Digital Filters

Second Edition

T0178457

Dietrich Schlichthärle

Digital Filters

Basics and Design

Second Edition

 Springer

Dr.-Ing. Dietrich Schlichthärle
Avaya GmbH&Co.KG
D-60326 Frankfurt, Main
Germany
e-mail: dietrich.schlichthaerle@ieee.org

ISBN 978-3-642-43285-9   ISBN 978-3-642-14325-0 (eBook)

DOI 10.1007/978-3-642-14325-0

Springer Heidelberg Dordrecht London New York

© Springer-Verlag Berlin Heidelberg 2000, 2011
Softcover re-print of the Hardcover 2nd edition 2011

Cover design: WMXDesign GmbH

Printed on acid-free paper

Springer is part of Springer Science+Business Media (www.springer.com)

To my parents

Preface

Today, there is an ever increasing number of applications that make use of digital filters or digital signal processing (DSP) algorithms in general. In electronic equipment of any kind, analog functions have been increasingly replaced by digital algorithms. This is true for instance for audio and video equipment, for communication, control and radar systems and in medical applications. Continuing improvement of CMOS technology has allowed higher complexity and increasing processing speed of dedicated integrated circuits and programmable signal processors. Higher clock speeds and parallel processing with the associated higher throughput opens the door for more and more real-time applications.

The advantages of digital signal processing are based on the fact that the performance of the applied algorithms is always predictable. There is no dependence on the tolerances of electrical components as in analog systems. This allows the implementation of interference-proof and long-term stable systems. The adjustment of components which is often necessary in analog circuits becomes obsolete. Also, extremely low frequencies can be processed without the problems which occur in analog systems.

In the field of consumer applications, digital signal processing has obtained wide application in the storage and reproduction of audio and video signals. In particular, CD technology, which has entered most households today, has to be mentioned. Compared with the previous vinyl technology, the reproduction of audio signals may take place arbitrarily often without any loss of quality. Besides the real-time processing of signals, simulation techniques which allow the imitation of physical processes in arbitrary time scales on computers gain more and more importance.

It is the aim of this textbook to give insight into the characteristics and the design of digital filters. The starting point for the design of recursive filters is the theory of continuous-time systems (Chap. 1) and the design methods for analog filters as introduced in Chap. 2 which are based on Butterworth, Chebyshev and Bessel polynomials as well as on elliptic functions. After the introduction of the theory of discrete-time systems (Chap. 3) and the basic structures of digital filters in Chap. 5, we derive the rules which allow the calculation of the coefficients of recursive digital filters from those of continuous-time reference filters in Chap. 6.

An important part of this textbook is devoted to the design of nonrecursive filters which cannot be treated with the classical filter design methods. Approximation criteria such as the minimisation of the mean-square or maximum error and the maximally flat design are presented in Chap. 7. Of foremost significance is the Remez exchange algorithm which is widely used in commercial filter design software.

The connecting point between the world of the physical signals which are considered continuous-time and the world of number series and mathematical algorithms which are discrete-time by nature is the sampling theorem, which is treated in Chap. 4.

Chapter 8 is devoted to all those effects which are due to the finite register lengths in real software or hardware implementations. Both signal samples and filter coefficients can be represented with limited accuracy only. The consequences are deviations from the design target, additive noise and—under certain circumstances—the occurrence of instabilities.

The final chapter gives an introduction to terms like oversampling and noise shaping. These are techniques which are primarily used to reduce the complexity of analog components in analog-to-digital and digital-to-analog converters.

In the Chap. 10, the textbook is completed by a selection of filter design tables for Butterworth, Chebyshev, Cauer and Bessel filters. Furthermore, the interested reader will find several subroutines written in Turbo Pascal which can be used in their own filter design programs. These may be easily translated into other dialects or programming languages.

The study of this book requires the knowledge of the theory of continuous-time systems. In this context, Chap. 1 gives only a short overview of terms and relations which are used in later chapters. A basic knowledge of algebra, infinitesimal calculus and the mathematics of complex numbers is also needed.

Maintal, February 2000   Dietrich Schlichthärle

Preface to the Second Edition

Besides numerous corrections and refinements of the existing material, the second edition of this textbook comprises substantial enhancements which round out the introduction of the structures and the design techniques of digital filters.

Chapter 2 (Analog filters) was supplemented by the consideration of ladder and lattice filters which form the basis of important classes of digital filters. These inherit the favourable characteristics of their analog counterparts.

Chapter 5 (Filter structures) received the most enhancements. Additional 50 pages are devoted to allpass filters and their applications. The allpass filter is a versatile building block which can be used so solve filter problems in numerous areas where phase shifting, fractional delays, or selective filtering with special properties are required. The outstanding feature of digital allpass filters is the possibility to achieve exactly unity gain independent of any coefficient inaccuracies. Due to this property, filters can be built featuring notably low coefficient sensitivity. A further completely new section deals with the transformation of filter block diagrams into software algorithms. This textbook cannot deeply discuss the topic of DSP programming. But some remarks concerning software structures of digital filters should not be missing in this chapter.

Chapter 6 (Design of IIR filters) was enhanced by the introduction of algorithms for the design of allpass filters based on given specifications.

Chapter 9 (Oversampling and Noise Shaping) was completely revised. The topic of CIC filters which attained great relevance in the area of upsampling and downsampling was added.

All program examples throughout this book were converted from Pascal to C which is the preferred programming language of digital signal processing.

Although the number of book pages increased substantially compared to the first edition, it is the belief of the author that this will be beneficial to readers who need more detailed information on any of the topics treated in this book.

Maintal, May 2010   Dietrich Schlichthärle

Contents

1 Continuous-Time Systems

The relationships between the input and output signals of analog systems which are introduced in this chapter are part of the theory of continuous-time systems. For a certain class of systems which have the properties of linearity and time-invariance (LTI systems), this theory provides relatively simple mathematical relationships. In the present textbook, we restrict ourselves to the consideration of filters which possess these two properties.

Linearity means that two superimposed signals pass through a system without mutual interference. Suppose that $y_1(t)$ is the response of the system to $x_1(t)$ and $y_2(t)$ the response to $x_2(t)$. Applying a linear combination of both signals $x = a\, x_1(t) + b\, x_2(t)$ to the input results in $y = a\, y_1(t) + b\, y_2(t)$ at the output.

Time-invariance means that the characteristics of the system do not change with time. A given input signal results in the same output signal at any time. So the property of time-invariance implies that if $y(t)$ is the response to $x(t)$, then the response to $x(t-t_0)$ is $y(t-t_0)$.

If these conditions are fulfilled, the characteristics of the system can be fully described by its reaction to simple test signals such as pulses, steps and harmonic functions.

1.1 Relations in the Time Domain

One of these possible test signals is the so-called delta function $\delta(t)$. The delta function is an impulse of unit area and infinitesimally short duration occurring at $t = 0$. The filter reacts to this signal with the impulse response $h(t)$. As we deal, in practice, with causal systems where the effect follows the cause in time, $h(t)$ will disappear for negative t. If the impulse response of a system is known, it is possible to calculate the reaction $y(t)$ of the system to any arbitrary input signal $x(t)$.

$$y(t) = \int_{-\infty}^{+\infty} x(t-\tau)h(\tau)\mathrm{d}\tau = \int_{-\infty}^{+\infty} x(\tau)h(t-\tau)\mathrm{d}\tau = x(t) * h(t) \qquad (1.1)$$

Relationship (1.1) is called the convolution integral: $y(t)$ is the convolution of $x(t)$ with $h(t)$. Both integrals are fully equivalent, as can be shown by a simple substitution of variables. In an abbreviated form, the convolution is usually denoted by an asterisk.

D. Schlichthärle, *Digital Filters*, DOI: 10.1007/978-3-642-14325-0_1,
© Springer-Verlag Berlin Heidelberg 2011

A further possible test signal is the step function $u(t)$. The reaction of the system to this signal is called the step response $a(t)$. The step function is defined as

$$u(t) = \begin{cases} 0 \text{ for } t < 0 \\ 1/2 \text{ for } t = 0 \\ 1 \text{ for } t > 0 \end{cases} .$$

For causal systems, also $a(t)$ will disappear for negative t.

$$a(t) = 0 \qquad \text{for } t < 0$$

If the step response of the system is known, the input and output signals of the system are related by the following equation:

$$y(t) = \int_{-\infty}^{+\infty} a(t-\tau)\frac{\mathrm{d}x(\tau)}{\mathrm{d}\tau}\mathrm{d}\tau = -\int_{-\infty}^{+\infty} a(\tau)\frac{\mathrm{d}x(t-\tau)}{\mathrm{d}\tau}\mathrm{d}\tau = a(t) * x'(t) . \tag{1.2}$$

In this case, the output signal is the convolution of the step response of the system with the time derivative of the input signal.

Delta function and unit step are related as

$$\delta(t) = \frac{\mathrm{d}u(t)}{\mathrm{d}t} \qquad\qquad \text{and} \tag{1.3a}$$

$$u(t) = \int_{-\infty}^{t} \delta(\tau)\,\mathrm{d}\tau \qquad\qquad \text{vice versa.} \tag{1.3b}$$

For LTI systems, similar relations exist between impulse and step response:

$$h(t) = \frac{\mathrm{d}a(t)}{\mathrm{d}t} \qquad\qquad \text{and} \tag{1.4a}$$

$$a(t) = \int_{-\infty}^{t} h(\tau)\,\mathrm{d}\tau \qquad\qquad \text{vice versa.} \tag{1.4b}$$

1.2 Relations in the Frequency Domain

In the frequency domain, the system behaviour can be analysed by applying an exponential function to the input of the form

$$x(t) = e^{pt} , \tag{1.5}$$

where p denotes the complex frequency. p consists of an imaginary part ω, which is the frequency of a harmonic oscillation, and a real part σ, which causes an increasing or decreasing amplitude, depending on the sign of σ.

$$p = \sigma + j\omega$$

The corresponding real oscillation is obtained by forming the real part of (1.5).

$$\text{Re } x(t) = e^{\sigma t} \cos \omega t$$

Substitution of (1.5) for the input signal in the convolution integral (1.1) gives the reaction of the system to the complex exponential stimulus.

$$y(t) = \int\limits_{-\infty}^{+\infty} e^{p(t-\tau)} h(\tau) d\tau$$

$$y(t) = e^{pt} \int\limits_{-\infty}^{+\infty} h(\tau) e^{-p\tau} d\tau \tag{1.6}$$

The integral in (1.6) yields a complex-valued expression which is a function of the complex frequency p.

$$H(p) = \int\limits_{-\infty}^{+\infty} h(t) e^{-pt} dt \tag{1.7}$$

Consequently, the output signal $y(t)$ is also an exponential function which we obtain by multiplying the input signal by the complex factor $H(p)$. Thus, input and output signals differ in amplitude and phase.

$$y(t) = e^{pt} H(p)$$

$$\text{Re } y(t) = |H(p)| e^{\sigma t} \cos[\omega t + \arg H(p)] \tag{1.8}$$

The function $H(p)$ is called the system or transfer function as it completely characterises the system in the frequency domain.

1.2.1 The Laplace Transform

The integral relationship (1.7) is called the Laplace transform. Thus, the transfer function $H(p)$ is the Laplace transform of the impulse response $h(t)$ of the system. In practice, a special form of the Laplace transform is used which is only defined on the interval $t \geq 0$, the so-called one-sided Laplace transform.

Table 1-1 Table of selected Laplace and Fourier transforms

Time domain	Laplace transform	Fourier transform
$\delta(t)$	1	1
$u(t)$	$\dfrac{1}{p}$	$\dfrac{1}{j\omega}+\pi\delta(\omega)$
$t\,u(t)$	$\dfrac{1}{p^2}$	$j\pi\dfrac{d\delta(\omega)}{d\omega}-\dfrac{1}{\omega^2}$
$e^{-\alpha t}u(t)$	$\dfrac{1}{p+\alpha}$	$\dfrac{1}{j\omega+\alpha}$
$(1-e^{-\alpha t})u(t)$	$\dfrac{\alpha}{p(p+\alpha)}$	$\pi\delta(\omega)+\dfrac{\alpha}{j\omega(j\omega+\alpha)}$
$e^{j\omega_0 t}u(t)$	$\dfrac{1}{p-j\omega_0}$	$\pi\delta(\omega-\omega_0)+\dfrac{1}{j\omega-\omega_0}$
$\cos\omega_0 t\,u(t)$	$\dfrac{p}{p^2+\omega_0{}^2}$	$\dfrac{\pi}{2}[\delta(\omega-\omega_0)+\delta(\omega+\omega_0)]+\dfrac{j\omega}{\omega_0{}^2-\omega^2}$
$\sin\omega_0 t\,u(t)$	$\dfrac{\omega_0}{p^2+\omega_0{}^2}$	$\dfrac{\pi}{2j}[\delta(\omega-\omega_0)-\delta(\omega+\omega_0)]+\dfrac{\omega_0}{\omega_0{}^2-\omega^2}$
$\cos\omega_0 t\,e^{-\alpha t}u(t)$	$\dfrac{p+\alpha}{(p+\alpha)^2+\omega_0{}^2}$	$\dfrac{\alpha+j\omega}{(\alpha+j\omega)^2+\omega_0{}^2}$
$\sin\omega_0 t\,e^{-\alpha t}u(t)$	$\dfrac{\omega_0}{(p+\alpha)^2+\omega_0{}^2}$	$\dfrac{\omega_0}{(\alpha+j\omega)^2+\omega_0{}^2}$

$$F(p)=\int_{0}^{+\infty}f(t)e^{-pt}\,dt \tag{1.9a}$$

If we apply this transform to the impulse response of a system, this does not impose any limitation as the impulse response disappears anyway for $t<0$ as pointed out in Sect. 1.1. When calculating the integral (1.9a), the real part of p (σ) is used as a parameter which can be freely chosen such that the integral converges. Table 1-1 shows some examples and Table 1-2 introduces the most important rules of the Laplace transform. The inverse transform from the frequency domain back into the time domain is expressed as

$$f(t)=\frac{1}{2\pi j}\int_{\sigma-j\infty}^{\sigma+j\infty}F(p)e^{pt}\,dp. \tag{1.9b}$$

Equation (1.9b) is a line integral in the complex p-plane in which the integration path is chosen with an appropriate σ such that the integral converges.

Table 1-2 Rules of the Laplace and Fourier transforms

Time domain	Laplace transform	Fourier transform
$f(t-t_0)$	$F(p)e^{-pt_0}$	$F(j\omega)e^{-j\omega t_0}$
$f(at)$	$\dfrac{1}{\|a\|}F(p/a)$	$\dfrac{1}{\|a\|}F(j\omega/a)$
$\dfrac{d\,f(t)}{dt}$	$pF(p)-f(0)$	$j\omega\,F(j\omega)$
$\displaystyle\int_0^t f(\tau)\,d\tau$	$\dfrac{1}{p}F(p)$	————
$\displaystyle\int_{-\infty}^t f(\tau)\,d\tau$	————	$\dfrac{F(j\omega)}{j\omega}+\pi\,F(0)\delta(\omega)$
$f(t)*g(t)$	$F(p)\,G(p)$	$F(j\omega)\,G(j\omega)$

In this context, the Laplace transform has an interesting property which is important for the characterisation of the behaviour of LTI systems in the frequency domain. The convolution of two functions $x_1(t) * x_2(t)$ in the time domain is equivalent to the multiplication of the Laplace transforms of both functions $X_1(p)X_2(p)$ in the frequency domain. For LTI systems, the output signal is the convolution of the input signal and the impulse response. In the frequency domain, the Laplace transform of the output signal is therefore simply the Laplace transform of the input signal multiplied by the system function $H(p)$.

$$Y(p) = H(p)\,X(p) \tag{1.10}$$

Hence (1.10) establishes a relation between input and output signal in the frequency domain and is therefore equivalent to (1.1), which relates input and output signals in the time domain.

1.2.2 The Fourier Transform

With $\sigma = 0$ or $p = j\omega$ in (1.7), we obtain a special case of the Laplace transform.

$$F(j\omega) = \int_{-\infty}^{+\infty} f(t)e^{-j\omega t}\,dt$$

$$\tag{1.11a}$$

This integral relationship is called the Fourier transform. Unlike the Laplace transform, (1.11a) does not contain a damping factor which enforces convergence. So the Fourier integral only converges if the signal is limited in time or decays fast enough. Otherwise, $F(j\omega)$ will contain singularities which can be treated using the concept of distributions. These singularities are frequently called spectral lines, which are mathematically represented as delta functions in the frequency domain. The concept of distributions removes the named limitations of the Fourier integral and extends its applicability to periodic functions with infinite energy. Table 1-1 shows some examples of Fourier-transformed functions and allows a direct comparison with the corresponding Laplace transform. Simple transition between Laplace transform $F(p)$ and Fourier transform $F(j\omega)$ by interchanging p and $j\omega$ is only possible if the Fourier integral converges for all ω, which means that no spectral lines exist in the frequency domain. Stable filters which we aim at in practical applications have transfer functions with this property, so that $H(j\omega)$ can be obtained from $H(p)$ by substituting $j\omega$ for p. The inverse Fourier transform is given by

$$f(t) = \frac{1}{2\pi} \int_{-\infty}^{+\infty} F(j\omega)e^{j\omega t} d\omega. \qquad (1.11b)$$

Table 1-2 shows some further properties of the Fourier transform.

The Fourier transform $H(j\omega)$ of the impulse response $h(t)$ is also referred to as the system or transfer function. Evaluation of (1.8) for $\sigma = 0$ or $p = j\omega$ shows that $H(j\omega)$ describes the behaviour of the system for the case of harmonic input signals in the steady state.

$$\operatorname{Re} y(t) = |H(j\omega)| \cos(\omega t + \arg H(j\omega))$$

$H(j\omega)$ is therefore the frequency response of the system. The magnitude of $H(j\omega)$ is called the amplitude or magnitude response which is often expressed in the form

$$a(\omega) = 20 \log |H(j\omega)| , \qquad (1.12a)$$

where $a(\omega)$ is the logarithmic gain of the system given in decibels (dB). The negative argument of the frequency response is the phase response of the system.

$$b(\omega) = -\arg H(j\omega) = -\arctan\left(\frac{\operatorname{Im} H(j\omega)}{\operatorname{Re} H(j\omega)}\right) \qquad (1.12b)$$

The phase response determines the phase shift between the harmonic oscillations at the input and at the output of the system with the frequency ω as a parameter. The derivative of the phase response is the group delay τ_g of the system.

$$\tau_g(\omega) = \frac{db(\omega)}{d\omega} \qquad (1.12c)$$

If the preservation of the temporal shape of the signal is an issue, the group delay has to be flat in the passband of the filter which is equivalent to linear-phase behaviour. This guarantees that all parts of the spectrum experience the same delay when passing through the system.

Example 1-1

Determine the amplitude, phase, and group delay responses of a low-pass filter characterised by the transfer function

$$H(p) = \frac{a}{a+p} \qquad \text{with } a > 0 \, .$$

Substitution of $j\omega$ for p yields the frequency response of the filter.

$$H(j\omega) = \frac{a}{a+j\omega} \qquad \text{with } a > 0$$

The magnitude of this expression is the amplitude response.

$$|H(j\omega)| = \frac{a}{|a+j\omega|} = \frac{a}{\sqrt{a^2 + \omega^2}} = \frac{1}{\sqrt{1 + \omega^2 / a^2}}$$

The phase response is the negative argument of the frequency response.

$$b(\omega) = -\arg H(j\omega) = \arg\frac{1}{H(j\omega)} = \arctan(\omega / a)$$

By differentiation of the phase response, we obtain the group delay of the filter.

$$\tau_g(\omega) = \frac{d\,b(\omega)}{d\omega} = \frac{1/a}{1 + \omega^2 / a^2}$$

1.3 Transfer Functions

The transfer behaviour of networks consisting of lumped elements such as inductors, capacitors, resistors, transformers and amplifiers is described by linear differential equations with real coefficients. Equation (1.13) shows the general form of this differential equation with $x(t)$ denoting the input signal and $y(t)$ the output signal of the system.

$$\begin{aligned}
a_N \frac{dy^{(N)}}{dt^N} &+ a_{N-1}\frac{dy^{(N-1)}}{dt^{N-1}} + \ldots + a_1\frac{dy}{dt} + a_0 y \\
&= b_M \frac{dx^{(M)}}{dt^M} + b_{M-1}\frac{dx^{(M-1)}}{dt^{M-1}} + \ldots + b_1\frac{dx}{dt} + b_0 x
\end{aligned} \qquad (1.13)$$

Applying the Laplace transform to (1.13) yields an algebraic equation which completely characterises the system in the frequency domain. The following relationship assumes that the system is at rest for $t = 0$. For an electrical network, this means that the stored energy in all capacitors and inductors is zero.

$$a_N p^N Y + a_{N-1} p^{N-1} Y + \ldots + a_1 p Y + a_0 Y$$
$$= b_M p^M X + b_{M-1} p^{M-1} X + \ldots + b_1 p X + b_0 X$$

Reordering the above relation gives the transfer function of the network, which is the quotient of the Laplace transforms of the output and the input signals.

$$H(p) = \frac{Y(p)}{X(p)} = \frac{b_M p^M + b_{M-1} p^{M-1} + \ldots + b_1 p + b_0}{a_N p^N + a_{N-1} p^{N-1} + \ldots + a_1 p + a_0} \qquad (1.14)$$

Equation (1.14) is a rational fractional function of the frequency variable p which has the following properties:

- The coefficients a_i and b_i are real.
- The degree of the denominator N which equals the order of the filter corresponds to the number of energy stores (capacitors and inductors) in the network.
- The degree of the numerator must be smaller than or equal to the degree of the denominator. Otherwise the system would contain ideal differentiators which are not realisable.

Beside the rational fractional representation, which is closely related to the differential equation of the system, there are two further forms of the transfer function which are of practical importance: pole/zero representation and partial-fraction expansion.

If the numerator and denominator polynomials in (1.14) are decomposed into the respective zero terms, we obtain the pole/zero representation.

$$H(p) = \frac{Y(p)}{X(p)} = \frac{b_M}{a_N} \frac{(p - p_{01}) \cdots (p - p_{0(M-1)})(p - p_{0M})}{(p - p_{\infty 1}) \cdots (p - p_{\infty(N-1)})(p - p_{\infty N})} \qquad (1.15)$$

The zeros of the numerator polynomial are the zeros of the transfer function, the zeros of the denominator polynomial are the poles. Poles and zeros have the following properties:

- Poles and zeros are either real or appear as conjugate-complex pairs. This property is forced by the real coefficients of the numerator and denominator polynomials.
- The real part of the poles has to be negative, which means that the poles of the system must lie in the left half of the p-plane. This condition guarantees stability of the system. It ensures that any finite input signal results in a finite output signal. When the input is set to zero, all internal signals and the output signal will converge to zero.

The pole/zero representation has special importance in cases where given transfer functions are realised as a serial connection of first- and second-order subsystems (Fig. 1-1a). First-order sections are used to realise real poles, second-order sections implement conjugate-complex pole pairs.

Assuming ideal noiseless amplifiers, the pole/zero pairing when creating second-order filter blocks and the sequence of these subfilters can be chosen arbitrarily. As the operational amplifiers used in practice generate noise and have a finite dynamic range, the pole/zero assignment and the choice of sequence are degrees of freedom within which to optimise the noise behaviour of the overall system.

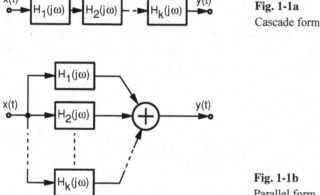

Fig. 1-1a
Cascade form

Fig. 1-1b
Parallel form

Example 1-2

Derive the pole/zero representation of the transfer function

$$H(p) = \frac{p^3 + 9p^2 + 20p + 16}{p^3 + 5p^2 + 8p + 6} \, . \tag{1.16a}$$

The roots of the numerator and denominator polynomial are determined using a standard math tool such as Matlab.
The roots of the numerator polynomial are -6.18 and $-1.41 \pm j0.78$.
The roots of the denominator polynomial are -3 and $-1 \pm j1$.
Insertion of the obtained poles and zeros in (1.15) yields

$$H(p) = \frac{(p - p_{01})(p - p_{02})(p - p_{03})}{(p - p_{\infty 1})(p - p_{\infty 2})(p - p_{\infty 3})}$$

$$H(p) = \frac{(p + 6.18)(p + 1.41 + j0.78)(p + 1.41 - j0.78)}{(p + 3)(p + 1 + j)(p + 1 - j)} \, . \tag{1.16b}$$

The complex pole and zero pairs in (1.16b) are combined to a second-order term with real coefficients. The given transfer function can therefore be realised by cascading a first-order and a second-order filter block.

$$H(p) = \frac{(p+6.18)}{(p+3)} \frac{(p^2+2.82p+2.46)}{(p^2+2p+2)} \qquad (1.16c)$$

Unlike the pole/zero representation, which consists of the product of first- and second-order terms, the partial-fraction expansion of the transfer function is composed of a sum of terms (1.17).

$$H(p) = \frac{A_1}{p - p_{\infty 1}} + \frac{A_2}{p - p_{\infty 2}} + \cdots + \frac{A_{N-1}}{p - p_{\infty(N-1)}} + \frac{A_N}{p - p_{\infty N}} \qquad (1.17)$$

This representation is closely related to a possible realisation of the transfer function as a parallel connection of first- and second-order systems (Fig. 1-1b).

The coefficients A_i of the partial-fraction expansion can be easily derived on condition that the poles are known. First of all, the transfer function $H(p)$ is multiplied by the pole term $(p - p_{\infty i})$. Then the limit of the resulting expression is taken for $p \rightarrow p_{\infty i}$, which gives the constant A_i as the result.

$$A_i = H(p)(p - p_{\infty i})\Big|_{p \rightarrow p_{\infty i}} \qquad (1.18)$$

It is obvious that this procedure yields the desired result if we apply (1.18) to (1.17). Another way to determine the partial-fraction coefficients, which avoids the forming of the limit in (1.18), is given by (1.19):

$$A_i = \frac{N(p_{\infty i})}{D'(p_{\infty i})}. \qquad (1.19)$$

$N(p)$ is the numerator polynomial and $D'(p)$ the derivative of the denominator polynomial [9].

When the coefficients A_i are determined, conjugate-complex terms can be combined to give realisable second-order transfer functions with real coefficients. It must be noted that the partial-fraction expansion requires the degree of the numerator to be less than the degree of the denominator. A transfer function with this property is referred to as strictly proper. This condition is not met for high-pass and bandstop filters for which the degrees of numerator and denominator are equal. In these cases, long division of the numerator polynomial by the denominator polynomial can be applied prior to the procedures described above. This results in the sum of a constant, which is equivalent to the gain of the system for $\omega \rightarrow \infty$, and a rational fractional function with the degree of the numerator reduced by one.

Applying an impulse to the input of the parallel combination of first- and second-order subsystems (Fig. 1-1b) results in an impulse response which is composed of the sum of all partial responses. These are decaying exponentials and exponentially decaying sinusoids. The decay rate and the frequency of these partial responses are determined by the poles of the system. For the design of digital filters which are intended to reproduce the properties of analog filters in the time domain, partial-fraction expansion of transfer functions is therefore of special interest.

Example 1-3

Calculate the partial fraction expansion of (1.16).

In the first step, we have to divide the numerator by the denominator polynomial, which is required as the degrees of numerator and denominator are equal. Equation (1.20a) shows the result of the long division.

$$H(p) = 1 + \frac{4p^2 + 12p + 10}{p^3 + 5p^2 + 8p + 6} \tag{1.20a}$$

We use (1.19) to calculate the partial fraction coefficients A_i.

$$A_i = \frac{N(p_{\infty i})}{D'(p_{\infty i})} = \frac{4p^2 + 12p + 10}{3p^2 + 10p + 8}\bigg|_{p = p_{\infty i}}$$

$$A_1 = \frac{4p^2 + 12p + 10}{3p^2 + 10p + 8}\bigg|_{p=-3} = \frac{4(-3)^2 + 12(-3) + 10}{3(-3)^2 + 10(-3) + 8} = 2$$

$$A_2 = \frac{4p^2 + 12p + 10}{3p^2 + 10p + 8}\bigg|_{p=-1+j} = \frac{4(-1+j)^2 + 12(-1+j) + 10}{3(-1+j)^2 + 10(-1+j) + 8} = 1$$

$$A_2 = \frac{4p^2 + 12p + 10}{3p^2 + 10p + 8}\bigg|_{p=-1-j} = \frac{4(-1-j)^2 + 12(-1-j) + 10}{3(-1-j)^2 + 10(-1-j) + 8} = 1$$

Insertion of these results in (1.17) yields the desired partial fraction representation.

$$H(p) = 1 + \frac{A_1}{p - p_{\infty 1}} + \frac{A_2}{p - p_{\infty 2}} + \frac{A_3}{p - p_{\infty 3}}$$

$$H(p) = 1 + \frac{2}{p+3} + \frac{1}{p+1-j} + \frac{1}{p+1+j} \tag{1.20b}$$

In order to obtain realisable subsystems, the pair of conjugate-complex terms is combined to a transfer function of second order (1.20c).

$$H(p) = 1 + \frac{2p+2}{p^2 + 2p + 2} + \frac{2}{p+3} \tag{1.20c}$$

Partial systems of first and second order are of special importance for the design of filters, as already mentioned. These are the smallest realisable units into which transfer functions can be decomposed. Chapter 8 explains in more detail the advantages of a design based on low-order partial systems. The transfer function of a general second-order filter can be expressed as

$$H(p) = \frac{b_0 + b_1 p + b_2 p^2}{a_0 + a_1 p + a_2 p^2} \ . \tag{1.21a}$$

In practice, this transfer function is often normalised such that $a_0 = 1$.

$$H(p) = \frac{b_0 + b_1 p + b_2 p^2}{1 + a_1 p + a_2 p^2} \tag{1.21b}$$

It is sometimes advantageous to factor the constant b_0 out of the numerator polynomial, which then appears as a gain factor G in front of the transfer function.

$$H(p) = G \frac{1 + b_1 p + b_2 p^2}{1 + a_1 p + a_2 p^2} \tag{1.21c}$$

It is easy to verify that G is the gain of the system at the frequency $p = 0$.
   The transfer function of a complex pole pair of the form

$$H(p) = \frac{1}{p^2 + a_1 p + a_0}$$

can be characterised in three ways:
- by the coefficients a_1 and a_0 of the denominator polynomial as above
- by the real part $\mathrm{Re}\, p_\infty$ and imaginary part $\mathrm{Im}\, p_\infty$ of the poles
- by the resonant frequency ω_0 and quality factor Q

$$H(p) = \frac{1}{p^2 + (\omega_0 / Q)p + \omega_0^2} = \frac{1}{p^2 - 2\,\mathrm{Re}\, p_\infty p + (\mathrm{Re}\, p_\infty)^2 + (\mathrm{Im}\, p_\infty)^2} \tag{1.22}$$

The quality factor Q has two illustrative interpretations:
- The quality factor is the ratio of the gain of the pole pair at the resonant frequency ω_0 to the gain at $\omega = 0$.

$$Q = |H(j\omega_0) / H(0)|$$

- The quality factor approximately corresponds to the reciprocal of the relative 3-dB bandwidth of the resonant peak of the pole pair.

$$Q \approx \omega_0 / \Delta\omega_{3dB}$$

From (1.22), we can derive a relationship between the quality factor and the location of the pole:

$$|\mathrm{Im}\, p_\infty / \mathrm{Re}\, p_\infty| = \sqrt{4Q^2 - 1} \ . \tag{1.23}$$

Equation (1.23) shows that poles of constant Q are located on straight lines in the complex p-plane as depicted in Fig. 1-2.

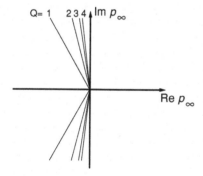

Fig. 1-2
Location of poles with constant quality
factor Q in the complex p-plane

1.4 Allpass Transfer Function

The poles of the transfer function of a stable system must always lie in the left
p-halfplane in order to guarantee a stable system. Zeros, in contrast, can lie
everywhere in the p-plane. Since the magnitude of a zero term of the form

$$j\omega - (\mathrm{Re}\, p_0 \pm j\,\mathrm{Im}\, p_0)$$

is identical with the magnitude of

$$j\omega - (-\mathrm{Re}\, p_0 \pm j\,\mathrm{Im}\, p_0) \; ,$$

we have the choice to place a zero in the right or left halfplane. There is no
difference in magnitude but in phase. A transfer function with all zeros located in
the left p-half-plane realises a given magnitude function with the lowest possible
phase shift as the frequency progresses from zero to infinity. Such transfer
functions are therefore called minimum-phase. A non-minimum-phase transfer
function can always be decomposed into the product of a minimum-phase and an
allpass transfer function as illustrated by (1.24) where the zeros in the right half-
plane are combined in $H_{0\mathrm{right}}(p)$.

$$H(p) = \underbrace{H_l(p)\, H_{0\mathrm{right}}(p)}_{non-minimum-phase} = \underbrace{H_l(p)\, H_{0\mathrm{left}}(p)}_{minimum-phase} \underbrace{\frac{H_{0\mathrm{right}}(p)}{H_{0\mathrm{left}}(p)}}_{allpass} \qquad (1.24)$$

The zeros in the right half-plane are replaced by corresponding zeros in the left
half-plane. The poles and zeros of the allpass transfer function form mirror-image
pairs which are symmetric about the imaginary axis (Fig. 1-3). Moreover complex
zeros and poles occur in complex-conjugate pairs. Therefore poles and zeros of an
allpass filter also form antisymmetric pairs about the origin.

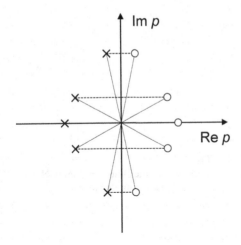

Fig. 1-3
Location of zeros and poles of an allpass transfer function

Because for every pole $p_{\infty i}$, there exists an antisymmetric zero p_{0i} with

$$p_{\infty i} = -p_{0i} \; ,$$

the transfer function of an allpass filter

$$H_{\text{Allp}}(p) = \frac{(p - p_{01})(p - p_{02}) \cdots (p - p_{0N})}{(p - p_{\infty 1})(p - p_{\infty 2}) \cdots (p - p_{\infty N})}$$

can be expressed as

$$H_{\text{Allp}}(p) = \frac{(p + p_{\infty 1})(p + p_{\infty 2}) \cdots (p + p_{\infty N})}{(p - p_{\infty 1})(p - p_{\infty 2}) \cdots (p - p_{\infty N})} \; .$$

Factoring out -1 from each of the zero terms in the numerator finally results in

$$H_{\text{Allp}}(p) = (-1)^N \frac{(-p - p_{\infty 1})(-p - p_{\infty 2}) \cdots (-p - p_{\infty N})}{(p - p_{\infty 1})(p - p_{\infty 2}) \cdots (p - p_{\infty N})} \; .$$

Thus numerator polynomial $N(p)$ and denominator polynomial $D(p)$ of the allpass filter are simply related as

$$N(p) = (-1)^N D(-p) \; .$$

The transfer function of an allpass filter is therefore of the form

$$H_{\text{Allp}}(p) = (-1)^N \frac{D(-p)}{D(p)} \; .$$

Due to the term $(-1)^N$, the sign of the gain of the allpass filter depends on the filter order. If this term is omitted, the gain is always $+1$ for $p = 0$.

1.5 The Squared Magnitude of Transfer Functions

We often need to calculate the squared magnitude of the frequency response $|H(j\omega)|^2$ from a given transfer function $H(p)$. The first possible approach is to replace p with $j\omega$ to obtain the frequency response $H(j\omega)$ of the system. From the mathematics of complex numbers we know that the squared magnitude can be obtained as the product of a given complex number and its complex-conjugate. Thus we can write

$$|H(j\omega)|^2 = H(j\omega)H^*(j\omega) \ .$$

Since $H(j\omega)$ is a rational function of $j\omega$ with real coefficients, it follows that

$$H^*(j\omega) = H(-j\omega) \ .$$

The squared magnitude can therefore be expressed as

$$|H(j\omega)|^2 = H(j\omega)H(-j\omega) \ . \qquad (1.25)$$

It is sometimes more convenient to calculate with functions in p instead of $j\omega$. By replacing $j\omega$ with p in (1.25), we can express the squared magnitude as

$$|H(j\omega)|^2 = H(p)H(-p) \qquad \text{for } p = j\omega$$

Thus we can alternatively start with multiplying $H(p)$ by $H(-p)$ and then perform the substitution of $j\omega$ for p as demonstrated in Example 1-4.

Example 1-4

Calculate the squared magnitude of the frequency response of a filter with the transfer function

$$H(p) = \frac{p^2 + 2}{p^3 + 2p^2 + 3p + 1} \ .$$

In the first step, we multiply $H(p)$ by $H(-p)$.

$$H(p)H(-p) = \frac{p^2 + 2}{p^3 + 2p^2 + 3p + 1} \ \frac{p^2 + 2}{-p^3 + 2p^2 - 3p + 1}$$

$$H(p)H(-p) = \frac{p^4 + 4p^2 + 4}{-p^6 - 2p^4 - 5p^2 + 1}$$

Substitution of $j\omega$ for p yields

$$|H(j\omega)|^2 = \frac{(j\omega)^4 + 4(j\omega)^2 + 4}{-(j\omega)^6 - 2(j\omega)^4 - 5(j\omega)^2 + 1} = \frac{\omega^4 - 4\omega^2 + 4}{\omega^6 - 2\omega^4 + 5\omega^2 + 1}$$

In the context of the mathematical treatment of filter transfer functions, it is sometimes required to derive a transfer function in p from the given squared magnitude of the frequency response

$$|H(j\omega)|^2 = \frac{|N(\omega)|^2}{|D(\omega)|^2} = \frac{f(\omega^2)}{g(\omega^2)} \; .$$

Both numerator and denominator are polynomials in ω^2. The problem is conveniently solved in the p-plane. By substituting p for $j\omega$ (or $-p^2$ for ω^2), we obtain

$$H(p)H(-p) = \frac{f(-p^2)}{g(-p^2)} = \frac{N(p)N(-p)}{D(p)D(-p)} \qquad \text{for } p = j\omega \; .$$

Next, we determine the zeros of $f(-p^2)$ and $g(-p^2)$. In both cases, we obtain zeros which are located symmetrically about the $j\omega$-axis. $f(-p^2)$ may also have pairs of zeros on the $j\omega$-axis.

The zeros of $g(-p^2)$ determine the poles of the transfer function which must, for a stable filter, lie in the left p-half-plane. So we choose all the zeros with negative real part to construct the polynomial $D(p)$. All the others are assigned to $D(-p)$.

For the construction of the numerator polynomial $N(p)$, however, we can choose the zeros of $f(-p^2)$ in the left or right half-plane or even a mixture of both. Zeros on the $j\omega$-axis occur in pairs and are evenly assigned to $N(p)$ and $N(-p)$. A unique solution of the problem only exists if we restrict ourselves to minimum-phase filters which have their zeros located in the left half-plane or on the $j\omega$-axis.

Example 1-5

Derive the minimum-phase transfer function of a filter whose squared magnitude is given as

$$|H(j\omega)|^2 = \frac{\omega^4 - 18\omega^2 + 81}{\omega^6 - 44\omega^4 + 484\omega^2 + 2704} \; .$$

First we substitute $-jp$ for ω.

$$H(p)H(-p) = \frac{(-jp)^4 - 18(-jp)^2 + 81}{(-jp)^6 - 44(-jp)^4 + 484(-jp)^2 + 2704}$$

$$H(p)H(-p) = \frac{p^4 + 18p^2 + 81}{-p^6 - 44p^4 - 484p^2 + 2704}$$

Calculation of the roots of numerator and denominator polynomial yields:

Numerator: $p_{01/2} = \pm j\,3$

                    $p_{03/4} = \pm j\,3$

Denominator: $p_{\infty 1/2} = 1 \pm j\,5$

$p_{\infty 3/4} = -1 \pm j\,5$

$p_{\infty 5/6} = \pm 2$

The zeros are all located on the $j\omega$-axis. The poles form mirror-image pairs about the $j\omega$-axis. Poles and zeros are now appropriately assigned to $D(p)$, $D(-p)$, $N(p)$, and $N(-p)$.

$$H(p)H(-p) = \frac{N(p)N(-p)}{D(p)D(-p)} = \frac{(p+j3)(p-j3)}{(p+1+j5)(p+1-j5)(p+2)}$$

$$\times \frac{(p+j3)(p-j3)}{(p-1+j5)(p-1-j5)(p-2)}$$

$$H(p) = \frac{N(p)}{D(p)} = \frac{(p+j3)(p-j3)}{(p+1+j5)(p+1-j5)(p+2)} = \frac{p^2+9}{p^3+4p^2+30p+52}$$

2 Analog Filters

In this chapter, we introduce the classical Butterworth, Chebyshev, Cauer, and Bessel filters which are intended to serve as the prototypes for the design of digital filters with related characteristics. For each of these types, the equations are derived to determine filter coefficients which meet the tolerance scheme of a desired frequency response. Some tables of filter coefficients can be found in the appendix which allow the filter design for a limited range of filter specifications. These, however, cannot replace commercially available filter design handbooks or filter design software.

We start by considering low-pass filters as the prototypes for the design of digital filters. After the introduction of the characteristics of the ideal low-pass filter, we throw some light on the filter types named above, each of which is, in a certain sense, an optimum approximation of the ideal low-pass characteristic.

Appropriate transformations allow the conversion of low-pass filters into high-pass, bandpass and bandstop filters with related characteristics. The corresponding transformation rules are also introduced in this chapter.

One possible approach to design digital filters is to use the transfer functions of analog reference filters as the basis. This results in the so-called direct-form structures which aim at realising a given transfer function. Another approach is to convert the structure of the analog filter to a related structure in the discrete-time domain. In this way, the good properties of passive analog filters in terms of stability and low coefficient sensitivity can also be made available to the discrete-time implementation. An important type of circuit in this context is the lossless network terminated in resistances at both ends. In Sect. 2-8, we derive the basic relations and properties of this kind of filter implementation and introduce the ladder and lattice structure in more detail. These have their counterparts in the discrete-time domain known as the digital ladder filter and the digital lattice filter respectively.

2.1 Filter Approximation

In the context of the characteristics of the ideal low-pass filter, we consider at first the distortionless system. If the waveform of a signal is to be preserved when passing through the system, multiplication of the signal by a constant factor k (frequency-independent gain) and a frequency-independent delay t_0 are the only allowed operations. This yields the following general relationship between input signal $x(t)$ and output signal $y(t)$ of the distortionless system in the time domain:

D. Schlichthärle, *Digital Filters*, DOI: 10.1007/978-3-642-14325-0_2,
© Springer-Verlag Berlin Heidelberg 2011

$$y(t) = k\, x(t - t_0)\,.$$

Applying the Fourier transform to this expression, we obtain a corresponding expression in the frequency domain:

$$Y(j\omega) = X(j\omega)\, k\, e^{-j\omega t_0} \qquad\qquad \text{and}$$

$$H(j\omega) = \frac{Y(j\omega)}{X(j\omega)} = k\, e^{-j\omega t_0}\,.$$

Magnitude, phase and group delay response of the distortionless system can be calculated using (1.12).

$$a(\omega) = 20 \log k$$

$$b(\omega) = \omega\, t_0$$

$$\tau_{\mathrm{g}}(\omega) = t_0$$

Magnitude and group delay response are frequency-independent, and the phase response is linear.

The ideal low-pass filter also features a linear phase in the passband. For $0 \le \omega < \omega_{\mathrm{c}}$, the gain is constantly unity. In the stopband $\omega > \omega_{\mathrm{c}}$, however, the filter cuts off completely (Fig. 2-1).

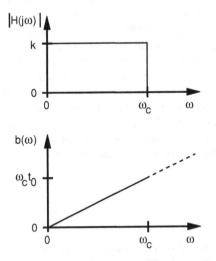

Fig. 2-1
Magnitude and phase response of the ideal low-pass filter

Such a filter is not realisable in practice. Approximations to the ideal behaviour have to be found which are implementable with a reasonable amount of hardware resources. Since the transfer functions of filters, made up of discrete RLC networks, are rational fractional functions of the frequency variable p or $j\omega$, it is obvious to use polynomials and rational functions for these approximations.

The lossless network terminated in resistances at both ends is the preferred network structure to realise a variety of well-known filter characteristics such as Butterworth, Chebyshev, Cauer and Bessel. In Sect. 2.8, we will show that this type of network is structurally-bounded. This means that the magnitude of the filter transfer function cannot exceed a certain maximum value. This fact is taken into account in the design process of these filters by assuming transfer functions with squared magnitude of the form

$$|H(j\omega)|^2 = \frac{1}{1+|K(j\omega)|^2} \quad \text{or}$$

$$H(p)H(-p) = \frac{1}{1+K(p)K(-p)} \quad \text{for } p = j\omega \tag{2.1}$$

as the starting point. Expression (2.1) can never exceed unity. The characteristic function $K(p)$ determines the kind of approximation to the ideal filter. For a low-pass filter, $K(j\omega)$ has the property of assuming low values between $\omega = 0$ and a certain cutoff frequency ω_c, but increasing rapidly above this frequency.

$K(p)$ may be a polynomial as in the case of Butterworth, Chebyshev and Bessel filters. For inverse Chebyshev and Cauer filters, $K(p)$ is a rational fractional function. In general, we can express the characteristic function as

$$K(p) = \frac{v(p)}{u(p)} . \tag{2.2}$$

Zeros of the numerator in (2.2) cause low values of $K(p)$. They determine therefore the behaviour of the filter in the passband. Zeros of the denominator polynomial lead to large values of $K(p)$. They determine the frequencies where the transfer function $H(p)$ has its zeros in the stopband. Substitution of (2.2) into (2.1) yields

$$H(p)H(-p) = \frac{1}{1+\dfrac{v(p)v(-p)}{u(p)u(-p)}} = \frac{u(p)u(-p)}{u(p)u(-p)+v(p)v(-p)} . \tag{2.3}$$

The poles of $H(p)$ are thus determined by both polynomials $u(p)$ and $v(p)$. The zeros of $H(p)$ are identical with the zeros of $u(p)$. A transfer function complementary to (2.3) is obtained by calculating

$$1 - H(p)H(-p) = \frac{v(p)v(-p)}{u(p)u(-p)+v(p)v(-p)} \tag{2.4}$$

In the theory of lossless networks terminated in resistances, (2.3) and (2.4) are referred to as the transmittance $t(p)$ and reflectance $\rho(p)$ of the network.

$$t(p)t(-p) = \frac{u(p)u(-p)}{u(p)u(-p)+v(p)v(-p)}$$

$$\rho(p)\rho(-p) = \frac{v(p)v(-p)}{u(p)u(-p)+v(p)v(-p)}$$

Transmittance and reflectance are related as

$$t(p)t(-p) + \rho(p)\rho(-p) = 1 \quad \text{for } p = j\omega \quad \text{or}$$
$$|t(j\omega)|^2 + |\rho(j\omega)|^2 = 1 \tag{2.5}$$

which is known as the Feldtkeller equation. $t(p)$ and $\rho(p)$ are said to be power-complementary.

Butterworth, Chebyshev and Cauer filters form a family of filter characteristics with distinct properties. All poles and zeros of $K(p)$ lie on the imaginary axis. The polynomials $u(p)$ and $v(p)$ are thus either even or odd depending on the order N and the type of filter as summarised in Table 2-1. This property results in certain symmetry conditions of the lossless filter network as we will show in Sect. 2.8.2.

Table 2-1 Characterisation of the polynomials $u(p)$ and $v(p)$ for Butterworth, Chebyshev and Cauer filters

Type of filter	Filter order N	Order of $u(p)$	Order of $v(p)$	r
Low-pass	N even	even	even	0
	N odd	even	odd	1
High-pass	N even	even	even	0
	N odd	odd	even	−1
Bandpass	$N/2$ even	even	even	0
	$N/2$ odd	odd	even	−1
Bandstop	$N/2$ even	even	even	0
	$N/2$ odd	even	odd	1

Due to the properties of the characteristic function according to Table 2-1, $K(p)$ can be generally expressed as

$$K(p) = \varepsilon\, p^r\, \frac{(p^2 + \omega_1^2)(p^2 + \omega_2^2)\dots(p^2 + \omega_n^2)}{(p^2 + \omega_{01}^2)(p^2 + \omega_{02}^2)\dots(p^2 + \omega_{0m}^2)} \tag{2.6}$$

The applicable values of r (1, 0, or −1) can also be found in Table 2-1. The parameter ε gives some control over the ripple of the magnitude response in the passband and/or stopband of the filter.

The desired magnitude characteristic of a low-pass filter is often specified in terms of a tolerance scheme, which has to be satisfied by the filter (Fig. 2-2). ω_p is the edge frequency of the passband, ω_s marks the beginning of the stopband. The region $\omega_p < \omega < \omega_s$ is the transition band between passband and stopband.

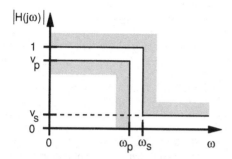

Fig. 2-2
Tolerance scheme of a low-pass filter

The relative width of this region,

ω_s / ω_p,

corresponds to the sharpness of the cutoff of the filter which, apart from other parameters, largely determines the complexity of the implementation of the filter. v_p specifies the allowable passband deviation of the gain with respect to the ideal case, while v_s corresponds to the maximum allowable stopband gain. Both figures are often specified in decibels (dB). The according tolerance scheme is open towards the bottom as the gain $|H(j\omega)| = 0$ approaches negative infinity on the logarithmic scale.

2.2 Butterworth Filters

With this type of filter, the squared magnitude response of the filter has the form

$$|H(j\omega)|^2 = \frac{1}{1 + \left(\dfrac{\omega}{\omega_c}\right)^{2N}} \, . \tag{2.7}$$

The Butterworth filter type features an optimally flat magnitude characteristic in the passband. All derivatives of the magnitude function are zero at $\omega = 0$ which leads to the named behaviour. All zeros of the filter occur at $\omega = \infty$. As a consequence, the magnitude monotonically approaches zero for $\omega \to \infty$. Figure 2-3 sketches a typical magnitude characteristic of a Butterworth filter plotted into a tolerance mask. The filter is specified as follows:

Order N :   4
Minimum gain in the passband v_p :   0.9
Maximum gain in the stopband v_s :   0.1

Normalising the frequency axis with respect to a reference frequency, in most cases the cutoff frequency of the filter, simplifies the mathematical treatment of filter transfer functions. Normalised frequencies are denoted by Ω or P.

Magnitude $|H(j\Omega)|$

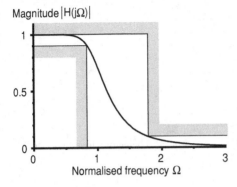

Normalised frequency Ω

Fig. 2-3
Magnitude characteristic of a fourth-order Butterworth filter

$$\Omega = \frac{\omega}{\omega_c} \qquad P = \frac{p}{\omega_c} \qquad\qquad (2.8)$$

In the normalised representation of the magnitude characteristic, the cutoff frequency of the filter is located at $\Omega = 1$.

$$|H(j\Omega)|^2 = \frac{1}{1+\Omega^{2N}} \qquad\qquad (2.9)$$

At the cutoff frequency ω_c ($\Omega = 1$), the gain amounts to $1/\sqrt{2}$, independently of the order of the filter, which corresponds in good approximation to a loss of 3 dB.

For the design of the filter and later on for the conversion to a digital filter, it is important to determine the poles and thus the transfer function $H(j\Omega)$ of the filter, for which we currently only know the magnitude (2.9). The procedure to derive the transfer function from the given squared magnitude is described in Sect. 1.5. In the first step, we replace Ω with $-jP$ to arrive at a function of the complex frequency P.

$$H(P)H(-P) = \frac{1}{1+(-jP)^{2N}} = \frac{1}{1+(-1)^N P^{2N}} = \frac{1}{1+P^N(-P)^N}$$

The characteristic function of the Butterworth filter can therefore be expressed as

$$K(P) = \frac{v(P)}{u(P)} = \frac{P^N}{1} = P^N$$

The desired poles of the transfer function are found by solving

$$1+(-1)^N P^{2N} = 0 \quad \text{or} \quad P^{2N} + (-1)^N = 0 \qquad\qquad (2.10)$$

for P and choosing those roots which lie in the left p-half-plane. The $2N$ roots have a magnitude of 1 and are equally spaced around the unit circle in the normalised P-plane. They can be calculated analytically. The $2N$ solutions are

$$P_{\infty v} = j e^{j\pi(2v+1)/2N} \qquad \text{with } 0 \le v \le 2N-1$$

or, after separation into the real and imaginary parts,

$$P_{\infty v} = -\sin[\pi(2v+1)/2N] + j\cos[\pi(2v+1)/2N]$$

$$\text{with } 0 \le v \le 2N-1 \; . \qquad (2.11)$$

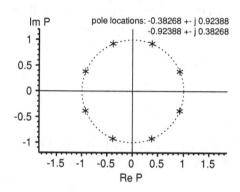

Im P   pole locations: -0.38268 +- j 0.92388
                              -0.92388 +- j 0.38268

Re P

Fig. 2-4
Location of the poles of a forth-order
Butterworth filter

Figure 2-4 sketches the location of the poles of the example shown in Fig. 2-3. Half of the poles lie in the right half, the other half in the left half of the P-plane. The transfer function of the filter is simply composed of the poles in the left half-plane as derived in Sect. 1.5.

$$H(P) = \frac{1}{(P - P_{\infty 1})(P - P_{\infty 2}) \; ... \; (P - P_{\infty N})} \qquad (2.12)$$

According to Sect. 1.3, this transfer function can be realised as a cascade arrangement of second-order sections and, in the case of an odd filter order, with an additional first-order filter block.

$$H(P) = \frac{1}{1 + A_{11}P + A_{21}P^2} \; ... \; \frac{1}{1 + A_{1k}P + A_{2k}P^2} \frac{1}{1 + A_{1(k+1)}P}$$

The filter design table in the appendix shows the coefficients A_{1i} and A_{2i} for filter orders up to $N = 12$. If N is even, $k = N/2$ second-order sections are needed to realise the filter. For odd orders, the filter is made up of one first-order and $k = (N-1)/2$ second-order sections. Up to now, we have only considered transfer functions in the normalised form. Substitution of p/ω_c for P yields the unnormalised transfer function $H(p)$.

By multiplying out the pole terms in (2.12), we obtain a polynomial in the denominator, which is called the Butterworth polynomial $B_n(P)$. For applications where this polynomial is needed in closed form, the coefficients can be obtained from the corresponding filter design tables.

$$H(P) = \frac{1}{B_n(P)} \tag{2.13}$$

The Butterworth filter is fully specified by two parameters, the cutoff frequency ω_c and the filter order N. These have to be derived from characteristic data of the tolerance scheme. A closer look at Fig. 2-3 shows that the magnitude graph just fits into the tolerance scheme if it passes through the edges (v_p, Ω_p) and (v_s, Ω_s). This yields two equations which can be used to determine the two unknown filter parameters in (2.7). As a result, the required filter order is specified by

$$N = \frac{\log \dfrac{\sqrt{1/v_s^2 - 1}}{\sqrt{1/v_p^2 - 1}}}{\log \dfrac{\omega_s}{\omega_p}} \ .$$

For the mathematical treatment of the filter types considered here and in later sections, it proved to be advantageous to introduce a normalised gain factor,

$$v_{\text{norm}} = v \Big/ \sqrt{(1 - v^2)} \ ,$$

which will simplify many relations. This applies, for instance, to the calculation of the filter order:

$$N = \frac{\log(v_{\text{pnorm}} / v_{\text{snorm}})}{\log(\omega_s / \omega_p)} \ . \tag{2.14a}$$

The filter order calculated using (2.14a) is not an integer and has to be rounded up in order to definitely comply with the prescribed tolerance scheme. Thus the order is in general a little higher than needed to fulfil the specification. This somewhat higher complexity can be used to improve the filter performance with respect to the slope, the passband ripple or the stopband attenuation.

Using (2.14b), the cutoff frequency of the low-pass filter can be determined such that the passband edge is still touched. The behaviour in the stopband, however, is better than that specified.

$$\omega_c = \omega_p \, v_{\text{pnorm}}^{1/N} \tag{2.14b}$$

Example 2-1

Determine the normalised poles and the transfer function of the 4th order Butterworth filter.

Evaluation of (2.11) for $N = 4$ yields the following roots of (2.10):

$$p_{\infty 1/2} = -0.9239 \pm j0.3827$$
$$p_{\infty 3/4} = -0.3827 \pm j0.9239$$
$$p_{\infty 5/6} = +0.3827 \pm j0.9239$$
$$p_{\infty 7/8} = +0.9239 \pm j0.3827$$

The resulting pole locations are plotted in Fig. 2-4. For the construction of the transfer function, we choose the poles in the left half-plane.

$$H(P) = \frac{1}{(P - P_{\infty 1})(P - P_{\infty 2})(P - P_{\infty 3})(P - P_{\infty 4})}$$

$$H(P) = \frac{1}{(P + 0.9239 + j0.3827)(P + 0.9239 - j0.3827)}$$

$$\times \frac{1}{(P + 0.3827 + j0.9239)(P + 0.3827 - j0.9239)}$$

$$H(P) = \frac{1}{P^2 + 1.8478P + 1} \frac{1}{P^2 + 0.7654P + 1}$$

$$H(P) = \frac{1}{P^4 + 2.6132P^3 + 3.4143P^2 + 2.6132P + 1} \tag{2.15}$$

The denominator polynomial in (2.15) is the 4th order Butterworth polynomial $B_4(P)$.

2.3 Chebyshev Filters

With this type of filter, the squared magnitude function is of the form

$$|H(j\Omega)|^2 = \frac{1}{1 + \varepsilon^2 T_N^2(\Omega)} . \tag{2.16}$$

$T_n(\Omega)$ is the Chebyshev polynomial of the first kind of the order n. There are various mathematical representations of this polynomial. The following is based on trigonometric and hyperbolic functions.

$$T_n(\Omega) = \begin{cases} \cos(n \arccos \Omega) & 0 \le \Omega \le 1 \\ \cosh(n \cosh^{-1} \Omega) & \Omega > 1 \end{cases} \tag{2.17}$$

Applying some theorems of hyperbolic functions, the following useful representation can be derived [9].

$$T_n(\Omega) = \frac{\left(\Omega + \sqrt{\Omega^2 - 1}\right)^n + \left(\Omega + \sqrt{\Omega^2 - 1}\right)^{-n}}{2} \tag{2.18}$$

For integer orders n, (2.18) yields polynomials with integer coefficients. The first polynomials for $n = 1 \ldots 5$ are given below.

$$T_1(\Omega) = \Omega$$

$$T_2(\Omega) = 2\Omega^2 - 1$$

$$T_3(\Omega) = 4\Omega^3 - 3\Omega$$
$$T_4(\Omega) = 8\Omega^4 - 8\Omega^2 + 1$$
$$T_5(\Omega) = 16\Omega^5 - 20\Omega^3 + 5\Omega$$

A simple recursive formula enables the computation of further Chebyshev polynomials.

$$T_{n+1}(\Omega) = 2\Omega T_n(\Omega) - T_{n-1}(\Omega) \tag{2.19}$$

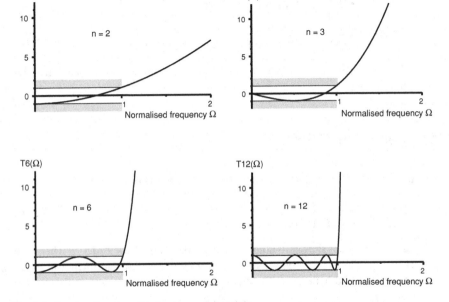

Fig. 2-5 Graphs of various Chebyshev polynomials

Figure 2-5 shows the graphs of some Chebyshev polynomials. These oscillate between +1 and −1 in the frequency range $0 \le \Omega \le 1$. The number of maxima and minima occurring is proportional to the degree n of the polynomial. For $\Omega > 1$ the functions monotonously approach infinity. The slope of the curves also increases proportionally to the order of the polynomial.

ε is another independent design parameter which can be used to influence the ripple of the magnitude in the passband, as can be seen upon examination of (2.16). The smaller ε is, the smaller is the amplitude of the ripple in the passband. At the same time, the slope of the magnitude curve in the stopband decreases. For a given filter order, sharpness of the cutoff and flatness of the magnitude characteristic are contradictory requirements. Comparison of Fig. 2-3 and Fig. 2-6 demonstrates that the achievable sharpness of the cutoff of the Chebyshev filter is higher compared to the Butterworth filter, if one assumes equal filter order, passband ripple and stopband attenuation.

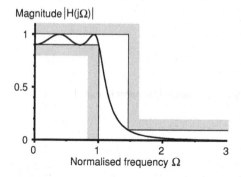

Fig. 2-6
Magnitude characteristic of a fourth-order Chebyshev filter
$v_p = 0.9$, $\varepsilon = 0.484$

In order to determine the poles of the Chebyshev filter, we start with a change to the complex frequency variable $P = j\Omega$.

$$|H(P)|^2 = \frac{1}{1 + \varepsilon^2 T_N^2(-jP)}$$

$$\text{for } P = j\Omega$$

The zeros of the denominator polynomial are the poles of the transfer function $H(P)$. So the poles can be determined by equating the denominator with zero.

$$1 + \varepsilon^2 T_N^2(-jP) = 0 \quad \text{or}$$
$$T_N(-jP) = \pm j / \varepsilon$$

By substitution of (2.17), we obtain the relationship

$$\cos[N \arccos(-jP)] = \pm j / \varepsilon \ .$$

Solving for P to obtain the poles is performed in two steps. We start with the introduction of a complex auxiliary variable $R + jI$.

$$R + jI = \arccos(-jP) \tag{2.20}$$

Thus the problem reduces, for the time being, to the solution of

$$\cos[N (R + jI)] = \pm j / \varepsilon \ .$$

Comparison of the real and imaginary parts on both sides of the equals sign yields two equations for determining R and I.

$$\cos NR \cosh NI = 0 \tag{2.21a}$$

$$\sin NR \sinh NI = \pm 1 / \varepsilon \tag{2.21b}$$

As cosh $x \neq 0$ for all real x, we get an equation from (2.21a) which can be used to directly determine the real part R of the auxiliary variable.

$$\cos NR = 0$$

$$NR = \pi/2 + v\pi$$

$$R_v = \frac{\pi}{2N} + v\frac{\pi}{N} \qquad (2.22a)$$

Substitution of (2.22a) into (2.21b) yields an equation which can be used to determine the imaginary part I.

$$(-1)^v \sinh NI = \pm 1/\varepsilon$$

$$I_v = \pm\frac{1}{N}\operatorname{arsinh}(1/\varepsilon) \qquad (2.22b)$$

The next step in calculating the poles of the Chebyshev filter is to solve for P in (2.20).

$$R + jI = \arccos(-jP_{\infty v})$$

$$-jP_{\infty v} = \cos(R + jI)$$

$$P_{\infty v} = j\cos R_v \cosh I_v + \sin R_v \sinh I_v$$

Substitution of the already determined auxiliary variable (2.22a,b) leads to the desired relationship.

$$P_{\infty v} = \sin[\pi(2v+1)/2N]\sinh[\operatorname{arsinh}(1/\varepsilon)/N]$$
$$+ j\cos[\pi(2v+1)/2N]\cosh[\operatorname{arsinh}(1/\varepsilon)/N] \qquad (2.23)$$
$$\text{with } 0 \leq v \leq 2N-1$$

$2N$ solutions are obtained, which are located on an ellipse in the P-plane. Compare this to the situation with Butterworth filters in the last section, where the poles were found to be on the unit circle. Figure 2-7 shows a plot of the pole locations for the fourth-order Chebyshev filter whose magnitude graph has already been sketched in Fig. 2-6. Using the procedure introduced in Sect. 1.5, the transfer function of the desired filter is simply composed of the N poles in the left half-plane. This leads again to a transfer function of the form

$$H(P) = \frac{1}{(P-P_{\infty 1})(P-P_{\infty 2})\ldots(P-P_{\infty N})} \qquad (2.24)$$

or, after separation into first- and second-order filter blocks, to a product representation of partial transfer functions.

$$H(P) = V\frac{1}{1+A_{11}P+A_{21}P^2}\ldots\frac{1}{1+A_{1k}P+A_{2k}P^2}\frac{1}{1+A_{1(k+1)}P}$$

The factor V normalises the maximum gain of the low-pass filter to unity and can be calculated as

$$V = \begin{cases} 1 & \text{for } N \text{ odd} \\ 1/\sqrt{1+\varepsilon^2} & \text{for } N \text{ even.} \end{cases}$$

Also for Chebyshev filters, we have included tables in the appendix to show the coefficients A_{1i} and A_{2i} for filter orders up to $N = 12$. If N is even, $k = N/2$ second-order sections are needed to realise the filter. For odd orders, the filter is made up of one first-order and $k = (N-1)/2$ second-order sections. Substitution of p/ω_c for P yields the unnormalised transfer function $H(p)$.

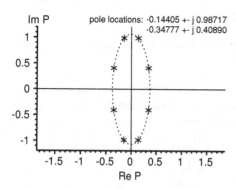

pole locations: -0.14405 +- j 0.98717
                -0.34777 +- j 0.40890

Fig. 2-7
Plot of the poles of a fourth-order Chebyshev filter (related magnitude graph is shown in Fig. 2-6)

The characteristics of a Chebyshev filter are determined by three parameters: the cutoff frequency ω_c, the parameter ε and the filter order N. According to (2.16), the parameter ε is directly related to the allowable deviation of the passband gain v_p.

$$v_p^2 = 1/(1+\varepsilon^2) \tag{2.25}$$

The cutoff frequency of the filter, to which the frequency axis is normalised, is not the point of 3-dB attenuation, as in the case of the Butterworth filter, but the edge frequency, where the curve of the magnitude leaves the passband in the tolerance scheme (Fig. 2-6).

$$\omega_c = \omega_p \quad \text{or} \quad \Omega_p = 1 \tag{2.26a}$$

The relation to calculate the required order of the filter is derived from the condition that the graph just touches the stopband edge (v_s, ω_s).

$$N = \frac{\cosh^{-1}(v_{pnorm}/v_{snorm})}{\cosh^{-1}(\omega_s/\omega_p)} \tag{2.26b}$$

The similarity to the corresponding relation for the Butterworth filter is striking. The logarithm is simply replaced by the inverse hyperbolic cosine. For definite compliance with the prescribed tolerance scheme, the result of (2.26b) has to be rounded up to the next integer.

By appropriate choice of the parameter ε, we can determine how the "surplus" filter order, which is a result of rounding up, will be used to improve the filter performance beyond the specification. With

$$\varepsilon = 1 / v_{\text{pnorm}} \qquad\qquad (2.26c)$$

we get a filter that has exactly the prescribed ripple in the passband, whereas the stopband behaviour is better than required. With

$$\varepsilon = \frac{1}{v_{\text{snorm}} \cosh\left(N \cosh^{-1}(\omega_s / \omega_p)\right)} \ , \qquad\qquad (2.26d)$$

on the other hand, the magnitude plot will just touch the stopband edge, but the passband ripple is smaller than specified.

Example 2-2

Determine the squared magnitude of the transfer function of a Chebyshev filter, satisfying the following specification:

Passband ripple	3 dB
Stopband attenuation	40 dB
Passband edge frequency	1 kHz
Stopband edge frequency	1.7 kHz

From the passband ripple and the stopband attenuation, we calculate the minimum gain in the passband and the maximum gain in the stopband.

$$v_p = 10^{-3/20} = 0.7079 \qquad v_{\text{pnorm}} = 1.0024$$
$$v_s = 10^{-40/20} = 0.01 \qquad v_{\text{snorm}} = 0.0100$$

The required order of the filter is obtained using (2.26b).

$$N = \frac{\cosh^{-1}(1.0024 / 0.01)}{\cosh^{-1}(1.7 / 1)} = 4.7191$$

The choice is $N = 5$. The filter parameter ε is calculated as

$$\varepsilon = 1 / v_{\text{pnorm}} = 0.9976 \ .$$

With these results, we can determine the desired squared magnitude of the transfer function using (2.16).

$$\left| H(j\Omega) \right|^2 = \frac{1}{1 + \varepsilon^2 T_5^2(\Omega)}$$

$$\left| H(j\Omega) \right|^2 = \frac{1}{1 + 0.9952 \left(16\Omega^5 - 20\Omega^3 + 5\Omega\right)^2} \qquad \text{with } \Omega = \frac{\omega}{2\pi \times 1\,\text{kHz}}$$

2.4 Inverse Chebyshev Filters

Occasionally, there are applications which require a regular ripple in the stopband and a monotonic shape of the magnitude in the passband. Such a filter can be easily derived from the Chebyshev filter of Sect. 2.3, which has a squared magnitude function (2.16) of the form

$$|H(j\Omega)|^2 = \frac{1}{1 + \varepsilon^2 T_N^2(\Omega)} \; .$$

In the first step, we apply the transformation $\Omega \rightarrow 1/\Omega$ to invert the frequency axis, which yields a high-pass response.

$$|H(j\Omega)|^2 = \frac{1}{1 + \varepsilon^2 T_N^2(1/\Omega)} \qquad (2.27)$$

In the second step, (2.27) is subtracted from unity which results in a low-pass magnitude function again. The ripple can now be found in the stopband.

$$|H(j\Omega)|^2 = 1 - \frac{1}{1 + \varepsilon^2 T_N^2(1/\Omega)}$$

$$|H(j\Omega)|^2 = \frac{\varepsilon^2 T_N^2(1/\Omega)}{1 + \varepsilon^2 T_N^2(1/\Omega)} \qquad (2.28)$$

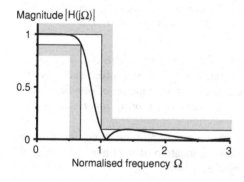

Fig. 2-8
Magnitude characteristic of a fourth-order inverse Chebyshev filter
$N = 4$
$\varepsilon = 0.1005$

Figure 2-8 shows the corresponding magnitude characteristic which is monotonic in the passband and shows ripples in the stopband. This behaviour requires a transfer function with finite zeros which are provided by the numerator polynomial in (2.28). These zeros can be calculated from the inverse of the zeros of the Chebyshev polynomial.

$$P_{0\mu} = \pm j \frac{1}{\cos[\pi(2\mu+1)/2N]} \qquad (2.29)$$

$$\text{with } 0 \leq \mu \leq M/2-1$$

The order of the numerator polynomial M is always even. It is related to the order N of the denominator polynomial as

$M = N$ for N even and

$M = N - 1$ for N odd.

The normalised poles of the inverse Chebyshev filter equal the reciprocal of the poles of the Chebyshev filter as introduced in Sect. 2.3. Figure 2-9 shows the location of the poles and zeros for a particular Chebyshev filter.

In the case of a cascade of first- and second-order filter blocks, we have the following product representation of partial transfer functions.

$$H(P) = \frac{1 + B_{21}P^2}{1 + A_{11}P + A_{21}P^2} \cdots \frac{1 + B_{2k}P^2}{1 + A_{1k}P + A_{2k}P^2} \frac{1}{1 + A_{1(k+1)}P} \qquad (2.30)$$

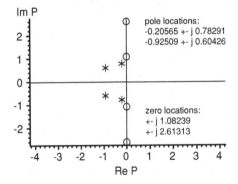

pole locations:
-0.20565 +- j 0.78291
-0.92509 +- j 0.60426

zero locations:
+- j 1.08239
+- j 2.61313

Fig. 2-9
Pole and zero plot of a fourth-order inverse Chebyshev filter according to the specification of Fig. 2-8

The filter design tables in the Appendix show the coefficients A_{1i}, A_{2i} and B_{2i} for filter orders up to $N = 12$. If N is even, $k = N/2$ second-order sections are needed to realise the filter. For odd orders, the filter is made up of one first-order and $k = (N-1)/2$ second-order sections. Substitution of p/ω_c for P yields the unnormalised transfer function $H(p)$.

The parameters ε and ω_c have a somewhat different meaning for inverse Chebyshev filters. ε determines the stopband attenuation of the filter and is related to the maximum gain v_s in the stopband by the equation

$$v_s^2 = \varepsilon^2 / (1 + \varepsilon^2).$$

The cutoff and normalisation frequency ω_c is now the edge frequency of the stopband.

$$\omega_c = \omega_s \quad \text{or} \quad \Omega_s = 1 \qquad (2.32a)$$

For the determination of the required filter order, (2.26b) is still valid.

$$N = \frac{\cosh^{-1}(v_{pnorm} / v_{snorm})}{\cosh^{-1}(\omega_s / \omega_p)} \tag{2.32b}$$

The parameter ε specifies the maximum normalised gain v_{snorm} and thus determines the ripple in the stopband.

$$\varepsilon = v_{snorm} \tag{2.32c}$$

The presented design rules result in frequency responses which have exactly the prescribed ripple in the stopband. The rounded up filter order leads to a better behaviour in the passband then specified.

For a given tolerance scheme, Chebyshev and inverse Chebyshev filters require the same filter order. The complexity of the inverse filter, however, is higher because the zeros have to be implemented, too, but consequently we obtain a better group delay behaviour in the passband than that of the Chebyshev filter.

2.5 Elliptic Filters

We showed that, compared to the Butterworth filter, a given tolerance scheme can be satisfied with a lower filter order if, with respect to the magnitude characteristic, a regular ripple is allowed in the passband (Chebyshev filter) or stopband (inverse Chebyshev filter). The next logical step is to simultaneously allow a ripple in the passband and in the stopband. Figure 2-10 shows the magnitude characteristic of a corresponding fourth-order filter.

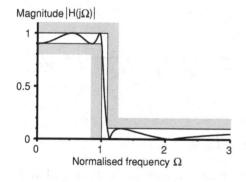

Fig. 2-10
Magnitude characteristic of a fourth-order elliptic filter
$N = 4$
$\varepsilon = 0.4843$
$k_1 = 0.0487$

The filter type with this property is called a Cauer or elliptic filter because elliptic functions are used to construct a magnitude response with the desired characteristics. The squared magnitude function of the elliptic filter is of the form

$$|H(u)|^2 = \frac{1}{1 + \varepsilon^2 \, sn^2(u, k_1)} \quad \text{with} \quad u = f(\Omega), \tag{2.33}$$

where $sn(u,k_1)$ is a Jacobian elliptic function [37]. u is a complex variable which is appropriately related to the frequency variable Ω. k_1 and ε are design parameters.

2.5.1 Properties of the Jacobian Elliptic Function

It has been known since the 1930s that low-pass filter transfer functions with equal-ripple response in both the passband and the stopband can be described exactly with Jacobian elliptic functions [16]. $w = sn(u,k)$ is one of the 12 known Jacobian functions [56]. It is the inverse function of the incomplete elliptic integral

$$u = \int_0^w \frac{dt}{\sqrt{(1-t^2)(1-k^2t^2)}} \ .$$

The Jacobian elliptic function is a function of the complex variable $u = u_1 + ju_2$, where u_1 is the real part and u_2 the imaginary part of u. In the directions of both the real u_1-axis and the imaginary ju_2-axis, this function is periodic (doubly periodic) and has periodically recurrent poles (×) and zeros (○) in the complex u-plane (Fig. 2-11).

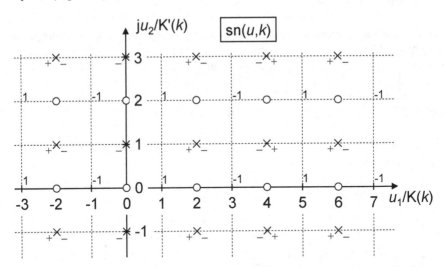

Fig. 2-11 Location of the poles and zeros of the Jacobi elliptic function

The periods in both directions are determined by the parameter k. The period $K(k)$ in the direction of the real axis is the complete elliptic integral of the first kind with the parameter k. The period in the direction of the imaginary axis is the complementary elliptic integral $K'(k)$.

$$K'(k) = K(k')$$

$$\text{with } k' = \sqrt{1-k^2}$$

Complete elliptic integrals can be obtained from tables or calculated with arbitrary precision by using appropriate series expansions. Fig. 2-12 shows the graph of K as a function of the parameter k. In particular, we have

$$K(0) = \frac{\pi}{2}.$$
(2.34)

As k approaches 1, $K(k)$ can be approximated [37] by

$$K(k) \approx \ln\left(\frac{4}{\sqrt{1-k^2}}\right) = \ln\left(\frac{4}{k'}\right).$$
(2.35)

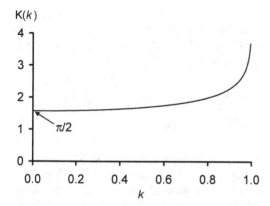

Fig. 2-12
Complete elliptic integral of the first kind

Along the dotted paths in Fig. 2-11, the Jacobian elliptic function $\mathrm{sn}(u,k)$ assumes real values. On a horizontal line through the zeros where $u = u_1 + j2nK'$ (n integer), $\mathrm{sn}(u,k)$ oscillates between -1 and $+1$ with the period $4K$. For small values of k, $\mathrm{sn}(u_1 + j2nK')$ can, in fact, be approximated with good accuracy by a sine wave:

$$\mathrm{sn}(u_1 + j2nK', k) \approx \sin\left(\frac{\pi u_1}{2K(k)}\right) \qquad n \text{ integer}.$$

Fig. 2-13 shows the shape of the periodic function for various values of k. If k approaches 1, the sine wave becomes more and more saturated.

Choosing a horizontal path through the poles with $u = u_1 + j(2n+1)K'$, the Jacobian elliptic function oscillates between $\pm 1/k$ and $\pm\infty$. A good approximation for $k \ll 1$ is given by

$$\mathrm{sn}\left(u_1 + j(2n+1)K', k\right) \approx \frac{1}{k \sin\left(\dfrac{\pi u_1}{2K(k)}\right)} \qquad n \text{ integer}.$$

sn(u_1,k)

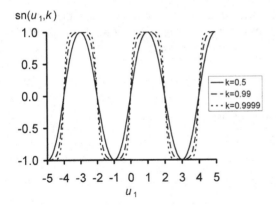

Fig. 2-13

The Jacobian elliptic function sn along the path $u = u_1 + j2nK'$

Finally we consider the vertical dotted paths in Fig. 2-11 which run in the middle between the poles and zeros. Here sn(u,k) oscillates between ± 1 and $\pm 1/k$ with the period 2K'. An approximation can be expressed as

$$sn\big((2n+1)K+ju_2,k\big) \approx (-1)^n \frac{1}{\sqrt{1-(1-k^2)\sin^2\left(\dfrac{\pi u_2}{2\,K'(k)}\right)}} \qquad n \text{ integer} .$$

Fig. 2-14 summarizes the shape of the Jacobian elliptic function sn for arguments u on the dotted grid in Fig. 2-11. The graph shows the absolute function values. It is interesting to note that these lie, for the three considered paths, in bands which do not overlap. So from the value of sn(u,k), it can be unambiguously decided if the argument u lies on a horizontal line through the zeros, on a horizontal line through the poles, or on a vertical line in the middle between poles and zeros.

sn(u_1), sn(K+ju_2), sn(u_1+jK')

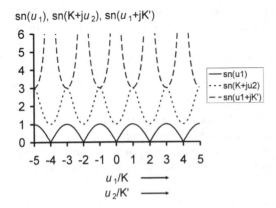

Fig. 2-14

The absolute value of sn(u,k) on the three considered paths, $k = 0.333$

As mentioned above, the Jacobian elliptic function sn(u,k) belongs to a family of further elliptic functions. In the context of the calculation of the poles and zeros of elliptic filters, we will make use of the functions cn(u,k) and dn(u,k) which are related to sn(u,k) by the following functional equations:

$$\text{cn}^2(u,k) = 1 - \text{sn}^2(u,k)$$

$$\text{dn}^2(u,k) = 1 - k^2 \text{sn}^2(u,k).$$

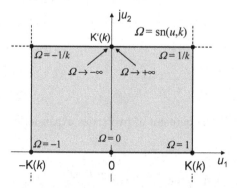

Fig. 2-15
The basic rectangle in the u-plane

The dotted lines in Fig. 2-11 subdivide the u-plane into a pattern of basic rectangles which all look similar showing a pole on one edge and a zero on the opposite edge. Fig. 2-15 depicts one of these rectangles in more detail. On the path $jK' \rightarrow -K + jK' \rightarrow -K \rightarrow 0 \rightarrow K \rightarrow K + jK' \rightarrow jK'$, $\Omega = \text{sn}(u,k)$ increases monotonically from $-\infty$ to $+\infty$ as illustrated in Fig. 2-16.

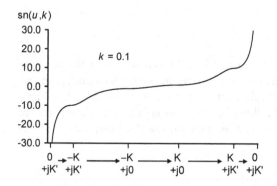

Fig. 2-16
Graph of the Jacobian elliptic function on the basic rectangle for $k = 0.1$

Conversely, the inverse Jacobian elliptic function $u = \text{sn}^{-1}(\Omega,k)$ maps the whole real axis $-\infty < \Omega < +\infty$ onto a rectangle in the u-plane or more precisely, as a consequence of the periodicity of the elliptic function, onto each of the basic rectangles in the u-plane. These have a width of $2K(k)$ and a height of $K'(k)$.

For some applications, it might be interesting to create a rectangle with arbitrary size as Ω proceeds from $-\infty$ to $+\infty$. With the parameter k, we can define the ratio between the width and the height since, with increasing k, $K(k)$ increases and K'(k) decreases and vice versa. By the introduction of an additional factor a, the rectangle can be scaled to the desired size:

$$u = a \operatorname{sn}^{-1}(\Omega, k) \qquad \text{with } -\infty < \Omega < +\infty . \tag{2.36}$$

Width W and height H of the rectangle can be calculated as

$$W = 2a\,\mathrm{K}(k) \tag{2.37a}$$

$$H = a\,\mathrm{K}'(k) . \tag{2.37b}$$

Division of (2.37b) by (2.37a) yields

$$\frac{\mathrm{K}'(k)}{\mathrm{K}(k)} = \frac{H}{W/2} .$$

The related value of k can be calculated as the quotient of two series expansions [37].

$$k = \left(\frac{2q^{1/4}(1 + q^2 + q^6 + q^{12} + \ldots)}{1 + 2q + 2q^4 + 2q^9 + \ldots} \right)^2 \qquad \text{with } q = e^{-\pi \mathrm{K}'/\mathrm{K}} = e^{-2\pi H/W} \tag{2.38}$$

The scale factor a is obtained from (2.37) as

$$a = \frac{W}{2\,\mathrm{K}(k)} = \frac{H}{\mathrm{K}'(k)} . \tag{2.39}$$

2.5.2 Elliptic Filter Design

If we follow the path $a \to b \to c \to d$ in Fig. 2-17 and substitute the resulting values of the Jacobian elliptic function in (2.33), we obtain the frequency response of a low-pass filter with equal ripple in the passband and stopband. The same frequency response results if we follow the path $a \to e \to f \to d$. The number of poles and zeros which lie on the closed rectangle in Fig. 2-17 corresponds to the order of the filter

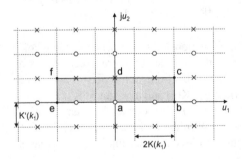

Fig. 2-17
Rectangle in the u-plane representing a third-order elliptic filter

In the passband, which is the region between the points a and b, the absolute value of the Jacobian elliptic function oscillates between 0 and 1. Accordingly, the square of the transfer function (2.33) assumes values between 1 and $1/(1+\varepsilon^2)$. The region between the points b and c is the transition band of the filter. In the region between the points c and d, the absolute value of the elliptic function oscillates between $1/k_1$ and ∞. The square of the transfer function thus assumes values between 0 and $1/(1+\varepsilon^2/k_1^2)$ which corresponds to the stopband of the filter. Fig. 2-18 shows an example of the frequency response of a third-order low-pass filter.

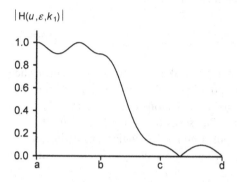

Fig. 2-18
Elliptic low-pass filter, $N = 3$, $\varepsilon = 0.48$, $k_1 = 0.049$

The minimum gain v_p in the passband can be calculated as

$$v_\mathrm{p} = 1\Big/\sqrt{1+\varepsilon^2} \ .$$

The maximum gain v_s in the stopband is obtained as

$$v_\mathrm{s} = 1\Big/\sqrt{1+\varepsilon^2/k_1^2} \ .$$

From these two equations, we can derive relations which enable us to determine the design parameters ε and k_1 of the elliptic filter from the characteristics of a desired tolerance mask.

$$\varepsilon = \sqrt{1/v_\mathrm{p}^2 - 1} = 1/v_\mathrm{pnorm} \tag{2.40a}$$

$$k_1 = \sqrt{1/v_\mathrm{p}^2 - 1}\Big/\sqrt{1/v_\mathrm{s}^2 - 1} = v_\mathrm{snorm}/v_\mathrm{pnorm} \tag{2.40b}$$

To complete the picture, we need to find a relation which maps the whole real frequency axis $-\infty < \Omega < +\infty$ onto the edge of the shaded rectangle in Fig. 2-17 in order to obtain a frequency response representation as a function of the frequency Ω rather than of a rectangular path in the u-plane. Based on the inverse Jacobian elliptic function, we derived in the previous section the desired relationship for a rectangle with arbitrary width W and height H. In our example of a third-order filter, the rectangle in Fig. 2-17 has the dimensions

$$W = 2N\,\mathrm{K}(k_1) = 6\,\mathrm{K}(k_1)$$
$$H = \mathrm{K'}(k_1)$$

The scale factor a in (2.36) is calculated using (2.39):

$$a = \frac{W}{2\,\mathrm{K}(k)} = \frac{H}{\mathrm{K'}(k)} = N\frac{\mathrm{K}(k_1)}{\mathrm{K}(k)} = \frac{\mathrm{K'}(k_1)}{\mathrm{K'}(k)} \tag{2.41}$$

The rectangular path in the u-plane is therefore obtained as

$$u_{\mathrm{W,H}} = N\frac{\mathrm{K}(k_1)}{\mathrm{K}(k)}\,\mathrm{sn}^{-1}(\Omega,k) \qquad \text{for } -\infty < \Omega < +\infty \ . \tag{2.42a}$$

The rectangle created by (2.42a) is symmetrical about the u_2-axis. But in the case of even filter orders, the corresponding pole/zero pattern cannot be arranged symmetrically about the ju_2-axis as demonstrated in Fig. 2-19. An offset of $\mathrm{K}(k_1)$ is therefore needed in (2.42a) so as to shift the rectangle to the appropriate position in relation to the poles and zeros which are to lie on its upper and lower edge. For even filter orders, the rectangle in the u-plane is therefore calculated as

$$u_{\mathrm{W,H}} = N\frac{\mathrm{K}(k_1)}{\mathrm{K}(k)}\,\mathrm{sn}^{-1}(\Omega,k) + \mathrm{K}(k_1) \qquad \text{for } -\infty < \Omega < +\infty \ . \tag{2.42b}$$

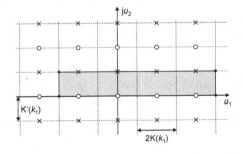

Fig. 2-19
Characteristic rectangle of a fourth-order elliptic filter in the u-plane

By substitution of (2.42) in (2.33), the squared magnitude response of the elliptic filter can now be expressed in closed form as a function of the frequency Ω:

$$|H(\Omega)|^2 = \frac{1}{1+\varepsilon^2\,\mathrm{sn}^2\!\left(N\dfrac{\mathrm{K}(k_1)}{\mathrm{K}(k)}\,\mathrm{sn}^{-1}(\Omega,k),k_1\right)} \qquad \text{for } N \text{ odd}$$

$$|H(\Omega)|^2 = \frac{1}{1+\varepsilon^2\,\mathrm{sn}^2\!\left(N\dfrac{\mathrm{K}(k_1)}{\mathrm{K}(k)}\,\mathrm{sn}^{-1}(\Omega,k)+\mathrm{K}(k_1),k_1\right)} \qquad \text{for } N \text{ even}$$

$$\tag{2.43}$$

The parameter k is still unknown. Simple rearrangement of (2.41) yields the relation

$$\frac{K'(k)}{K(k)} = \frac{1}{N}\frac{K'(k_1)}{K(k_1)}$$

(2.44)

which can be used to calculate k with the aid of (2.38). Based on relationship (2.42), we can assign concrete frequency values to the points a to f in Fig 2-17 as shown in Fig. 2-20.

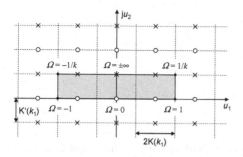

Fig. 2-20
Characteristic rectangle of a third-order elliptic filter in the u-plane, parameterised with frequency values

Like k_1, k is also a parameter which is related to the tolerance mask of the elliptic filter. The bottom right corner of the rectangle in Fig. 2-20 corresponds to the passband edge of the filter. Here the normalised frequency variable Ω assumes a value of 1. This means that the frequency axis is normalised to the passband edge frequency. The upper right corner corresponds to the stopband edge of the filter. Here the frequency variable assumes a value of $1/k$. With $\Omega_p = 1$ and $\Omega_s = 1/k$, the parameter k is related as

$$k = \frac{\Omega_p}{\Omega_s} = \frac{\omega_p}{\omega_s}$$

(2.45)

with the edge frequencies of the filter. Thus k defines the slope of the filter.

(2.44) relates three important characteristics of the filter, namely order N, stopband attenuation and passband ripple expressed by k_1, and slope of the filter expressed by k. If two of these parameters are given, the third one is uniquely determined. If k and k_1 are prescribed in a tolerance scheme, (2.46) provides a relation to determine the required filter order.

$$N = \frac{K'(k_1)/K(k_1)}{K'(k)/K(k)}$$

(2.46)

Since complete elliptic integrals cannot be expressed by elementary functions, (2.46) is a somewhat unwieldy relation, which makes it necessary to use tables or computer support to determine the required filter order. A good approximation of (2.46) can be derived with the aid of the approximations of the complete elliptic integral as given by (2.34) and (2.35). In practical filter design applications, k is close to unity while $k_1 \ll 1$. $K(k)$, $K'(k)$, $K(k_1)$, and $K'(k_1)$ can therefore be approximated as follows:

$$K(k) \approx \ln\left(\frac{4}{\sqrt{1-k^2}}\right) = \frac{1}{2}\ln\left(\frac{16}{1-k^2}\right) = \frac{1}{2}\ln\left(\frac{16}{(1-k)(1+k)}\right) \approx \frac{1}{2}\ln\left(\frac{8}{1-k}\right)$$

$$K'(k) = K\left(\sqrt{1-k^2}\right) \approx \pi/2$$

$$K(k_1) \approx \pi/2$$

$$K'(k_1) = K\left(\sqrt{1-k_1^2}\right) \approx \ln\left(\frac{4}{k_1}\right)$$

This simplifies (2.46) to

$$N \approx \frac{2}{\pi^2}\ln\frac{4}{k_1}\ln\frac{8}{1-k}. \qquad (2.47)$$

It is mandatory that k, k_1, and N satisfy equation (2.46). If this is not the case, the rectangle defined by (2.42) will not match the rectangle in Fig. 2-17 or Fig. 2-19 which will inevitably lead to deviations from the desired optimum frequency response. In practical cases, k and k_1 are derived from the tolerance mask of a filter design problem. The order of the filter as calculated with (2.46) or (2.47) will in general not be integer and has to be rounded up. It is essential that k or k_1 or both are recalculated based on the final value of the filter order so as to satisfy (2.46).

2.5.3 Calculation of the Poles and Zeros

For the calculation of the poles and zeros of the elliptic filter, we apply the procedure introduced in Sect. 1.5. In the first step, $-jP$ is substituted for Ω in the squared magnitude response (2.43).

$$|H(P)|^2 = \frac{1}{1+\varepsilon^2 \, \mathrm{sn}^2\left(N\dfrac{K(k_1)}{K(k)}\mathrm{sn}^{-1}(-jP,k),k_1\right)} \qquad \text{for } N \text{ odd}$$

$$\qquad\qquad (2.48)$$

$$|H(P)|^2 = \frac{1}{1+\varepsilon^2 \, \mathrm{sn}^2\left(N\dfrac{K(k_1)}{K(k)}\mathrm{sn}^{-1}(-jP,k)+K(k_1),k_1\right)} \qquad \text{for } N \text{ even}$$

The poles and zeros of (2.48) are calculated while taking into account that the poles of the denominator are the zeros of the transfer function and vice versa. The obtained poles and zeros are finally appropriately assigned to $H(P)$ and $H(-P)$.

By introducing the intermediate variable u with

$$u = N \frac{K(k_1)}{K(k)} \operatorname{sn}^{-1}(-jP,k) \quad \text{for } N \text{ odd } \text{ or}$$

$$u = N \frac{K(k_1)}{K(k)} \operatorname{sn}^{-1}(-jP,k) + K(k_1) \quad \text{for } N \text{ even respectively,}$$

$$(2.49)$$

the poles and zeros of (2.48) can be first determined in the u-plane by evaluation of the denominator $1+\varepsilon^2 \operatorname{sn}^2(u,k_1)$. With the aid of (2.49), the poles and zeros are then transformed into the P-plane.

Calculation of the zeros:
The zeros of (2.48) coincide with the poles of the elliptic function in the denominator. Since the elliptic function appears in the square, all calculated zeros are double and can be equally assigned to $H(P)$ and $H(-P)$. We see by inspection of Fig. 2-17 and 2-19 that the poles of the denominator are located at

$$u_{0i} = 2i\,K(k_1) + jK'(k_1)$$

$$\text{with } i = -(N-1)/2 \ ... \ +(N-1)/2 \quad \text{for } N \text{ odd}$$

$$\text{and } i = -N/2+1 \ ... \ +N/2 \qquad \text{for } N \text{ even.}$$

The complex zero frequencies P_0 are calculated using the transform equation (2.49).

$$u_{0i} = N \frac{K(k_1)}{K(k)} \operatorname{sn}^{-1}(-jP_0,k) = 2i\,K(k_1) + jK'(k_1) \quad \text{for } N \text{ odd and}$$

$$u_{0i} = N \frac{K(k_1)}{K(k)} \operatorname{sn}^{-1}(-jP_0,k) + K(k_1) = 2i\,K(k_1) + jK'(k_1) \quad \text{for } N \text{ even.}$$

By rearrangement and substitution of (2.46), one obtains

$$P_0 = j\,\operatorname{sn}\!\left(\frac{2i}{N}K(k) + j\,K'(k),k\right) \quad \text{for } N \text{ odd and}$$

$$P_0 = j\,\operatorname{sn}\!\left(\frac{2i-1}{N}K(k) + j\,K'(k),k\right) \quad \text{for } N \text{ even.}$$

Using an identity of elliptic functions, we finally arrive at the expression

$$P_{0i} = j\frac{1}{k\,\operatorname{sn}\!\left(\dfrac{2i}{N}K(k),k\right)} \qquad \text{with } i = -(N-1)/2 \ ... \ +(N-1)/2$$
$$\text{for } N \text{ odd and}$$

$$P_{0i} = j\frac{1}{k\,\operatorname{sn}\!\left(\dfrac{2i-1}{N}K(k),k\right)} \qquad \text{with } i = -N/2+1 \ ... \ +N/2$$
$$\text{for } N \text{ even.}$$

or in more compact form

$$P_{0i} = j\frac{1}{k\,\mathrm{sn}\left(\dfrac{m}{N}K(k),k\right)} \quad \text{with } m = -(N-1) \text{ to } N-1 \text{ in steps of } 2 \, . \quad (2.50)$$

Calculation of the poles:
The poles of (2.48) coincide with the zeros of the denominator. The location of the zeros in the u-plane are calculated by equating the denominator with zero.

$$1 + \varepsilon^2 \,\mathrm{sn}^2(u_\infty, k_1) = 0$$

$$1 + \varepsilon^2\left(1 - \mathrm{cn}^2(u_\infty, k_1)\right) = 0$$

$$1 + \varepsilon^2\left(1 - \frac{1}{\mathrm{cn}^2(ju_\infty, k_1')}\right) = 0$$

$$1 + \varepsilon^2\left(1 - \frac{1}{1 - \mathrm{sn}^2(ju_\infty, k_1')}\right) = 0$$

$$\mathrm{sn}^2(ju_\infty, k_1') = \frac{1}{1 + \varepsilon^2}$$

$$u_{\infty lm} = \pm j\,\mathrm{sn}^{-1}\left(\frac{1}{\sqrt{1+\varepsilon^2}}, k_1'\right) + 2l\,K(k_1) + j2m\,K'(k_1)$$

The pole pairs occur periodically in the u-plane. N of these pairs are associated with the rectangle which characterizes the elliptic filter of order N.

$$u_{\infty i} = 2i\,K(k_1) \pm jU_\infty, \quad U_\infty = \mathrm{sn}^{-1}\left(\frac{1}{\sqrt{1+\varepsilon^2}}, k_1'\right)$$

with $i = -(N-1)/2 \ \dots \ +(N-1)/2$ for N odd   (2.51)
and $i = -N/2+1 \ \dots \ +N/2$ for N even

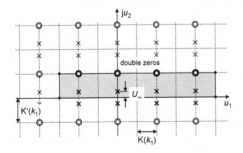

Fig. 2-21
Location of the zeros and poles of a fourth-order elliptic filter in the u-plane

The locations of the poles in the P-plane are calculated using the transform equation (2.49).

$$u_{\infty i} = N \frac{K(k_1)}{K(k)} sn^{-1}(-jP_{\infty i},k) = 2i\,K(k_1) \pm jU_{\infty} \quad \text{for } N \text{ odd and}$$

$$u_{\infty i} = N \frac{K(k_1)}{K(k)} sn^{-1}(-jP_{\infty i},k) + K(k_1) = 2i\,K(k_1) \pm jU_{\infty} \quad \text{for } N \text{ even.}$$

$$(2.52)$$

Solving (2.52) for P yields the desired relations to calculate the poles in the P-plane.

$$P_{\infty i} = j\, sn\left(\frac{2i}{N}K(k) \pm j\frac{K'(k)}{K'(k_1)}U_{\infty},k\right)$$

with $i = -(N-1)/2 \dots +(N-1)/2$ for N odd and

$$P_{\infty i} = j\, sn\left(\frac{2i-1}{N}K(k) \pm j\frac{K'(k)}{K'(k_1)}U_{\infty},k\right)$$

with $i = -N/2+1 \dots +N/2$ $\qquad$ for N even

These expressions can be simplified to a form which is likewise valid for even and odd filter orders N:

$$P_{\infty i} = j\, sn\left(\frac{m}{N}K(k) \pm j\frac{K'(k)}{K'(k_1)}U_{\infty},k\right)$$

with $m = -(N-1)$ to $N-1$ in steps of 2.

Using the functional relation

$$sn(x+jy,k) = \frac{sn(x,k)\,dn(y,k') + j\,sn(y,k')\,cn(y,k')\,cn(x,k)\,dn(x,k)}{1 - dn^2(x,k)sn^2(y,k')},$$

real and imaginary part of the poles can be calculated separately. With

$$u_r = \frac{m}{N}K(k) \quad \text{and} \quad u_i = \pm\frac{K'(k)}{K'(k_1)}U_{\infty}$$

we finally arrive at

$$\text{Re } P_{\infty i} = \frac{\pm sn(u_i,k')\,cn(u_i,k')\,cn(u_r,k)\,dn(u_r,k)}{1 - dn^2(u_r,k)sn^2(u_i,k')}$$

$$\text{Im } P_{\infty i} = \frac{sn(u_r,k)\,dn(u_i,k')}{1 - dn^2(u_r,k)sn^2(u_i,k')}$$

$$(2.53)$$

with $m = -(N-1)$ to $N-1$ in steps of 2

The poles with the negative sign in the real part (stable poles) are assigned to $H(P)$, those with the positive real part to $H(-P)$.

The Pascal routine **Cauer** in the Appendix is an implementation of the equations derived in this section. It calculates the poles and zeros based on (2.50) and (2.53). In case a filter is specified in terms of a tolerance mask, the subroutine **CauerParam** can be used to determine the desired filter parameters k, k_1, ε, and N. As an example, Figure 2-16 illustrates the location of the poles and zeros of the filter specified in Figure 2-10.

Some tables of elliptic filter coefficients are included in the Appendix. But these can only cover a very limited range of the possibilities of elliptic filters because, compared to the Chebyshev filter, another design parameter (k_1) is available which offers an additional degree of freedom for characterizing the desired behaviour.

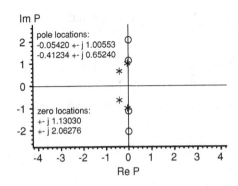

Fig. 2-22
Pole and zero plot of a fourth-order elliptic filter according to the specification of Fig. 2-10

For the cascade implementation of partial filters of first and second order, we have the following transfer function:

$$H(P) = V \frac{1 + B_{21}P^2}{1 + A_{11}P + A_{21}P^2} \cdots \frac{1 + B_{2k}P^2}{1 + A_{1k}P + A_{2k}P^2} \frac{1}{1 + A_{1(k+1)}P} \quad .$$

The factor V normalises the maximum gain of the low-pass filter to unity. V is given by

$$V = \begin{cases} 1 & \text{for } N \text{ odd} \\ 1/\sqrt{1+\varepsilon^2} & \text{for } N \text{ even} \end{cases} \quad .$$

The degree of the numerator polynomial M is always even. It is related to the degree N of the denominator polynomial as

$$M = N \qquad \text{for } N \text{ even} \quad \text{and}$$
$$M = N - 1 \qquad \text{for } N \text{ odd} \quad .$$

Example 2-3

Design a Cauer filter as specified below and show the pole/zero plot of the resulting filter (make use of the filter design tables in the appendix).

Passband ripple:	1 dB
Stopband attenuation:	40 dB
Passband edge frequency:	1 kHz
Stopband edge frequency:	2 kHz

The given specification translates to the following characteristic values:

$$v_p = 10^{-2/20} = 0.891$$

$$v_s = 10^{-40/20} = 0.01$$

$$\omega_p = 2\pi \times 1000 \ s^{-1}$$

$$\omega_s = 2\pi \times 2000 \ s^{-1}$$

According to (2.45), the parameter k of the elliptic function is calculated as

$$k = \frac{\omega_p}{\omega_s} = 0.5 \ .$$

The parameter k_1 is obtained using (2.40b).

$$k_1 = \sqrt{1/v_p^2 - 1} \Big/ \sqrt{1/v_s^2 - 1} = 0.00510$$

The required filter order can be approximately calculated using (2.47).

$$N \approx \frac{2}{\pi^2} \times \ln\frac{4}{k_1} \times \ln\frac{8}{1-k}$$

$$N \approx \frac{2}{\pi^2} \times \ln\frac{4}{0.00510} \times \ln\frac{8}{1-0.5} = 3.74$$

We round the filter order up to the next integer and choose $N = 4$. The filter coefficients are obtained from the filter design table in the appendix.

$$H(P) = 0.8913 \frac{1+0.3860P^2}{1+0.2109P+1.0015P^2} \frac{1+0.0805P^2}{1+2.0139P+2.7642P^2}$$

The roots of the numerator and denominator polynomials are the zeros and poles of the transfer function.

$$1+0.3860P^2 = 0 \qquad \Rightarrow \quad P_{01/2} = \pm j1.61$$

$$1+0.0805P^2 = 0 \qquad \Rightarrow \quad P_{03/4} = \pm j3.53$$

$$1+0.2109P+1.0015P^2 = 0 \quad \Rightarrow \quad P_{\infty 1/2} = -0.11 \pm j0.99$$

$$1+2.0139P+2.7642P^2 = 0 \quad \Rightarrow \quad P_{\infty 3/4} = -0.36 \pm j0.48$$

The poles and zeros occur in complex-conjugate pairs. The zeros are located on the imaginary axis. The poles lie in the left half-plane which guarantees stability of the filter. The figure shows the resulting pole/zero plot of the filter.

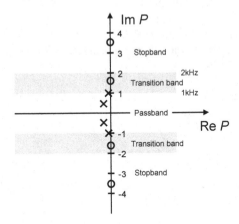

Pole/zero plot of the transfer function of Example 2-3

2.6 Bessel Filters

Bessel low-pass filters are used if the waveform of a given input signal is to be preserved after filtering. Impulse deformations are mainly due to group delay distortion which occurs especially in the vicinity of the cutoff frequency of selective filters. Bessel filters have poles with a very low quality factor close to aperiodic damping, which avoids any overshot in the output signal if, for instance, a step function is applied. The group delay characteristic is optimally flat, comparable to the magnitude response of the Butterworth filter. Thus Bessel filters can also be used to delay signals. In this case, the passband of the filter has to be chosen according to the bandwidth of the signal to be delayed. The larger the product of the bandwidth of the signal and the delay time is the higher is the filter order needed.

We begin the following considerations with a low-pass transfer function of the form

$$H(p) = \frac{1}{1 + a_1 p + a_2 p^2 + a_3 p^3 + \ldots a_N p^N} \, . \tag{2.53}$$

We obtain the corresponding frequency response by substituting $j\omega$ for p.

$$H(j\omega) = \frac{1}{1 + ja_1\omega - a_2\omega^2 - ja_3\omega^3 + \ldots + a_N(j\omega)^N} \tag{2.54}$$

The coefficient a_1 has a special meaning. For ω approaching zero we have

$$H(j\omega) \approx \frac{1}{1 + ja_1\omega} \quad .$$

Calculation of the phase response yields for low frequencies

$$b(\omega) \approx \arctan\frac{a_1\omega}{1} = \arctan(a_1\omega) \quad .$$

The corresponding group delay characteristic of the filter takes the form

$$\tau_g(\omega) = \frac{d}{d\omega}\arctan(a_1\omega) = \frac{a_1}{1 + (a_1\omega)^2} \quad .$$

It appears that the coefficient a_1 equals the group delay τ_0 for ω approaching zero.

$$a_1 = \tau_g(0) = \tau_0$$

Starting from (2.53), we arrive at a normalised representation of the transfer function if we use the definitions $P = p\,\tau_0$ and $A_i = a_i\,/\,\tau_0^i$.

$$H(P) = \frac{1}{1 + P + A_2 P^2 + A_3 P^3 + \ldots + A_N P^N} \tag{2.55}$$

Using the definition $\Omega = \omega\,\tau_0$, the corresponding normalised frequency response reads

$$H(j\Omega) = \frac{1}{1 + j\Omega - A_2\Omega^2 - jA_3\Omega^3 + \ldots + A_N(j\Omega)^N} \quad . \tag{2.56}$$

Using (1.12b), the resulting phase response is of the form

$$b(\Omega) = \arctan\frac{\Omega - A_3\Omega^3 + A_5\Omega^5 - \ldots}{1 - A_2\Omega^2 + A_4\Omega^4 - \ldots} \quad . \tag{2.57}$$

We shall next determine the coefficients A_i in such a way that we obtain a phase response that comes as close as possible to the ideal linear-phase case

$$b(\omega) = \omega\,\tau_0 \qquad \text{or} \qquad b(\Omega) = \Omega \quad .$$

τ_0 is the desired constant delay time we are aiming at. The A_i should therefore satisfy the following relationship over a large range of frequencies:

$$\Omega \approx \arctan\frac{\Omega - A_3\Omega^3 + A_5\Omega^5 - \ldots}{1 - A_2\Omega^2 + A_4\Omega^4 - \ldots} \tag{2.58a}$$

or

$$\tan\Omega \approx \frac{\Omega - A_3\Omega^3 + A_5\Omega^5 - \ldots}{1 - A_2\Omega^2 + A_4\Omega^4 - \ldots} \quad . \tag{2.58b}$$

(2.58b) suggests choosing the A_i as the coefficients of truncated series expansions of the sine in the numerator and of the cosine in the denominator. These are well known and read as

$$A_i = 1/i! \ .$$

(2.58b) would then take the form

$$\tan \Omega \approx \frac{\Omega - \frac{1}{6}\Omega^3 + \frac{1}{120}\Omega^5 - \dots}{1 - \frac{1}{2}\Omega^2 + \frac{1}{24}\Omega^4 - \dots} \ . \tag{2.59}$$

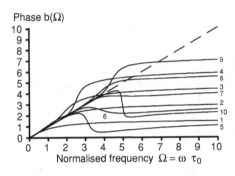

Fig. 2-23
Phase response of low-pass filters with coefficients based on truncated series expansions of sine and cosine, $N = 1 \dots 10$

This choice of low-pass coefficients does not yield satisfactory results, however. The linear phase response is only achieved in a very small range at low frequencies (Fig. 2-23) because the sine and cosine representations used here are series expansions about the point $\Omega = 0$. For higher frequencies, we obtain very different graphs of the phase response, depending on the filter order. The same is true for the magnitude characteristic, as shown in Fig. 2-24.

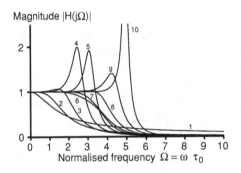

Fig. 2-24
Magnitude response of low-pass filters with coefficients based on the truncated series expansion of sine and cosine, $N = 1 \dots 10$

The chosen approach to approximate the tangent by the quotient of truncated series expansions of sine and cosine is apparently not suitable to solve the present problem. For degrees of approximation $N > 4$ the poles and zeros do not occur at real values of Ω, as would be expected from the tangent function. Furthermore, stability is not guaranteed with these coefficients, as the corresponding transfer functions have poles in the right half of the P-plane. Usable sets of coefficients are obtained, however, if the rational fractional function (2.59) is expanded into a continued fraction. Continued-fraction expansion is a mathematical method that is used, for instance, for the synthesis of passive two-terminal networks from given rational fractional impedance or admittance functions [58]. Applying this method to our problem, we have the following representation of the tangent function:

$$\tan \Omega \approx \cfrac{1}{\cfrac{1}{\Omega} - \cfrac{1}{\cfrac{3}{\Omega} - \cfrac{1}{\cfrac{5}{\Omega} - \cfrac{1}{\cfrac{7}{\Omega} - \cdots}}}} \ .$$

According to the desired degree of approximation, this infinite continued fraction is truncated after the Nth partial quotient. Conversion back to the rational fractional form after truncation yields the following approximations for $N = 1, 2$ and 3:

$$N = 1 \qquad \tan \Omega \approx \Omega,$$

$$N = 2 \qquad \tan \Omega \approx \frac{3\Omega}{3 - \Omega^2} \quad \text{and} \qquad\qquad (2.60)$$

$$N = 3 \qquad \tan \Omega \approx \frac{15\Omega - \Omega^3}{15 - 6\Omega^2} \ .$$

This truncated continued fraction shows the tendency to preserve important properties of the tangent function:

- All poles and zeros are real.
- Poles and zeros alternate.
- The coefficients lead to stable low-pass filters for all orders.

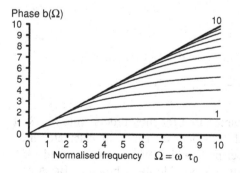

Phase b(Ω)

Normalised frequency $\Omega = \omega\,\tau_0$

Fig. 2-25
Phase response of low-pass filters with coefficients based on the truncated continued fraction of the tangent, $N = 1 \dots 10$

Figure 2-25 demonstrates the much better behaviour with respect to the phase response compared to Fig. 2-23. By differentiation we obtain the corresponding group delay characteristic, which is constant over a wide range of frequencies (Fig. 2-26). The magnitude (Fig. 2-27) shows a monotonically decreasing shape, similar to the Butterworth filter but with a lower slope.

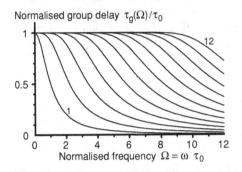

Normalised group delay $\tau_g(\Omega)/\tau_0$

Fig. 2-26
Group delay characteristics of low-pass filters with coefficients based on the truncated continued-fraction expansion of the tangent,
$N = 1 \dots 12$

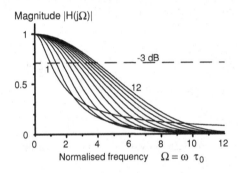

Magnitude $|H(j\Omega)|$

Fig. 2-27
Magnitude response of Bessel filters with the frequency axis normalised to τ_0,
$N = 1 \dots 12$

Inserting the coefficients of the tangent approximation (2.60) into (2.55) we obtain the following transfer functions:

$$N = 1 \qquad H(P) = \frac{1}{1+P}$$

$$N = 2 \qquad H(P) = \frac{3}{3+3P+P^2} \qquad\qquad (2.61)$$

$$N = 3 \qquad H(P) = \frac{15}{15+15P+6P^2+P^3} .$$

In Table 2-2, the coefficients are summarised for filter orders up to $N = 8$. It is not needed to evaluate continued fraction expansions to calculate these coefficients. Simple relations of Bessel polynomials can be used instead [75]. For inclusion in filter design programs, we provide the subroutine **Bessel** in the Appendix, which is an implementation of (2.62). The Bessel filter coefficients can be expressed as

$$a_v = \frac{(2n-v)!}{(n-v)!v!2^{n-v}}.$$ (2.62)

The calculation of the poles is not possible in closed form as it was the case for the filter types treated up to now. The only way is to calculate the coefficients as described above and to numerically determine the roots of the denominator polynomial in (2.55). For a cascade implementation of partial filters of first and second order, we have transfer functions of the form

$$H(P) = \frac{1}{1+A_{11}P+A_{21}P^2} \cdots \frac{1}{1+A_{1k}P+A_{2k}P^2} \frac{1}{1+A_{1(k+1)}P}.$$ (2.63)

Table 2-2 Coefficients of Bessel filters

n	a_0	a_1	a_2	a_3	a_4	a_5	a_6	a_7	a_8
1	1	1							
2	3	3	1						
3	15	15	6	1					
4	105	105	45	10	1				
5	945	945	420	105	15	1			
6	10395	10395	4725	1260	210	21	1		
7	135135	135135	62370	17325	3150	378	28	1	
8	2027025	2027025	945945	270270	51975	6930	630	36	1

The design table in the Appendix shows the coefficients A_{1i} and A_{2i} for filter orders up to $N=12$. Substitution of $p\,\tau_0$ for P yields the unnormalised transfer function $H(p)$ whose coefficients can be obtained by the following relationships:

$$a_{1i} = A_{1i}\,\tau_0 \qquad \text{and} \qquad a_{2i} = A_{2i}\,\tau_0^2.$$ (2.64)

In our considerations so far, the time delay τ_0 was the design parameter which determined, together with the filter order, the characteristics of the Bessel filter. It is of interest, of course, to know the corresponding 3-dB cutoff frequency as well. For $N>3$, the following equation which can only be solved numerically can be used to determine the cutoff frequency Ω_{3dB}.

$$|H(j\Omega)|^2 = \frac{1}{(1-A_2\Omega^2+A_4\Omega^4 - \ldots)^2 + (\Omega-A_3\Omega^3+A_5\Omega^5 - \ldots)^2} = \frac{1}{2}$$

This equation has N solutions one of which is real and positive. This solution corresponds to the desired cutoff frequency. Table 2-3 shows the results up to the order $N=12$.

The determination of the 3-dB cutoff frequency provides an interesting relationship, which is valid for Bessel filters in general. The normalised and unnormalised cutoff frequencies are related as

$$\Omega_{3dB}(N) = \omega_{3dB}\,\tau_0.$$ (2.65)

N	$\Omega_{3dB}(N)$
1	1.000
2	1.361
3	1.756
4	2.114
5	2.427
6	2.703
7	2.952
8	3.180
9	3.392
10	3.591
11	3.780
12	3.959

Table 2-3
Normalised 3-dB cutoff frequencies of the Bessel filter

From (2.65) it follows that, for a given filter order N, the product of the group delay and the cutoff frequency is a constant. The higher the cutoff frequency is, the lower the attainable group delay, and vice versa. Equation (2.65) can also be used to estimate the filter order required to delay a signal of bandwidth $\omega_B \approx \omega_{3dB}$ at an amount of τ_0 with low distortion.

Finally, (2.65) opens up the possibility of designing Bessel filters with a prescribed 3-dB cutoff frequency. By substitution into (2.64), we obtain relationships determining the corresponding unnormalised filter coefficients:

$$a_{1i} = A_{1i}\, \Omega_{3dB} / \omega_{3dB} \qquad \text{and} \qquad a_{2i} = A_{2i}\, \Omega_{3dB}^2 / \omega_{3dB}^2 \, . \qquad (2.66a)$$

From (2.66a) we can derive a set of normalised coefficients A'_{1i} and A'_{2i} such that the transfer function (2.56) is no longer normalised to the time delay τ_0, but to the 3-dB cutoff frequency ω_{3dB}.

$$A'_{1i} = A_{1i}\, \Omega_{3dB} \qquad\qquad A'_{2i} = A_{2i}\, \Omega_{3dB}^2 \qquad\qquad (2.66b)$$

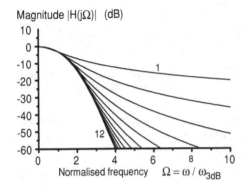

Magnitude $|H(j\Omega)|$ (dB)

Normalised frequency $\quad \Omega = \omega / \omega_{3dB}$

Fig. 2-28
Magnitude response of Bessel filters normalised to the 3-dB cutoff frequency, $N = 1 \dots 12$

These coefficients can also be found in a table in the Appendix for filter orders up to 12. When using these coefficients, the normalised frequency P has to be replaced by p/ω_{3dB} in (2.63) in order to obtain the unnormalised transfer function. Figure 2-28 shows the corresponding logarithmic magnitude responses.

It is interesting to note that, up to the cutoff frequency, the magnitude response is almost independent of the chosen filter order. According to [75], the attenuation characteristic can be approximated by a square function in this region.

$$a(\omega/\omega_{3dB}) \approx 3\,(\omega/\omega_{3dB})^2 \qquad \text{(dB)}$$

Similarly, the width of the transition band between passband and stopband is scarcely influenced by the filter order. However, for higher frequencies the filter order significantly influences the slope of the graph, which shows an asymptotic behaviour with a decay of $6N$ dB/octave.

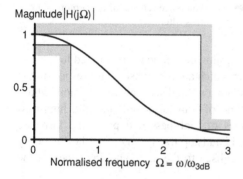

Magnitude $|H(j\Omega)|$

Normalised frequency $\Omega = \omega/\omega_{3dB}$

Fig. 2-29
Magnitude characteristic of a fourth-order Bessel filter

For comparison with Butterworth, Chebyshev and elliptic filters, we show the graph of the magnitude characteristic of a fourth-order Bessel filter plotted into a tolerance mask (Fig. 2-29). The shape is monotonic, similar to the behaviour of the Butterworth filter. The width of the transition band is larger, however. This is the price that has to be paid to enforce linearity of the phase response.

2.7 Filter Transformations

In the previous sections, we only considered low-pass filters with different characteristics in the passband and stopband. The coefficient tables in the Appendix also refer to low-pass filters only. The preferred analysis of this filter type is justified by the fact that the transition to other filter types such as high-pass, bandpass and bandstop filters is possible through relatively simple transformations of the frequency variable p. In the first step, the tolerance scheme of the filter to be designed has to be converted to the specification of the prototype low-pass filter. Afterwards the approximated prototype filter is converted back to the desired filter type using appropriate frequency transformations. In this section, we introduce the transformations for high-pass, bandpass and bandstop filters. Passband ripple and stopband attenuation are directly adopted into the tolerance scheme of the prototype low-pass filter (Fig. 2-30). The slope ω_s/ω_p of the prototype filter is derived from the edge frequencies of the tolerance scheme of the filter to be designed.

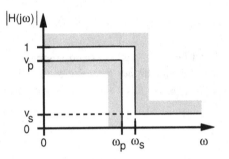

Fig. 2-30
Tolerance scheme of the prototype low-pass filter

The transformations introduced in the following sections are applied to transfer functions of the prototype low-pass filter which are normalised with respect to the frequency axis. As a result, we again obtain a normalised transfer function, whose reference frequency depends on the type of transformation. For the high-pass transformation, the reference frequencies of high-pass and prototype low-pass filter are identical. For the bandpass and bandstop transformations, the frequency variable of the transfer function is normalised to the mid-band frequency of the passband or stopband respectively.

Figure 2-31 shows the magnitude characteristic of a fourth-order elliptic filter. This filter will be used in the following sections to demonstrate the effect of the various transformations on the frequency response.

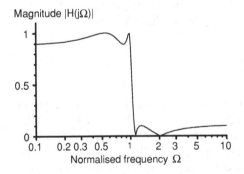

Fig. 2-31
Magnitude characteristic of a fourth-order elliptic filter
Passband ripple: 1 dB
Stopband attenuation: 20 dB

2.7.1 Low-Pass High-Pass Transformation

In this case, the slope of the low-pass prototype filter is derived from the edge frequencies of the high-pass tolerance scheme in Fig. 2-32 by the following relationship:

$$\omega_s / \omega_p = \omega_2 / \omega_1 \; . \tag{2.67}$$

After the design of the prototype filter, the high-pass transfer function complying with the tolerance scheme in Fig. 2-32 is obtained by the following transformation:

$$P \;\; \rightarrow \;\; 1 / P \; . \tag{2.68}$$

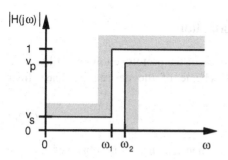

Fig. 2-32
Prescribed tolerance scheme of the high-pass filter

In the transfer function of the prototype low-pass filter, the normalised frequency variable P has simply to be replaced by its reciprocal $1/P$. It is easy to see that this transformation reverses the frequency axis, which turns a low-pass into a high-pass filter. Figure 2-33 shows the magnitude characteristic of the high-pass filter which results from Fig. 2-31 by the transformation (2.68).

For Chebyshev and elliptic filters, the passband edge is the frequency to which the transfer functions of the prototype low-pass filter and the desired high-pass filter are normalised. The 3-dB cutoff frequency is the reference for Butterworth filters and it is also preserved after the transformation. If the Butterworth filter is not specified by the cutoff frequency but by the edge frequencies of a tolerance scheme, the coefficients of the prototype low-pass filter have to be transformed in such a way that the transfer function is normalised to the passband edge frequency. This modification of the coefficients, with the help of (2.69a), guarantees that the given tolerance scheme is satisfied.

$$B_2' = B_2 \, f^2 \qquad\qquad A_2' = A_2 \, f^2 \qquad\qquad A_1' = A_1 \, f \tag{2.69a}$$

with f defined as

$$f = \frac{\omega_p}{\omega_{3dB}} = \left(\frac{1}{v_p^2} - 1 \right)^{1/(2N)} \tag{2.69b}$$

Magnitude $|H(j\Omega)|$

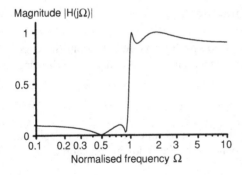

Normalised frequency Ω

Fig. 2-33
Magnitude characteristic of a fourth-
order high-pass filter:
Passband ripple: 1 dB
Stopband attenuation: 20 dB
(derived from Fig. 2-31 by transformation of the frequency variable)

2.7.2 Low-Pass Bandpass Transformation

Equation (2.70) is used to calculate the slope of the prototype low-pass filter from the edge frequencies of the tolerance scheme of the desired bandpass filter (Fig. 2-34).

$$\omega_s / \omega_p = (\omega_4 - \omega_1)/(\omega_3 - \omega_2) \qquad (2.70)$$

For the specification of the tolerance scheme, it has to be observed that the applied frequency transformation leads to symmetrical bandpass filters which feature equal relative width of both transition bands. Hence we have

$$\omega_1 \omega_4 = \omega_2 \omega_3. \qquad (2.71)$$

Three edge frequencies can be chosen arbitrarily. The fourth one is determined by (2.71).

$|H(j\omega)|$

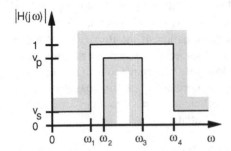

Fig. 2-34
Tolerance scheme of the bandpass filter

After the design of the prototype filter, the bandpass transfer function complying with the tolerance scheme in Fig. 2-34 is obtained by the following transformation:

$$P \rightarrow (P + 1/P)/B . \qquad (2.72a)$$

B is the bandwidth of the filter normalised to the mid-band frequency.

$$B = (\omega_3 - \omega_2) / \omega_m \qquad (2.72b)$$

ω_m is the mid-band frequency of the filter, which is related to the edge frequencies of the tolerance scheme as

$$\omega_m = \sqrt{\omega_1 \omega_4} = \sqrt{\omega_2 \omega_3} \; . \qquad (2.72c)$$

The bandpass transfer function that we obtain by (2.72a) is normalised to this frequency. For Butterworth filters, we have to apply (2.69) to the prototype low-pass filter prior to the transformation, in order to comply with the given tolerance scheme. If we omit this step, the bandwidth B would be the difference between the two 3-dB points, and not between the passband edges.

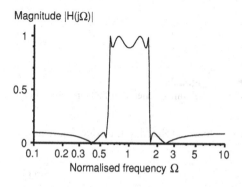

Fig. 2-35
Magnitude characteristic of an eighth-order bandpass filter:
Passband ripple: 1 dB
Stopband attenuation: 20 dB
Bandwidth $B = 1$
(created by LP to BP transformation from Fig. 2-31)

   Transformation (2.72) doubles the filter order. This means that the bandpass filter is of twice the order of the prototype low-pass filter. If we replace P by $j\Omega$ in (2.72), it is easy to see that the transformation turns a low-pass filter into a bandpass filter:

$$\Omega \;\rightarrow\; \Omega' = (\Omega - 1/\Omega) / B \; .$$

If Ω passes through the range $0 \le \Omega \le 1$, the low-pass characteristic is passed in the reverse direction from the stopband to the passband ($\Omega' = -\infty \dots 0$). If Ω passes through the range $1 \le \Omega \le \infty$, we start in the passband and end up in the stopband ($\Omega' = 0 \dots \infty$). Both ranges together yield a bandpass behaviour. Figure 2-35 shows the magnitude characteristic of the eighth-order bandpass filter, which results from the low-pass prototype in Fig. 2-31.

2.7.3 Low-Pass Bandstop Transformation

Equation (2.73) is used to calculate the slope of the prototype low-pass filter from the edge frequencies of the tolerance scheme of the desired bandstop filter (Fig. 2-36).

$$\omega_s / \omega_p = (\omega_4 - \omega_1) / (\omega_3 - \omega_2) \qquad (2.73)$$

The same symmetry property as expressed by (2.71), which imposes constraints on the choice of the edge frequencies, applies to the bandstop filter. After the design of the prototype filter, the bandstop transfer function complying with the tolerance scheme in Fig. 2-36 is obtained by the transformation

$$P \rightarrow B/(P+1/P) \,. \tag{2.74a}$$

B is the bandwidth of the filter normalised to the mid-band frequency.

$$B = (\omega_4 - \omega_1)/\omega_m \tag{2.74b}$$

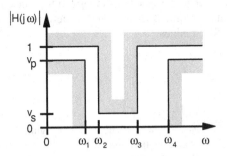

Fig. 2-36
Tolerance scheme of the bandstop filter

ω_m is the mid-band frequency of the filter, which is related to the edge frequencies of the tolerance scheme as

$$\omega_m = \sqrt{\omega_1 \omega_4} = \sqrt{\omega_2 \omega_3} \,. \tag{2.74c}$$

All the comments that we made in the previous section concerning the normalisation of the transfer function, the special handling of Butterworth filters, and the doubling of the filter order, also apply to the bandstop filter discussed here. With the same argumentation as presented there, it is easy to understand that the low-pass characteristic is passed in the first place from the passband to the stopband and afterwards in the opposite direction, which yields a bandstop characteristic. Figure 2-37 shows the magnitude characteristic of the eighth-order bandstop filter that results from the low-pass prototype in Fig. 2-31.

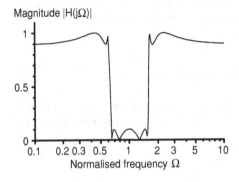

Fig. 2-37
Magnitude characteristic of an eighth-order bandstop filter:
Passband ripple: 1 dB
Stopband attenuation: 20 dB
Bandwidth B = 1
(created by LP to BS transformation from Fig. 2-31)

Example 2-4

Determine the transfer function of a Chebyshev bandpass filter with the following specification:

Filter order:	4
Passband ripple:	1 dB
Midband frequency:	1 kHz
Bandwidth:	500 Hz

The prototype for the design of the specified bandpass filter is a 2nd order Chebyshev low-pass filter with 1 dB ripple in the passband. From the design table in the appendix, we obtain the following transfer function for this filter:

$$H(P) = 0.89125 \frac{1}{1 + 0.99567P + 0.90702P^2}$$

Substitution of $(P + 1/P) / B$ for P yields the desired transfer function of the band-pass filter. B is the relative bandwidth

$$B = \frac{\Delta\omega}{\omega_m} = \frac{2\pi \times 500\,\text{Hz}}{2\pi \times 1\,\text{kHz}} = 0.5$$

of the filter.

$$H(P) = 0.89125 \frac{1}{1 + \dfrac{0.99567}{0.5}\left(P + \dfrac{1}{P}\right) + \dfrac{0.90702}{0.5^2}\left(P + \dfrac{1}{P}\right)^2}$$

$$H(P) = \frac{0.89125\,P^2}{P^2 + 1.99134\,(P^3 + P) + 3.62808\,(P^2 + 1)^2}$$

$$H(P) = \frac{0.24565\,P^2}{P^4 + 0.54887\,P^3 + 2.27563\,P^2 + 0.54887\,P + 1}$$

The following figure sketches the magnitude response of the filter.

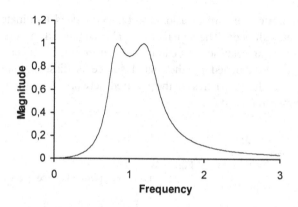

2.8 Passive Filter Networks

In the preceding paragraphs, we introduced a number of selective filter characteristics which make use of different classes of polynomials and rational functions. At the times when the theory of these filters was developed, realisation was mainly based on lossless LC networks terminated in resistors at one or both ends of the circuit. Since the filter characteristics strongly depend on the values of the terminating resistors, filter networks were often embedded between tube amplifier stages to decouple these networks from the impedances of signal source and signal sink. Later on, transistors and operational amplifiers took over this function. The move to active realisations had the goal to avoid the inductors present in filter networks because these were and still are expensive, hard to manufacture with high accuracy, and unwieldy especially in low-frequency applications. Tunable filters is another field of applications where active filters have some advantages. In many cases, only one potentiometer is required to change a filter parameter where a number of inductors and capacitors must be tuned in a passive LC-filter realisation.

Though LC networks are only seldom used in filter implementations today, the underlying concepts of passiveness and more specifically of losslessness have survived in the design of active analog filters and in digital signal processing. Inherent stability and low coefficient sensitivity are outstanding attributes of resistively terminated lossless networks. The basic approach to preserve these attributes is to convert the Kirchhoff node and mesh equations of the circuits into signal flow graphs which can then be translated to various forms of implementations. In the following, we will introduce some basic relations in the context of lossless networks terminated in resistors at both ends. These will be used in Chap. 5 to derive structures of digital filters which inherit the named good properties.

2.8.1 The Lossless Two-port

Figure 2-38 shows the basic structure of a lossless LC-network N terminated in resistances R_1 and R_2 at both ends. The signal source E is connected to port 1 of the network. Some interesting relationships can be derived from the fact that N is lossless. So all the power launched by the signal source is dissipated in the terminating resistors. The characteristics of the inserted network determine how this power is shared between R_1 and R_2.

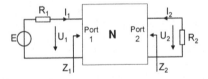

Fig. 2-38
Lossless two-port inserted between resistive terminations

A first important quantity is the squared magnitude of the transmittance t which is defined as the ratio between the power P_2 dissipated in R_2 and the maximum (available) power P_a that can be delivered by the source E with its internal resistance R_1.

$$P_2 = \frac{|U_2|^2}{R_2} \qquad\qquad P_a = \frac{|E|^2}{4R_1}$$

$$|t|^2 = \frac{P_2}{P_a} = 4\frac{R_1}{R_2}\left|\frac{U_2}{E}\right|^2$$

$$t = 2\sqrt{\frac{R_1}{R_2}}\frac{U_2}{E} = 2\sqrt{\frac{R_1}{R_2}}H \tag{2.75}$$

The transmittance t corresponds, apart from a constant factor, to the transfer function H of the filter circuit between the signal source E and the output of the resistively terminated network U_2. Clearly, the magnitude of t can never exceed unity by definition. t reaches this maximum if

1. the network is lossless and thus the whole input power $\mathrm{Re}(U_1\,I_1)$ is passed to the resistor R_2 and
2. the network transforms the real resistance R_2 into the real internal resistance of the source R_1 (matching between source and load).

This results, according to (2.75), in a well defined upper bound of the magnitude of the transfer function H.

$$|H|_{max} = \frac{1}{2}\sqrt{\frac{R_2}{R_1}}$$

Networks with this property are called structure-induced bounded [72]. Let us now assume that a network has the named property, and for a given frequency ω_k we have

$$|t(j\omega_k)| = 1 \ .$$

If the ith component of the network changes its value V_{i0} due to ageing or temperature changes for instance, the magnitude of the transmittance t and of the related transfer function H can only decrease. The left curve in Fig. 2-39 shows a corresponding characteristic which has a maximum at V_{i0} equivalent with a zero slope. Thus the first-order sensitivity of the transmittance with respect to tolerances of the network elements vanishes around the nominal value.

$$\left.\frac{\partial|t(j\omega_k)|}{\partial V_i}\right|_{V_i = V_{i0}} = 0 \qquad\qquad \forall i, \forall k \tag{2.76}$$

Each element of the network shows the described behaviour. A sensitivity characteristic as indicated by the right curve in Fig. 2-39 cannot occur in a structure-induced bounded network. Selective filters often have several maxima with $|t|=1$ distributed over the passband. Between these maxima, the gain only deviates marginally from unity. We can therefore expect that structure-induced bounded networks exhibit excellent sensitivity properties in the whole passband because the derivative (2.76) assumes low values or vanishes completely in this frequency range.

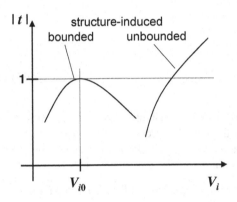

Fig. 2-39
Examples of sensitivity characteristics

The reflectance ρ is complementary to the transmittance. Its square magnitude is the share of the available power which is not transferred to R_2 due to an imperfect match between the signal source and the terminating resistance. This share of P_a is also referred to as the reflected power P_r.

$$|\rho|^2 = \frac{P_a - P_2}{P_a} = \frac{P_r}{P_a} = 1 - \frac{P_2}{P_a} = 1 - |t|^2 \tag{2.77}$$

Transmittance and reflectance are thus related as

$$|\rho|^2 + |t|^2 = 1 \ , \tag{2.78}$$

which is known as the Feldtkeller equation.

The input impedance Z_1 of the lossless network terminated in R_2 as shown in Fig. 2-38 is an important quantity, which is directly related to the transfer behaviour of the filter network. First we observe that the power fed into port 1 of the network N must be equal to the power dissipated in R_2 since N is lossless.

$$\frac{|I_1|^2}{\operatorname{Re} Z_1} = \frac{|U_2|^2}{R_2} \tag{2.79}$$

The input current of the network I_1 can be calculated as

$$I_1 = \frac{E}{R_1 + Z_1} \ . \tag{2.80}$$

(2.79) and (2.80) substituted in (2.75) yields

$$|t|^2 = \frac{4R_1 \operatorname{Re} Z_1}{|R_1 + Z_1|^2} . \tag{2.81}$$

Transmittance and input impedance of the network are thus related as

$$t = 2\frac{\sqrt{R_1 \operatorname{Re} Z_1}}{R_1 + Z_1} .$$

The numerator of (2.81) can be expanded to the form [75]

$$|t|^2 = \frac{(R_1 + \operatorname{Re} Z_1)^2 + (\operatorname{Im} Z_1)^2 - (R_1 - \operatorname{Re} Z_1)^2 - (\operatorname{Im} Z_1)^2}{|R_1 + Z_1|^2}$$

which finally results in

$$|t|^2 = \frac{|R_1 + Z_1|^2 - |R_1 - Z_1|^2}{|R_1 + Z_1|^2} = 1 - \frac{|R_1 - Z_1|^2}{|R_1 + Z_1|^2} = 1 - |\rho|^2 .$$

The reflectance is thus related to the input impedance Z_1 of the network as

$$\rho = \frac{R_1 - Z_1}{R_1 + Z_1} \quad \text{and} \quad Z_1 = R_1 \frac{1 - \rho}{1 + \rho} , \tag{2.82}$$

which is the well-known form of a reflection coefficient.

The problem of finding an appropriate network realising a given transfer function or transmittance can thus be converted to the problem of finding a resistively terminated LC network that realises a given driving point impedance.

The transfer impedance Z_{21} of the network terminated in R_2 is another useful quantity, which is closely related to the transmittance. It is defined as the ratio of the output voltage U_2 to the input current I_1 of the lossless network.

$$Z_{21} = \frac{U_2}{I_1} = \frac{U_2}{E} \frac{E}{I_1}$$

Substitution of (2.75) and (2.80) yields

$$Z_{21} = \frac{1}{2} t \sqrt{\frac{R_2}{R_1}} (R_1 + Z_1) \tag{2.83}$$

Using (2.82), we obtain

$$Z_{21} = \frac{1}{2} t \sqrt{\frac{R_2}{R_1}} \left(R_1 + R_1 \frac{1 - \rho}{1 + \rho} \right)$$

$$Z_{21} = \sqrt{R_1 R_2} \frac{t}{1+\rho} \qquad (2.84a)$$

Solving (2.83) for t, the transmittance can be expressed in terms of the input impedance and the transfer impedance.

$$t = 2\sqrt{\frac{R_1}{R_2}} \frac{Z_{21}}{R_1 + Z_1} \qquad (2.84b)$$

Example 2-5
Determine Z_1 for the implementation of a third-order Chebyshev low-pass filter with a passband ripple of 1 dB and a passband cutoff frequency at 1 kHz. Both ends of the LC network are to be terminated in a resistance $R = R_1 = R_2$ of 1 kΩ.

From the Chebyshev filter table in the appendix, we get the desired transfer function

$$t(P) = \frac{1}{1+0.497P+1.006\ P^2} \frac{1}{1+2.024\ P} \ .$$

$$t(P) = \frac{1}{2.035\ P^3 + 2.012\ P^2 + 2.521\ P + 1}$$

The frequency variable P is here normalised with respect to the passband edge frequency. In the following, we denote the numerator polynomials of transmittance and reflectance as $u(P)$ and $v(P)$ respectively and the denominator polynomial as $d(P)$. So the squared magnitude of the transmittance can be written in the form

$$|t|^2 = \frac{u(P)u(-P)}{d(P)d(-P)} \qquad \text{for } p = j\omega \ .$$

Using the Feldtkeller equation (2.78), the reflectance is calculated as

$$|\rho|^2 = 1 - |t|^2 = 1 - \frac{u(P)u(-P)}{d(P)d(-P)} = \frac{d(P)d(-P) - u(P)u(-P)}{d(P)d(-P)}$$

$$|\rho|^2 = \frac{v(P)v(-P)}{d(P)d(-P)} = \frac{d(P)d(-P) - u(P)u(-P)}{d(P)d(-P)}$$

The numerator polynomial of the reflectance is thus related to the known polynomials of the transmittance as

$$v(P)v(-P) = d(P)d(-P) - u(P)u(-P) \ .$$

With

$$u(P) = 1 \quad \text{and}$$

$$d(P) = 2.035P^3 + 2.012\ P^2 + 2.521\ P + 1$$

we get

$$v(P)v(-P) = -4.143\ P^6 - 6.214\ P^4 - 2.330\ P^2$$
$$= -4.143\ P^2(P^2 + 0.75)(P^2 + 0.75)$$
$$= 2.035\ P(P^2 + 0.75) \times 2.035\ (-P)((-P)^2 + 0.75)\ .$$

We take the left part of the last expression as the desired numerator polynomial of the reflectance.

$$\rho = \frac{v(P)}{d(P)} = \frac{2.035\ P(P^2 + 0.75)}{2.035\ P^3 + 2.012\ P^2 + 2.521\ P + 1}$$

Substitution of ρ in (2.82) yields the desired input impedance Z_1 of the network.

$$Z_1 = R\frac{2.012\ P^2 + 0.994\ P + 1}{4.071\ P^3 + 2.012\ P^2 + 4.047P + 1}$$

$$\text{with } R = 1\ k\Omega \text{ and } P = p/(2\pi \times 1\ kHz)$$

At low frequencies, Z_1 approaches R which provides a perfect match to the impedance of the source. The maximum possible power is transferred to the load resistance. For high frequencies, Z_1 approaches zero equivalent with a short circuit. So all the available power is dissipated in the source resistance R_1 and nothing is transferred to the load resistance at the output. The input impedance of the network thus realises a transmittance with low-pass characteristic.

For some types of network structures, Z_1 is the appropriate starting point for the calculation of the values of the capacitors and inductors realising the desired transfer behaviour. In other cases, it is more advantageous to use the elements of the z-matrix of the lossless network for this purpose.

2.8.2 Two-port Equations

The I/O-relations of two-ports are often described in the form of matrices. Widely used are impedance, admittance, chain, and scattering matrices. The latter will be used extensively in the context of wave digital filters which will be introduced in Sect. 5.7. For the mathematical treatment of the lossless filter networks terminated in resistances, we choose the impedance matrix or z-matrix. The elements of the z-matrix describe the relationship between voltages and currents for the unterminated or open-circuit case. Voltages and currents of the two-port are related as

$$\begin{pmatrix} U_1 \\ U_2 \end{pmatrix} = \begin{pmatrix} z_{11} & z_{12} \\ z_{21} & z_{22} \end{pmatrix} \begin{pmatrix} I_1 \\ I_2 \end{pmatrix} \quad \text{or} \quad \begin{matrix} U_1 = z_{11}I_1 + z_{12}I_2 \\ U_2 = z_{21}I_1 + z_{22}I_2 \end{matrix} \ . \tag{2.85}$$

z_{11} and z_{22} are the open-circuit driving-point impedances of the ports.

$$z_{11} = \left. \frac{U_1}{I_1} \right|_{I_2=0} \qquad z_{22} = \left. \frac{U_2}{I_2} \right|_{I_1=0}$$

z_{12} and z_{21} are transfer impedances which describe the relation between voltages and currents at different ports.

$$z_{21} = \left. \frac{U_2}{I_1} \right|_{I_2=0} \qquad z_{12} = \left. \frac{U_1}{I_2} \right|_{I_1=0}$$

The elements of the z-matrix are not independent of each other. For all networks composed of inductors, capacitors and resistors, the transfer impedances z_{21} and z_{12} are equal. This is a consequence of the reciprocity of RLC networks. The reciprocity theorem states the following: Choose any two ports of a given RLC network. Inject a certain current into the first port and measure the voltage at the second port with a high-impedance voltmeter. Then inject the same current into the second port and measure the voltage at the first port. Both measured voltages are the same.

We investigate now the case that port 2 is terminated in a resistance R_2. Considering the direction of the arrowheads in Fig. 2-38, voltage and current at port 2 are related as

$$U_2 = -R_2 I_2 .$$

Substitution in (2.85) yields the following relationships for the input impedance Z_1 and the transfer impedance Z_{21} of the terminated network

$$Z_1 = \left. \frac{U_1}{I_1} \right|_{U_2=-R_2 I_2} = z_{11} - \frac{z_{21}^2}{z_{22} + R_2} = \frac{\det z + R_2 z_{11}}{z_{22} + R_2} \qquad (2.86)$$

$$Z_{21} = \left. \frac{U_2}{I_1} \right|_{U_2=-R_2 I_2} = \frac{z_{21} R_2}{z_{22} + R_2} \qquad (2.87)$$

In the following, we denote the numerator of the transmittance as $u(P)$ and the numerator of the reflectance as $v(P)$. The denominator $d(P)$ is the same for both quantities. Making use of the relationships derived in this section and further properties known from network theory [75], all the afore mentioned impedance quantities can be expressed in terms of $u(P)$, $v(P)$ and $d(P)$.

$$Z_1 = R_1 \frac{d - v}{d + v} \qquad (2.88)$$

$$Z_{21} = \sqrt{R_1 R_2} \, \frac{u}{d + v} \qquad (2.89)$$

The elements of the z-matrix are calculated as summarised in Table 2-4. The relations depend if u and v are even or odd. The functions even(x) and odd(x) denote the even and odd parts respectively of the polynomial x.

Table 2-4 Relations to determine the elements of the z-matrix

	u even	u odd
z_{11}	$R_1 \dfrac{\text{even}(d) - \text{even}(v)}{\text{odd}(d) + \text{odd}(v)}$	$R_1 \dfrac{\text{odd}(d) - \text{odd}(v)}{\text{even}(d) + \text{even}(v)}$
z_{22}	$R_2 \dfrac{\text{even}(d) + \text{even}(v)}{\text{odd}(d) + \text{odd}(v)}$	$R_2 \dfrac{\text{odd}(d) + \text{odd}(v)}{\text{even}(d) + \text{even}(v)}$
z_{21}	$\sqrt{R_1 R_2} \dfrac{u}{\text{odd}(d) + \text{odd}(v)}$	$\sqrt{R_1 R_2} \dfrac{u}{\text{even}(d) + \text{even}(v)}$

In many practical cases, the driving point impedances z_{11} and z_{22} are not independent of each other. Provided that the lossless network is terminated with equal resistances $R_1 = R_2$ at both ends, Table 2-4 reveals that z_{11} and z_{22} are equal under the following conditions:

- u is even and v is odd, or
- u is odd and v is even,

which results in both cases in an odd characteristic function

$$K(p) = \frac{v(p)}{u(p)} .$$

The corresponding squared magnitude of the transfer function has the form

$$|t|^2 = \frac{1}{1 + |K(j\omega)|^2} = \frac{1}{1 + (\text{odd rational function of } \omega)^2} . \qquad (2.90)$$

Symmetric two-ports have the property that the ports can reverse roles without affecting the transfer characteristics of the filter. Table 2-1 shows for which types of Butterworth, Chebyshev and Cauer filters this symmetry condition applies:

- odd-order high-pass and low-pass filters
- bandpass and bandstop filters derived from odd-order low-pass reference filters

Filters with even characteristic functions and a squared magnitude response of the form

$$|t|^2 = \frac{1}{1 + (\text{even rational function of } \omega)^2} \qquad (2.91)$$

can be realised with antisymmetrical (antimetrical) structures for which z_{11} and z_{22} are related as

$$z_{11} z_{22} = R_1 R_2 + z_{21}^2 . \qquad (2.92)$$

Example 2-6

Determine the elements of the impedance matrix for the Chebyshev filter specified in Example 2-5.

In Example 2-5, we calculated the polynomials u, v, and d as

$$u = 1$$

$$v = 2.035\ P(P^2 + 0.75) = 2.035\ P^3 + 1.526\ P$$

$$d = 2.035\ P^3 + 2.012\ P^2 + 2.521\ P + 1$$

u is even and v is odd. Substitution in the relations in the first column of Table 2-4 yields the desired rational fractional functions in the normalised frequency variable P.

$$z_{11} = 0.494\ R\ \frac{P^2 + 0.497}{P(P^2 + 0.994)}$$

$$z_{22} = 0.494\ R\ \frac{P^2 + 0.497}{P(P^2 + 0.994)}$$

$$z_{21} = 0.246\ R\ \frac{1}{P(P^2 + 0.994)}$$

2.8.3 Network Configurations

There is a wide variety of network configurations to realise the desired input impedance Z_1 or the matrix impedances z_{11}, z_{22}, and z_{21} for a filter to be designed. Some basic structures are the T, Π, bridged-T, twin-T, lattice, and ladder as shown in Fig. 2-40. Each of the boxes in these diagrams may represent a single capacitor or inductor or a complex lossless network made up of these elements.

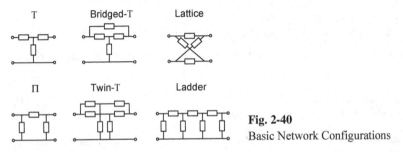

Fig. 2-40
Basic Network Configurations

In the first step of a filter design, a basic network structure is chosen and the impedances of the elements of the structure are calculated based on the given matrix impedances. As an example, Figure 2-41 shows the relationships to calculate the network elements of the T-network.

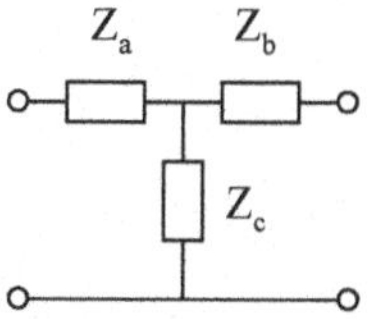

$$Z_a = z_{11} - z_{21}$$
$$Z_b = z_{22} - z_{21}$$
$$Z_c = z_{21}$$

Fig. 2-41

Relationships to determine the elements of the T-structure

Then it has to be checked if the calculated impedances are realisable with an LC-network. Each of the basic network configurations has certain restrictions concerning the realisability of a given transfer function. Realisable impedances of lossless LC-networks must fulfil the following requirement:

- Z must be an odd function in the frequency variable P. This means that the order of the numerator polynomial must be even and the order of the denominator polynomial odd or vice versa.
- The order of numerator and denominator must differ by one.
- Poles and zeros of Z are simple, lie on the imaginary axis, and alternate on the imaginary axis.

Example 2-7

Check if the filter calculated in Example 2-5 and Example 2-6 can be realised with the T-structure.

Using the relations in Fig. 2-41 and the matrix impedances calculated in Example 2-6, the impedances Z_a, Z_b and Z_c of the T-network are calculated as

$$Z_a = z_{11} - z_{21} = 0.494\,R\frac{P}{P^2 + 0.994}$$

$$Z_b = z_{22} - z_{21} = 0.494\,R\frac{P}{P^2 + 0.994}$$

$$Z_c = z_{21} = 0.246\,R\frac{1}{P(P^2 + 0.994)}$$

Z_a and Z_b fulfil the requirements and can be easily realised with parallel resonant circuits. Z_c, however, is not realisable since the orders of numerator and denominator differ by 3. The filter, therefore, cannot be realised as a T-network.

In the third step, the impedances of the network elements have to be realised with an appropriate LC circuit. Most common are the canonical structures of Foster and Cauer shown in Fig. 2-42. The elements of the Foster network are obtained by partial-fraction expansion of the respective impedance/admittance functions, those of the Cauer network by continued-fraction expansion [75].

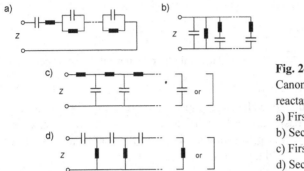

Fig. 2-42

Canonical realisations of reactance functions

a) First Foster network

b) Second Foster network

c) First Cauer network

d) Second Cauer network

In the following we consider in more detail the ladder and lattice structures which are often used as the analog prototypes for the realisation of digital filters (digital ladder filters, digital lattice filters).

2.8.4 Ladder Filters

Figure 2-43 shows the basic structure of a ladder filter which consists of a cascade of alternating series and shunt impedances. Each impedance z_i may represent a single component (capacitor or inductor) or a combination of several of these components. The ladder structure may be looked upon as a cascade of T and Π networks. Unlike the bridged-T, twin-T or lattice, the ladder has only one path between input and output. So zeros of the transfer function cannot be realised by cancellation of signals within the network. Zeros must rather be realised by zeros (short circuits at certain frequencies) in the shunt branches or by poles (open circuits at certain frequencies) in the series branches of the structure. Since the poles and zeros of the driving point impedances of LC networks lie on the imaginary axis, the ladder structure as a whole can only create zeros on the imaginary axis.

For the realisation of inverse Chebyshev and Cauer filters, this is not a restriction, because all these types of filters have their zeros on the imaginary axis as we showed in this chapter.

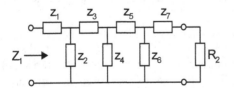

Fig. 2-43

The ladder structure

In order to calculate the input impedance of the circuit shown in Fig. 2-43, we start at the output terminal. The impedance of the last branch plus the terminating resistor is

$$z_7 + R_2 \ .$$

This impedance is connected in parallel with z_6 which results in the admittance

$$1/z_6 + \frac{1}{z_7 + R_2} \ .$$

Series connection of z_5 yields the impedance

$$z_5 + \cfrac{1}{1/z_6 + \cfrac{1}{z_7 + R_2}} \ .$$

Proceeding in this way back to the input terminal, we obtain the input impedance Z_1 as

$$Z_1 = z_1 + \cfrac{1}{1/z_2 + \cfrac{1}{z_3 + \cfrac{1}{1/z_4 + \cfrac{1}{z_5 + \cfrac{1}{1/z_6 + \cfrac{1}{z_7 + R_2}}}}}} \ .$$

This form is called a continued fraction expansion which explicitly shows the impedances or admittances of the individual series and shunt branches of the ladder as the partial quotients. In the next example, we will show how these partial quotients are calculated from the rational fractional function of a given impedance.

Example 2-8

Determine the values of the inductors and capacitors of a ladder implementation of the filter specified in Example 2-5.

The input impedance Z_1 of the filter network was calculated in Example 2-5 as

$$Z_1 = R \frac{2.012\ P^2 + 0.994\ P + 1}{4.071\ P^3 + 2.012\ P^2 + 4.047 P + 1}$$

The values of the components of the ladder are derived by continued-fraction expansion of the impedance function.

$$Z_1 = 0 + \cfrac{1}{\cfrac{4.071\ P^3 + 2.012\ P^2 + 4.047 P + 1}{R\ (2.012\ P^2 + 0.994\ P + 1)}}$$

The first iteration step yields $z_1 = 0$. This means that the first series branch is a simple short circuit. Long division of the polynomials yields

$$(4.071\,P^3 + 2.012\,P^2 + 4.047P + 1)/R\,(2.012\,P^2 + 0.994\,P + 1)$$

$$= 2.023\,P/R - (4.071\,P^3 + 2.012\,P^2 + 2.023\,P)$$

Remainder : $2.024\,P + 1$

$$Z_1 = 0 + \cfrac{1}{\cfrac{2.023P}{R} + \cfrac{2.024P+1}{R(2.012P^2 + 0.994P + 1)}}$$

$$Z_1 = 0 + \cfrac{1}{\cfrac{2.023\,P}{R} + \cfrac{1}{\cfrac{R\,(2.012\,P^2 + 0.994\,P + 1)}{2.024\,P + 1}}}$$

$1/z_2 = 2.023P/R$ obviously represents a capacitor in the first shunt branch.

$$z_2 = \frac{R}{2.023\,P} = \frac{R\omega_p}{2.023\,p} = \frac{1}{pC_2}$$

$$C_2 = \frac{2.023}{R\omega_p} = \frac{2.023}{1000\,\Omega \times 2\pi \times 1000\,\text{Hz}} = 0.322\,\mu\text{F}$$

We perform again a long division resulting in

$$R\,(2.012\,P^2 + 0.994\,P + 1)/(2.024\,P + 1)$$

$$= 0.994\,PR - R\,(2.012\,P^2 + 0.994\,P)$$

Remainder : R

After the next iteration, the continued fraction thus assumes the following form:

$$Z_1 = 0 + \cfrac{1}{\cfrac{2.023\,P}{R} + 0.994\,PR + \cfrac{R}{2.024\,P + 1}}$$

$$Z_1 = 0 + \cfrac{1}{\cfrac{2.023\,P}{R} + 0.994\,PR + \cfrac{1}{\cfrac{2.024\,P}{R} + \cfrac{1}{R}}}$$

$z_3 = 0.994\,PR$ is the impedance of an inductor which forms the next series branch.

$$z_3 = 0.994 \, PR = \frac{0.994 \, pR}{\omega_p} = pL_3$$

$$L_3 = \frac{0.994 \times 1000 \, \Omega}{2\pi \times 1000 \, \text{Hz}} = 158 \, \text{mH}$$

The next shunt branch consists of the parallel connection of a capacitor and the terminating resistor. The value of this capacitor is equal to the value of C_2. The minor difference is due to the calculation with 3 decimals. This result confirms our earlier statement that odd-order Chebyshev filters are symmetrical. Fig. 2-44 shows the circuit diagram of the Chebyshev filter which we used as an example in the previous calculations.

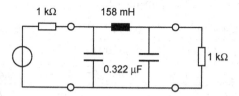

Fig. 2-44
3rd order Chebyshev low-pass filter, 1 dB ripple, passband cutoff frequency 1 kHz

2.8.5 Lattice Filter

Though all four impedances of a lattice network may be different, the lattice designation is commonly used for the symmetrical case as shown in Fig. 2-45. All symmetrical impedance matrices with $z_{11} = z_{22}$ can be realised with a lattice network. The relations to derive the lattice impedances Z_a and Z_b are quite simple:

$$Z_a = z_{11} - z_{21}$$
$$Z_b = z_{11} + z_{21}$$

(2.93)

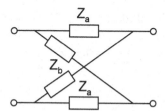

Fig. 2-45
The symmetric lattice structure

The lattice network exhibits, unlike the ladder, two signal paths between input and output. This leads to a different way of implementing transmission zeros. Where appropriate poles in the series branches or zeros in the shunt branches are provided for this purpose in T, Π, and ladder networks, signal cancellation is the corresponding mechanism in lattice networks. This also lifts the restriction that zeros must lie on the imaginary axis.

Example 2-9

Realise the filter specified in Example 2-5 as a lattice network and determine the values of the needed components.

We have seen above, that the filter to be realised possesses a symmetrical z-matrix. Thus it is realisable as a lattice. We take z_{11} and z_{21} from Example 2-6.

$$z_{11} = 0.494\, R \frac{P^2 + 0.497}{P(P^2 + 0.994)} = R \frac{0.494\, P^2 + 0.246}{P(P^2 + 0.994)}$$

$$z_{21} = 0.246\, R \frac{1}{P(P^2 + 0.994)} = R \frac{0.246}{P(P^2 + 0.994)}$$

Substitution in (2.93) yields the lattice impedances.

$$Z_a = z_{11} - z_{12} = R \frac{0.494\, P^2}{P(P^2 + 0.994)} = R \frac{0.494\, P}{P^2 + 0.994}$$

$$Z_b = z_{11} + z_{12} = R \frac{0.494\, P^2 + 0.491}{P(P^2 + 0.994)} = R \frac{0.494\,(P^2 + 0.994)}{P(P^2 + 0.994)} = R \frac{0.494}{P}$$

Z_a is the impedance of a parallel connection of a capacitor and an inductor as can be easily seen by rearrangement of the impedance term.

$$Z_a = \frac{0.494\, PR}{P^2 + 0.994} = \frac{1}{\dfrac{P^2 + 0.994}{0.494\, PR}} = \frac{1}{\dfrac{P}{0.494\,\omega_p R} + \dfrac{\omega_p}{0.497\, pR}} = \frac{1}{pC_a + \dfrac{1}{pL_a}}$$

Z_b is the impedance of a capacitor.

$$Z_b = \frac{0.494\, R}{P} = \frac{0.494\,\omega_p R}{p} = \frac{1}{pC_b}$$

The values of the components can now be easily calculated as

$$C_a = C_b = \frac{1}{0.494\,\omega_p R} = \frac{1}{0.494 \times 2\pi \times 1000\ \text{Hz} \times 1000\ \Omega} = 0.322\,\mu\text{F}$$

and

$$L_a = \frac{0.497 R}{\omega_p} = \frac{0.497 \times 1000\ \Omega}{2\pi \times 1000\ \text{Hz}} = 79\ \text{mH}\ .$$

At low frequencies, the inductors act as a short circuit. The load resistor is directly connected to the source resistor, and the maximum power is transferred to the load. With increasing frequency, two mechanisms provide an increasing attenuation. First of all, the source is shorted by the capacitors. Additionally, the lossless network more and more turns into a bridge type of network made up of four equal capacitors which provides additional attenuation. A perfect cancellation due to the bridged configuration, however, requires four exactly equal capacitors which is hardly to implement with analog components.

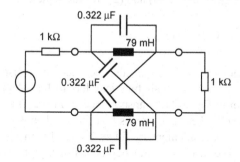

Fig. 2-46
Lattice implementation of a
3rd order Chebyshev filter

Interesting relationships can be found if we express the transmittance and reflectance of the network in terms of the impedances Z_a and Z_b of the lattice implementation. We start with (2.84b)

$$t = 2\sqrt{\frac{R_1}{R_2}}\frac{Z_{21}}{R_1 + Z_1}$$

and substitute (2.88) and (2.89). We assume a symmetrical network with $z_{11} = z_{22}$ and $R_1 = R_2 = R$.

$$t = 2\frac{z_{21}R}{(R + z_{11})(R + z_{11}) - z_{21}^2} = 2\frac{z_{21}R}{(R + z_{11} - z_{21})(R + z_{11} + z_{21})}$$

Substitution of (2.93) yields

$$t = \frac{R(Z_b - Z_a)}{(R + Z_a)(R + Z_b)}.$$

By partial fraction expansion, we finally obtain

$$t = \frac{1}{2}\left[\frac{R - Z_a}{R + Z_a} - \frac{R - Z_b}{R + Z_b}\right] = \frac{1}{2}[A_1(p) - A_2(p)]. \tag{2.94}$$

The symmetrical lattice thus realises the given transfer function as the difference of two allpass transfer functions $A_1(P)$ and $A_2(P)$, where $A_1(P)$ is exclusively determined by the lattice impedance Z_a and $A_2(P)$ by the lattice impedance Z_b. The lattice structure is not very useful for the implementation of analog filters. Zeros are realised by cancellation of signals, which requires highly accurate components in order to achieve the desired stopband attenuation. Digital allpass filters, however, can be realised with exact unity gain for all frequencies, which makes the lattice approach very attractive for digital implementations. These inherit the low parameter sensitivity of resistively terminated lossless networks from the analog reference filters.

The reflectance ρ can also be expressed in terms of the lattice impedances Z_a and Z_b. Starting with (2.82) and substituting (2.88) and (2.93) finally yields

$$\rho = \frac{1}{2}\left[\frac{R-Z_a}{R+Z_a} + \frac{R-Z_b}{R+Z_b}\right] = \frac{1}{2}[A_1(p) + A_2(p)] \; . \tag{2.95}$$

The same allpass transfer functions as before can also be used to realise the reflectance of the filter. Merely the sum of both functions has to be calculated instead of the difference as shown in Fig. 2-47. Transmittance and reflectance are complementary with respect to each other. This means that the passbands of one match the stopbands of the other. Complementary filters are useful in various signal processing applications where different frequency bands of a given signal are to be processed separately. Examples are spectral analysis, voice compression or crossover networks in audio applications.

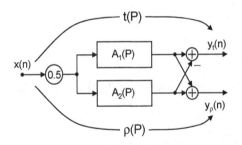

Fig. 2-47
Transmittance and reflectance realised as the difference and sum of two allpass transfer functions

By combination of (2.94) and (2.95), the allpass transfer functions $A_1(p)$ and $A_2(p)$ can be expressed in terms of the transmittance and reflectance of the filter.

$$\begin{aligned} A_1(p) &= \rho(p) + t(p) \\ A_2(p) &= \rho(p) - t(p) \end{aligned} \tag{2.96}$$

Filters as realised in Fig. 2-47 are doubly complementary. On the one hand, the sum of transmittance and reflectance is an allpass transfer function with unity magnitude for all frequencies. Transfer functions with this property are called allpass-complementary.

$$t(j\omega) + \rho(j\omega) = A_1(j\omega) \quad \text{and} \quad |t(j\omega) + \rho(j\omega)| = 1$$

Besides being allpass-complementary, transmittance and reflectance are related by the Feldtkeller equation. So t and ρ are also power-complementary.

$$|t(j\omega)|^2 + |\rho(j\omega)|^2 = 1$$

For allpass-complementary filters, there are only constraints concerning the sum or difference of their transfer functions. Power-complementary filters, in contrast, require the magnitude of both filters to be limited to unity.

Example 2-10

Calculate the allpass transfer functions derived from the lattice implementation of the filter specified in Example 2-5.

From the given specification, transmittance and reflectance were calculated in Example 2-5 as

$$t(p) = \frac{1}{(1+0.497P+1.006\ P^2)(1+2.024\ P)}$$

$$\rho(p) = \frac{2.035P\ (P^2+0.75)}{(1+0.497P+1.006\ P^2)(1+2.024\ P)}$$

Substitution of $t(P)$ and $\rho(P)$ in (2.96) yields the desired allpass transfer functions.

$$A_1(P) = \frac{2.035\ P^3+1.526\ P+1}{(1+0.497P+1.006\ P^2)(1+2.024\ P)}$$

$$A_1(P) = \frac{(1-0.497P+1.006\ P^2)(1+2.024\ P)}{(1+0.497P+1.006\ P^2)(1+2.024\ P)} = \frac{1-0.497P+1.006\ P^2}{1+0.497P+1.006\ P^2}$$

$$A_2(P) = \frac{2.035\ P^3+1.526\ P-1}{(1+0.497P+1.006\ P^2)(1+2.024\ P)}$$

$$A_2(P) = -\frac{(1+0.497P+1.006\ P^2)(1-2.024\ P)}{(1+0.497P+1.006\ P^2)(1+2.024\ P)} = -\frac{1-2.024\ P}{1+2.024\ P}$$

The following steps can be summarised to derive the allpass decomposition of a given transfer function $t(P) = u(P)/d(P)$.

- Check if $u(P)$ is an even or odd polynomial. If this is the case,
- calculate $\rho(P)$ from the given $t(P)$ as demonstrated in Example 2-5.
- Check if $v(P)$ is even while $u(P)$ is odd or vice versa. If this is the case,
- calculate $A_1(p)$ and $A_2(p)$ using (2.96).
- Cancel common factors in the numerators and denominators of $A_1(p)$ and $A_2(p)$.

We have seen that all transfer functions whose squared magnitude can be written in the form of (2.90) can be expressed as the difference of two allpass transfer functions. Relation (2.97) summarises the relationships between the involved numerator and denominator polynomials.

$$t(P) = \frac{u(P)}{d(P)} = \frac{1}{2}(A_1(P)-A_2(P)) = \frac{1}{2}\left(\frac{d_1(-P)}{d_1(P)} \pm \frac{d_2(-P)}{d_2(P)}\right)$$

$A_2(P)$ is of the form $-d_2(-P)/d_2(P)$ if $u(P)$ is even and of the form $d_2(-P)/d_2(P)$ if $u(P)$ is odd.

$$t(P) = \frac{0.5[d_2(P)d_1(-P) \pm d_1(P)d_2(-P)]}{d_1(P)d_2(P)} \tag{2.97}$$

(2.97) reveals that the allpass decomposition of a transfer function can be reduced to the problem of separating the denominator into two appropriate polynomials. This finally means that the poles must be distributed conveniently among the two allpass transfer functions. For odd-order Butterworth, Chebyshev and Cauer low-pass and high-pass filters, there is a simple rule for this problem [25]. Fig. 2-48 shows the typical pole pattern of an odd-order filter. These poles must be alternatingly allocated to the allpass transfer functions $A_1(P)$ and $A_2(P)$.

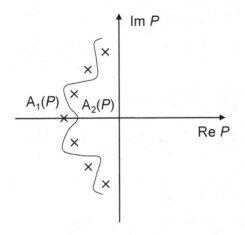

Fig. 2-48
Alternating distribution of poles

Example 2-11
Determine the allpass transfer functions of the allpass decomposition of the Chebyshev filter specified in Example 2-5 using the alternating distribution of poles.

The transfer function of the filter to be designed is given as

$$t(P) = \frac{1}{1 + 0.497P + 1.006P^2} \frac{1}{1 + 2.024P}$$

Figure 2-49 shows the location of the three poles in the P-plane
The denominator polynomials $d_1(P)$ and $d_2(P)$ are thus obtained as

$$d_1(P) = P + 0.494$$
$$d_2(P) = (P + 0.246 + j0.966)(P + 0.246 - j0.966) = P^2 + 0.494P + 0.994$$

The given transfer function $t(P)$ and the reflectance $\rho(P)$ can now be expressed as the sum and difference of the two allpass transfer functions

$$A_1(P) = \frac{d_1(-P)}{d_1(P)} \quad \text{and} \quad A_2(P) = -\frac{d_2(-P)}{d_2(P)} \; ,$$

which yields the same result as obtained in Example 2-10. The sign of $A_2(P)$ is chosen negative since $u(P)$ is even.

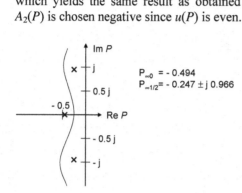

$P_{\infty 0} = -0.494$
$P_{\infty 1/2} = -0.247 \pm j\, 0.966$

Figure 2-49
Location of the poles of the
3rd order Chebyshev filter

3 Discrete-Time Systems

In Chap. 1, we introduced the relations between the input and output signals of continuous-time LTI systems. In the time domain, the output signal $y(t)$ is calculated by convolution of the input signal $x(t)$ with the impulse response $h(t)$ of the system (1.1). In the frequency domain, the spectrum of the output signal $Y(j\omega)$ is obtained by multiplication of the spectrum of the input signal $X(j\omega)$ by the transfer function $H(j\omega)$ (1.10). In this chapter we derive the corresponding relations for discrete-time systems, which look quite similar.

Let us first consider the context of discrete-time systems. Input and output signals are no longer physical quantities, but sequences of dimensionless numbers. Having in mind a physical meaning behind a discrete-time signal, the physical quantity can always be recovered by multiplying the numbers of the sequence by a reference value of this quantity. This is exactly what digital-to-analog converters do. In the following, the input sequence of the system is denoted by $x(n)$, and the output sequence by $y(n)$ where n points to the nth element of the sequence. If the sequence is related to a real-time signal or process, an association with the actual time axis can be made by writing $x(nT)$ and $y(nT)$ where T is the quantisation period of the time axis. A discrete-time system is the implementation of a mathematical algorithm, which transforms the input sequence $x(n)$ into the output sequence $y(n)$. The algorithm may be realised in dedicated hardware or in software running on Digital Signal Processors (DSPs) or on general purpose processors. The processing of signal sequences may take place in real-time, which requires synchronisation with other processes like A/D and D/A conversion happening at the same time. On the other hand, signal data may be processed off-line without any relationship to the time axis. An example is processing of data that are available on disk and that are stored back to disk after processing.

The terms "discrete-time" and "digital" are often used synonymously. But strictly speaking, discrete-time means quantisation of the time axis only, whereas digital is related to a real technical implementation in which signal levels have to be quantised too, as they can only be represented by a finite number of bits. The effects resulting from the limited resolution of signal levels are discussed in detail in Chap. 8.

In the discrete-time domain, we also deal with transfer systems that have the properties of linearity and time-invariance. Instead of time-invariance, however, we prefer to use the term "shift-invariance", since the processing of data may take place without any relation to the real time axis, as stated above. Thus we speak of linear shift-invariant (LSI) systems.

D. Schlichthärle, *Digital Filters*, DOI: 10.1007/978-3-642-14325-0_3,
© Springer-Verlag Berlin Heidelberg 2011

Linearity means that two superimposed signals pass through a system without mutual interference. Let $y_1(n)$ be the response of the system to $x_1(n)$ and $y_2(n)$ the response to $x_2(n)$. Applying a linear combination of both individual signals $x(n) = a\, x_1(n) + b\, x_2(n)$ to the input results in $y(n) = a\, y_1(n) + b\, y_2(n)$ at the output.

The property of shift-invariance can be described as follows: Let $y(n)$ be the reaction of the system to the input sequence $x(n)$. If one applies the same sequence to the input but delayed for n_0 cycles of the algorithm or n_0 clock cycles of the hardware, then the response to $x(n-n_0)$ is simply $y(n-n_0)$.

If both properties apply, the characteristics of the system can be fully described by the reaction to simple test signals such as pulses, steps and harmonic sequences.

3.1 Discrete-Time Convolution

The unit-sample sequence $\delta(n)$ adopts the role of the delta function as a stimulus to analyse the behaviour of the system in the discrete-time domain. The unit-sample sequence is defined as

$$\delta(n) = \begin{cases} 1 & \text{for } n = 0 \\ 0 & \text{for } n \neq 0 . \end{cases} \tag{3.1}$$

The response of the system to $\delta(n)$ is the unit-sample response $h(n)$. Provided that the system is linear and shift-invariant, we can derive a general relationship between input and output signal. We start by considering the reaction of the system to a single sample out of the whole of the input sequence. The n_0th pulse of the input sequence $x(n)$ can be written in the form

$$x(n_0)\,\delta(n - n_0) .$$

This single pulse leads to the output signal

$$x(n_0)\,h(n - n_0) .$$

To obtain this result, we made use of the properties of linearity and shift-invariance. If we now superimpose the responses of the individual input samples by summation over all possible n_0, we obtain the response of the system to the whole input sequence $x(n)$, again making use of the linear property.

$$\sum_{n_0=-\infty}^{+\infty} x(n_0)h(n - n_0) = y(n) \tag{3.2}$$

According to (3.2), we can calculate the output sequence $y(n)$ as a response to any arbitrary input sequence $x(n)$ if the unit-sample response $h(n)$ is known. Relation (3.2) is called the discrete convolution. It is the analogous relation to (1.1). In the discrete-time case, a summation takes the place of the integral. $x(n)$ and $h(n)$ can reverse roles in (3.2), which yields the following two representations:

$$y(n) = \sum_{k=-\infty}^{+\infty} x(k)h(n-k) = \sum_{k=-\infty}^{+\infty} h(k)x(n-k) \ . \tag{3.3}$$

As in the continuous-time case, convolution is denoted with an asterisk (*). So (3.3) can be written in the abbreviated form

$$y(n) = x(n)* h(n) = h(n)* x(n) \ .$$

The following example shall illustrate the method of discrete convolution.

Example 3-1

A unit-step sequence, which is defined as

$$x(n) = 1 \quad \text{for } n \geq 0 \qquad \text{and} \qquad x(n) = 0 \quad \text{for } n < 0$$

is applied to the input of the system under consideration. The unit-sample response of this system is specified as follows:

$$h(n) = \begin{cases} 10 & \text{for } n = 0 \\ 4 & \text{for } n = 1 \\ 2 & \text{for } n = 2 \\ 1 & \text{for } n = 3 \\ 0 & \text{otherwise} \ . \end{cases}$$

Determine the output sequence $y(n)$ of the system.

In order to perform the convolution according to (3.3), we write down the sequences $h(n)$ and $x(n)$ in the form of a table. Below we demonstrate the calculation of the output sample $y(n=2)$.

k	-3	-2	-1	0	1	2	3	4	5	6	7
x(k)	0	0	0	1	1	1	1	1	1	1	1
h(2-k)	0	0	1	2	4	10	0	0	0	0	0
y(2)											

y(2) = 0 + 0 + 0 + 2 + 4 + 10 + 0 + 0 + 0 + 0 + 0 = **16**

The values of $h(n)$ and $x(n)$ that appear in the table one below the other have to be multiplied. Afterwards all products are summed to get the output sample $y(2)$. To calculate other output samples, the position of the unit-sample response, which appears in reversed order in the 3rd line, has to be shifted accordingly. For other values of n, the convolution yields the following output samples:

n	-2	-1	0	1	2	3	4	5	6	7	8	9 .....
x(n)	0	0	1	1	1	1	1	1	1	1	1	1 .....
y(n)	0	0	10	14	16	17	17	17	17	17	17	17 .....

As a reaction to the unit-step sequence, we observe a slow rise up to the final value at the output, which is a low-pass behaviour.

The response to the unit-step sequence fully characterises the systems in the time domain as well. The unit-step function is defined as

$$u(n) = \begin{cases} 1 & \text{for } n \geq 0 \\ 0 & \text{for } n < 0 \end{cases} . \tag{3.4}$$

Substitution of (3.4) in the convolution sum (3.3) gives the following relationship between unit-sample response $h(n)$ and unit-step response $a(n)$ of the system:

$$a(n) = \sum_{m=-\infty}^{+\infty} u(n-m)h(m)$$

$$a(n) = \sum_{m=-\infty}^{n} h(m) . \tag{3.5a}$$

Conversely, the unit-sample response can be expressed as the difference of two consecutive samples of the unit-step response.

$$h(n) = a(n) - a(n-1) \tag{3.5b}$$

Digital filters, in principle, perform discrete convolution according to (3.3). Numerous algorithms and filter structures are available for the practical implementation of the convolution which differ in complexity and performance. For the best choice, the respective application has to be considered. In Chap. 5, we introduce the most important filter structures and highlight the characteristics of the various realisations.

3.2 Relations in the Frequency Domain

In order to describe the properties of the discrete-time system in the frequency domain, we choose a complex harmonic oscillation as a test function, as we did previously in the continuous-time case. The discrete-time harmonic sequence is obtained by the substitution $t = nT$ in the continuous-time form:

$$x(t) = e^{j\omega_0 t}$$

$$x(n) = e^{j\omega_0 nT} . \tag{3.6}$$

T is the period at which the complex exponential is sampled. The reciprocal of T is commonly referred to as the sampling frequency f_s.

$$f_s = 1/T$$

The corresponding angular frequency is defined as

$$\omega_s = 2\pi f_s = 2\pi/T .$$

In the case of digital real-time applications, where signal processing is synchronised with other processes in the continuous-time domain, such as A/D and D/A conversion, frequency f or ω and period T have the usual meaning of "cycles per time unit" and "period of sampling". But if the samples $x(n)$ are available in stored form, on a hard disk for instance, it is no longer possible to conclude how these samples have been generated. A 200 Hz sinusoid sampled at 1 kHz yields the same sequence as a 1000 Hz sinusoid sampled at 5 kHz. Hence it seems reasonable to use the normalised frequency ωT which specifies how the angle progresses between two successive samples of the harmonic oscillation. Using normalised frequencies allows the treatment of discrete-time systems independent of the real time axis.

No special symbol has been introduced in the literature for the frequency normalised to the sampling period. Some authors use ω for this purpose, as well, and indicate a different meaning in the course of the text. Using this style, the samples of a harmonic exponential oscillation can be expressed as

$$x(n) = e^{j\omega n} .$$

Throughout this text, the combination ωT will be used exclusively. We leave it to the reader to assume either processes in real time, with the corresponding meaning of ω and T, or signal processing without any relation to the real time axis, where the interpretation of ωT as a normalised frequency is more appropriate.

Convolution of (3.6) with the unit-sample response yields the response of the system to the complex exponential.

$$y(n) = \sum_{m=-\infty}^{+\infty} e^{j\omega_0 T(n-m)} h(m)$$

$$y(n) = e^{j\omega_0 Tn} \sum_{m=-\infty}^{+\infty} e^{-j\omega_0 Tm} h(m) \tag{3.7}$$

The summation term in (3.7) results in a complex variable $H(e^{j\omega_0 T})$, which is a function of the normalised frequency $\omega_0 T$. The relation

$$H(e^{j\omega_0 T}) = \sum_{m=-\infty}^{+\infty} e^{-j\omega_0 Tm} h(m)$$

is commonly referred to as the discrete-time Fourier transform (DTFT). $H(e^{j\omega_0 T})$ is the discrete-time Fourier transform of the sequence $h(m)$. Thus (3.7) becomes

$$y(n) = e^{j\omega_0 Tn} H(e^{j\omega_0 T}) .$$

Table 3-1 Some selected DTFT pairs

Time domain	Frequency domain
$\delta(n)$	1
$u(n)$	$\dfrac{1}{1-e^{-j\omega T}}+\pi\sum_{n=-\infty}^{+\infty}\delta(\omega T-2\pi n)$
$nu(n)$	$\dfrac{e^{-j\omega T}}{(1-e^{-j\omega T})^2}+j\pi\sum_{n=-\infty}^{+\infty}\delta'(\omega T-2\pi n)$
$e^{j\omega_0 Tn}\,u(n)$	$\dfrac{1}{1-e^{j\omega_0 T}e^{-j\omega T}}+\pi\sum_{n=-\infty}^{+\infty}\delta(\omega T-\omega_0 T-2\pi n)$
$z_0^n\,u(n)$	$\dfrac{1}{1-z_0\,e^{-j\omega T}}\qquad \mathrm{Re}\,z_0<1$

Table 3-1 shows some selected DTFT transform pairs. In the cases where the DTFT sum does not converge, periodic Dirac pulses occur in the frequency domain.

$H(e^{j\omega_0 T})$ modifies the complex exponential applied to the input with respect to amplitude and phase and therefore represents the gain and phase shift of the system specified by its unit-sample response $h(m)$. The DTFT of the unit-sample response evaluated at arbitrary frequencies ω is thus the frequency response of the system.

$$H(e^{j\omega T})=\sum_{n=-\infty}^{+\infty}e^{-j\omega Tn}h(n) \tag{3.8}$$

The magnitude of the frequency response is the magnitude response or amplitude response $A(\omega)$ of the discrete-time system.

$$A(\omega)=\left|H(e^{j\omega T})\right|$$

The amplitude response gives the gain of the system as a function of frequency. Gains are often expressed on a logarithmic scale in units of dB.

$$a(\omega)=20\log\left|H(e^{j\omega T})\right|=20\log A(\omega)\quad\text{in dB}$$

The frequency response of a system has often the form of a rational fractional function of $e^{j\omega T}$.

$$H(e^{j\omega T})=\frac{N(e^{j\omega T})}{D(e^{j\omega T})}$$

For the determination of the amplitude response, the magnitude of numerator and denominator can be calculated separately.

$$A(\omega) = \frac{\left|N(e^{j\omega T})\right|}{\left|D(e^{j\omega T})\right|} = \frac{A_N(\omega)}{A_D(\omega)}$$

The phase response $b(\omega)$ of the system is the negative argument of the frequency response.

$$b(\omega) = -\arg H(e^{j\omega T}) = -\arctan \frac{\operatorname{Im} H(e^{j\omega T})}{\operatorname{Re} H(e^{j\omega T})}$$

The phase response gives the phase shift, which a sine wave of frequency ω experiences when passing through the system, as a function of frequency. The frequency response can be expressed by magnitude and phase as

$$H(e^{j\omega T}) = A(\omega)e^{-jb(\omega)} \tag{3.9}$$

If the frequency response is given in the form of a rational fractional function, the phase of numerator and denominator can again be calculated separately.

$$H(e^{j\omega T}) = \frac{A_N(\omega)e^{-jb_N(\omega)}}{A_D(\omega)e^{-jb_D(\omega)}} = \frac{A_N(\omega)}{A_D(\omega)}e^{-j(b_N(\omega)-b_D(\omega))}$$

In this case, the overall phase of the system is given as

$$b(\omega) = b_N(\omega) - b_D(\omega). \tag{3.10}$$

The group delay response, finally, is defined as the derivative of phase with respect to frequency.

$$\tau_g(\omega) = \frac{d b(\omega)}{d\omega}$$

The group delay determines the delay of the envelope of a narrowband signal, centred around ω, when passing through the system. For a rational fractional frequency response function, the group delay is calculated as

$$\tau_g(\omega) = \tau_{gN}(\omega) - \tau_{gD}(\omega)$$

which can be easily derived by differentiation of (3.10).

The group delay can be determined directly from the frequency response while avoiding the intermediate calculation of the phase response. Starting point is (3.9).

$$H(e^{j\omega T}) = A(\omega)e^{-jb(\omega)}$$

$$e^{-jb(\omega)} = \frac{H(e^{j\omega T})}{A(\omega)}$$

Taking the natural logarithm on both sides yields

$$-jb(\omega) = \ln H(e^{j\omega T}) - \ln A(\omega)$$

$$b(\omega) = j\ln H(e^{j\omega T}) - j\ln A(\omega)$$

$$\tau_g(\omega) = \frac{db(\omega)}{d\omega} = j\frac{H'(e^{j\omega T})}{H(e^{j\omega T})} - j\frac{A'(\omega)}{A(\omega)}$$

$$\tau_g(\omega) = j\left(\mathrm{Re}\frac{H'(e^{j\omega T})}{H(e^{j\omega T})} + j\,\mathrm{Im}\frac{H'(e^{j\omega T})}{H(e^{j\omega T})} \right) - j\frac{A'(\omega)}{A(\omega)}$$

$$\tau_g(\omega) = -\mathrm{Im}\frac{H'(e^{j\omega T})}{H(e^{j\omega T})} + j\left(\mathrm{Re}\frac{H'(e^{j\omega T})}{H(e^{j\omega T})} - \frac{A'(\omega)}{A(\omega)} \right) .$$

Since the group delay is a real function, the imaginary part must disappear. So we finally have

$$\tau_g(\omega) = -\mathrm{Im}\frac{H'(e^{j\omega T})}{H(e^{j\omega T})} = -\mathrm{Im}\frac{dH(e^{j\omega T})/d\omega}{H(e^{j\omega T})} . \tag{3.11}$$

Example 3-2

Determine the frequency response of a discrete-time system which delays the input signal by three sampling periods. Calculate the group delay of the system using (3.11).

The impulse response of the delay system can be written as

$$h(n) = \delta(n-3) .$$

The impulse $\delta(n)$ applied to the input of the delayer appears three sampling periods later at the output. The frequency response is calculated as the Fourier transform of the impulse response.

$$H(e^{j\omega T}) = \sum_{n=-\infty}^{+\infty} h(n)e^{-j\omega nT} = \sum_{n=-\infty}^{+\infty} \delta(n-3)e^{-j\omega nT}$$

Only the summand for $n = 3$ contributes the sum which results in the response

$$H(e^{j\omega T}) = e^{-3j\omega T} .$$

The group delay is calculated using (3.11) as

$$\tau_g(\omega) = -\mathrm{Im}\frac{H'(e^{j\omega T})}{H(e^{j\omega T})} = -\mathrm{Im}\frac{-3jT\,e^{-3j\omega T}}{e^{-3j\omega T}} = \mathrm{Im}(j3T) = 3T ,$$

which is the expected result.

According to (3.8), the frequency response of discrete-time systems is a function of the term $e^{j\omega T}$. As a consequence, the frequency response is periodic with the period $2\pi/T = \omega_s$. This result is closely related to the fact that the sequences

$$x(n) = e^{j(\omega + k\omega_s)nT} = e^{jn(\omega T + 2\pi k)}$$

are identical for all integer values of k. Since the input sequences are identical, the reactions $y(n)$ of the system will also be identical for all k. Consequently, the frequency response of the system assumes the same complex values at frequencies $\omega + k\,\omega_s$, which means periodicity with period ω_s.

Application of the DTFT to the convolution (3.3) yields the starting point for the derivation of a general relationship between discrete-time input and output signals in the frequency domain.

$$Y(e^{j\omega T}) = \sum_{n=-\infty}^{+\infty} e^{-j\omega Tn}\, y(n)$$

$$Y(e^{j\omega T}) = \sum_{n=-\infty}^{+\infty} e^{-j\omega Tn} \sum_{k=-\infty}^{+\infty} x(k)h(n-k)$$

Changing the order of the two summations gives

$$Y(e^{j\omega T}) = \sum_{k=-\infty}^{+\infty} x(k)e^{-j\omega Tk} \sum_{n=-\infty}^{+\infty} h(n-k)e^{-j\omega T(n-k)} \quad .$$

With $m = n-k$ it follows that

$$Y(e^{j\omega T}) = \sum_{k=-\infty}^{+\infty} x(k)e^{-j\omega Tk} \sum_{m=-\infty}^{+\infty} h(m)e^{-j\omega Tm}$$

$$Y(e^{j\omega T}) = X(e^{j\omega T})H(e^{j\omega T}) \quad . \tag{3.12}$$

Thus the DTFT of the output signal is the product of the DTFT of the input signal and the frequency response. This relationship is analogous to the one for continuous-time systems.

As already mentioned, the discrete-time Fourier transform of a sequence $f(n)$ can be written in the form

$$F(e^{j\omega T}) = \sum_{n=-\infty}^{+\infty} e^{-j\omega Tn}\, f(n) \, . \tag{3.13}$$

$F(e^{j\omega T})$ has the meaning of the spectrum of the sequence of numbers $f(n)$. Conversely, if the spectrum is known, the sequence $f(n)$ can be calculated. For the derivation of the inverse relationship, we start by multiplying both sides of (3.13) by $e^{j\omega kT}$ and integrating over one period of the periodic spectrum.

$$\int_{-\omega_s/2}^{+\omega_s/2} F(e^{j\omega T}) e^{j\omega kT}\, d\omega = \int_{-\omega_s/2}^{+\omega_s/2} \sum_{n=-\infty}^{+\infty} f(n) e^{-j\omega nT} e^{j\omega kT}\, d\omega$$

$$\int_{-\omega_s/2}^{+\omega_s/2} F(e^{j\omega T}) e^{j\omega kT} \, d\omega = \sum_{n=-\infty}^{+\infty} f(n) \int_{-\omega_s/2}^{+\omega_s/2} e^{j\omega(k-n)T} \, d\omega$$

The integral on the right side is nonzero for $n = k$ only, since the integral is otherwise evaluated over whole periods of the complex exponential.

$$\int_{-\omega_s/2}^{+\omega_s/2} F(e^{j\omega T}) e^{j\omega kT} \, d\omega = f(k) \int_{-\omega_s/2}^{+\omega_s/2} 1 \, d\omega = f(k) \, \omega_s$$

$$f(k) = \frac{1}{\omega_s} \int_{-\omega_s/2}^{+\omega_s/2} F(e^{j\omega T}) e^{j\omega kT} \, d\omega = \frac{1}{2\pi} \int_{-\pi}^{+\pi} F(e^{j\omega T}) e^{j\omega kT} \, d\omega T \qquad (3.14)$$

Relation (3.14) constitutes the inverse discrete-time Fourier transform of a discrete-time signal. If the sequence $f(n)$ is real, the spectrum $F(e^{j\omega T})$ has to fulfil the symmetry condition

$$F(e^{j\omega T}) = F^*(e^{-j\omega T}) \, . \qquad (3.15)$$

The raised asterisk denotes the complex conjugate. Only if (3.15) is met, will the integral (3.14) yield real values for $f(k)$.

3.3 The z-Transform

3.3.1 Definition of the z-Transform

In the context of the FT of discrete-time signals, we come across the same problems that we have already encountered with continuous-time signals. Assuming idealised signals such as the unit-step sequence or a sinusoid sequence extending from $n = -\infty$ to $n = +\infty$, the Fourier sum (3.10a) does not converge. On the other hand, these functions are often used in theoretical considerations. If the FT is applied to calculate the spectrum of such discrete-time signals, the only way out is to treat the occurring singularities using the concept of distributions. The latter allows the characterisation of the energy, which is concentrated on one frequency or distributed over several or an infinite number of frequencies, by delta functions. We already used the same approach in the context of continuous-time signals. In many applications, however, a detour is made via the frequency domain to take advantage of the much simpler multiplication instead of the equivalent convolution in the time domain. In these cases, an approach similar to the Laplace transform is available in the discrete-time domain.

First of all, we only consider sequences whose elements are zero for $n < 0$ (right-sided transform). The samples to be transformed are multiplied by a damping term of the form $e^{-\sigma nT}$. σ is a real parameter that can be freely chosen such that the Fourier sum converges in any case.

$$F(e^{j\omega T}, \sigma) = \sum_{n=0}^{+\infty} f(n) e^{-\sigma nT} e^{-j\omega nT}$$

Combining the exponentials yields

$$F(e^{j\omega T}, \sigma) = \sum_{n=0}^{+\infty} f(n) e^{-(\sigma + j\omega)nT} \quad .$$

The term $e^{(\sigma + j\omega)T}$ is commonly abbreviated by the letter z.

$$F(z) = \sum_{n=0}^{+\infty} f(n) z^{-n} \tag{3.16}$$

Equation (3.16) constitutes the so-called right-sided z-transform (ZT). With σ approaching zero, equivalent to replacing z by $e^{j\omega T}$, the z-transform turns into the Fourier transform. This transition requires caution, however. The substitution is only possible without any problems if the Fourier sum of the sequence $f(n)$ converges for all frequencies ω!

3.3.2 Properties of the z-Transform

The properties of the z-transform are discussed in this section only in so far as they are needed for the understanding of the following chapters. Emphasis is put on the convolution and shift theorems whose application is illustrated in a number of examples. Detailed analysis of the z-transform, also with methods of the theory of complex functions, can be found in [27, 55].

The Convolution Theorem

Let $c(n)$ be the result of the convolution of $a(n)$ and $b(n)$.

$$c(n) = \sum_{k=-\infty}^{+\infty} a(k) b(n-k)$$

According to (3.12) the z-transform of $c(n)$ can be calculated as

$$C(z) = \sum_{n=0}^{\infty} \left(\sum_{k=-\infty}^{+\infty} a(k) b(n-k) \right) z^{-n} \quad .$$

Interchanging the order of summations yields

$$C(z) = \sum_{k=-\infty}^{+\infty} a(k) \sum_{n=0}^{\infty} b(n-k) z^{-n} \ .$$

The substitution $m = n - k$ results in

$$C(z) = \sum_{k=-\infty}^{+\infty} a(k) \sum_{m=-k}^{+\infty} b(m) z^{-(m+k)} \ .$$

As we only consider right-sided sequences, there are no contributions to the sums over m and k for $m < 0$ and $k < 0$.

$$C(z) = \sum_{k=0}^{\infty} a(k) \left(\sum_{m=0}^{\infty} b(m) z^{-m} \right) z^{-k}$$

If we choose σ, the real part of z, such that the two sums converge, we can express $C(z)$ as the product of two sums:

$$C(z) = \sum_{k=0}^{\infty} a(k) z^{-k} \sum_{m=0}^{\infty} a(m) z^{-m}$$

and hence

$$C(z) = A(z) B(z) \ .$$

It turns out that the property of the Fourier transform, that convolution in the time domain is equivalent to multiplication in the frequency domain, is also valid in the context of the z-transform. Applied to discrete-time systems, this means that the product of the z-transform of the input $X(z)$ and the unit-sample response of the system $H(z)$ yields the z-transform of the output $Y(z)$.

$$Y(z) = X(z) H(z) \tag{3.17}$$

$H(z)$ is called the system or transfer function. If the Fourier sum of $h(n)$ converges, as stated earlier, z may be replaced by $e^{j\omega T}$ to get the frequency response of the system. Figure 3-1 summarises again the relationships between the input and output signals of discrete-time systems.

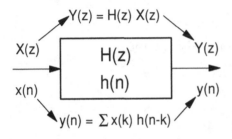

Fig. 3-1
Relationships between the input and output signals of discrete-time systems

The Shift Theorem

Let $F(z)$ be the z-transform of the sequence $f(n)$ and $F_d(z)$ the z-transform of the delayed sequence $f_d(n) = f(n-n_0)$ with $n_0 > 0$.

$$F_d(z) = \sum_{n=0}^{\infty} f(n-n_0)z^{-n}$$

We shall show how $F(z)$ and $F_d(z)$ are related. Substitution of m for $n-n_0$ yields

$$F_d(z) = \sum_{m=-n_0}^{\infty} f(m)\, z^{-(m+n_0)}$$

$$F_d(z) = z^{-n_0} \sum_{m=-n_0}^{\infty} f(m)\, z^{-m} \ .$$

Since $f(m)$ is zero for $m < 0$, there is no contribution to the sum for negative m. Hence we can write

$$F_d(z) = z^{-n_0} \sum_{m=0}^{\infty} f(m)\, z^{-m} = z^{-n_0} F(z) \ . \tag{3.18}$$

The delay of a sequence by n_0 sample periods in the time domain leads to a multiplication by z^{-n_0} in the frequency domain. Multiplication by the term z^{-1}, in particular, means a delay by one sample period. The term z^{-1} is therefore called the unit-delay operator. The unit delay is of great importance as it is a basic building block of digital signal processing algorithms, so it will often be found in the following chapters. Table 3-2 summarises the most important rules of the z-transform.

Time domain	Frequency domain
$f(n-n_0)$	$F(z)z^{-n_0}$
$f(n)z_0^n$	$F(z/z_0)$
$nf(n)$	$-zF'(z)$
$f_1(n) * f_2(n)$	$F_1(z)\,F_2(z)$

Table 3-2
A selection of the most important rules of the z-transform

Inverse z-Transform

For completeness, we introduce in this section the inverse z-transform, which can be mathematically expressed as

$$f(n) = \frac{1}{2\pi j} \oint_C F(z)\, z^{n-1}\, dz \ . \tag{3.19}$$

The integration path C is a contour in the convergence region of $F(z)$, which encircles the origin and is passed through counter-clockwise. In the case of the right-sided z-transform considered here, C encloses all poles of $F(z)$. The integration according to (3.19) needs not be carried out explicitly very often. Extensive tables are available from which the most important z-transform pairs needed in practice can be obtained. Most problems in the context of first- and second-order systems may be solved with the transform pairs shown in Table 3-3.

Table 3-3
z-transforms of selected signal sequences

Time domain	z-transform
$\delta(n)$	1
$u(n)$	$\dfrac{z}{z-1}$
$n\,u(n)$	$\dfrac{z}{(z-1)^2}$
$z_0^n\,u(n)$	$\dfrac{z}{z-z_0}$
$(1-z_0^n)\,u(n)$	$\dfrac{z(1-z_0)}{(z-1)(z-z_0)}$
$\cos(\omega_0 n)\,u(n)$	$\dfrac{z^2 - z\cos\omega_0}{z^2 - 2z\cos\omega_0 + 1}$
$\sin(\omega_0 n)\,u(n)$	$\dfrac{z\sin\omega_0}{z^2 - 2z\cos\omega_0 + 1}$
$z_0^n\cos(\omega_0 n)\,u(n)$	$\dfrac{z^2 - zz_0\cos\omega_0}{z^2 - 2zz_0\cos\omega_0 + z_0^2}$
$z_0^n\sin(\omega_0 n)\,u(n)$	$\dfrac{zz_0\sin\omega_0}{z^2 - 2zz_0\cos\omega_0 + z_0^2}$

Example 3-3

Determine the z-transform of the unit-sample sequence $\delta(n)$.

$$f(n) = \delta(n)$$

$$F(z) = \sum_{n=0}^{\infty} \delta(n)\, z^{-n}$$

According to the definition of $\delta(n)$, there is only one contribution to the above sum for $n = 0$. For all other values of n, the unit-sample sequence is zero.

$$F(z) = 1 \times z^{-0} = 1$$

Example 3-4

Determine the z-transform of the unit-step sequence $u(n)$.

All samples of the unit-step sequence assume the value 1 for $n \geq 0$. Hence

$$F(z) = \sum_{n=0}^{\infty} u(n) z^{-n} = \sum_{n=0}^{\infty} z^{-n} \ .$$

This geometric progression converges for $|z^{-1}| < 1$, which is equivalent to choosing $\sigma > 0$, and yields

$$F(z) = \frac{1}{1-z^{-1}} = \frac{z}{z-1} \ . \tag{3.20}$$

Example 3-5

Determine the z-transform of the decaying exponential $e^{-\alpha nT} u(n)$.

$$f(n) = e^{-\alpha nT} u(n) \quad \text{with } \alpha > 0$$

$$F(z) = \sum_{n=0}^{\infty} e^{-\alpha nT} z^{-n} = \sum_{n=0}^{\infty} (e^{-\alpha T} z^{-1})^n$$

Again we have to deal with the sum of an infinite geometric progression. Convergence is assured provided that $|e^{-\alpha T} z^{-1}| < 1$ or $\sigma > -\alpha$. The infinite summation yields

$$F(z) = \frac{1}{1 - e^{-\alpha T} z^{-1}} = \frac{z}{z - e^{-\alpha T}} \ . \tag{3.21}$$

Example 3-6

Let a discrete-time system be characterised by an exponentially decaying unit-sample response of the form $h(n) = e^{-\alpha nT} u(n)$. As the input signal we apply the unit-step sequence $u(n)$. Determine the output sequence $y(n)$.

The output sequence $y(n)$ may be obtained in the time domain by convolution of $h(n)$ and $u(n)$.

$$y(n) = \sum_{k=-\infty}^{+\infty} u(k) h(n-k)$$

But in this example, we prefer to make use of the convolution theorem and perform our calculations in the frequency domain, where the convolution is replaced by the multiplication of the z-transforms.

$$Y(z) = X(z) H(z)$$

With the results of the previous examples, we can directly write down the z-transform of the output signal which is simply the product of (3.20) and (3.21).

$$Y(z) = \frac{z}{z-1} \frac{z}{z-e^{-\alpha\ T}} = \frac{z^2}{(z-1)(z-e^{-\alpha\ T})}$$

Partial fraction expansion yields

$$Y(z) = \frac{1}{1-e^{-\alpha T}} \frac{z}{z-1} - \frac{e^{-\alpha T}}{1-e^{-\alpha T}} \frac{z}{z-e^{-\alpha T}} \ .$$

Again using the results from these examples, it is easy to return to the time domain.

$$y(n) = \frac{1}{1-e^{-\alpha T}} u(n) - \frac{e^{-\alpha T}}{1-e^{-\alpha T}} e^{-\alpha nT} u(n)$$

$$y(n) = \frac{1}{1-e^{-\alpha T}} \left(1 - e^{-\alpha(n+1)T}\right) u(n)$$

Figure 3-2 shows the input signal, the unit-sample response and the output signal of this example.

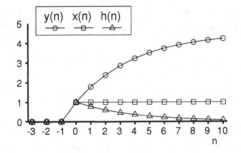

Fig. 3-2
Unit-step response of a filter with the unit-sample response
$h(n) = e^{-\alpha nT} u(n)$,
$\alpha T = 0.25$

Example 3-7

Determine the magnitude response of the transfer function

$$H(z) = \frac{z}{z-e^{-\alpha T}} \ .$$

The corresponding frequency response is obtained by substituting $e^{j\omega T}$ for z.

$$H(e^{j\omega T}) = \frac{1}{1-e^{-\alpha T} e^{-j\omega T}} \tag{3.22}$$

We multiply (3.22) by $H(e^{-j\omega T})$ to get the squared magnitude of the transfer function.

$$\left|H(e^{j\omega T})\right|^2 = \frac{1}{1-e^{-\alpha T}e^{-j\omega T}}\frac{1}{1-e^{-\alpha T}e^{+j\omega T}}$$

$$\left|H(e^{j\omega T})\right|^2 = \frac{1}{1+e^{-2\alpha T}-e^{-\alpha T}(e^{+j\omega T}+e^{-j\omega T})}$$

$$\left|H(e^{j\omega T})\right|^2 = \frac{1}{1+e^{-2\alpha T}-2e^{-\alpha T}\cos\omega T}$$

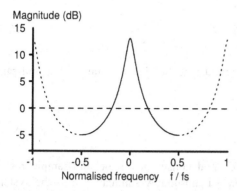

Fig. 3-3
Logarithmic magnitude of a first
order low-pass filter
$h(n) = e^{-\alpha Tn} u(n)$, $\alpha T = 0.25$

Figure 3-3 shows the logarithmic magnitude in dB, which is periodic with period f_s. The magnitude shows a low-pass characteristic, as could be expected from the slow rise in the reaction of the system to the unit-step sequence.

3.4 Stability Criterion in the Time Domain

There are a number of definitions of stability. One approach is the definition of stability by Lyapunov [43], who considers the behaviour of the system starting from an initial state with vanishing input signal. We come back to this definition later in Chap. 8. In this discussion we assume the so-called BIBO-stability (Bounded Input – Bounded Output), which requires that a limited input signal always results in a limited output signal. The requirement can be formulated mathematically as follows:

$$|y(n)| = \left|\sum_{k=-\infty}^{+\infty}x(n-k)h(k)\right| < B . \tag{3.23}$$

We are looking for constraints imposed on $h(k)$ which guarantee that the magnitude of the convolution sum stays below a certain upper boundary B whenever the input sequence is bounded. By application of the triangle inequality to the magnitude of the convolution sum, we obtain an expression which is guaranteed to be equal or larger than the term in (3.23). If this expression is bounded, then also (3.23) is surely fulfilled.

$$\sum_{k=-\infty}^{+\infty} |x(n-k)| \, |h(k)| < B$$

Replacing the magnitude of the input signal $|x(n)|$ by its maximum x_{max} again yields an expression whose magnitude is equal to or larger than the previous one.

$$\sum_{k=-\infty}^{+\infty} x_{max} |h(k)| = x_{max} \sum_{k=-\infty}^{+\infty} |h(k)| < B \qquad (3.24)$$

In order to fulfil (3.24), it is required that the absolute sum of the unit-sample response is bounded.

$$\sum_{k=-\infty}^{+\infty} |h(k)| < \infty \qquad (3.25)$$

The inequality (3.25) is the desired constraint on the unit-sample response guaranteeing BIBO-stability. If $h(k)$ is absolutely summable, then the system is stable in the discussed sense.

Inequality (3.25) is a criterion in the time domain. As the behaviour of systems is mostly characterised by the transfer function in the frequency domain, it would be desirable to have available a stability criterion in the frequency domain, as well. We come back to that problem later in Chap. 5, where we investigate the structures of digital filters and their transfer functions.

3.5 Further Relations of Interest

3.5.1 Magnitude at $\omega = 0$ and $\omega = \pi/T$

The absolute sum of the unit-sample response has an important meaning with respect to the stability of the system, as shown in the previous section. Occasionally, the sum of the sequence and the sum of the squared sequence are of interest, too.

Let us start with considering (3.8) for the special case $\omega = 0$. It turns out that the sum of the unit-sample response corresponds to the gain of the system at the frequency $\omega = 0$.

$$\sum_{n=-\infty}^{+\infty} h(n) = H(0) \qquad (3.26a)$$

For $\omega = \pi/T$, the exponential in (3.8) alternates between $+1$ and -1. Hence we have

$$\sum_{n=-\infty}^{+\infty} h(n)(-1)^n = H(\pi/T) \ . \tag{3.26b}$$

Equations (3.26a,b) can be used to normalise the unit-sample response $h(n)$ to an arbitrary system gain with respect to $\omega = 0$ or $\omega = \pi/T$.

3.5.2 Parseval's Theorem

The starting point of the following calculations is relation (3.8) again.

$$H(e^{j\omega T}) = \sum_{n=-\infty}^{+\infty} e^{-j\omega Tn} h(n) \tag{3.27}$$

The conjugate-complex frequency response can be written as

$$H^*(e^{j\omega T}) = \sum_{m=-\infty}^{+\infty} e^{+j\omega Tm} h(m) \ . \tag{3.28}$$

Integration of the product of (3.27) and (3.28) over a whole period $0 \dots 2\pi/T$ of the periodic frequency response yields the following relationship. We have directly taken into account that the product of H and H^* corresponds to the squared magnitude of the frequency response.

$$\int_0^{2\pi/T} \left| H(e^{j\omega T}) \right|^2 d\omega = \int_0^{2\pi/T} \sum_{n=-\infty}^{+\infty} h(n) e^{-j\omega Tn} \sum_{m=-\infty}^{+\infty} h(m) e^{+j\omega Tm} \, d\omega \tag{3.29a}$$

$$\int_0^{2\pi/T} \left| H(e^{j\omega T}) \right|^2 d\omega = \sum_{n=-\infty}^{+\infty} h(n) \sum_{m=-\infty}^{+\infty} h(m) \int_0^{2\pi/T} e^{j\omega T(m-n)} \, d\omega \tag{3.29b}$$

When interchanging summation and integration in (3.29a), we tacitly assumed that the system under consideration is stable. The integral on the right side of (3.29b) is only nonzero for $m = n$ because the integration otherwise extends over whole periods of the periodic complex exponential.

$$\int_0^{2\pi/T} \left| H(e^{j\omega T}) \right|^2 d\omega = \sum_{n=-\infty}^{+\infty} h(n) \sum_{m=-\infty}^{+\infty} h(m) \ 2\pi\delta(n-m)/T = 2\pi/T \sum_{n=-\infty}^{+\infty} h^2(n)$$

$$\sum_{n=-\infty}^{+\infty} h^2(n) = \frac{T}{2\pi} \int_0^{2\pi/T} \left| H(e^{j\omega T}) \right|^2 d\omega = \frac{T}{\pi} \int_0^{\pi/T} \left| H(e^{j\omega T}) \right|^2 d\omega \tag{3.30}$$

The sum over the squared unit-sample response provides information about the mean-square gain of the system. Equation (3.30) is known in the literature as one of the forms of Parseval's theorem.

The sum of a squared sequence is commonly referred to as the energy of the sequence. Equation (3.30) opens up the possibility of calculating the energy of the signal in the frequency domain by integrating the squared magnitude of the spectrum.

Example 3-8

Determine the sum over the squared unit-sample response of an ideal low-pass filter with cutoff frequency ω_c.

The gain of the ideal low-pass filter is unity up to the cutoff frequency ω_c. Beyond this frequency, up to $\omega = \pi/T$, the gain is zero. Evaluation of (3.30) with these data yields the following result:

$$\sum_{n=-\infty}^{+\infty} h^2(n) \approx \frac{T}{\pi} \int_0^{\omega_c} 1 \, d\omega = \frac{\omega_c T}{\pi} = 2 f_c T$$

Thus, the energy of the unit-sample response is proportional to the cutoff frequency of the low-pass filter if the gain is constant in the passband.

3.5.3 Group Delay Calculation in the z-Domain

It was shown in Sect. 3.2 that the group delay of a system can be calculated based on the derivative of the frequency response with respect to the frequency ω (3.11). This calculation can also be performed in the z-domain. Using

$$\frac{d e^{j\omega T}}{d\omega} = j T e^{j\omega T} \, ,$$

(3.11) can be expressed in the form

$$\tau_g = -\operatorname{Im} jT \frac{e^{j\omega T} \, d H(e^{j\omega T})/d e^{j\omega T}}{H(e^{j\omega T})} \, .$$

With $z = e^{j\omega T}$ we get

$$\tau_g = -T \operatorname{Im} j \frac{z \, d H(z)/d z}{H(z)} \Bigg|_{z=e^{j\omega T}}$$

$$\tau_g = -T \operatorname{Im} j \left(\operatorname{Re} \frac{z \, d H(z)/d z}{H(z)} + j \operatorname{Im} \frac{z \, d H(z)/d z}{H(z)} \right) \Bigg|_{z=e^{j\omega T}}$$

$$\tau_g = -T\,\mathrm{Re}\,\frac{z\,\mathrm{d}\,H(z)/\mathrm{d}\,z}{H(z)}\bigg|_{z=e^{j\omega T}} = -T\,\mathrm{Re}\,\frac{z\,H'(z)}{H(z)}\bigg|_{z=e^{j\omega T}} \qquad (3.31)$$

Example 3-9

Determine the group delay of the transfer function

$$H(z) = (1 + z^{-1})/2$$

using (3.31). This filter calculates the average value of two consecutive samples of the discrete-time signal.

Substitution of the transfer function in (3.31) yields

$$\tau_g = -T\,\mathrm{Re}\,\frac{z(-z^{-2})/2}{(1+z^{-1})/2}\bigg|_{z=e^{j\omega T}} = T\,\mathrm{Re}\,\frac{z^{-1}}{1+z^{-1}}\bigg|_{z=e^{j\omega T}}.$$

Substitution of $e^{j\omega T}$ for z gives

$$\tau_g = T\,\mathrm{Re}\,\frac{e^{-j\omega T}}{1+e^{-j\omega T}}.$$

Multiplication of numerator and denominator by the term $e^{j\omega T/2}$ results in the expression

$$\tau_g = T\,\mathrm{Re}\,\frac{e^{-j\omega T/2}}{e^{+j\omega T/2}+e^{-j\omega T/2}} = T\,\mathrm{Re}\,\frac{\cos(\omega T/2)-j\sin(\omega T/2)}{2\cos(\omega T/2)}.$$

Taking the real part of this expression yields the result

$$\tau_g = T\,\frac{\cos(\omega T/2)}{2\cos(\omega T/2)} = \frac{T}{2}.$$

4 Sampling Theorem

4.1 Introduction

The sampling theorem, popularised by Shannon in 1948 [67], is one of the most important laws of signal theory. It constitutes the connection between the domain of the continuous-time physical signals and the abstract digital world, where signals are represented by sequences of numbers and the behaviour of electrical, mechanical or hydraulic systems is imitated by mathematical algorithms. The sampling theorem is a prerequisite for the pulse code modulation (PCM) which forms the basis for time multiplex techniques and digital measurement and control, as well as for the possibility of high quality transmission, processing and storage of signals in digital form.

The sampling theorem states that it is, under certain circumstances, sufficient to know the values of a continuous-time signal at discrete times in order to reconstruct the original continuous-time signal completely.

The following considerations will highlight some important relationships between continuous-time and discrete-time signals.

4.2 Sampling

After periodic sampling of a continuous-time signal $f_a(t)$, we only know the values $f(n) = f_a(nT)$ of the original signal at equidistant instants nT. We will show how the missing information about the behaviour of the signal between the samples manifests itself in the frequency domain.

Time function $f_a(t)$ and spectrum $F_a(j\omega)$ of analogue signals are related by the transform pair of the Fourier transform. Provided that we are only interested in the signal values at the instants $t = nT$, the inverse transform (1.11b) takes on the form

$$f(n) = f_a(nT) = \frac{1}{2\pi} \int_{-\infty}^{+\infty} F_a(j\omega) e^{j\omega Tn} \, d\omega \ . \tag{4.1}$$

Decomposition of integral (4.1) into frequency intervals of $2\pi / T$ and summation over all partial integrals yields

D. Schlichthärle, *Digital Filters*, DOI: 10.1007/978-3-642-14325-0_4,
© Springer-Verlag Berlin Heidelberg 2011

$$f(n) = \frac{1}{2\pi} \sum_{k=-\infty}^{+\infty} \int_{-\pi/T}^{+\pi/T} F_a(j\omega + jk2\pi/T) e^{j(\omega Tn + 2\pi kn)} \, d\omega \, .$$

Using the relation $\omega_s = 2\pi/T$, and taking into account that k and n are integers, it follows

$$f(n) = \frac{1}{2\pi} \sum_{k=-\infty}^{+\infty} \int_{-\omega_s/2}^{+\omega_s/2} F_a(j\omega + jk\omega_s) e^{j\omega Tn} \, d\omega \, .$$

By interchanging integration and summation we obtain

$$f(n) = \frac{1}{2\pi} \int_{-\omega_s/2}^{+\omega_s/2} e^{j\omega Tn} \sum_{k=-\infty}^{+\infty} F_a[j(\omega + k\omega_s)] \, d\omega \, . \tag{4.2a}$$

The sum under the integral in (4.2a) corresponds to the periodically continued spectrum of the original analog signal. For the calculation of the sample values from the spectrum of the analog signal, we have, according to (4.2a), to calculate the inverse discrete-time Fourier transform over one period of the periodically continued spectrum. Figure 4-1 shows a corresponding example. It is obvious from (4.2a) that the bold detail in the interval $-\omega_s/2 \leq \omega \leq +\omega_s/2$ of the periodically continued spectrum contains the complete information of the samples $f(n)$.

Spectrum

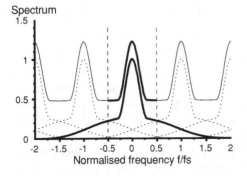

Fig. 4-1
Periodic continuation of the spectrum
(Shannon condition not met)

Comparison of (4.2a) with (3.14) yields a relation between the DTFT of $f(n)$ and the spectrum of the original analog signal $f_a(t)$.

$$\frac{1}{2\pi} \sum_{k=-\infty}^{+\infty} F_a[j(\omega + k\omega_s)] = \frac{1}{\omega_s} F(e^{j\omega T})$$

Thus sampling of the analog signal $f_a(t)$ with the period T leads to the following equivalent relations in the time and frequency domains:

time domain : $f(n) = f_a(nT)$

frequency domain : $F(e^{j\omega T}) = \dfrac{1}{T} \displaystyle\sum_{k=-\infty}^{+\infty} F_a[j(\omega + k\omega_s)]$.   (4.3)

The Fourier transform of the sequence $f(n)$ is, apart from the constant factor $1/T$, identical to the periodically continued spectrum of the analog signal $f_a(t)$. By insertion of (3.13) in (4.3), we obtain the inverse relation to (4.2a), which provides information about the analog spectrum derived from the samples $f(n)$.

$$\sum_{k=-\infty}^{+\infty} F_a[j(\omega + k\omega_s)] = T \sum_{n=-\infty}^{+\infty} f(n)e^{-j\omega Tn} \qquad (4.2b)$$

From (4.2b) it can be seen that the loss of information about the behaviour of the analog signal $f_a(t)$ between the samples $f_a(nT)$ has the consequence that only the periodically continued spectrum is available. The original spectrum of the analog signal, in general, can not be recovered any more. The reason is that we cannot identify how the bold part of the periodical spectrum in Fig. 4-1 is composed of the periodically continued parts of the original spectrum. There is an infinite number of possibilities for decomposing this spectrum into partial spectra. The equivalent statement in the time domain is that there is an infinite number of possible analog functions $f_a(t)$ that could interpolate the given samples $f_a(nT)$.

4.3 Band-Limited Signals

Normally, the original spectrum $F_a(j\omega)$ cannot be recovered, if only samples of the analog signal $f_a(t)$ are available. The only exception is the special situation where the analog spectrum $F_a(j\omega)$ is limited to the frequency range $0 \leq |\omega| < \omega_s/2$. No overlapping or "aliasing" takes place if the spectrum is periodically continued (Fig. 4-2). The original spectrum and hence the original analog signal can be recovered by simply removing the periodically continued parts of the spectrum of the discrete-time signal, which is accomplished by low-pass filtering at a cutoff frequency of $\omega_s/2$. This low-pass filtering interpolates the discrete-time samples and yields the original analog signal. Hence no information is lost if a signal is sampled at a frequency which is at least double the highest frequency contained in the spectrum of the signal. This is the essential message of the sampling theorem.

The mathematical relations between the samples $f(n)$ and the spectrum of the analog signal $F_a(j\omega)$ become considerably simpler if the condition of the sampling theorem is fulfilled. We can omit the periodic continuation in (4.2a) because there is no aliasing, and the spectrum in the interval $-\omega_s/2 < \omega < +\omega_s/2$ is identical to the original spectrum of the analog signal.

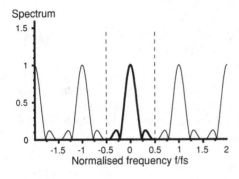

Fig. 4-2
Periodic continuation of the spectrum
without aliasing

$$f(n) = \frac{1}{2\pi} \int_{-\omega_s/2}^{+\omega_s/2} F_a(j\omega) e^{j\omega Tn} \, d\omega \qquad (4.4a)$$

The absence of aliasing also simplifies the inverse relation (4.2b). The spectrum $F_a(j\omega)$ can be directly calculated from the samples $f(n)$.

$$F_a(j\omega) = \begin{cases} T \sum_{n=-\infty}^{+\infty} f(n) e^{-j\omega Tn} = T F(e^{j\omega T}) & \text{for } |\omega| < \pi/T \\ 0 & \text{for } |\omega| \geq \pi/T \end{cases} \qquad (4.4b)$$

In (4.4b) the spectrum of the analog signal $F_a(j\omega)$ is formally calculated as the inverse Fourier transform of the samples $f(n)$, setting to zero all spectral components outside the range $-\omega_s/2 < \omega < +\omega_s/2$. The latter corresponds to an ideal low-pass filtering at a cutoff frequency of $\omega_s/2$.

4.4 Signal Conversion

4.4.1 Analog-to-Digital Conversion

The process of A/D-conversion consists in practice of three essential steps, as shown in Fig. 4-3:

- low-pass filtering
- sample and hold
- the actual analog-to-digital conversion

The low-pass filter limits the spectrum of the input signal $f_a(t)$ to half the sampling frequency. This avoids any overlapping of spectral components if the spectrum is periodically continued by the subsequent sampling process. Half the sampling frequency is often referred to as the Nyquist frequency. In the following, we will use the term baseband for the frequency range up to the Nyquist frequency, which corresponds to the frequency band usable for user information if the signal is digitised at a given sample rate. Low-pass filtering has to be performed carefully. Alias spectral components that are mirrored into the baseband cannot be removed later. It is impossible to distinguish after sampling which part of the spectrum is the original one and which part stems from the periodic continuation. Without having the original spectrum available, it is also impossible to restore the signal later in the time domain.

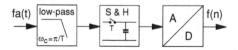

Fig. 4-3
Functionalities of analog-to-digital conversion

The usable range of the baseband is further reduced by the transition band of the low-pass filter, which cannot be made arbitrarily small in real implementations. The effort that has to be spent on filtering depends on the efficiency with which the baseband is to be used and on quality requirements with respect to the conservation of the original form of the signal. For the latter aspect, the group delay distortions in the vicinity of the cutoff frequency are an important factor.

Samplers that sample a given signal at only double the highest frequency contained in the signal are called Nyquist samplers. In Chap. 9 we will describe the concept of oversampling, which uses sampling rates much higher than needed to fulfil the sampling theorem. Conversion techniques based on this concept allow us to shift the complexity of filtering and interpolation into the digital domain, to reduce the costs of A/D and D/A converters and to improve the overall performance of signal processing by linear-phase filtering, which is not possible in the continuous-time domain.

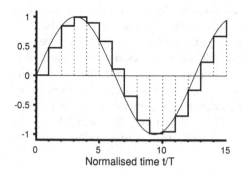

Normalised time t/T

Fig. 4-4
Sampling using a "sample and hold" circuit

In practice, sampling of signals is mostly performed using "sample and hold" circuits. The result is a stepped signal $f_s(t)$ which keeps the signal value at the sampling instant $f_a(nT)$ over a whole sampling period T (Fig. 4-4). This is necessary because most of the A/D converter types require the sampled value for a while to determine the corresponding numerical value in digital form. Some types of converters need multiple access to the sample value. If the value changes during the conversion process, the result would be incorrect.

The optimal choice of an A/D converter architecture depends on the respective applications. Factors such as accuracy, resolution, the nature of the analog input signal, conversion speed, and environmental conditions have to be considered in the process of selection. Most of the converters on the market are based on one of the following principles or on variations of these:

- flash
- sub-ranging
- successive approximation
- dual slope
- voltage-to-frequency conversion (VFC)
- delta-sigma

Common to all architectures is the function to convert the rounded or truncated ratio of the analog input voltage to a certain reference voltage into a numerical value represented by a number of bits. Delta-sigma converters are often used in combination with the concept of oversampling, which will be explained in more detail in Chap. 9.

4.4.2 Digital-to-Analog Conversion

4.4.2.1 Implementational Aspects

Relation (4.4b) shows that the reconstruction of the analog signal $f_a(t)$ is conceptionally achieved by low-pass filtering of the spectrum $F(e^{j\omega T})$ of the corresponding discrete-time sequence $f(n)$. This procedure interpolates the values of the signal between the digital samples. What does it mean in a real physical implementation, however, to interpolate a sequence of numbers by means of an analog low-pass filter as suggested by (4.4b)? We need an intermediate step, in which the digital sample $f(n)$ is represented by some analog equivalent which can then be filtered using real analog hardware. This is the actual function of the D/A converter, which generates at its output a sequence of electrical pulses, at a rate of $1/T$, whose amplitude is proportional to the value of the corresponding digital samples $f(n)$. Figure 4-5 shows the essential steps of digital-to-analog conversion, which include the actual D/A conversion, deglitching and low-pass filtering.

Fig. 4-5
Functionalities of digital-to-analog conversion

The most frequently used principle of D/A conversion is to sum the current of switched current sources. If digital samples with N-bit resolution are to be converted, the converter provides N current sources whose current is graded in binary exponential steps. Let the least significant bit (LSB) of the digital word be represented by a current of I_0. The subsequent sources exhibit current values of $2\,I_0$, $4\,I_0$, $8\,I_0$, etc. The source representing the most significant bit (MSB) thus exhibits a value of $2^{N-1}\,I_0$. Given a digital word to be converted, the current sources are switched on or off depending on the value of the corresponding bits. The current thus accumulated is then converted to a proportional voltage, which appears at the output of the converter. The advantage of using current sources in contrast to voltage sources is the higher achievable conversion speed.

In practical implementations, it cannot be guaranteed that all current sources are switched at the same instant in the transition period from one digital input value to the next one. This results in a period in which the analog output value is uncertain. A so-called deglitcher is used to sample the output of the converter at an instant when it has reached a stable voltage and to provide an impulse at this voltage with a well-defined length. The function is thus similar to the sample and hold circuit used in A/D conversion. It is common to choose the length of the pulses equal to the sampling period T, which results in a stepped signal. The influence that the choice of the pulse length has on the conversion process will be discussed in more detail below in the mathematical analysis of interpolation by low-pass filtering.

The low-pass filter removes the periodically continued parts of the spectrum and interpolates the digital samples. Depending on the application, more or less effort is spent to come as close as possible to the behaviour of the ideal low-pass filter. If the best possible preservation of the form of a signal is an issue, oversampling techniques are superior, which allow linear-phase filtering in the digital domain.

4.4.2.2 Mathematical Analysis of the D/A Conversion Process

Figure 4-6 illustrates the principle of the recovery of analog signals from sequences of digital sample values $f(n)$. The two main blocks are a pulse modulator and a low-pass filter. The pulse modulator generates a sequence of pulses at the sampling rate $1/T$. The amplitude of the pulses is proportional to the value of the respective sample to be reproduced. The low-pass filter removes the periodically continued parts of the spectrum which means interpolation and smoothing of the pulse stream in the time domain.

The pulse modulator is characterized by the temporal shape $p(t)$ of the generated pulses or equivalently by their spectral shape $P(j\omega)$. The sequence of pulses $f_s(t)$, weighted according to the values of the related digital samples $f(n)$, can be mathematically expressed as

$$f_s(t) = \sum_{n=-\infty}^{+\infty} f(n)\, p(t - nT) \,. \tag{4.5}$$

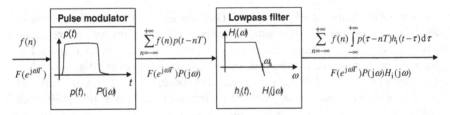

Fig. 4-6   Principle of digital-to-analog conversion

(4.5) implements the transition from a discrete-time sequence of sample values to a real physical analog signal. The spectrum of $f_s(t)$ is obtained by taking the Fourier transform of (4.5).

$$F_s(j\omega) = \int_{t=-\infty}^{+\infty} \sum_{n=-\infty}^{+\infty} f(n)\,p(t-nT)\,e^{-j\omega t}\,dt$$

Interchanging summation and integration yields

$$F_s(j\omega) = \sum_{n=-\infty}^{+\infty} f(n) \int_{t=-\infty}^{+\infty} p(t-nT)\,e^{-j\omega t}\,dt \; .$$

The integral is the Fourier transform of the shifted pulse $p(t)$. Applying the shift theorem of the Fourier transform gives

$$F_s(j\omega) = \sum_{n=-\infty}^{+\infty} f(n)\,e^{-j\omega nT} \int_{t=-\infty}^{+\infty} p(t)\,e^{-j\omega T}\,dt \; .$$

The summation term is nothing else but the discrete-time Fourier transform $F(e^{j\omega T})$ of the sequence $f(n)$. The integral corresponds to the Fourier transform and thus to the spectrum $P(j\omega)$ of the pulse $p(t)$. The spectrum of the pulse stream can therefore be expressed as

$$F_s(j\omega) = F(e^{j\omega T})\,P(j\omega) \; . \tag{4.6}$$

The subsequent interpolation low-pass filter is characterized by its impulse response $h_i(t)$ and the corresponding frequency response $H_i(j\omega)$. The reconstructed analog signal $f_r(t)$ after interpolation is obtained by convolution of $f_s(t)$ and the impulse response $h_i(t)$ of the low-pass filter.

$$f_r(t) = \int_{\tau=-\infty}^{+\infty} f_s(\tau)\,h_i(t-\tau)\,d\tau$$

$$f_r(t) = \int_{\tau=-\infty}^{+\infty} \sum_{n=-\infty}^{+\infty} f(n)\,p(\tau-nT)\,h_i(t-\tau)\,d\tau$$

Interchanging summation and integration yields

$$f_r(t) = \sum_{n=-\infty}^{+\infty} f(n) \int_{\tau=-\infty}^{+\infty} p(\tau - nT) h_i(t-\tau) \, d\tau \; . \tag{4.7}$$

The corresponding relation in the frequency domain reads as

$$F_r(j\omega) = F(e^{j\omega T}) P(j\omega) H_i(j\omega) \tag{4.8}$$

4.4.2.3 Storage and Playout of Analog Signals

In many technical fields of application, there is a need to digitally record analog signals, to store them on media such as hard disk, flash prom, or DVD, and to playout these signals later, e.g. for listening, viewing, or for further analysis. These applications are not limited to audio or video which immediately cross one's mind. In many technical areas, continuous-time physical quantities such as pressure, displacement, acceleration, voltage, current, or light intensity need to be recorded for later evaluation. The sampling theorem must be fulfilled in all these cases by adapting the sampling rate to the rate of change of the respective physical signals or by low-pass filtering of the measured signal to half the sampling frequency before sampling. In this way, misinterpretation of the measurement results can be avoided. Audio and video engineers are aware of this because aliasing errors are immediately audible or visible. In other technical fields, the consequences of undersampling are often ignored which may lead to inexplicable or misleading results.

Sampling an analog signal at double the highest frequency of interest is also known as Nyquist sampling (as opposed to oversampling). A step-by-step analysis in the frequency domain will reveal which factors influence the true-to-original reproduction of the analog signal. At first, the analog signal $f_a(t)$ passes an anti-aliasing filter with frequency response $H_a(j\omega)$ which is intended to remove the frequency components above half the sampling frequency $\omega_s/2 = \pi/T$. This filtered signal is then sampled and digitized. The latter means that the sample values are digitally represented by a given limited number of bits. This digitisation adds a noise spectrum $Q(e^{j\omega T})$ to the signal whose nature will be discussed in detail in Chap. 8. The signal is now ready to be stored. Using (4.3), the spectrum of the stored signal, expressed by the Fourier transform $F(e^{j\omega T})$ of the sample sequence $f(n)$, can be written as

$$F(e^{j\omega T}) = \frac{1}{T} \sum_{n=-\infty}^{+\infty} F_a[j(\omega + n\omega_s)] H_a[j(\omega + n\omega_s)] + Q(e^{j\omega T}) \; . \tag{4.9}$$

Substitution of (4.9) in (4.8) results in the following relation between the original signal $F_a(j\omega)$ and the reproduced signal $F_r(j\omega)$.

$$F_r(j\omega) = \left[\frac{1}{T}\sum_{n=-\infty}^{+\infty}F_a[j(\omega+n\omega_s)]H_a[j(\omega+n\omega_s)]+Q(e^{j\omega T})\right]P(j\omega)H_i(j\omega) \quad (4.10)$$

All the efforts required to store and play back the signal resides in the analog domain. Nyquist sampling does not involve any digital signal processing. The characteristics of the anti-aliasing and interpolation filters and the chosen resolution of the digital signal representation are the deciding factors for the quality of the reproduced signal.

The noise term in (4.10) can be made arbitrarily small by choosing an appropriate wordlength for the digital signal representation.

- PCM transmission in telephony uses a compressed coding with 8 bit per sample. The achievable signal-to-noise ratio (SNR) is about 35 dB.
- CD audio uses 16 bit linear PCM with a maximum SNR of 98 dB.
- Studio technology uses 24 or even 32 bits. Each additional bit improves the SNR by 6 dB

Ignoring the noise term in (4.10), we obtain the following relationship between original and the recovered analog signal:

$$F_r(j\omega) = \frac{1}{T}P(j\omega)H_i(j\omega)\sum_{n=-\infty}^{+\infty}F_a[j(\omega+n\omega_s)]H_a[j(\omega+n\omega_s)] \quad (4.11)$$

The choice of appropriate filter characteristics for the anti-aliasing and interpolation filters is the most critical aspect in the case of Nyquist sampling. On the one hand, the optimum usage of the available frequency range $0 \le \omega < \omega_s/2$ requires steep-slope low-pass filters. Such filters, however, are known to create considerable group delay distortion in the vicinity of the cutoff frequency which may not be tolerable in many applications. For digital filters, it is quite easy to realise linear-phase filters. Analog filters, however, can only approximate a linear phase response. Allpass networks can be used, for instance, to compensate more or less perfectly the phase response of the actual low-pass filter. In practical applications, sufficient bandwidth is reserved for the transition band of the involved filters. Public switched telephony (PSTN) uses the frequency band from 300 Hz to 3.6 kHz. The commonly used sampling rate in PCM transmission is 8 kHz ($f_s/2 = 4$ kHz). A band of 400 Hz is thus available for the filter slope. CD audio uses a sampling rate of 44.1 Khz ($f_s/2 = 22.05$ kHz). Since the highest audible frequency is about 20 kHz for young persons, about 2 kHz are available for the frequency response to drop down to sufficient low gain values.

To complete the picture, we finally have to consider the term $P(j\omega)/T$ in equation (4.11) which may also create unwanted frequency distortion. $P(j\omega)$ is the spectrum of the pulses generated by the pulse modulator in the digital-to-analog converter. In practice, these pulses are rectangular-shaped with an amplitude proportional to the values of the digital sample sequence. In most practical cases, the pulse width is chosen equal to the sampling period T which results in a stepped signal similar to the "sample and hold" signal shown in Fig. 4-4. In the following, however, we will discuss the general case of a rectangular pulse with arbitrary width $\tau \leq T$.

$P(j\omega)$ is obtained as the Fourier transform of the rectangular pulse $p(t)$ which is defined as

$$p(t) = \begin{cases} 0 & \text{for } t < 0 \\ 1/2 & \text{for } t = 0 \\ 1 & \text{for } 0 < t < \tau \\ 1/2 & \text{for } t = \tau \\ 0 & \text{for } t > \tau \end{cases}$$

$$P(j\omega) = \int_{t=-\infty}^{+\infty} p(t)e^{-j\omega t}dt = \int_{t=0}^{\tau} e^{-j\omega t}dt = \frac{e^{-j\omega\tau}-1}{-j\omega} = e^{-j\omega\tau/2}\frac{e^{-j\omega\tau/2}-e^{+j\omega\tau/2}}{-j\omega}$$

$$P(j\omega) = \tau e^{-j\omega\tau/2}\frac{\sin(\omega\tau/2)}{\omega\tau/2} = \tau e^{-j\omega\tau/2}\text{sinc}(\omega\tau/2)$$

$$P(j\omega)/T = \frac{\tau}{T}e^{-j\omega\tau/2}\text{sinc}(\omega\tau/2) \tag{4.12}$$

The distortion term (4.12) is composed of three terms which are considered in the following in more detail.

$e^{-j\omega\tau/2}$ simply represents a delay at an amount of half the pulse width τ, which is more or less caused by the sluggish behaviour of the interpolating low-pass filter.

τ/T is a gain factor which attenuates the output signal of the filter according to the duty cycle of the pulse sequence. This is a result of the averaging behaviour of the low-pass filter, which spreads the energy of the pulses in time to yield a continuous analog signal.

The third term, a sinc function, requires a closer look, because it introduces a frequency-dependent distortion, as depicted in Fig. 4-7.

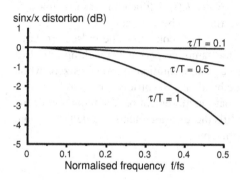

Fig. 4-7
Frequency response of the sinx/x-distortion (τ/T =1, 0.5 and 0.1)

The frequency response of this distortion which is sometimes referred to as $\sin x/x$-distortion or sinc distortion can be expressed as

$$H_{\text{sinc}}(j\omega) = \frac{\sin(\omega\tau/2)}{\omega\tau/2} \ .$$

At first glance, the best solution seems to be the choice of narrow pulses. With decreasing τ, the sinc function approaches unity, which avoids the need for any further correction of the frequency response. The disadvantage of this approach is that the gain factor τ/T becomes smaller and smaller, which must be compensated for by additional amplification of the signal. Assuming a certain amount of noise in the environment of the D/A conversion circuitry, this means deterioration in the achievable signal-to-noise ratio. In practical implementations, it is harder to reduce the environmental noise by 20 dB, for instance, than to incorporate a correction into the frequency response of the interpolation filter that amounts to about 4 dB at half the sampling frequency if τ is chosen equal to T (Fig. 4-7). So most of the D/A converters on the market deliver a stepped signal with $\tau = T$ requiring the previously mentioned compensation of the $\sin x/x$-distortion which can take place in the digital or in the analog domain.

4.4.2.4 Ideal Signal Reconstruction

In this section, we consider the ideal reconstruction of the analog signal $f_r(t)$ from a given sequence of samples $f(n)$. In doing so, we have to make assumptions which are not realisable in practice. But the results are nevertheless useful for general interpolation problems and point to the right way of implementing appropriate realisable filters.

Equation (4.7) is the starting point to derive a relation describing the ideal signal reconstruction.

$$f_r(t) = \sum_{n=-\infty}^{+\infty} f(n) \int_{\tau=-\infty}^{+\infty} p(\tau - nT) h_i(t - \tau) \, d\tau \tag{4.13}$$

Pulse shape $p(t)$ and low-pass impulse response $h_i(t)$ will be chosen in such a way that all the previously mentioned imperfections can be avoided.

In case of the stepped output signal of the D/A converter, the pulses $p(t)$ have unity amplitude and a duration equal to the sampling period T, resulting in pulses with an area of T. We define now a pulse whose duration approaches zero while the area is still T. This avoids on the one hand sinc distortion. On the other hand, the pulse power remains unchanged so that the aforementioned attenuation of the signal can be avoided. A pulse which fulfils these requirements is a delta function, which has unit area by definition, multiplied by T.

$$p(t) = T \, \delta(t)$$

Substitution in (4.13) yields

$$f_r(t) = \sum_{n=-\infty}^{+\infty} f(n)\,T \int_{\tau=-\infty}^{+\infty} \delta(\tau-nT)h_i(t-\tau)\,d\tau$$

$$f_r(t) = \sum_{n=-\infty}^{+\infty} f(n)\,T\,h_i(t-nT) \tag{4.14}$$

For $h_i(t)$ we assume the impulse response of an ideal low-pass filter with a cutoff frequency at $\omega_s/2$. $h_i(t)$ is calculated as the inverse Fourier transform of the ideal low-pass frequency response which features unity gain in the frequency range $-\omega_s/2 < \omega < +\omega_s/2$ and zero gain outside this range.

$$h_i(t) = \frac{1}{2\pi} \int_{\omega=-\infty}^{+\infty} H_i(j\omega)e^{j\omega t}\,d\omega = \frac{1}{2\pi} \int_{\omega=-\omega_s/2}^{+\omega_s/2} e^{j\omega t}\,d\omega$$

$$h_i(t) = \frac{1}{2\pi}\frac{e^{j\omega_s t/2}-e^{-j\omega_s t/2}}{jt} = \frac{1}{T}\frac{T}{2\pi}\frac{\sin\omega_s t/2}{t/2} = \frac{1}{T}\frac{\sin\omega_s t/2}{\omega_s t/2}$$

$$h_i(t) = \frac{1}{T}\frac{\sin\pi t/T}{\pi t/T} = \frac{1}{T}\operatorname{sinc}\pi t/T \tag{4.15}$$

Substitution of (4.15) in (4.14) yields the desired relationship.

$$f_r(t) = \sum_{n=-\infty}^{+\infty} f(n)\frac{\sin\pi(t-nT)/T}{\pi(t-nT)/T} = \sum_{n=-\infty}^{+\infty} f(n)\operatorname{sinc}\pi(t/T-n) \tag{4.16}$$

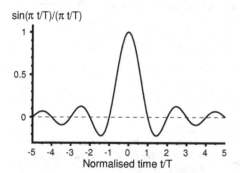

sin(π t/T)/(π t/T)

Normalised time t/T

Fig. 4-8
Graph of the function sinc(π t/T)

Relation (4.16) shows that the ideal band-limited interpolation of the samples is realized by means of the sinc function. The sinc function has the special property that it becomes unity at $t=0$ and vanishes at all other sampling instants $t=nT$ ($n \neq 0$) (Fig. 4-8). If we assign to each sampling instant a sinc function with an amplitude according to the value of the respective sample, we obtain the desired interpolated signal by summation of all shifted and weighted sinc functions

(Fig. 4-9). Because of the previously mentioned property, the sinc functions do not interfere at the sampling instants so that the interpolated signal assumes exactly the same values at the sampling instants as the samples that have been interpolated. If the original analog signal $f_a(t)$ was band-limited to half the sampling frequency, interpolation of the samples $f_a(nT)$ by sinc functions yields the perfect reconstruction of the original signal.

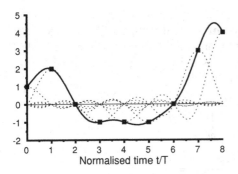

Fig. 4-9
Interpolation of samples by means of the sinc function

As mentioned earlier, the ideal low-pass filter is not realisable, and (4.15) also shows that the impulse response is not causal. Real filters have to be used, with all their associated imperfections with respect to the magnitude and group delay characteristics. It is evident that interpolation will not be done by means of sinc functions in practical implementations. If the preservation of the form of a signal is of importance, we should consider reconstructing the signal exactly at least at the sampling instants. This can be accomplished by means of interpolating impulse responses with properties similar to the sinc function (all values but one on a grid with period T are zero). An example of such an impulse response is shown in Fig. 4-10.

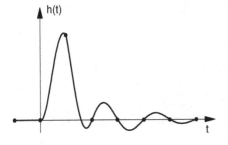

Fig. 4-10
Impulse response of a causal inter-polation filter preserving the values at the sampling instants

Filters with this property fulfil the so-called Nyquist criterion, which plays an important role in digital data transmission. The overall impulse response of a digital transmission system which is determined by the pulse shaper on the transmitter side, by the characteristics of the transmission medium and by the frequency response of the equaliser at the receiver side, should approximate as close as possible a frequency response which fulfils the Nyquist criterion. This minimizes interference between neighbouring pulses on the transmission line (intersymbol interference, ISI) and thus reduces the risk of wrong decisions at the receiving side concerning the amplitude of the recovered pulses. Such interference-free low-pass filters exhibit an interesting behaviour in the frequency domain, as sketched in Fig. 4-11: The slope in the transition band is symmetrical about the frequency $f_s/2$. If periodically continued, the overall frequency response sums up to a constant. This is the Nyquist criterion expressed in the frequency domain.

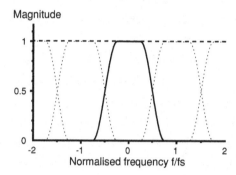

Fig. 4-11
Example of a frequency response that fulfils the Nyquist criterion

By shifting the discrete-time axis appropriately, sampling an impulse response $h(t)$ which meets the Nyquist criterion yields a discrete-time impulse response which consists of only one nonzero sample.

$$h(n) = h(nT) = \delta(n)$$

The corresponding Fourier transform is a constant.

$$H(e^{j\omega T}) = 1$$

Equation (4.3) directly provides the condition which a continuous-time frequency response has to meet in order to result in a constant discrete-time frequency response upon sampling.

$$\frac{1}{T} \sum_{k=-\infty}^{+\infty} H_a[j(\omega+k\omega_s)] = H(e^{j\omega T}) = 1$$

$$\sum_{k=-\infty}^{+\infty} H_a[j(\omega+k\omega_s)] = T$$

The periodically continued frequency response of the reconstruction filter must be a constant. The raised-cosine filter [26] is a well known example for a Nyquist filter which meets the above condition.

5 Filter Structures

5.1 Graphical Representation of Discrete-Time Networks

5.1.1 Block diagram and flowgraph

Digital filter algorithms are usually represented in the form of block diagrams and flowgraphs. Both representations are based on the following three basic elements:
- Addition of two signal sequences
- Multiplication of a sequence by a constant
- Delay of a sequence by one sample period T

The last-mentioned function corresponds to the operation of a D flip-flop in digital circuits. Triggered by the rising or falling edge of the applied clock signal, the information at the input is stored and passed to the output. This leads to the situation that we find a sample value at the output of the delay element that was applied to the input during the previous clock period. Shift-register structures, in which sample sequences can be stored and shifted, are realised by serial arrangement of the named delay elements.

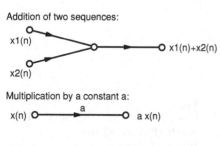

Addition of two sequences:

$x1(n)$

$x1(n)+x2(n)$

$x2(n)$

Multiplication by a constant a:

$x(n)$ —— a —— $a\,x(n)$

Delay by one sample period T:

$x(n)$ —— z^{-1} —— $x(n-1)$

Fig. 5-1
Elements of the flowgraph

Flowgraphs are comparable to the graphs used for the analysis of electrical networks. The topology of the network is described by means of nodes and branches. Figure 5-1 shows the elements of a flowgraph. The delay element is denoted with the unit-delay operator z^{-1}. Multiplication by a constant a is

D. Schlichthärle, *Digital Filters*, DOI: 10.1007/978-3-642-14325-0_5,
© Springer-Verlag Berlin Heidelberg 2011

represented by a branch denoted with a. The addition of two signals has no special symbol. Signals running up to a common node are added at this node.

Block diagrams depict the complexity of a digital filter. The amount of memory and the number of mathematical operations such as additions and multiplications required to realise the filter become visible. Figure 5-2 shows the symbols used.

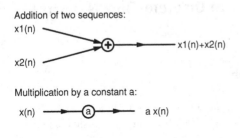

Addition of two sequences:

Multiplication by a constant a:

Delay by one sample period:

Fig. 5-2
Elements of the block diagram

As an introductory example, we consider a digital filter that calculates the average of two successive samples. Figure 5-3 shows the corresponding flowgraph and block diagram. The algorithm of the filter can be expressed as

$$y(n) = 0.5 \left[x(n) + x(n-1) \right] .$$

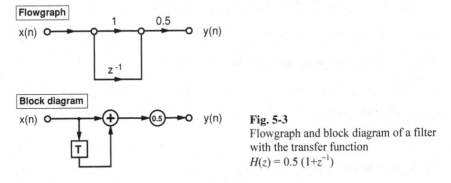

Fig. 5-3
Flowgraph and block diagram of a filter with the transfer function
$H(z) = 0.5 \, (1 + z^{-1})$

The transfer function of this filter can be obtained by taking the z-transform of the algorithm.

$$Y(z) = 0.5 \left(X(z) + X(z) z^{-1} \right)$$

$$Y(z) = 0.5 (1 + z^{-1}) X(z)$$

This results in the transfer function

$$H(z) = \frac{Y(z)}{X(z)}$$

$$H(z) = 0.5(1 + z^{-1}) = 0.5\frac{z+1}{z} \quad .$$

The inverse transform of $H(z)$ into the time domain yields the unit-sample response of the filter. With the results of Sect. 3.3 we obtain:

$$H(z) = 0.5 \times 1 + 0.5 \times 1 \times z^{-1}$$

$$h(n) = 0.5\,\delta(n) + 0.5\,\delta(n-1) \quad .$$

The filter reacts to the unit-sample sequence with two unit-samples of half the amplitude (Fig. 5-4). Inspection of the flowgraph leads to the same result.

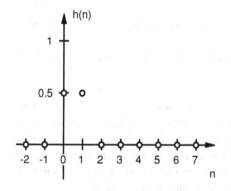

Fig. 5-4
Unit-sample response of the filter with the transfer function $H(z) = 0.5\,(1+z^{-1})$

For the hardware realisation of this filter, a memory is required that keeps the current input sample for calculating the average with the following input sample. Furthermore we need an adder. The multiplication by 0.5 can be simply realised by a shift operation in binary arithmetic.

Instead of developing special hardware, digital filters are often realised in the form of mathematical algorithms implemented on microcomputers or signal processors. These solutions are more flexible and are especially appropriate for applications in which the filter parameters have to be modified often, or during operation. Block diagrams can be used as the basis for the development of the corresponding program code. The following C pseudo-code is implementing the algorithm of our example.

```
/* X : current input sample
   X1: memory to keep the current input sample
   Y : currently calculated output sample
   ********************************************* */
X1 = 0;  // initialisation of the memory
```

```
do {
X = inputSample;
/* In real-time applications, the inputSample is provided by
    an A/D converter or input port of the hardware. Otherwise
    it is taken from a data array or from hard disk. */
Y = 0.5*(X+X1);
X1 = X; // keep input sample
outputSample = Y;
/* The result is stored back to memory or output via a port or
    a D/A converter. */
} while (!finished);
```

5.1.2 Realisability

Mathematical algorithms that describe the behaviour of discrete-time systems may be represented in the form of flowgraphs. In order to be realisable, a flowgraph has to meet the following two conditions:

- The sequence of the mathematical operations that are to be performed must be clearly identifiable.
- It must be guaranteed that all operations can recur periodically with the frequency $f_s = 1/T$.

The first condition leads to the requirement that there must not exist any delay-free directed loops in the flowgraph. A loop is called directed if the arrowheads in all branches have the same orientation with respect to a given orientation of the loop. Loop I in Fig. 5-5 is directed and does not contain any delay element. In this case it is not possible to clearly identify in which sequence the mathematical operations have to be performed. Thus the flowgraph is not realisable.

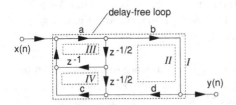

Fig. 5-5
Examples of realisable and unrealisable loops in a flowgraph

The condition of periodicity of all operations leads to the requirement that the total delay in any loop (directed or not) must be equal to an integer multiple (zero, positive or negative) of T. The total delay is calculated as the sum of the delays in all branches of the loop. The delay in a branch that has the same orientation as the loop is assumed positive, otherwise it is assumed negative. Delays of fractions of T may occur in a flowgraph if, for instance, a number of mathematical operations has to be performed sequentially within one period T. The total delay in a loop, however, must be in any case an integer multiple of T. Examples are shown again in Fig. 5-5. Loop III has a total delay of $3/2\ T$, the delay in loop IV amounts to $-T/2$. Both loops are not realisable by definition. The only loop that fulfils the realisability condition is loop II with a total delay of $-T$.

All filter structures that we have treated up to now fulfil the realisability condition. Common to all these structures is that the input signal to each delay element in the recursive part of the filter is always calculated as a linear combination of the output signals of the delay elements. Thus all directed loops contain delays (e.g. Fig. 5-22 or Fig. 5-36). The time period T between the instants at which the input signals of the delay elements are switched to the output of the delay elements must be sufficient to perform all the mathematical operations needed to calculate these linear combinations.

Transformations that convert continuous-time systems into discrete-time systems have to meet certain conditions to come to a useful result: one is stability, which is assured by having poles within the unit circle in the z-plane. In this section we have become acquainted with a second important condition. The discrete-time system that we obtain by a transformation must be realisable in the sense that we have discussed above.

5.2 FIR Filters

5.2.1 The Basic Structure of FIR Filters

In the case of FIR (Finite Impulse Response) filters, the output sequence is calculated as a linear combination of the current and M past input samples. Figure 5-6 shows a flowgraph and block diagram representing this type of filter. The input data pass through an M-stage shift register. After each shift operation, the outputs of the shift register are weighted with the coefficients b_r and summed up. The sum is the output value of the filter associated with the respective clock cycle.

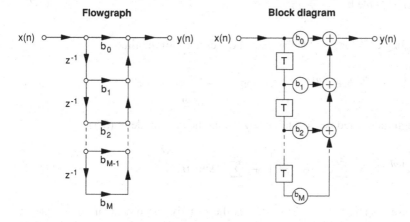

Fig. 5-6 Flowgraph and block diagram of the FIR filter

The weighting coefficients b_r, which determine the characteristics of the filter, have an obvious meaning. Let us consider the unit-sample response of the filter, which we obtain by applying a unit-sample sequence $\delta(n)$ to the input. This sequence has only one nonzero sample with the value 1. This unit sample propagates step-by-step through the shift register. It is evident that the coefficients b_r will appear at the output one after the other starting with b_0. This means that the coefficients b_r are the samples of the unit-sample response $h(r)$ of the filter. From the flowgraph we can derive the following algorithm:

$$y(n) = b_0 x(n) + b_1 x(n-1) + \ldots + b_M x(n-M)$$

or expressed in compact form

$$y(n) = \sum_{r=0}^{M} x(n-r)b_r \ . \tag{5.1}$$

Relation (5.1) is the convolution sum that we derived in Sect. 3.1, because b_r can be replaced by $h(r)$ as in the previous discussion.

$$y(n) = \sum_{r=0}^{M} x(n-r)h(r)$$

The FIR filter, therefore, has a structure that realises the convolution sum in direct form. The length of the shift register determines the length of the unit-sample response, which is finite and gives the filter its name. So the summation in (5.1) covers a finite number of elements which is in contrast to the general convolution sum (3.3). The transfer function of the FIR filter is obtained by taking the z-transform of the unit-sample response $h(n)$.

$$H(z) = \sum_{r=0}^{M} h(r)z^{-r} = \sum_{r=0}^{M} b_r z^{-r} \tag{5.2}$$

Substituting $e^{j\omega T}$ for z yields the frequency response of the filter.

$$H(e^{j\omega T}) = \sum_{r=0}^{M} b_r e^{-j\omega rT} = \sum_{r=0}^{M} b_r \cos \omega rT - j \sum_{r=0}^{M} b_r \sin \omega rT \tag{5.3}$$

The magnitude response is obtained by taking the magnitude of (5.3).

$$\left| H(e^{j\omega T}) \right|^2 = \left(\sum_{r=0}^{M} b_r \cos \omega rT \right)^2 + \left(\sum_{r=0}^{M} b_r \sin \omega rT \right)^2 \tag{5.4}$$

The phase response of the FIR filter is the negative argument of the frequency response.

$$b(\omega) = -\arg H(e^{j\omega T}) = \arctan \frac{\displaystyle\sum_{r=0}^{M} b_r \sin \omega rT}{\displaystyle\sum_{r=0}^{M} b_r \cos \omega rT} \tag{5.5}$$

The group delay response is calculated as the derivative of the phase response with respect to the frequency ω.

$$\tau_g = T \frac{\displaystyle\sum_{r=0}^{M} b_r \cos r\omega T \sum_{r=0}^{M} rb_r \cos r\omega T + \sum_{r=0}^{M} b_r \sin r\omega T \sum_{r=0}^{M} rb_r \sin r\omega T}{\left|H(e^{j\omega T})\right|^2} \tag{5.6}$$

An alternative filter structure can be found by transposition of the flowgraph. There are transformation rules that can be applied to flowgraphs without changing the transfer function. A transposed filter is obtained by reversing the direction of signal flow in all branches and interchanging the input and output. Figure 5-7 shows the flowgraph and block diagram of the transposed structure, as derived from the corresponding representations in Fig. 5-6. While the original structure was characterised by a continuous shift register, the transposed structure features single memory cells separated by adders.

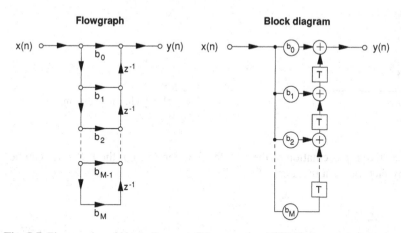

Fig. 5-7 Flowgraph and block diagram of the transposed FIR filter

The complexity of FIR filters may be quite high in standard filter applications, because many multiplications and additions are required to reproduce the complete impulse response in the time domain. The complexity is determined by the number of coefficients and consequently by the length of the impulse response to be realised. The kind of filter characteristic and the accuracy of the approximation of the frequency response highly determine the number of coefficients. Filters with a hundred coefficients and more can be often found in practice. However, the FIR filter has some interesting properties that make it attractive in realisation and application:

1. The FIR filter has a highly regular structure which is advantageous for the implementation.
2. The FIR filter is definitely stable since the stability criterion (3.21) is always fulfilled.
3. The filter always quiets down if the input signal vanishes. Uncontrolled oscillations, which may occur with recursive systems due to numerical problems, are excluded.
4. There are filter applications where the marked phase distortions occurring in the vicinity of the cutoff frequency are not tolerable, and hence a linear phase filter is required. One possibility to solve this problem is to linearise the phase response of a given filter by means of complex all-pass networks which can be accomplished only approximately. A more elegant solution is to use FIR filters. Linear-phase filters have an impulse response which is symmetrical about a sampling instant n_0 (Fig. 5-8). As this symmetry condition can be met exactly by FIR filters, the resulting frequency response is exactly linear-phase. With other filter structures, linear phase can only be approximated at the cost of complex implementations.

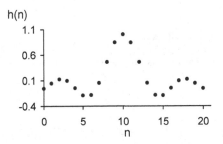

Fig. 5-8
Example of a linear-phase impulse response ($n_0 = 10$)

An alternative representation of the transfer function (5.2) of the FIR filter can be found by factoring out the term z^{-M}.

$$H(z) = \frac{\sum_{r=0}^{M} b_r z^{M-r}}{z^M} \tag{5.7}$$

This leads to positive exponents in the sum. In the numerator we get a polynomial in z of degree M which possesses M zeros. The denominator of (5.7) represents an Mth-order pole at $z = 0$.

$$H(z) = b_0 \frac{(z - z_{01})(z - z_{02}) \cdots (z - z_{0M})}{z^M} \tag{5.8}$$

Figure 5-9 shows the possible locations of poles and zeros in the z-plane. Since we consider here real filters which feature real filter coefficients and real impulse responses, the zeros z_{0i} in (5.8) must be real or occur in complex-conjugate pairs.

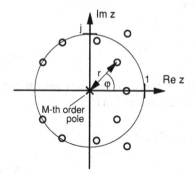

Fig. 5-9
Possible location of poles and zeros of
an Mth-order FIR filter

For a further analysis of magnitude, phase, and group delay of the FIR filter, we divide each factor in the numerator of (5.8) by z which results in the following representation of the transfer function:

$$H(z) = b_0(1 - z_{01}z^{-1})(1 - z_{02}z^{-1})...(1 - z_{0M}z^{-1})$$

$$H(z) = b_0 \prod_{n=1}^{M}(1 - z_{0n}z^{-1}) \tag{5.9}$$

Substitution of z by $e^{j\omega T}$ yields the frequency response of the filter.

$$H(e^{j\omega T}) = b_0 \prod_{n=1}^{M}\left(1 - z_{0n}e^{-j\omega T}\right) \tag{5.10}$$

It is advantageous for the following calculations to express the position of the zeros in polar coordinates.

$$z_{0n} = r_{0n}e^{j\varphi_{0n}}$$

While real and imaginary part of the zero have no direct physical meaning, the polar coordinates r_0 and φ_0 are closely related to the characteristics of the frequency response. The angle φ_0 determines the centre frequency ω_0 where the zero becomes effective.

$$\varphi_0 = \omega_0 T = 2\pi f_0 T$$

The radius r_0 determines the shape of the resonance peak. With z_0 expressed in polar coordinates, (5.10) assumes the form

$$H(e^{j\omega T}) = b_0 \prod_{n=1}^{M}\left(1 - r_{0n}e^{j\varphi_{0n}}e^{-j\omega T}\right) = b_0 \prod_{n=1}^{M}\left(1 - r_{0n}e^{-j(\omega T - \varphi_{0n})}\right) \tag{5.11}$$

In the following, we will analyse in detail how the zero terms of the form

$$H_{zero}(e^{j\omega T}) = 1 - r_0\,e^{-j(\omega T - \varphi_0)} = 1 - r_0\cos(\omega T - \varphi_0) + j r_0\sin(\omega T - \varphi_0) \quad (5.12)$$

contribute to the overall response of the filter.

5.2.2 Magnitude Response

We start with considering the magnitude response of the zero term which is obtained by taking the magnitude of the complex expression (5.12).

$$\left|H_{zero}(e^{j\omega T})\right|^2 = \left(1 - r_0\cos(\omega T - \varphi_0)\right)^2 + \left(r_0\sin(\omega T - \varphi_0)\right)^2$$

$$\left|H_{zero}(e^{j\omega T})\right|^2 = 1 + r_0^2 - 2r_0\cos(\omega T - \varphi_0) \quad (5.13)$$

The minimum of the magnitude occurs at $\omega = \varphi_0/T$. The gain is $1 - r_0$ in the minimum and $1 + r_0$ in the maximum. Figure 5-10 shows the magnitude responses of a zero for various radii r_0. The overall magnitude response of an Mth order FIR filter can be expressed as

$$\left|H(e^{j\omega T})\right|^2 = b_0 \prod_{n=1}^{M}\left(1 + r_{0n}^2 - 2r_{0n}\cos(\omega T - \varphi_{0n})\right) \quad (5.14)$$

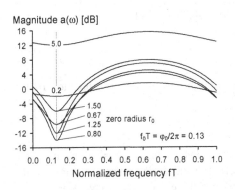

Magnitude a(ω) [dB]

Fig. 5-10
Magnitude of a zero with the radius r_0 as a parameter

Relation (5.13) and Fig. 5-10 show that zeros close to or on the unit circle yield high attenuation around the frequency φ_0/T, while zeros close to the origin achieve low attenuation with more or less variation of the magnitude. Thus it can be expected that, as a result of FIR filter design, zeros in the stopband will be located close to the unit circle while the passband features zeros close to or within the origin. Figure 5-11 shows an example of the frequency response of a filter with twelve zeros equally spaced in angle. The maximum gain is normalised to unity. Six zeros are located in the frequency range $-0.25 < fT < 0.25$ at a radius of

$r_0 = 0.4$ (low attenuation). The remaining zeros have been placed onto the left half of the unit circle (high attenuation). As a whole, the described placing of zeros results in a low-pass behaviour.

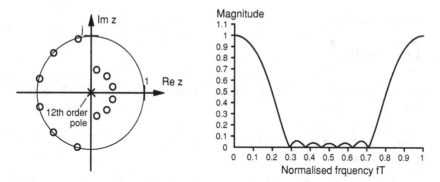

Fig. 5-11 Pole/zero chart and frequency response of a twelfth-order FIR low-pass filter

It is interesting to note that for zeros having reciprocal radii, the magnitude curves in Fig. 5-10 are identical apart from a constant factor. So we obviously have a degree of freedom to realise a given magnitude response: we can chose the zeros to lie within the unit circle ($r_0 < 1$) or outside the unit circle ($r_0 > 1$) in the z-plane. It will be shown later that the two variants differ dramatically with respect to the phase and group delay characteristics.

5.2.3 Reciprocal Conjugate Zeros

The relationships between a polynomial in z and the corresponding polynomial having zeros with reciprocal radii are of interest in a number of applications. Examples are linear-phase filters and allpass filters. A zero $z_0^{(r)}$ with reciprocal radius can be expressed as

$$z_0^{(r)} = \frac{1}{r_0} e^{j\varphi_0} = \frac{1}{r_0\, e^{-j\varphi_0}} = \frac{1}{z_0^*} \tag{5.15}$$

z_0 and $z_0^{(r)} = 1/z_0^*$ are referred to as a reciprocal-conjugate pair. Substituting (5.15) in (5.9) yields the transfer function of the polynomial with reciprocal zero radii.

$$H^{(r)}(z) = b_0 \prod_{n=1}^{M}\left(1 - \frac{1}{z_{0n}^*} z^{-1}\right)$$

$$H^{(r)}(z) = b_0 \prod_{n=1}^{M}(-1)\frac{z^{-1}}{z_{0n}^*}\left(1 - z_0^* z\right)$$

$$H^{(\mathrm{r})}(z) = b_0(-1)^M \frac{z^{-M}}{\displaystyle\prod_{n=1}^{M} z_{0n}^*} \prod_{n=1}^{M}\left(1 - z_0^* z\right) \tag{5.16}$$

The product of all zeros is

$$\prod_{n=1}^{M} z_{0n}^* = (-1)^M \frac{b_M}{b_0}$$

which can be easily verified by calculating the coefficient of z^{-M} in (5.9). Substitution in (5.16) yields

$$H^{(\mathrm{r})}(z) = \frac{b_0^2}{b_M} z^{-M} \prod_{n=1}^{M}\left(1 - z_0^* z\right)$$

Due to the assumption of a real filter, for each z_0 in the product term also a factor with the complex-conjugate counterpart will be present. So we can replace z_0^* by z_0 without changing anything.

$$H^{(\mathrm{r})}(z) = \frac{b_0^2}{b_M} z^{-M} \prod_{n=1}^{M}(1 - z_0 z) = \frac{b_0}{b_M} z^{-M} H(1/z) \tag{5.17}$$

Substitution of z by $e^{\mathrm{j}\omega T}$ yields the corresponding relationships for the frequency response.

$$H^{(\mathrm{r})}(e^{\mathrm{j}\omega T}) = \frac{b_0}{b_M} e^{-\mathrm{j}M\omega T} H(e^{-\mathrm{j}\omega T}) = \frac{b_0}{b_M} e^{-\mathrm{j}M\omega T} H^*(e^{\mathrm{j}\omega T}) \tag{5.18}$$

Taking the magnitude of (5.18) reveals that $H^{(\mathrm{r})}(z)$ and $H(z)$ have, apart from a constant factor, the same magnitude response.

$$\left|H^{(\mathrm{r})}(e^{\mathrm{j}\omega T})\right| = \left|\frac{b_0}{b_M}\right|\left|H^*(e^{\mathrm{j}\omega T})\right| = \left|\frac{b_0}{b_M}\right|\left|H(e^{\mathrm{j}\omega T})\right|$$

The coefficients of a polynomial with reciprocal zeros have an interesting property which can be demonstrated by substituting (5.2) in (5.17).

$$H^{(\mathrm{r})}(z) = \frac{b_0}{b_M} z^{-M} H(1/z) = \frac{b_0}{b_M} z^{-M} \sum_{r=0}^{M} b_r (1/z)^{-r} = \frac{b_0}{b_M} z^{-M} \sum_{r=0}^{M} b_r z^r$$

$$H^{(\mathrm{r})}(z) = \frac{b_0}{b_M} \sum_{r=0}^{M} b_r z^{-(M-r)}$$

Compared to the original polynomial $H(z)$, the coefficients appear in reversed order when reciprocal radii are used for the zeros. $H(z)$ and $H^{(\mathrm{r})}(z)$ are said to form a mirror-image pair.

5.2.4 Phase Response

The phase response of a zero is calculated as the negative argument of (5.12).

$$b_{zero}(\omega) = -\arg H_{zero}(e^{j\omega T}) = -\arctan\frac{r_0 \sin(\omega T - \varphi_0)}{1 - r_0 \cos(\omega T - \varphi_0)} \qquad (5.19)$$

The shapes of the phase response curves differ considerably depending on where the zeros are located in the z-plane. Figure 5-12 illustrates the case where the radius of the zeros is less than unity, i.e. the zeros are lying within the unit circle. The phase response oscillates symmetrically about zero. The overall phase shift is zero as the frequency goes from zero to the sampling frequency.

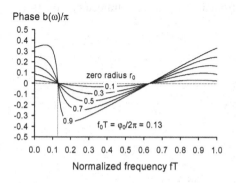

Fig. 5-12
Phase shift of the zero term for radii less than unity

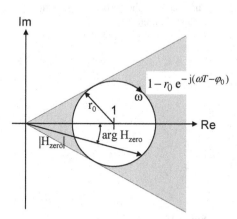

Fig. 5-13
Location curve of the zero term for radii less than unity

The location curve of the zero term (5.12) in the complex plane is a circle with radius r_0. The centre of this circle is located at $1 + j0$ on the real axis. If ω goes from 0 to $2\pi/T$, the complex values of (5.12) complete a full circle as depicted in Fig. 5-13. The length of the trajectory from the origin to the locations on the circle represent the magnitude of the zero term for a given frequency while the angle between the real axis and the trajectory reflects the phase of the complex zero term. The swing of the phase is bounded by the upper and lower tangents to the circle. The phase response is limited according to the relation

$$|b(\omega)| \leq \arcsin r_0 .$$

Fig. 5-14 shows the phase response for radii greater than unity which is equivalent to zeros lying outside the unit circle. In contrast to the previous case, the phase response curves are monotonously increasing. The phase does not return back to the starting value when the frequency goes from 0 to the sampling frequency $2\pi/T$. A linear-phase term is superimposed to the phase characteristic which leads to an overall phase shift of 2π in the frequency range $0 \leq \omega \leq 2\pi/T$. This is also illustrated in Fig. 5-15. If we move along the circle with increasing frequency, the angle between the trajectory and the real axis increases monotonously. If the frequency goes from 0 to $2\pi/T$, the trajectory covers an overall phase shift of 2π.

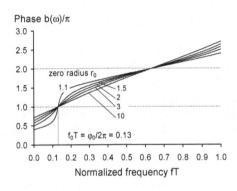

Fig. 5-14
Phase shift of the zero term for radii greater than unity

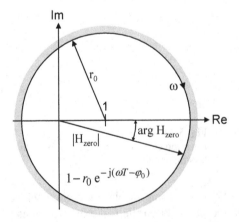

Fig. 5-15
Location curve of the zero term for radii greater than unity

Each zero of the filter lying outside the unit circle contributes this linear phase shift to the overall phase response of the filter. A filter whose zeros are all lying outside the unit circle is therefore called a maximum-phase filter. This choice of the zero locations leads to the largest phase shift for a given magnitude response. If, in contrast, all zeros are lying within the unit circle, the phase response oscillates about zero. This location of the zeros leads to the smallest possible phase shift for a given magnitude response. Such a filter is therefore called minimum-phase.

The phase response of an Mth-order FIR filter is the sum of the phase shifts of all contributing zeros.

$$b(\omega) = -\sum_{n=1}^{M} \arctan \frac{r_{0n} \sin(\omega T - \varphi_{0n})}{1 - r_{0n} \cos(\omega T - \varphi_{0n})} \qquad (5.20)$$

Also in the context of the phase response, it is worth considering the relationships between filters having mirror-image coefficient sets. Starting point is (5.18).

$$H^{(\mathrm{r})}(e^{j\omega T}) = \frac{b_0}{b_M} e^{-jM\omega T} H^*(e^{j\omega T})$$

Taking the argument on both sides of the equation yields

$$\arg H^{(\mathrm{r})}(e^{j\omega T}) = -M\omega T - \arg H(e^{j\omega T}) \ .$$

Since the phase response $b(\omega)$ is the negative argument of the frequency response, we finally have

$$b^{(\mathrm{r})}(\omega) = M\omega T - b(\omega) \ . \qquad (5.21)$$

If all zeros of $H(e^{j\omega T})$ lie inside the unit circle in the z-plane, the phase response $b(\omega)$ will oscillate about zero. The change in phase is zero as ω goes from 0 to $2\pi/T$. The corresponding filter with reciprocal zeros (mirror-image coefficients) behaves differently. The phase response $b^{(\mathrm{r})}(\omega)$ exhibits a linear-phase term $M\omega T$, which is equivalent with a delay of M sampling periods. The graph of the phase response oscillates about this linear-phase shift as depicted in Fig. 5-14. The change in phase is $2\pi M$ as ω goes from 0 to $2\pi/T$.

5.2.5 Group Delay

The group delay response of a zero is calculated as the derivative of the phase response (5.19) with respect to ω.

$$b_{\mathrm{zero}}(\omega) = -\arctan \frac{r_0 \sin(\omega T - \varphi_0)}{1 - r_0 \cos(\omega T - \varphi_0)}$$

$$\frac{d b_{\mathrm{zero}}(\omega)}{d\omega} = \tau_{\mathrm{gzero}}(\omega) = T \frac{r_0^2 - r_0 \cos(\omega T - \varphi_0)}{1 + r_0^2 - 2r_0 \cos(\omega T - \varphi_0)} \qquad (5.22)$$

Group delay τ_g/T

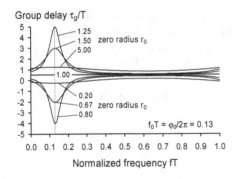

Fig. 5-16
Group delay of the zero term with the
radius r_0 as a parameter

Figure 5-16 shows graphs of the group delay response for various radii r_0. For zeros outside the unit circle, the group delay is always positive. For zeros inside the unit circle, the group delay assumes negative values around the centre frequency φ_0/T. The extremum at this frequency assumes a value of

$$\tau_{g\,zero}(\varphi_0\,/T)=T\frac{r_0}{r_0-1}\ .$$

The relationship between the group delay characteristics of FIR filters with mirror-image coefficient sets is simply obtained by differentiation of (5.21).

$$\tau_g^{(r)}(\omega)=MT-\tau_g(\omega) \tag{5.23}$$

Example 5-1
Given a first-order FIR filter as shown in the following figure.

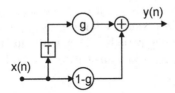

Modify the block diagram in such a way that only one coefficient multiplier is needed. Sketch the magnitude response and the group delay of the filter for g assuming the values $0, 0.25, 0.5, 0.75,$ and 1.
    From the given block diagram, we can derive the following relationship between input sequence $x(n)$ and output sequence $y(n)$:

$$y(n)=(1-g)x(n)+gx(n-1)$$

By rearrangement of this expression, we obtain a form with only one coefficient multiplication.

$$y(n)=x(n)+g\big(x(n-1)-x(n)\big)$$

A corresponding block diagram is shown below.

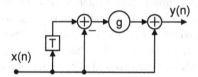

Taking the z-transform yields the transfer function of the filter.

$$Y(z) = X(z) + g\left(X(z)z^{-1} - X(z)\right)$$

$$H(z) = \frac{Y(z)}{X(z)} = 1 + g(z^{-1} - 1)$$

The frequency response is obtained by substituting $e^{j\omega T}$ for z.

$$H(e^{j\omega T}) = 1 + g(e^{-j\omega T} - 1) = 1 + g(\cos\omega T - j\sin\omega T - 1)$$

$$H(e^{j\omega T}) = 1 + g(\cos\omega T - 1) - jg\sin\omega T$$

Magnitude:

$$\left|H(e^{j\omega T})\right| = \sqrt{(1 + g(\cos\omega T - 1))^2 + g^2\sin^2\omega T}$$

$$\left|H(e^{j\omega T})\right| = \sqrt{1 + 2g(1-g)(\cos\omega T - 1)}$$

Group delay:

$$b(\omega) = -\arctan\frac{-g\sin\omega T}{1 + g(\cos\omega T - 1)} = \arctan\frac{g\sin\omega T}{1 + g(\cos\omega T - 1)}$$

$$\tau_g(\omega) = \frac{db(\omega)}{d\omega} = \frac{g + g(1-g)(\cos\omega T - 1)}{1 + 2g(1-g)(\cos\omega T - 1)}$$

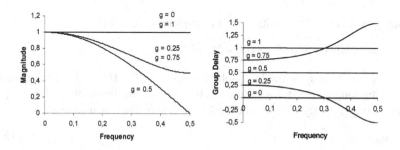

At low frequencies, up to about $0.2\omega_s$, this relative simple structure may be used as an adjustable delay line. The magnitude response is almost constant. The group delay is almost proportional to the coefficient g.

It is interesting to note that the group delay of our example may assume negative values. The occurrence of negative group delays often leads to confusion. It is sometimes argued that negative group delay is a contradiction to causality.

Let us assume that we apply a narrowband pulse to the input of a filter. The mid-band frequency of the pulse lies in the range where the filter shows a negative group delay. What is actually happening is that the centroid of the envelope of the narrowband pulse is slightly shifted towards the beginning of the pulse. This gives the impression that the output signal precedes the input signal. But observing the beginning of the input and output signals shows, of course, that causality is not violated.

5.2.6 Linear-Phase Filters

With the results of the previous sections, it is easy to build linear-phase filters. The series connection of a FIR filter of order M_1 with the transfer function $H_1(z)$ and a corresponding filter with mirror-image coefficients $H_1^{(r)}(z)$ yields the desired behaviour.

$$H(z) = H_1(z) H_1^{(r)}(z) \tag{5.24}$$

$H(z)$ is a filter of even order M with $M = 2M_1$. The magnitude response of $H(z)$ is the square of the magnitude response of $H_1(z)$.

$$\left| H(e^{j\omega T}) \right| = \frac{b_0}{b_{M_1}} \left| H_1(e^{j\omega T}) \right|^2$$

The phase response of (5.24) is calculated as the sum of the phase of both filter sections.

$$b(\omega) = b_1(\omega) + b_1^{(r)}(\omega)$$

Substitution of (5.21) results in the linear-phase expression

$$b(\omega) = b_1(\omega) + M_1 \omega T - b_1(\omega) = M_1 \omega T .$$

The group delay of the linear-phase filter is $M_1 T = MT/2$ which corresponds to half the length of the filter. The coefficients of the linear-phase filter have an interesting property. The set of coefficients is symmetrical, meaning that

$$b_n = b_{M-n} . \tag{5.25}$$

This is best demonstrated with an example.

$$H_1(z) = 1 + 3z^{-1} - 2z^{-2}$$
$$H_1^{(r)}(z) = -2 + 3z^{-1} + z^{-2}$$

$$H(z) = (1 + 3z^{-1} - 2z^{-2})(-2 + 3z^{-1} + z^{-2})$$
$$H(z) = -2 - 3z^{-1} + 14z^{-2} - 3z^{-3} - 2z^{-4}$$

This property of the coefficients leads to symmetrical impulse responses as depicted in Fig. 5-8.

In the following, we will show that this symmetry condition really results in a linear-phase characteristic. An Mth-order FIR filter has $M+1$ coefficients and the following frequency response:

$$H(e^{j\omega T}) = \sum_{n=0}^{M} b_n e^{-j\omega nT} = e^{-j\omega MT/2} \sum_{n=0}^{M} b_n e^{-j\omega(n-M/2)T} .$$

The sum is split into 3 parts.

$$H(e^{j\omega T}) = e^{-j\omega MT/2} \left[\sum_{n=0}^{\frac{M}{2}-1} b_n e^{-j\omega(n-M/2)T} + \sum_{n=\frac{M}{2}+1}^{M} b_n e^{-j\omega(n-M/2)T} + b_{M/2} \right]$$

Substitution of $M-m$ for n in the second sum yields

$$H(e^{j\omega T}) = e^{-j\omega MT/2} \left[\sum_{n=0}^{\frac{M}{2}-1} b_n e^{-j\omega(n-M/2)T} + \sum_{m=0}^{\frac{M}{2}-1} b_m e^{+j\omega(m-M/2)T} + b_{M/2} \right] .$$

$$H(e^{j\omega T}) = e^{-j\omega MT/2} \left[2 \sum_{n=0}^{\frac{M}{2}-1} b_n \cos\omega(n-M/2)T + b_{M/2} \right]$$

By introduction of a new set of coefficients B_n, the frequency response of the linear-phase FIR filter can be expressed as

$$H(e^{j\omega T}) = e^{-j\omega MT/2} \sum_{n=0}^{M/2} B_n \cos\omega nT \quad \text{with} \quad B_n = 2b_{M/2-n} \quad \text{for } n = 1 \ldots M/2 .$$
$$\text{and} \quad B_0 = b_{M/2}$$

The exponential term in front of the sum represents a linear-phase transfer function that causes the already mentioned delay of $MT/2$. This term has no influence on the magnitude. The sum is purely real and represents a zero-phase filter.

$$H_0(\omega) = \sum_{n=0}^{M/2} B_n \cos\omega nT \qquad (5.26)$$

The zero-phase frequency response $H_0(\omega)$ must not be confused with the magnitude response because $H_0(\omega)$ may also assume negative values.

Up to now, we discussed linear-phase FIR filters with even order M and even symmetry of the coefficients and the impulse response as defined by (5.25). We call this a case 1 filter which is the most versatile type of linear-phase filters. There are no restrictions concerning the filter characteristics to be implemented. Particularly at $\omega = 0$ and $\omega = \pi/T$, $H_0(\omega)$ can assume any arbitrary value. So low-pass, high-pass, bandpass, and bandstop filter types may be realized likewise.

In order to obtain odd-order linear-phase FIR filters, a first-order linear-phase term needs to be added to the case 1 filter. There are only two such terms which could be used for this purpose:

$$H(z) = 1 + z^{-1} \quad \text{and} \quad H(z) = 1 - z^{-1} \ .$$

These represent real zeros on the unit circle with $z_0 = 1$ and $z_0 = -1$ respectively. The frequency response of the first term can be expressed as

$$H(e^{j\omega T}) = 1 + e^{-j\omega T} = e^{-j\omega T/2}(e^{j\omega T/2} + e^{-j\omega T/2})$$
$$H(e^{j\omega T}) = 2 e^{-j\omega T/2} \cos \omega T/2$$

The term adds a delay of half the sampling period to the case 1 filter characteristics. Due to the cosine function, the magnitude response is forced to zero at $\omega = \pi/T$. So high-pass filters and bandstop filters cannot be realized with this case 2 type of filters. If the z-transfer function of a case 1 filter is multiplied by $1 + z^{-1}$, we again obtain a filter with even-symmetric coefficients. Expressed the other way round, it is always possible to extract the term $1 + z^{-1}$ from a case 2 transfer function. The zero-phase frequency response of a case 2 filter can be written as

$$H_0(\omega) = 2 \cos \omega T/2 \sum_{n=0}^{\frac{M-1}{2}} B_n \cos n\omega T \ .$$

Note that the increase of the filter order by one does not increase the degree of freedom for the filter design because $1 + z^{-1}$ is a fixed term. The number of available coefficients is not increased. The choice of case 2 filters is advantageous in those cases where the target magnitude response is anyway zero at $\omega = \pi/T$.

The frequency response of the alternative term $1 - z^{-1}$ can be expressed as

$$H(e^{j\omega T}) = 1 - e^{-j\omega T} = e^{-j\omega T/2}(e^{j\omega T/2} - e^{-j\omega T/2})$$
$$H(e^{j\omega T}) = 2 j e^{-j\omega T/2} \sin \omega T/2 = 2 e^{j\pi/2} e^{-j\omega T/2} \sin \omega T/2$$

This term adds a delay of half a sampling period and a constant phase shift of $\pi/2$ to the filter characteristics. Due to the sine function, the magnitude response is forced to zero at $\omega = 0$. So low-pass filter and bandstop filters cannot be realized with this case 3 type of linear-phase FIR filters. If a case 1 transfer function is multiplied by the term $1-z^{-1}$, we obtain a filter with odd-symmetric coefficients.

$$b_n = -b_{M-n} \qquad\qquad (5.27)$$

Figure 5-17 shows the basic shape of the impulse response of linear-phase filters with odd-symmetric coefficients.

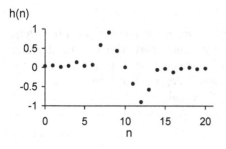

h(n)

Fig. 5-17
Example of an odd-symmetric linear-phase impulse response ($n_0 = 10$)

It is always possible to extract the term $1-z^{-1}$ from the transfer function of case 3 filters. The zero-phase frequency response can be written as

$$H_0(\omega) = 2\sin \omega T / 2 \sum_{n=0}^{\frac{M-1}{2}} B_n \cos n\omega T \;.$$

The additional constant phase shift of $\pi/2$ is the outstanding property of odd-symmetric FIR filters which opens up the possibility to design differentiators, Hilbert transformers, and any kind of quadrature filters.

To complete the picture, we finally consider the case 4 type featuring odd-symmetric coefficients and even filter order. Such a filter can be obtained by multiplying the transfer function of a case 1 filter by $1+z^{-1}$ and $1-z^{-1}$ which can be combined to the multiplication by $1-z^{-2}$. The frequency response of $1-z^{-2}$ can be expressed as

$$H(e^{j\omega T}) = 1 - e^{-2j\omega T} = e^{-j\omega T}(e^{j\omega T} - e^{-j\omega T})$$

$$H(e^{j\omega T}) = 2 j e^{-j\omega T} \sin \omega T = 2 e^{j\pi/2} e^{-j\omega T} \sin \omega T$$

This term adds a delay of one sampling period and a constant phase shift of $\pi/2$ to the filter characteristics. Due to the sine function, the magnitude response is forced to zero at $\omega = 0$ and at $\omega = \pi/T$. So only a bandpass type of filter can be realized with this option. By multiplication of a case 1 transfer function by $1-z^{-2}$, we obtain an even-order polynomial with odd-symmetric coefficients. Due to the odd symmetry, the coefficient $b_{M/2}$ must always be zero. This will be illustrated with a simple example where the second-order transfer function of a case 1 filter is multiplied by $1-z^{-2}$. The resulting coefficient b_2 is zero.

$$H(z) = 1 + 3z^{-1} + z^{-2} \quad \text{(case 1)}$$

$$H(z)(1 - z^{-2}) = 1 + 3z^{-1} + 0z^{-2} - 3z^{-3} - z^{-4} \quad \text{(case 4)}$$

The zero-phase frequency response of the case 4 filter can be expressed as

$$H_0(\omega) = 2\sin \omega T \sum_{n=0}^{\frac{M}{2}-1} B_n \cos n\omega T$$

Figure 5-18 depicts the possible locations of zeros for linear-phase FIR filters. Zeros within the unit circle always have a counterpart with reciprocal radius outside the unit circle and vice versa. This guarantees linear phase. Zeros on the unit circle are per se linear-phase. Complex zeros always have a complex-conjugate counterpart in order to make the filter real.

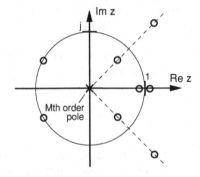

Fig. 5-18
Possible locations of poles and zeros for linear-phase FIR filters

Table 5-1 summarizes the characteristics of the various available options for the design of linear-phase FIR filters.

Table 5-1 Characteristics of linear-phase FIR filters

	order	symmetry	magnitude at $\omega = 0$	magnitude at $\omega = \pi/T$	Transfer function
case 1	even	even	any	any	$H_{C1}(z)$
case 2	odd	even	any	0	$H_{C2}(z) = (1+z^{-1})\,H_{C1}(z)$
case 3	odd	odd	0	any	$H_{C3}(z) = (1-z^{-1})\,H_{C1}(z)$
case 4	even	odd	0	0	$H_{C4}(z) = (1-z^{-2})\,H_{C1}(z)$

Because of the symmetry of the filter coefficients, the complexity of the filter realisation can be reduced considerably. All coefficients (for odd filter order) or all but one (for even filter order) occur in equal pairs so that the number of multiplications can be halved as shown in the block diagrams in Fig. 5-19. For filters with odd symmetry, the difference must be taken prior to coefficient multiplication instead of the sum.

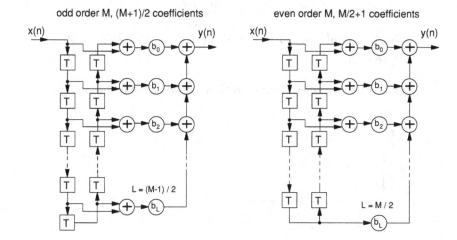

odd order M, (M+1)/2 coefficients

even order M, M/2+1 coefficients

Fig. 5-19 Optimised structures for linear-phase FIR filters

Example 5-2

Given a fourth-order FIR filter with the unit-sample response

$$h(0) = -1$$
$$h(1) = -2$$
$$h(2) = 0$$
$$h(3) = 2$$
$$h(4) = 1 \quad .$$

Calculate the phase response and the group delay of the filter and sketch the magnitude response.

The frequency response is obtained using (3.23):

$$H(e^{j\omega T}) = \sum_{n=0}^{4} h(n) e^{-j\omega nT}$$

$$H(e^{j\omega T}) = -1 - 2e^{-j\omega T} + 2e^{-j3\omega T} + e^{-j4\omega T}$$

$$H(e^{j\omega T}) = e^{-j2\omega T}(-e^{j2\omega T} - 2e^{j\omega T} + 2e^{-j\omega T} + e^{-j2\omega T})$$

$$H(e^{j\omega T}) = -je^{-j2\omega T} \, 2(2\sin\omega T + \sin 2\omega T)$$

$$H(e^{j\omega T}) = -j(\cos 2\omega T - j\sin 2\omega T)2(2\sin\omega T + \sin 2\omega T)$$

$$\left| H(e^{j\omega T}) \right| = 2|2\sin\omega T + \sin 2\omega T|$$

The phase response is the argument of the complex frequency response

$$b(\omega) = -\arg H(\mathrm{e}^{j\omega T}) = -\arctan\frac{-\cos 2\omega T}{-\sin 2\omega T} = -\arctan(\cot 2\omega T)$$

$$b(\omega) = \arctan(-\cot 2\omega T) = \arctan(\tan(2\omega T + \pi/2)) = 2\omega T + \pi/2$$

$$\tau_g(\omega) = 2T$$

The filter has a bandpass characteristic as shown in the following figure.

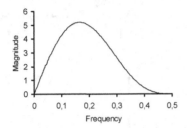

The phase response is linear. Due to the odd symmetry of the unit-sample response, the phase includes a constant phase shift of $\pi/2$. The delay corresponds to half the length of the FIR filter.

5.3 IIR Filters

5.3.1 An Introductory Example

Impulse responses with infinite length can be realised with a limited number of coefficients by also using delayed output values to calculate the current output sample. Remember that the FIR filter only uses the current and delayed input values. A simple first-order low-pass filter with exponentially decaying unit-sample response may serve as an example:

$$h(n) = \mathrm{e}^{-\alpha n}\,u(n) = (\mathrm{e}^{-\alpha})^n\,u(n)\ .$$

Taking the z-transform of the unit-sample response yields the transfer function of the filter. From Table 3-2 we see that

$$H(z) = \frac{z}{z - \mathrm{e}^{-\alpha}} = \frac{1}{1 - \mathrm{e}^{-\alpha}\,z^{-1}}\ .$$

The transfer function is the quotient of the z-transforms of the input and output sequences $X(z)$ and $Y(z)$.

$$H(z) = \frac{Y(z)}{X(z)} = \frac{1}{1 - \mathrm{e}^{-\alpha}\,z^{-1}}$$

$$Y(z)(1-e^{-\alpha}z^{-1}) = X(z)$$
$$Y(z)-e^{-\alpha}Y(z)z^{-1} = X(z)$$

Transformation back into the time domain using the rules derived in Chap. 3 yields an algorithm that relates input sequence $x(n)$ and output sequence $y(n)$.

$$y(n)-e^{-\alpha}y(n-1) = x(n)$$
$$y(n) = x(n)+e^{-\alpha}y(n-1)$$

The resulting filter merely requires one multiplier and one adder. The calculation of the current output value is based on the current input value and the previous output value. Figure 5-20 shows the flowgraph of this filter. The use of delayed output values gives the filter a feedback or recursive structure.

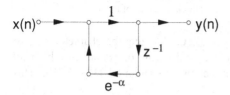

Fig. 5-20
Flow graph of an IIR filter with the transfer function $H(z)=z/(z-e^{-\alpha})$

5.3.2 Direct Form Filters

In a general recursive filter realisation, the current and M delayed values of the input are multiplied by the coefficients b_k, N delayed values of the output are multiplied by the coefficients a_k, and all the resulting products are added. Sorting input and output values to both sides of the equals sign yields the following sum representation (5.28):

$$\sum_{r=0}^{M}b_r x(n-r) = \sum_{i=0}^{N}a_i y(n-i). \tag{5.28}$$

This relation, commonly referred to as a difference equation, is analogous to the differential equation (1.13) which describes the input/output behaviour of continuous-time systems in the time domain. It is usual to normalise (5.28) such that $a_0 = 1$. This leads to the following filter algorithm:

$$y(n) = \sum_{r=0}^{M}b_r x(n-r) - \sum_{i=1}^{N}a_i y(n-i). \tag{5.29}$$

Figure 5-21 shows a block diagram representing (5.29). In contrast to the FIR filter, a second shift register appears that is used to delay the output sequence $y(n)$.

The structure with the coefficients b_r on the left of Fig. 5-21 is the well-known FIR filter structure, which represents the non-recursive part of the filter. The structure on the right is the recursive part of the filter, whose characteristics are determined by the coefficients a_i.

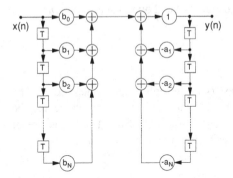

Fig. 5-21
Block diagram of an IIR filter

Both filter parts can be moved together without changing the characteristics of the filter. In the middle of the structure we find a row of adders in which all delayed and weighted input and output values are summed (Fig. 5-22). This structure is called the direct form I.

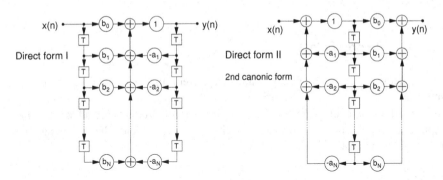

Fig. 5-22 Direct form I   **Fig. 5-23** Direct form II

The recursive and non-recursive parts in Fig. 5-21 can be considered as two independent filters arranged in series. The sequence of these filters can be reversed without changing the overall function. This rearrangement leads to two parallel shift registers that can be combined into one because both are fed by the same signal in this constellation. The resulting structure, as depicted in Fig. 5-23, manages with the minimum possible amount of memory. This structure is called the direct form II. Because of the minimum memory requirement, it is also referred to as a canonical form.

Transposed structures also exist for IIR filters. These are derived by reversing all arrows and exchanging input and output in the corresponding flowgraphs.

Figure 5-24 shows the transposed direct form II, which also features the minimum memory requirement. This structure is therefore canonical, too. With the transposed direct form I as depicted in Fig. 5-25, we have introduced all possible direct forms.

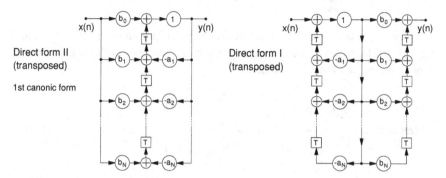

Fig. 5-24 Transposed direct form II   **Fig. 5-25** Transposed direct form I

The transfer function of the IIR filter is obtained by taking the z-transform of (5.28). On this occasion we make use of the shift theorem that we introduced in Sect. 3.3.2.

$$\sum_{r=0}^{M} b_r X(z) z^{-r} = \sum_{i=0}^{N} a_i Y(z) z^{-i}$$

$$X(z) \sum_{r=0}^{M} b_r z^{-r} = Y(z) \sum_{i=0}^{N} a_i z^{-i}$$

$$H(z) = \frac{Y(z)}{X(z)} = \frac{\sum_{r=0}^{M} b_r z^{-r}}{\sum_{i=0}^{N} a_i z^{-i}} \qquad (5.30)$$

Replacing z with $e^{j\omega T}$ yields the frequency response of the filter.

$$H(e^{j\omega T}) = \frac{\sum_{r=0}^{M} b_r e^{-j\omega rT}}{\sum_{i=0}^{N} a_i e^{-j\omega iT}} = \frac{\sum_{r=0}^{M} b_r \cos \omega rT - j \sum_{r=0}^{M} b_r \sin \omega rT}{\sum_{i=0}^{N} a_i \cos \omega iT - j \sum_{i=0}^{N} a_i \sin \omega iT} \qquad (5.31)$$

The magnitude response of the IIR filter is calculated by taking the magnitude of the expressions in the numerator and in the denominator of (5.31).

$$\left|H(e^{j\omega T})\right|^2 = \frac{\left(\displaystyle\sum_{r=0}^{M} b_r \cos \omega rT\right)^2 + \left(\displaystyle\sum_{r=0}^{M} b_r \sin \omega rT\right)^2}{\left(\displaystyle\sum_{i=0}^{N} a_i \cos \omega iT\right)^2 + \left(\displaystyle\sum_{i=0}^{N} a_i \sin \omega iT\right)^2} \tag{5.32}$$

The phase response of the filter is the negative argument of the frequency response (5.31). It can be calculated as the sum of the phase of the numerator and the negative phase of the denominator.

$$b(\omega) = \arctan \frac{\displaystyle\sum_{r=0}^{M} b_r \sin \omega rT}{\displaystyle\sum_{r=0}^{M} b_r \cos \omega rT} - \arctan \frac{\displaystyle\sum_{i=0}^{N} a_i \sin \omega iT}{\displaystyle\sum_{i=0}^{N} a_i \cos \omega iT} \tag{5.33}$$

The group delay of numerator and denominator can also be calculated separately using (5.6).

$$\tau_g(\omega) = \tau_{gnum} - \tau_{gden}$$

The filter structures depicted in Fig. 5-21 to 5-25 show that the filter coefficients are identical to the coefficients of the numerator and denominator polynomials of the transfer function. Since these filters are directly derived from the transfer function, they are also referred to as direct-form filters. These have the advantage that they manage with the lowest possible number of multipliers, which is the same as the number of coefficients of the transfer function. Moreover, choosing a canonical structure leads to a filter that realises a given transfer function with the lowest possible complexity in terms of memory and arithmetical operations.

5.3.3 Poles and Zeros

The transfer function (5.30) is a rational fractional function of the variable z. If we factor the term z^M out of the numerator polynomial and z^N out of the denominator polynomial, we obtain a form in which z appears with positive exponents in both sums.

$$H(z) = z^{N-M} \frac{\displaystyle\sum_{r=0}^{M} b_r z^{M-r}}{\displaystyle\sum_{i=0}^{N} a_i z^{N-i}}$$

The numerator polynomial possesses M zeros, the denominator polynomial N zeros respectively. The zeros of the denominator polynomial are commonly referred to as the poles of the transfer function which we will denote by z_∞ in the following. The zeros of the numerator are the zeros of the transfer function which we will denote by z_0. At $z = 0$ we find an $|N{-}M|$ th-order pole or zero depending on the difference between the degrees of numerator and denominator.

$$H(z) = z^{N-M}\,\frac{b_0(z - z_{01})(z - z_{02})\dots(z - z_{0M})}{a_0(z - z_{\infty1})(z - z_{\infty2})\dots(z - z_{\infty N})} \tag{5.34}$$

For a detailed analysis of the contribution of each single zero and pole to the overall characteristics of the filter, it is convenient to rewrite (5.34) in the form

$$H(z) = \frac{b_0\displaystyle\prod_{r=1}^{M}(1 - z_{0r}z^{-1})}{a_0\displaystyle\prod_{i=1}^{N}(1 - z_{\infty i}z^{-1})} = \frac{b_0\displaystyle\prod_{r=1}^{M}(1 - r_{0r}\,e^{j\varphi_{0r}}z^{-1})}{a_0\displaystyle\prod_{i=1}^{N}(1 - r_{\infty i}\,e^{j\varphi_{\infty i}}z^{-1})}\;.$$

The numerator is the transfer function of a FIR filter. Magnitude, phase, and group delay of the zero term $1-z_0z^{-1}$ were discussed in detail in Sect. 5-2. The corresponding relations for the pole term $1/(1-z_\infty z^{-1})$ can be directly derived from these results. The magnitude of the poles is the reciprocal of the magnitude of the zeros (5.13). Logarithmic magnitude, phase and group delay have inverse signs with respect to the corresponding expressions (5.19) and (5.22) for the zeros.

$$\left|H_{\text{pole}}(e^{j\omega T})\right|^2 = \frac{1}{1 + r_\infty^2 - 2r_\infty \cos(\omega T - \varphi_\infty)} \tag{5.35}$$

$$b_{\text{pole}}(\omega) = \arctan\!\left(\frac{r_\infty \sin(\omega T - \varphi_\infty)}{1 - r_\infty \cos(\omega T - \varphi_\infty)}\right) \tag{5.36}$$

$$\tau_{\text{g pole}}(\omega) = -T\,\frac{r_\infty^2 - r_\infty \cos(\omega T - \varphi_\infty)}{1 + r_\infty^2 - 2r_\infty \cos(\omega T - \varphi_\infty)} \tag{5.37}$$

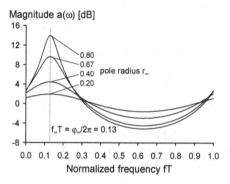

Fig. 5-26
Magnitude response of a pole with the radius r_∞ as a parameter

Figure 5-26 shows the graph of the magnitude response of a pole according to (5.35). This graph exhibits a more or less pronounced resonance peak depending on the distance of the poles from the unit circle. The closer the poles approach the unit circle, the higher is the quality factor.

In contrast to the diagrams which illustrate magnitude, phase, and group delay of the zeros, the corresponding diagrams for the poles only show graphs for $r_\infty < 1$. It is theoretically possible to calculate graphs for $r_\infty \geq 1$, as well. These would not be of practical value, however, because poles on or outside the unit circle are not stable, as will be shown in the next section. The zeros may still be placed inside, onto, or outside the unit circle as in the case of the FIR filter.

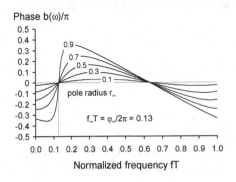

Fig. 5-27
Phase response of a pole with the radius r_∞ as a parameter

Figure 5-27 shows the phase response of a single pole with the radius as a parameter. The closer the radius approaches unity, the higher is the slope of the phase curve at the pole frequency. The group delay response (Fig 5-28) shows a peak at the pole frequency. The height of this peak becomes higher the closer the pole approaches the unit circle. The maximum and minimum values of the group delay curves are calculated as

$$\tau_{g\,max} = T\frac{r_\infty}{1-r_\infty} \quad \text{and} \quad \tau_{g\,min} = -T\frac{r_\infty}{1+r_\infty} \, .$$

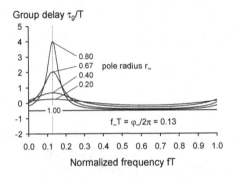

Fig. 5-28
Phase response of a pole with the radius r_∞ as a parameter

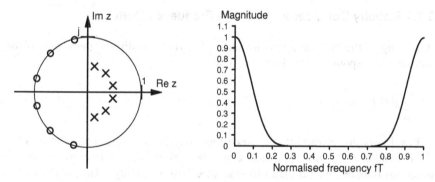

Fig. 5-29 Pole/zero plot and frequency response of a sixth-order IIR low-pass filter

In case of selective filters, the poles are distributed over the passband, since these can realise the much higher gain needed in the passband in contrast to the stopband. Zeros, if present, are positioned in the stopband, where they support the named behaviour by providing additional attenuation. Figure 5-29 shows the example of a sixth-order filter which possesses six poles at $r_\infty = 0.5$ and six zeros on the unit circle, all equally spaced in angle.

For reasons we will deal with in more detail in Chap. 8, it is desirable to cut up a filter into smaller subsystems with as low an order as possible. For real poles and zeros, first-order sections are the appropriate solution. Complex-conjugate pole and zero pairs are combined into second-order sections in order to obtain real filter coefficients. Figure 5-30 shows a corresponding cascade arrangement of second-order filter sections.

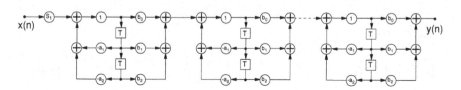

Fig. 5-30 A cascade of second-order filter sections

The scaling coefficient s_1 at the input of the cascade is chosen such that no overflow occurs in the first filter section. Such scaling factors also need to be inserted between the following sections. These, however, can be directly comprised in the non-recursive coefficients b_0, b_1 and b_2. On the one hand, overflow has to be avoided by these scaling factors, on the other hand the available dynamic range of the signal path has to be used in an optimised way to obtain the best possible signal-to-noise ratio. Together with the pairing of poles and zeros and the sequence of subsystems, these scaling coefficients are degrees of freedom, which have no direct influence on the frequency response but which decisively determine the noise behaviour and thus the quality of the filter.

5.3.4 Stability Considerations in the Frequency Domain

According to the considerations in Sect. 3.4, BIBO stability is guaranteed if the unit-sample response is absolutely summable.

$$\sum_{k=-\infty}^{+\infty} |h(k)| < \infty \qquad (5.38)$$

FIR filters are always stable, since their unit-sample responses consist of a finite number of samples. Which condition has to be fulfilled by a rational fractional transfer function $H(z)$ to guarantee BIBO stability? An analysis of the unit-sample response $h(n)$, which we obtain by inverse z-transform of $H(z)$ into the time domain, will answer this question. Partial-fraction expansion of $H(z)$ leads to an expression that allows a direct transformation into the time domain.

$$H(z) = \sum_{r=0}^{N} A_r \frac{z}{z - z_{\infty r}}$$

We obtain by inverse z-transform

$$h(n) = \sum_{r=0}^{N} A_r z_{\infty r}^{n} u(n).$$

Each summand that contributes to the impulse response has to fulfil condition (5.38). This means that the absolute sum of each term has to stay below a certain boundary S to assure stability.

$$\sum_{n=0}^{\infty} \left| A_r z_{\infty r}^{n} \right| < S$$

$$|A_r| \sum_{n=0}^{\infty} \left| z_{\infty r}^{n} \right| < S \qquad (5.39)$$

Expression (5.39) is the sum of an infinite geometric progression which converges if $|z_{\infty r}| < 1$. A plot of the possible locations in the complex z-plane shows that stable poles must lie within the unit circle (Fig. 5-31). Since the coefficients of the denominator polynomial of $H(z)$ are real, the poles have to be real or occur in complex-conjugate pairs.

There is an analogous condition in the theory of continuous-time systems. In the case of stable filters, the poles must lie within the left half of the p-plane with p being the complex frequency variable in the continuous-time domain. If we aim at imitating the behaviour of an analog filter with a digital filter, this is commonly accomplished by transformations that map the p-plane into the z-plane. In order to assure that a stable analog filter leads to a stable digital filter, the part of the p-plane that contains the poles of the analog filter has to be mapped into the interior of the unit circle in the z-plane. For a more general approach, the whole left half of the p-plane should be mapped into the unit circle of the z-plane.

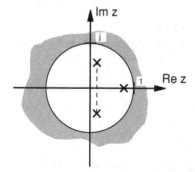

Fig. 5-31
Stability region of poles in the z-domain

Because of the condition that the poles of a stable discrete-time system have to lie within the unit circle, the coefficients a_i of the denominator polynomial cannot assume any arbitrary value. For elementary first- and second-order filter sections, the admissible range of values can be easily specified. A first-order pole is expressed as

$$\frac{1}{1+a_1z^{-1}} = \frac{z}{z+a_1} .$$

A stable system has to fulfil the condition

$$|a_1| < 1 . \tag{5.40}$$

In the case of second-order filter sections, the coefficients a_1 and a_2 determine the location of the poles.

$$\frac{1}{1+a_1z^{-1}+a_2z^{-2}} = \frac{z^2}{z^2+a_1z+a_2}$$

The triangle in Fig. 5-32 shows the possible combinations of coefficients in the a_1/a_2-plane that result in stable filters. A parabola subdivides the triangle into two regions. Coefficient pairs in the dark upper area lead to complex-conjugate pole pairs while coefficients in the lower part lead to two real poles.

In practice, there are further limitations concerning the choice of filter coefficients that we will consider in detail in Chap. 8. In the edge area of the triangle particularly, there is a potential danger of numerical instabilities. Another natural limitation follows from the fact that the coefficients can only be represented with limited precision. The possible combinations of the coefficients a_1 and a_2 form a grid in the stability triangle.

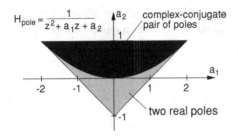

Fig. 5-32
Stability region of the coefficients a_1 and a_2 for second-order poles

5.4 State-Space Structures

Direct-form filters, as treated in the previous sections, are often used in practice because these structures realise a given transfer function with minimum cost. This filter type has the disadvantage that there are almost no possibilities to optimise all the effects that are caused by the finite precision of the coefficient and signal representation. These are:

- Sensitivity of the frequency response with respect to coefficient inaccuracies.
- Quantisation noise that results from the limitation of the word length after mathematical operations such as multiplication or floating point addition.
- Unstable behaviour, i.e. overflow oscillations and quantisation limit cycles.

Higher flexibility and far better possibilities to influence the named effects are offered by the so-called state-space structures.

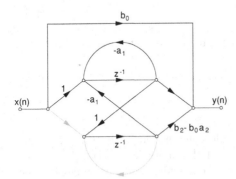

Fig. 5-33
Redrawn second-order direct form filter

In the case of state-space structures, a direct delayless path between input and output, as characterised by the coefficient b_0 of the transfer function, is realised in the form of a bypass. If the orders of the denominator and numerator polynomials are equal in a given application, the numerator polynomial first has to be divided by the denominator polynomial. This results in a constant and a new rational fractional transfer function that has no constant term in the numerator.

$$\frac{b_0 + b_1 z^{-1} + b_2 z^{-2}}{1 + a_1 z^{-1} + a_2 z^{-2}} = b_0 + \frac{(b_1 - b_0 a_1)z^{-1} + (b_2 - b_0 a_2)z^{-2}}{1 + a_1 z^{-1} + a_2 z^{-2}}$$

This constant b_0 determines the gain of the direct path between input and output of the filter. After splitting off the direct path, the second-order direct-form filter can be easily redrawn into the form of Fig. 5-33.

If this figure is completed by the grey paths, we obtain the general form of a second-order state-space structure (Fig. 5-34) which features feedback paths from each memory output $z_i(n)$ to each memory input $z_i(n+1)$. The coefficients in these feedback paths are a_{ij}. The input signal is multiplied by the coefficients b_i and added to the feedback signals at the input of the memories. The output signal of the filter is obtained by a linear combination of the memory outputs. The coefficients of this linear combination are c_i.

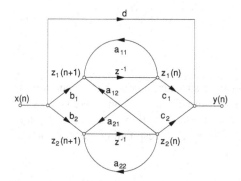

Fig. 5-34
Structure of a second-order state-space filter

The described behaviour can be mathematically expressed as

$$z_1(n+1) = a_{11} \, z_1(n) + a_{12} \, z_2(n) + b_1 \, x(n) \tag{5.41a}$$

$$z_2(n+1) = a_{21} \, z_1(n) + a_{22} \, z_2(n) + b_2 \, x(n) \tag{5.41b}$$

$$y(n) = c_1 \, z_1(n) + c_2 \, z_2(n) + d \, x(n) \ . \tag{5.41c}$$

While the rational fractional transfer function of a general direct-form filter has five coefficients which fully characterise the transfer behaviour, the corresponding state-space structure has nine coefficients. This means higher implementation complexity, but offers many degrees of freedom to optimise the filter performance. Let us now relate the coefficients of the transfer function

$$H(z) = \beta_0 + \frac{\beta_1 z + \beta_2}{z^2 + \alpha_1 z + \alpha_2} = \beta_0 + \frac{\beta_1 z^{-1} + \beta_2 z^{-2}}{1 + \alpha_1 z^{-1} + \alpha_2 z^{-2}}$$

to the coefficients of the state-space representation (5.41).

$$\beta_0 = d$$
$$\beta_1 = c_1\, b_1 + c_2\, b_2$$
$$\beta_2 = c_1\, b_2\, a_{12} + c_2\, b_1\, a_{21} - c_1\, b_1\, a_{22} - c_2\, b_2\, a_{11}$$
$$\alpha_1 = -(a_{11} + a_{22})$$
$$\alpha_2 = a_{11}\, a_{22} - a_{12}\, a_{21}$$

Relation (5.41) can be advantageously written in vector form.

$$z(n+1) = A\, z(n) + b\, x(n) \tag{5.42a}$$

$$y(n) = c^{\mathrm{T}}\, z(n) + d\, x(n) \tag{5.42b}$$

The vectors and matrices printed in bold have the following meaning:

$$z(n) = \begin{pmatrix} z_1(n) \\ z_2(n) \end{pmatrix} \qquad b = \begin{pmatrix} b_1 \\ b_2 \end{pmatrix} \qquad c = \begin{pmatrix} c_1 \\ c_2 \end{pmatrix} \text{ or } c^{\mathrm{T}} = \begin{pmatrix} c_1 & c_2 \end{pmatrix}$$

$$A = \begin{pmatrix} a_{11} & a_{12} \\ a_{21} & a_{22} \end{pmatrix}.$$

In this vector representation, the state-space structure can be easily generalised towards higher filter orders. For a general Nth-order filter, z, b and c are vectors of dimension N, and A is a $N \times N$ matrix. With these definitions, (5.42) is valid for any arbitrary filter order. Figure 5-35 shows a corresponding general block diagram.

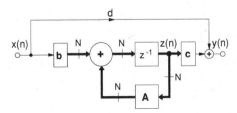

Fig. 5-35
General structure of an Nth-order state-space filter

As a consequence of the $N \times N$ matrix, the number of coefficients increases with the square of the filter order. A direct-form filter requires $2N+1$ multipliers, while a state-space filter is fully specified by $(N+1)^2$ coefficients. This is why higher-order state-space filters are not used very often in practice. Their use is limited, in general, to cascade realisations of second-order filter blocks. Compared to the straightforward implementation of an Nth-order state-space structure, however, the cascade realisation yields only sub-optimal results with regard to the optimisation of the filter performance, but is a good compromise with regard to implementation complexity. From the vector form (5.42) of the state-space algorithm, we can directly derive the transfer function $H(z)$ and the unit-sample response $h(n)$.

The transfer function can be expressed as

$$H(z) = c^T (z\mathbf{I} - A)^{-1} b + d \ . \tag{5.43}$$

$\mathbf{I}$ is the identity matrix. The exponent -1 means inversion of the matrix expression in parenthesis. The unit-sample response is expressed as

$$h(n) = 0 \qquad \text{for } n < 0 \tag{5.44a}$$

$$h(n) = d \qquad \text{for } n = 0 \tag{5.44b}$$

$$h(n) = c^T A^{n-1} b \quad \text{for } n > 0 \ . \tag{5.44c}$$

It can be shown mathematically that (5.44c) converges for $n \to \infty$ only if the magnitudes of the eigenvalues of the matrix A are less then unity. This condition is equivalent to the stability criterion with respect to the location of the poles in the z-plane, because the eigenvalues of A equal the poles of the transfer function (5.43).

We already mentioned that the state-space structure can realise a given transfer function in many ways, since this structure has more coefficients available than are needed to specify the transfer behaviour. We want to show now that a given filter structure can be easily transformed into an equivalent one if we apply simple vector and matrix transformations to the transfer function (5.43). We start by expanding (5.43) on both sides of the matrix expression in parenthesis by the term $T T^{-1}$.

$$H(z) = c^T T T^{-1} (z\mathbf{I} - A)^{-1} T T^{-1} b + d$$

T is an arbitrary nonsingular matrix. Since the product $T T^{-1}$ yields the identity matrix, the transfer function is not changed by this manipulation. The following assignment results in a new set of vectors and matrices:

$$A' = T^{-1} A T$$

$$b' = T^{-1} b \tag{5.45}$$

$$c'^T = c^T T \ .$$

It is easy to see that we obtain a filter with totally different coefficients but exactly the same transfer function as the original one.

When we introduced the state-space structure, we showed that the direct-form filter is a special case of this structure. Furthermore we became acquainted with four different variants of the direct form which have identical transfer behaviour. We will show that the state-space coefficients of one variant can be obtained by applying transformation (5.45) to the coefficients of an other variant. Let us consider direct form II and its transposed version as an example. Figure 5-36 shows the block diagram of a second-order direct form II filter in two ways: standard direct form and equivalent state-space representation.

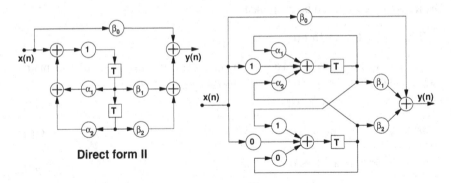

Fig. 5-36 Direct form II in standard and state-space representation

The following state-space coefficients can be derived from Fig. 5-36:

$$b = \begin{pmatrix} 1 \\ 0 \end{pmatrix} \qquad c^T = (\beta_1 \quad \beta_2)$$

$$A = \begin{pmatrix} -\alpha_1 & -\alpha_2 \\ 1 & 0 \end{pmatrix}.$$

The block diagram of the corresponding transposed structure is shown in Fig. 5-37.

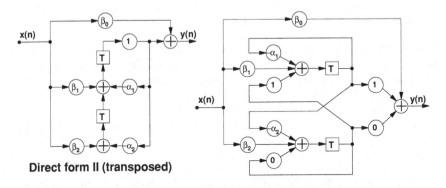

Fig. 5-37 Transposed direct form II in standard and state-space representation

Figure 5-37 yields the following state-space coefficients:

$$b' = \begin{pmatrix} \beta_1 \\ \beta_2 \end{pmatrix} \qquad c'^T = (1 \quad 0)$$

$$A' = \begin{pmatrix} -\alpha_1 & 1 \\ -\alpha_2 & 0 \end{pmatrix}.$$

It can easily be shown that the coefficients of the transposed structure are obtained by the transformation

$$T = \frac{1}{\beta_1\beta_2\alpha_1 - \beta_1^2\alpha_2 - \beta_2^2} \begin{pmatrix} \beta_2\alpha_1 - \beta_1\alpha_2 & -\beta_2 \\ -\beta_2 & \beta_1 \end{pmatrix}$$

$$T^{-1} = \begin{pmatrix} \beta_1 & \beta_2 \\ \beta_2 & \beta_2\alpha_1 - \beta_1\alpha_2 \end{pmatrix}$$

from the coefficients of the original direct form II.

5.5 The Normal Form

The filter structure that we deal with in this section is also referred to as the "coupled-loop" structure according to Gold and Rader in the literature.

A well-tried means to reduce the coefficient sensitivity of higher-order rational fractional transfer functions is the decomposition into first- and second-order partial transfer functions. It is obvious to improve the performance of the filter by further decomposition of the second-order sections. A second-order filter that realises a complex-conjugate pole pair cannot be easily subdivided into first-order partial systems because the resulting filter coefficients may be complex (5.46).

$$\frac{1}{z^2 + a_1 z + a_2} = \frac{1}{z - (\alpha + j\beta)} \frac{1}{z - (\alpha - j\beta)}$$

$$= \frac{z^{-1}}{1 - (\alpha + j\beta)z^{-1}} \frac{z^{-1}}{1 - (\alpha - j\beta)z^{-1}} \tag{5.46}$$

with $a_1 = -2\alpha$ and $a_2 = \alpha^2 + \beta^2$

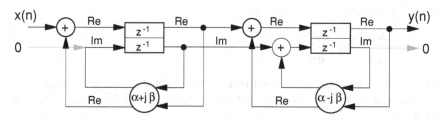

Fig. 5-38 Decomposition of a second-order pole into first-order sections

Figure 5-38 shows the block diagram of a second-order pole realised as a cascade of two first-order complex-conjugate poles. The input and output sequences of this filter must still be real. Within the structure, however, we find complex signals as a consequence of the complex coefficients. In Fig. 5-38, these complex signal paths are depicted as two parallel lines, one representing the real part, the other the imaginary part, of the respective signal. Restructuring and simplification yields the structure shown in Fig. 5-39. On this occasion, we have taken into account that some of the imaginary paths are not needed because the input and output signals of the overall second-order block are real. This structure requires four coefficients to realise the complex-conjugate pole pair instead of two in the case of the direct form. The coefficients equal the real part α and the imaginary part β of the pole pair. These are, derived from (5.46), related to the coefficients of the transfer function a_1 and a_2 as

$$\alpha = -a_1/2 \qquad \beta = \sqrt{a_2 - a_1^2/4} \qquad\qquad a_1 = -2\alpha \qquad a_2 = \alpha^2 + \beta^2 .$$

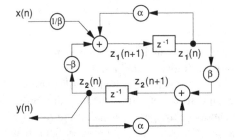

Fig. 5-39
The "coupled-loop" structure according to Rader and Gold

The normal form can be easily converted into the state-space form according to Fig. 5-34, which yields the following state-space coefficients:

$$b = \begin{pmatrix} 1/\beta \\ 0 \end{pmatrix} \qquad\qquad c^T = (0 \quad 1)$$

$$A = \begin{pmatrix} \alpha & -\beta \\ \beta & \alpha \end{pmatrix} .$$

Introduction of a polynomial into the numerator of (5.46) leads to a more general second-order transfer function.

$$H(z) = \frac{b_0 + b_1 z^{-1} + b_2 z^{-2}}{1 + a_1 z^{-1} + a_2 z^{-2}}$$

State-space vector b and system matrix A are not affected by this amendment. The vector c^T and the coefficient d take the non-recursive part of the transfer function into account.

$$c^T = \left(\beta b_1 + 2\alpha\beta b_0 \quad b_2 + b_1\alpha + (\alpha^2 - \beta^2)b_0 \right)$$

The normal filter can also be derived from the direct form II by simple application of the transformation

$$T = \begin{pmatrix} \beta & \alpha \\ 0 & 1 \end{pmatrix} \qquad\qquad T^{-1} = \begin{pmatrix} 1/\beta & -\alpha/\beta \\ 0 & 1 \end{pmatrix} .$$

5.6 Digital Ladder Filters

The filter structures, treated in this chapter up to now, are more or less direct implementations of the rational fractional transfer function in z. This is obvious especially in the case of the direct form since the coefficients of the transfer function are identical with the filter coefficients. With increasing filter order, however, the direct realisation of the transfer function becomes more and more critical with respect to noise performance, stability and coefficient sensitivity. We already showed two possibilities to avoid these problems:

- The decomposition of the transfer function into lower order partial systems, preferably of first and second order.
- The transition to state-space structures which require a higher implementation complexity but offer more flexibility with respect to the control of the named drawbacks.

The filter structures that we introduce in the following act another way. They are not guided by the realisation of transfer functions in z which in turn are derived from the transfer functions of analog reference filters in p. These structures aim at directly imitating the structure of analog reference filters in the discrete-time domain. In Sect. 2.8.4, we introduced the analog ladder filter which exhibits excellent properties with respect to stability and coefficient sensitivity. This makes them very appropriate to act as a prototype for digital filter implementations.

A starting point for the preservation of the good properties of the analog reference filter in the discrete-time domain is to convert the electrical circuit diagram of the analog filter, whose behaviour is governed by Kirchhoff's current and voltage laws, into a signal flowgraph or a more implementation-oriented block diagram. The signal quantities appearing in these flowgraphs or block diagrams may be voltages, currents or even wave quantities. In the next step, mathematical operators such as integrators, for instance, are replaced by corresponding discrete-time equivalents, and continuous-time properties of elements such as capacitors, inductors and resistors are converted into appropriate discrete-time representations.

5.6.1 Kirchhoff Ladder Filters

By means of a concrete example, we want to illustrate step by step how we can, starting with the Kirchhoff current and voltage laws and a resulting continuous-time signal flowgraph, finally obtain a block diagram of a discrete-time realisation of the filter. Fig. 5-40 shows a third-order ladder low-pass filter. We identify a lossless two-port inserted between resistive terminations with the well-known good sensitivity properties.

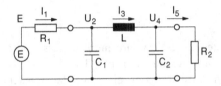

Fig. 5-40
Third-order ladder low-pass filter

The following current and voltage equations can be read from Fig. 5-40:

$$I_1 = \frac{E - U_2}{R_1} \qquad\qquad I_5 = \frac{U_4}{R_2} \qquad\qquad U_2 = \frac{I_1 - I_3}{j\omega C_1}$$

$$I_3 = \frac{U_2 - U_4}{j\omega L} \qquad\qquad\qquad U_4 = \frac{I_3 - I_5}{j\omega C_2} \ .$$

Figure 5-41a shows a translation of these equations into a flowgraph and a block diagram. Currents and voltages can now be interpreted as signal quantities which are related by additions and transfer functions in the connecting branches. This representation is independent of the actual physical implementation of the desired transfer behaviour. If we replace current and voltage by physical quantities like force and velocity, for instance, and elements like capacitors and inductors by masses and springs, this flowgraph could also be derived from a system that represents a third-order mechanical low-pass filter.

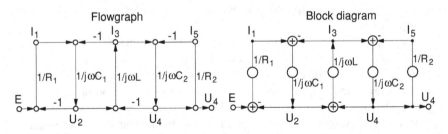

Fig. 5-41a Flowgraph and block diagram derived from a third-order ladder low-pass filter

If we normalise the impedances and admittances in Fig. 5-41a to an arbitrary resistance R, we obtain a flowgraph in which only voltages appear as signal quantities and all coefficients and transfer functions are dimensionless, which is advantageous for the further numerical treatment of the structure. This normalisation is achieved by multiplying all coefficients in the ascending branches by R and dividing all coefficients in the descending branches by R. This procedure does not change the transfer behaviour at all. The resulting block diagram is shown in Fig. 5-41b. Both lateral branches in Fig. 5-41b contain pure multiplications by dimension-less constants. The three inner branches contain integrators, as can be concluded from the frequency response that is inverse proportional to the frequency ($\sim 1/j\omega$). Thus 4 adders, 3 integrators and 2 coefficient multipliers are required to realise the low-pass filter.

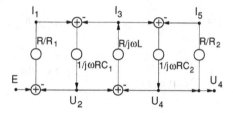

Fig. 5-41b
Normalised flowgraph of the third-order
low-pass filter

The coefficients and time constants of the integrators in the resulting block
diagram Fig. 5-41c can be calculated as

$$k_1 = R / R_1 \quad \tau_2 = RC_1$$
$$k_5 = R / R_2 \quad \tau_3 = L / R$$
$$\tau_4 = RC_2$$

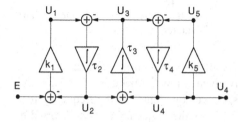

Fig. 5-41c
Active filter implementation of the
third-order low-pass filter

Block diagram Fig. 5-41c can be used as the basis for the design of an active
filter in which the integrators are realised by operational amplifiers. For the
transition from the continuous-time to a discrete-time realisation, we have to find
a discrete-time approximation of the integrator.

The frequency response of an integrator features the following two properties
that have to be reproduced as closely as possible:
• The phase is constantly 90°.
• The magnitude is inversely proportional to the frequency.
Transfer function and frequency response of the continuous-time integrator can be
expressed as

$$H(p) = \frac{1}{p\tau} \quad \rightarrow \quad H(j\omega) = \frac{1}{j\omega\tau} \; .$$

τ denotes the time constant of the integrator. The constant phase can be easily
realised by a discrete-time algorithm, while the reproduction of a magnitude
inversely proportional to the frequency causes problems, since the magnitude
function must be periodical with the sampling frequency and representable by a
trigonometric function.

Fig. 5-42 shows two possible candidates for the discrete-time realisation of the
integrator block. Transfer function and frequency response of these algorithms can
be expressed as

$$H(z) = \frac{T}{\tau} \frac{z^{-1/2}}{1-z^{-1}} \qquad \rightarrow \qquad H(e^{j\omega T}) = \frac{T}{2\tau} \frac{1}{j\sin(\omega T/2)} \qquad (5.47a)$$

$$H(z) = \frac{T}{2\tau} \frac{1+z^{-1}}{1-z^{-1}} \qquad \rightarrow \qquad H(e^{j\omega T}) = \frac{T}{2\tau} \frac{1}{j\tan(\omega T/2)} . \qquad (5.47b)$$

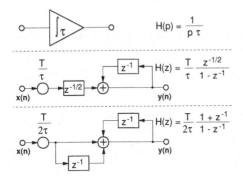

$$H(p) = \frac{1}{p\,\tau}$$

$$H(z) = \frac{T}{\tau} \frac{z^{-1/2}}{1-z^{-1}}$$

$$H(z) = \frac{T}{2\tau} \frac{1+z^{-1}}{1-z^{-1}}$$

Fig. 5-42
Possible discrete-time realisations of an integrator

Both realisations possess a constant phase of 90°. A good approximation of the magnitude of the analog integrator is only given at low frequencies where sine and tangent can be approximated by their respective arguments. As the frequency ω is replaced by the sine and tangent terms in (5.47), the frequency axis is distorted in both cases: (5.47a) means stretching, (5.47b) compression of the frequency axis. A way to circumvent this problem is to predistort the frequency response of the analog reference filter such that the discrete-time implementation has exactly the desired cutoff frequencies.

Starting from the circuit diagram of the analog implementation, we have mathematically found a way to derive a discrete-time block diagram of a filter which has a ladder filter as a model. At first glance, the only drawback of the discrete-time solution seems to be that the frequency axis is distorted which can be defused by proper measures.

More serious, however, is the fact that replacing the integrators in Fig. 5-41c by either discrete-time approximation depicted in Fig. 5-42 does not result in realisable systems in the sense discussed in Sect. 5.1.2. In one case, there exist loops in the block diagram with a total delay that is not an integer multiple of T. In the other case we find delayless loops. This result can be generalised: flowgraphs and block diagrams which are based on voltage, current and the corresponding Kirchhoff laws lead to unrealisable discrete-time block diagrams. We will show in the next section that this problem can be solved by the transition to wave quantities. The only way out in the present approach is to add or remove delays to guarantee realisability. This operation, however, will lead to unavoidable deviations of the frequency response, and the good coefficient sensitivity behaviour will be lost. We will demonstrate this by means of an example.

Example 5-3

Derive the discrete-time structure of a third-order Chebyshev filter with the following specification:

Filter order: $N = 3$
Passband ripple: 1.256 dB
Cutoff frequency: $f_c T = 0.1$

Use the continuous-time ladder structure according to Fig. 5-41c as the basis and replace the integrator with an appropriate discrete-time equivalent.

Replacing the analog integrator in Fig. 5-41c with the discrete-time version according to (5.47a) leads to the following block diagram.

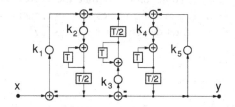

Fig. 5-43a
Block diagram of a third-order ladder low-pass filter

The filter coefficients in this representation are calculated as follows:

$$k_1 = R / R_1$$
$$k_2 = T / \tau_1 = T/RC_1$$
$$k_3 = T / \tau_2 = TR/L$$
$$k_4 = T / \tau_3 = T/RC_2$$
$$k_5 = R / R_2 \ .$$

(5.48)

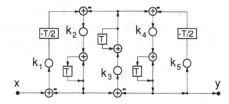

Fig. 5-43b
Block diagram of the discrete-time ladder filter after equivalent rearrangement

A number of equivalent rearrangements in Fig. 5-43a lead to the form of Fig. 5-43b. The lateral branches in this block diagram contain delays of $-T/2$. This leads to loops that violate the realisability conditions.

If the sampling frequency is high compared to the cutoff frequency of the low-pass filter, we can add delays of $T/2$ in both lateral branches (see Fig. 5-43c) without changing the frequency response too dramatically [4]. As a consequence, the left branch only contains the multiplication by the coefficient k_1, the right branch the multiplication by k_5. This modification leads to a realisable filter structure.

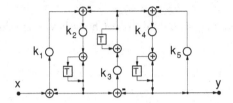

Fig. 5-43c
Block diagram after removal of
the delays in the lateral branches

The five filter coefficients of the block diagram in Fig. 5-43c can be determined using (5.48). We assume that the analogue reference filter is terminated with equal resistances $R_1 = R_2$ at both ends. To simplify the block diagram, we choose the normalisation resistance R equal to R_1 and R_2. The coefficients k_1 and k_5 become unity so that the corresponding multiplications can be avoided. The missing values for the capacitors and the inductor can be obtained from filter design tables such as [57]:

$$C_1 = C_2 = 2.211315 / (2\pi f_c R)$$
$$L = 0.947550 \, R / 2\pi f_c \ .$$

Substitution in (5.48) leads to the following coefficients of the discrete-time filter:

$$k_1 = k_5 = 1$$
$$k_2 = k_4 = 2\pi \, 0.1 / 2.211315 = 0.284138$$
$$k_3 = 2\pi \, 0.1 / 0.947550 = 0.663098 \ .$$

Figure 5-44 compares the magnitude characteristics of the analog and the discrete-time realisation. The deviations in the passband are caused by the manipulations of the block diagram which assure realisability. The property of boundedness of the magnitude gets lost since the curve visibly exceeds unity in the vicinity of the cutoff frequency. The frequency axis is stretched as a consequence of the frequency response of the used integrator approximation (5.47a).

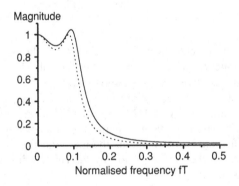

Fig. 5-44
Magnitude of a third-order
ladder filter (Chebyshev
1.256 dB ripple, cutoff
frequency $f_c T = 0.1$)
dashed line: analog filter
solid line: digital ladder filter

The deviation of the magnitude response with respect to the analog reference filter as depicted in Fig. 5-44 is relatively small for low cutoff frequencies. With increasing cutoff frequency, however, the removal of the delays in the block diagram becomes more and more noticeable. In the range of $f_cT = 0.25$, the original frequency response of the analog filter can not be recognised any more.

5.6.2 Wave Digital Filters

The considerations in the previous section showed that flowgraphs based on current and voltage as signal parameters (see Fig. 5-41a) result in non realisable discrete-time systems. By the modifications to remove this drawback, the resulting digital filter looses the good properties of the reference filter, and the frequency response more or less deviates from the specification. The transition to wave parameters can solve the problem. Fettweis [21, 23, 24] showed that this approach does not only result in realisable systems but also leads to filters with excellent stability properties. Under certain circumstances, passivity and thus stability can be guaranteed even if the unavoidable rounding operations are taken into account.

Wave parameters have their origin in transmission line theory, and form, in a somewhat modified form, the basis of wave digital filters. Because of their great importance for the understanding of wave digital filters, wave parameters are considered in more detail in the following section.

It must be noted that we cannot treat all aspects of wave digital filter theory in this textbook. The low-pass filter according to Fig. 5-40 is to serve as an example again in to order to finally arrive at a realisable discrete-time flowgraph. A detailed overview of theory and practice of wave digital filters can be found in [23] which has almost the extent of a book.

5.6.2.1 Wave Parameters

Wave parameters have their origin in transmission line theory. The voltage u at an arbitrary position on the line can be assumed to result from the superposition of a voltage wave propagating in the positive x-direction u_+ and a voltage wave in opposite direction u_-. The same is true for current waves on the line (see Fig. 5-45).

$$u = u_+ + u_-$$
$$i = i_+ + i_-$$

(5.49)

$$u_+ = i_+ R_L$$
$$u_- = -i_- R_L$$

$$u = u_+ + u_-$$
$$i = i_+ + i_-$$

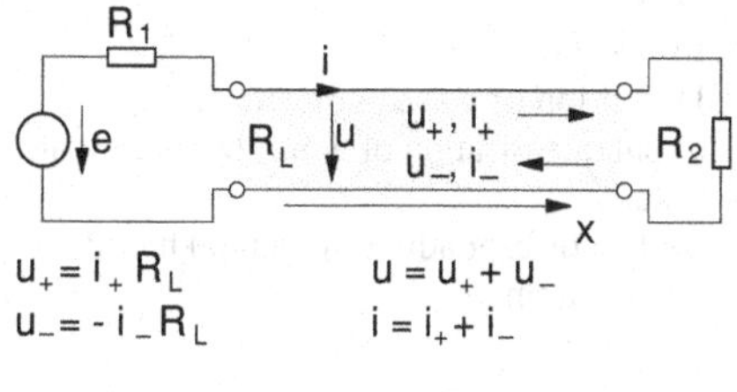

Fig. 5-45
Definition of wave parameters derived from the theory of transmission lines

Current and voltage waves are related by the characteristic impedance R_L of the line.

$$u_+/i_+ = R_L \qquad\qquad u_-/i_- = -R_L \qquad\qquad (5.50)$$

Using (5.49) and (5.50), we can calculate the voltage waves propagating in both directions at an arbitrary position on the line from the voltage and the current measured at this position.

$$u_+ = (u + R_L i)/2 \qquad\qquad u_- = (u - R_L i)/2 \qquad\qquad (5.51)$$

These relations are independent of the length of the line. In principle they are also valid for a line length equalling zero. In this case, R_L is no more the characteristic impedance of a line but a parameter that can be freely chosen. u_+ and u_- have the meaning of quantities that describe the electrical state of a port in the same way as the pair u and i. For our purpose we define the wave parameters for simplicity without the factor ½ and denote them by the letters a and b. Instead of R_L we write simply R.

$$a = u + Ri \qquad\qquad b = u - Ri \qquad\qquad (5.52a)$$

$$u = (a + b)/2 \qquad\qquad i = (a - b)/2R \qquad\qquad (5.52b)$$

R denotes the so-called port resistance. The parameters a and b defined in this way are called the voltage wave parameters since they have the dimension of a voltage. Similarly we can define current wave parameters

$$a = u/R + i \qquad\qquad b = u/R - i$$

or power wave parameters

$$a = u/\sqrt{R} + i\sqrt{R} \qquad\qquad b = u/\sqrt{R} - i\sqrt{R}$$

from which we can calculate the power of the incident and reflected waves by squaring.

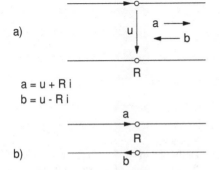

a)

$a = u + R\,i$
$b = u - R\,i$

b)

Fig. 5-46
Definition of wave parameters:
a) characterisation of a port by current and voltage
b) the corresponding flowgraph based on wave parameters

All three parameter definitions are suitable for the following considerations, but it has become custom to use voltage waves. Figure 5-46a depicts a port with the assignment of current, voltage, port resistance and the wave parameters a and b. Figure 5-46b shows the equivalent wave flowgraph. It is important to note that the indication of the port resistance in the flowgraph is absolutely mandatory, as otherwise the parameters a and b are not defined. Without the knowledge of the port resistance it is impossible to calculate back the electrical state of the port in terms of voltage u and current i from the wave parameters a and b.

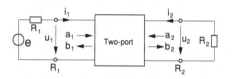

Fig. 5-47
Wave parameters in the context of a two-port

Figure 5-47 shows a two-port network inserted between two resistive terminations. In the following we calculate the wave parameters associated with the ports of this network. The waves propagating towards the network are denoted as a_i, the waves reflected by the network as b_i. The choice of the port impedances R_i is arbitrary in principle. In practice, however, one will select these particular parameter values of R_i such that they lead to the simplest overall expressions for our problem. This is the case if the port impedances are chosen equal to the terminating resistances. Based on this assumption, we calculate the four incident and reflected waves as depicted in Fig. 5-47 using (5.52).

wave a_1:
$$a_1 = u_1 + R_1 i_1$$
$$i_1 = (e - u_1)/R_1$$
$$a_1 = e \qquad (5.53a)$$

The incident wave at port 1 equals the voltage e of the source.

wave b_1:
$$b_1 = u_1 - R_1 i_1$$
$$i_1 = (e - u_1)/R_1$$
$$b_1 = 2u_1 - e \qquad (5.53b)$$

The reflected wave at port 1 depends on the input impedance of the corresponding network port. If this impedance equals R_1, which is the matching case, then $u_1 = e/2$, and the reflected wave b_1 vanishes. Otherwise b_1 is terminated in the internal resistor of the source which acts as a sink in this case.

wave a_2:
$$a_2 = u_2 + R_2 i_2$$
$$i_2 = -u_2/R_2$$
$$a_2 = 0 \qquad (5.53c)$$

Because of the matching termination, there is no incident wave at port 2.

wave b_2:

$$b_2 = u_2 - R_2 i_2$$
$$i_2 = -u_2/R_2$$
$$b_2 = 2u_2 \qquad\qquad (5.53d)$$

The reflected wave at port 2 is the part of the source signal that is transmitted over the network. Its value is the double of the voltage at port 2. Figure 5-48 summarises the calculated wave flows and depicts a flowgraph which is equivalent to the circuit diagram Fig. 5-47.

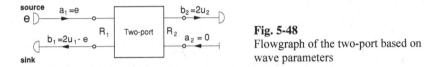

Fig. 5-48
Flowgraph of the two-port based on wave parameters

Fig. 5-48 introduces two new symbols, a wave source and a wave sink. The equivalent of a voltage source contains both elements, since it generates an outgoing wave and terminates an incoming wave in its internal resistor. The terminating resistance at the right side of the two-port is characterised by a sink only.

If the two-port is a selective filter and we consider the stopband of the frequency response, then the incident wave a_1 is, in essence, reflected at port 1. In the passband, the two-port network matches the internal resistor of the source R_1 more or less perfectly with the terminating resistor R_2. As a consequence, the wave a_1 is transmitted without, or with low, attenuation over the network and is terminated as wave b_2 in the resistor R_2, whereas b_1 vanishes. Thus the two-port shares the incident wave a_1 among the reflected waves b_1 and b_2.

Incident and reflected waves of a two-port are related by so-called scattering matrices:

$$\begin{pmatrix} b_1 \\ b_2 \end{pmatrix} = \begin{pmatrix} S_{11} & S_{12} \\ S_{21} & S_{22} \end{pmatrix} \begin{pmatrix} a_1 \\ a_2 \end{pmatrix}. \qquad\qquad (5.54)$$

The elements S_{ii} are reflection coefficients, which are also called the reflectances. They define the share of the incident wave a_i at port i that is reflected at this port. The elements S_{ij} describe the transfer behaviour from port j to port i. They are called the transmittances.

$$S_{11} = \frac{b_1}{a_1}\bigg|_{a_2 = 0} \qquad\qquad S_{12} = \frac{b_1}{a_2}\bigg|_{a_1 = 0}$$

$$S_{21} = \frac{b_2}{a_1}\bigg|_{a_2 = 0} \qquad\qquad S_{22} = \frac{b_2}{a_2}\bigg|_{a_1 = 0} \qquad\qquad (5.55)$$

The parameter S_{21} has a special meaning. If the network is properly terminated ($a_2 = 0$), then (5.53a) and (5.53d) apply and we get

$$S_{21} = 2\frac{u_2}{e} . \tag{5.56}$$

Apart from the factor 2, S_{21} is the transfer function of the two-port network terminated by R_1 and R_2. Furthermore, apart from the factor $\sqrt{(R2/R1)}$, (5.56) is identical to the definition of the transmittance (2.75). The difference is due to the fact that (2.75) is based on the power balance and therefore on power waves, whereas (5.56) is derived from voltage waves.

If the two-port is lossless, the sum of the power of the incident waves must equal the sum of the power of the reflected waves.

$$|a_1|^2 / R_1 + |a_2|^2 / R_2 = |b_1|^2 / R_1 + |b_2|^2 / R_2$$

Using (5.54) yields the following relations between the coefficients of the scattering matrix:

$$|S_{11}|^2 + (R_1/R_2)|S_{21}|^2 = 1 \tag{5.57a}$$

$$|S_{22}|^2 + (R_2/R_1)|S_{12}|^2 = 1 \tag{5.57b}$$

$$S_{11}S_{12}^*/R_1 + S_{21}S_{22}^*/R_2 = 0 \tag{5.57c}$$

Equation (5.57a) points to an interesting property of the reflectance of a lossless two-port. Transmittance S_{21} and reflectance S_{11} are complementary with respect to each other. Thus S_{11} can be considered a transfer function, too, which has its passband in that frequency range, where S_{21} has its stopband and vice versa. The output b_1 in the wave flowgraph Fig. 5-48 can therefore be used as the output of a filter with a complementary frequency response. The same applies to the pair S_{22} and S_{12}. From (5.57a) and (5.57b) we can derive further interesting properties of the scattering parameters for the lossless case.

$$\begin{aligned} |S_{11}| &\le 1 & |S_{21}| &\le \sqrt{R_2/R_1} \\ |S_{22}| &\le 1 & |S_{12}| &\le \sqrt{R_1/R_2} \end{aligned} \tag{5.58}$$

All frequency responses derived from the scattering parameters are bounded. It can therefore be expected that filters based on scattering parameters of lossless networks (inserted between resistive terminations) possess the excellent properties concerning coefficient sensitivity discussed in Sect. 2.8.1. Additionally, taking into account (5.57c) yields the following interesting relations:

$$|S_{11}| = |S_{22}| \tag{5.59a}$$

$$|S_{21}| = \frac{R_2}{R_1}|S_{12}| . \tag{5.59b}$$

Apart from a constant factor, the magnitudes of S_{12} and S_{21} are identical. Thus port 1 and port 2 can reverse roles.

Example 5-4

Derive the flowgraph of a simple voltage divider circuit (Fig. 5-49) based on voltage waves.

Since the two-port network directly connects source and terminating resistor, voltage and current at both ports are related as follows:

$$u_1 = u_2 \qquad\qquad i_1 = -i_2 \ .$$

Furthermore we have according to (5.52b)

$$u_1 = (a_1 + b_1)/2 \qquad\qquad i_1 = (a_1 - b_1)/2R_1$$
$$u_2 = (a_2 + b_2)/2 \qquad\qquad i_2 = (a_2 - b_2)/2R_2 \ .$$

From these relations we can determine the reflected waves b_1 and b_2 in terms of the incident waves a_1 and a_2.

$$b_1 = \frac{2R_1}{R_1 + R_2} a_2 - \frac{R_1 - R_2}{R_1 + R_2} a_1$$

$$b_2 = \frac{2R_2}{R_1 + R_2} a_1 - \frac{R_1 - R_2}{R_1 + R_2} a_2$$

$$(5.60)$$

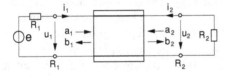

Fig. 5-49
Two-port representation of the voltage divider circuit

Using the abbreviation

$$r = \frac{R_1 - R_2}{R_1 + R_2} \ ,$$

(5.60) can be written as

$$b_1 = -ra_1 + (1 + r)a_2$$
$$b_2 = (1 - r)a_1 + ra_2$$

$$(5.61)$$

which leads to the following scattering matrix describing the wave-flow behaviour of the simple two-port in Fig. 5-49.

$$S = \begin{pmatrix} -r & 1+r \\ 1-r & r \end{pmatrix}$$

Figure 5-50 shows a possible realisation of (5.61) in the form of a block diagram. This structure is called a two-port adapter. It links ports with different port resistances R_i. If the internal resistance of the source R_1 and the terminating resistance R_2 are matched ($r = 0$), no reflections occur at the ports. Incident wave a_1 leaves the two-port as b_2, and a_2 passes the two-port accordingly and comes out as b_1.

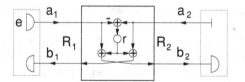

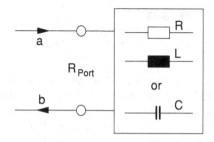

Fig. 5-50
Block diagram representation of the voltage divider circuit

The transfer behaviour of the voltage divider is described by the scattering parameter S_{21} as can be seen from (5.56).

$$S_{21} = 2\frac{u_2}{e} \quad \text{or} \quad \frac{u_2}{e} = \frac{1}{2}S_{21}$$

With $S_{21} = 1 - r$ from above, we can write

$$\frac{u_2}{e} = \frac{1}{2}(1-r) = \frac{1}{2}\left(1 - \frac{R_1 - R_2}{R_1 + R_2}\right) = \frac{R_2}{R_1 + R_2} \quad ,$$

which is the expected result.

5.6.2.2 Decomposition of the Analog Reference Network into One-Ports and Multi-Ports

The starting point for the design of the Kirchhoff ladder filter has been the establishment of the corresponding node and branch equations of the filter network. The theory of wave digital filters uses a different approach. In the first step, all elements of the electrical reference network are described by their properties with respect to wave parameters. Each element is considered as a one-port with an associated port resistance R_{Port} (see Fig. 5-51).

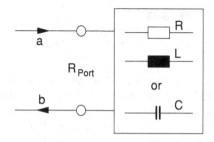

Fig. 5-51
Definition of the wave one-port

The circuit element is now characterised by its reflectance, which is the ratio of reflected to incident wave. In case of a resistor, we for simplicity choose $R_{Port} = R$.

The reflectance, and consequently the reflected wave, become zero (see the example of R_2 in Fig 5-49 and Fig. 5-50). The reflectance of capacitors and inductors is complex, frequency dependent and does not vanish for any frequency. For the choice of R_{Port} we will have to consider relations between the sampling period T of the target digital filter and the time constants $\tau = R_{Port}C$ or $\tau = L/R_{Port}$, as we will show in the next section.

In the next step, all network elements are linked together by adapters to form the wave flowgraph of the whole filter network. We have already introduced the two-port adapter (Fig. 5-50) which links together ports with different port resistances (see the example in the previous section where the adapter describes the matching behaviour of a voltage source with internal resistor R_1 and a terminating resistor R_2). For more complex networks, however, we need adapters with three or more ports. We attempt now to decompose the ladder filter according to Fig. 5-40 into one-ports that represent the elements of the circuit and adapters that describe how these elements are linked.

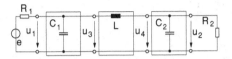

Fig. 5-52
Decomposition of a ladder filter into partial two-ports.

Figure 5-52 depicts the ladder filter modified into the form of a cascade of two-ports. Each two-port contains exactly one element, which is arranged either as a series or as a shunt element. For each of these two-ports, we can give a scattering matrix. In [76] these scattering matrices are used as the basis for the design of RC-active filters with low component sensitivity. For our purpose, however, we want to go one step further. We modify the flowgraph in such a way that the network elements appear in the form of one-ports. This can be accomplished by rearrangement as shown in Fig. 5-53. Note that we still have the same ladder filter that we started with.

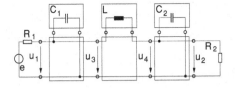

Fig. 5-53
Representation of the ladder filter by means of one-ports and three-ports

The circuit diagram now consists of a series connection of three-ports. The remaining unattached ports are terminated by one-ports which represent the voltage source with internal resistance R_1, the terminating resistance R_2, and the components of the lossless filter network C_1, C_2 and L. If we convert Fig. 5-53 into a wave flowgraph, we have to distinguish between two types of three-port adapters: C_1 and C_2 are linked to the connecting network via parallel adapters, the inductor L via a serial adapter. Figure 5-54 introduces the two new symbols that represent these adapter types and shows the result of the conversion.

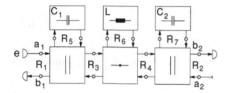

Fig. 5-54
Wave flowgraph of the ladder filter with three-ports as basic building blocks

Each of the adapters in Fig. 5-54 has to link three ports with different port resistances. After the interconnection of all blocks, we identify seven independent port resistances that have to be specified:

- R_1 and R_2 are determined by the internal resistance of the source and the terminating resistance of the network.
- R_5, R_6 and R_7 will be determined in such a way that the discrete-time realisation of the frequency dependent reflectance of the capacitors and of the inductor becomes as simple as possible.
- The choice of R_3 and R_4 seems to be arbitrary at first glance. We will see, however, that there are constraints with respect to the realisability of the discrete-time filter structure that have to be observed.

The general relation between incident and reflected waves of a three-port can be expressed as:

$$b_1 = S_{11}\, a_1 + S_{12}\, a_2 + S_{13}\, a_3$$
$$b_2 = S_{21}\, a_1 + S_{22}\, a_2 + S_{23}\, a_3 \qquad\qquad (5.62)$$
$$b_3 = S_{31}\, a_1 + S_{32}\, a_2 + S_{33}\, a_3 \; .$$

In the following, we determine the scattering coefficients of both three-port adapter types and give possible realisations of these adapters in the form of block diagrams. We proceed in the same way as we did for the derivation of the two-port adapter in the previous section. We start with the parallel adapter.

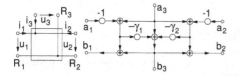

Fig. 5-55
Three-port parallel adapter

As an example, we show the calculation of the scattering coefficient S_{23} in more detail. This coefficient describes the situation that a voltage source is connected to port 3, and port 1 and port 2 are terminated with R_1 and R_2 respectively. So there are no incident waves at port 1 and port 2. From (5.62) it follows:

$$S_{23} = \frac{b_2}{a_3}\bigg|_{a_1 = 0,\, a_2 = 0} \; . \qquad\qquad (5.63a)$$

From the definition of the wave parameters (5.52) we have:

$$b_2 = u_2 - R_2 i_2$$
$$a_3 = u_3 + R_3 i_3 \ .$$

(5.63b)

According to Fig. 5-55, we can write down the following node and mesh equations for the parallel adapter.

$$u_1 = u_2 = u_3$$
$$i_1 + i_2 + i_3 = 0$$

(5.63c)

The terminations with R_1 and R_2 yield the following additional equations:

$$u_1 = -R_1 i_1 \qquad \text{and} \qquad u_2 = -R_2 i_2 \ .$$

(5.63d)

The set of equations (5.63a,b,c,d) finally results in:

$$S_{23} = \frac{2G_3}{G_1 + G_2 + G_3} \ .$$

G_n are the reciprocals of the port resistances R_n. In the same way we can calculate the remaining scattering coefficient.

$$b_1 = \frac{G_1 - G_2 - G_3}{G_1 + G_2 + G_3} a_1 + \frac{2G_2}{G_1 + G_2 + G_3} a_2 + \frac{2G_3}{G_1 + G_2 + G_3} a_3$$

$$b_2 = \frac{2G_1}{G_1 + G_2 + G_3} a_1 + \frac{G_2 - G_1 - G_3}{G_1 + G_2 + G_3} a_2 + \frac{2G_3}{G_1 + G_2 + G_3} a_3$$

$$b_3 = \frac{2G_1}{G_1 + G_2 + G_3} a_1 + \frac{2G_2}{G_1 + G_2 + G_3} a_2 + \frac{G_3 - G_1 - G_2}{G_1 + G_2 + G_3} a_3$$

Using the abbreviations

$$\gamma_1 = \frac{2G_1}{G_1 + G_2 + G_3} \qquad\qquad \gamma_2 = \frac{2G_2}{G_1 + G_2 + G_3}$$

we can simplify these relations. The transfer behaviour of the three-port parallel adapter is fully specified by the two coefficients γ_1 and γ_2.

$$b_1 = (\gamma_1 - 1)a_1 + \gamma_2 a_2 + (2 - \gamma_1 - \gamma_2)a_3$$

$$b_2 = \gamma_1 a_1 + (\gamma_2 - 1)a_2 + (2 - \gamma_1 - \gamma_2)a_3$$

(5.64a)

$$b_3 = \gamma_1 a_1 + \gamma_2 a_2 + (1 - \gamma_1 - \gamma_2)a_3$$

The block diagram in Fig. 5-55 shows a possible realisation of (5.64a). Two multipliers and six adders are required. From (5.64a) we can derive the following scattering matrix:

$$S = \begin{pmatrix} \gamma_1 - 1 & \gamma_2 & 2 - \gamma_1 - \gamma_2 \\ \gamma_1 & \gamma_2 - 1 & 2 - \gamma_1 - \gamma_2 \\ \gamma_1 & \gamma_2 & 1 - \gamma_1 - \gamma_2 \end{pmatrix} . \tag{5.64b}$$

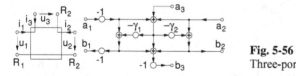

Fig. 5-56
Three-port serial adapter

The three-port serial adapter can be treated in the same way. According to Fig. 5-56 we can write down the following node and mesh equations:

$$i_1 = -i_2 = -i_3$$
$$u_1 - u_2 - u_3 = 0 .$$

Again using the definition of the wave parameters (5.52), the equations describing the respective port terminations and the abbreviations

$$\gamma_1 = \frac{2R_1}{R_1 + R_2 + R_3} \qquad\qquad \gamma_2 = \frac{2R_2}{R_1 + R_2 + R_3}$$

finally result in the following relations between the incident and reflected waves:

$$b_1 = (1 - \gamma_1)a_1 + \gamma_1 a_2 + \gamma_1 a_3$$
$$b_2 = \gamma_2 a_1 + (1 - \gamma_2)a_2 - \gamma_2 a_3 \tag{5.65a}$$
$$b_3 = (2 - \gamma_1 - \gamma_2)a_1 - (2 - \gamma_1 - \gamma_2)a_2 - (1 - \gamma_1 - \gamma_2)a_3 .$$

The transfer behaviour of the adapter is fully determined again by the two coefficients γ_1 and γ_2. The block diagram in Fig. 5-56 depicts a possible implementation of (5.65a). Two multipliers and six adders are required. The following scattering matrix can be derived from (5.65a):

$$S = \begin{pmatrix} 1 - \gamma_1 & \gamma_1 & \gamma_1 \\ \gamma_2 & 1 - \gamma_2 & -\gamma_2 \\ 2 - \gamma_1 - \gamma_2 & -2 + \gamma_1 + \gamma_2 & -1 + \gamma_1 + \gamma_2 \end{pmatrix} . \tag{5.65b}$$

5.6.2.3 The Reflectance of the Capacitor and the Inductor

What is missing in the wave flowgraph of Fig. 5-54 to complete the picture are the one-ports that describe the reflection behaviour of the capacitors and the inductor. The current and voltage of these elements are related by the respective impedances $j\omega L$ and $1/j\omega C$.

$$u_L = j\omega L i_L \qquad\qquad i_C = j\omega C u_C$$

Using (5.52a) and assuming an arbitrary port resistance, R_{Port} yields the following ratios of reflected to incident waves:

$$\frac{b_L}{a_L} = \frac{u_L - R_{Port} i_L}{u_L + R_{Port} i_L} = \frac{j\omega - R_{Port}/L}{j\omega + R_{Port}/L} \qquad\qquad (5.66a)$$

$$\frac{b_C}{a_C} = \frac{u_C - R_{Port} i_C}{u_C + R_{Port} i_C} = \frac{1/(R_{Port}C) - j\omega}{1/(R_{Port}C) + j\omega} \; . \qquad\qquad (5.66b)$$

The transfer functions (5.66) describe the behaviour of first-order all-pass filters. The magnitude of these transfer functions is unity. The phase can be expressed as:

$$b_L(\omega) = 2\arctan(\omega L / R_{Port}) + \pi \qquad\qquad (5.67a)$$

$$b_C(\omega) = 2\arctan(\omega R_{Port}C) \; . \qquad\qquad (5.67b)$$

(Note: In the literature, reflected waves and the phase response of a system are both denoted by the letter b!)

The group delay of the all-pass filter is the derivative of the phase with respect to the frequency ω.

$$\tau_L(\omega) = \frac{db_L(\omega)}{d\omega} = \frac{2L/R_{Port}}{1 + (\omega L / R_{Port})^2} \qquad\qquad (5.68a)$$

$$\tau_C(\omega) = \frac{db_C(\omega)}{d\omega} = \frac{2R_{Port}C}{1 + (\omega R_{Port}C)^2} \qquad\qquad (5.68b)$$

The ladder filter of Fig. 5-40 can therefore be realised using three first-order all-pass filters that are linked by three-port parallel and serial adapters (Fig. 5-57). In order to convert this continuous-time flowgraph into a discrete-time realisation of the filter, we have to find a discrete-time approximation of the all-pass filter.

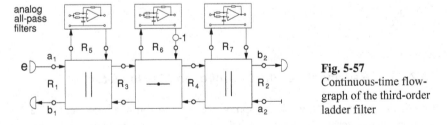

Fig. 5-57
Continuous-time flow-
graph of the third-order
ladder filter

The group delay according to (5.68) is relatively constant up to the cutoff frequency $\omega_c = 1/(R_{Port}C)$ or $\omega_c = R_{Port}/L$. At $\omega = 0$, the group delay amounts to

$\tau_g = 2R_{\text{Port}}C$ or $\tau_g = 2L/R_{\text{Port}}$. In the discrete-time domain, it is very easy to realise a constant delay corresponding to the sample period T. The port resistances R_{Port} are chosen now in such a way that the all-pass filters have a group delay of T at $\omega = 0$.

$$R_{\text{Port}L} = 2L/T \tag{5.69a}$$

$$R_{\text{Port}C} = T/2C \tag{5.69b}$$

If we replace the continuous-time all-pass filter by a delay of T, we obtain a discrete-time realisation of the ladder filter (see Fig. 5-58). A capacitor is thus represented by a delay in the flowgraph, an inductor by a delay plus an inverter.

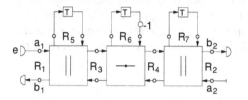

Fig. 5-58
Block diagram of the third-order wave digital filter

Which deviation with respect to the frequency response do we expect if the all-pass filter is replaced by a simple delay? The magnitude is constant and equals unity in both cases, but there is a difference with respect to the phase.

$$b_{\text{delay}}(\omega) = \omega T = 2\pi f T \tag{5.70a}$$

$$b_{\text{allpass}}(\omega) = 2\arctan(\omega T/2) = 2\arctan(\pi f T) \tag{5.70b}$$

Figure 5-59 compares both phase responses.

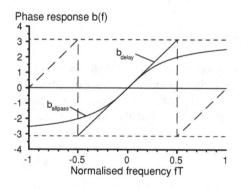

Phase response b(f)

Normalised frequency fT

Fig. 5-59
Phase of a first-order all-pass filter compared to the phase of a delay T

While the all-pass filter requires the whole frequency axis $-\infty \le f \le +\infty$ to pass over the phase range $-\pi$ to $+\pi$, the same is accomplished with the delay T in the frequency range $-f_s/2 \le f \le +f_s/2$. It can therefore be expected in the discrete-time realisation according to Fig. 5-58 that the frequency response of the analog reference filter is compressed to the range $-f_s/2 \le f \le +f_s/2$ and periodically continued. By comparison of (5.70a) and (5.70b) it is easy to understand that we obtain the frequency response of the discrete-time filter by replacing the frequency variable ω with the term

$$\omega' = \frac{2}{T}\tan(\omega T / 2) \tag{5.71}$$

Because of this distortion of the frequency axis, it is important to design the analog reference filter in such a way that characteristic frequencies, as the passband or stopband edges for instance, are shifted towards higher frequencies using (5.71). After compression of the frequency response in the discrete-time realisation, these frequencies will assume exactly the desired values. Fig. 5-60 compares the magnitude of the analog reference filter and of the corresponding wave digital filter implementation. The passband ripple remains unchanged. Because of the compression of the frequency axis, the cutoff frequency is shifted to a lower frequency and the slope of the magnitude plot in the transition band becomes steeper.

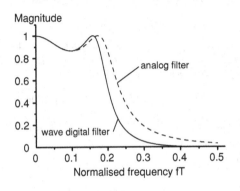

Fig. 5-60
Magnitude of a third-order ladder filter (analog and wave digital filter) Chebyshev 1.256 dB ripple, cutoff frequency $0.2 f_s$

5.6.2.4 The Role of the Bilinear Transform

Application of the so-called bilinear transform (5.72), which transforms a transfer function in the continuous-time frequency variable p into a corresponding transfer function in z, leads exactly to the behaviour described in the previous section: capacitors and inductors are converted into delays, and the frequency axis is distorted according to (5.71).

$$p = \frac{2}{T}\frac{z-1}{z+1} \tag{5.72}$$

This transform also plays an important role in the design of IIR filters, as we will show in Chap. 6.

Let us first apply this transform to an inductance L. Complex current and voltage are related as

$$U = pLI \ .$$

Substitution of (5.72) leads to the following relation in the z-domain:

$$U = \frac{2L}{T} I \frac{z-1}{z+1} \tag{5.73a}$$

$$U = RI\Psi . \tag{5.73b}$$

The term $2L/T$ has the dimension of a resistance and is determined by the value of the inductance L. The term $(z-1)/(z+1)$ is denoted by the letter ψ, which represents a new dimensionless frequency variable. Formally, we can now define the impedance of the inductor in the discrete-time domain as

$$Z_L = \frac{U}{I} = R\Psi = R\frac{z-1}{z+1} . \tag{5.74a}$$

With $R = T/2C$, we obtain for the capacitor

$$Z_C = \frac{U}{I} = R/\Psi = R\frac{z+1}{z-1} . \tag{5.74b}$$

In the following, we calculate the reflectance of the capacitor and inductor using the impedance definition (5.74). From (5.52a) we have

$$\frac{b}{a} = \frac{U - R_{\text{Port}}I}{U + R_{\text{Port}}I} = \frac{U/I - R_{\text{Port}}}{U/I + R_{\text{Port}}} = \frac{Z - R_{\text{Port}}}{Z + R_{\text{Port}}} .$$

Substitution of (5.74) yields

$$\frac{b_L}{a_L} = \frac{R\frac{z-1}{z+1} - R_{\text{Port}}}{R\frac{z-1}{z+1} + R_{\text{Port}}} = \frac{z\frac{R - R_{\text{Port}}}{R + R_{\text{Port}}} - 1}{z - \frac{R - R_{\text{Port}}}{R + R_{\text{Port}}}}$$

$$\frac{b_C}{a_C} = \frac{R\frac{z+1}{z-1} - R_{\text{Port}}}{R\frac{z+1}{z-1} + R_{\text{Port}}} = \frac{z\frac{R - R_{\text{Port}}}{R + R_{\text{Port}}} + 1}{z + \frac{R - R_{\text{Port}}}{R + R_{\text{Port}}}} .$$

If we choose R_{Port} equal to R in the impedance definition (5.74), we obtain the following results:

$$\frac{b_L}{a_L} = -z^{-1} \qquad \frac{b_C}{a_C} = z^{-1} .$$

Thus by appropriate choice of the port impedance, the bilinear transform turns capacitors and inductors into simple delays in the flowgraph. We consider now the frequency response of the impedances Z_L and Z_C. We replace z by $e^{j\omega T}$ in (5.74) and obtain

$$Z_L(\omega) = R\frac{e^{j\omega T} - 1}{e^{j\omega T} + 1} = jR\tan(\omega T / 2)$$

$$Z_C(\omega) = R\frac{e^{j\omega T} + 1}{e^{j\omega T} - 1} = -jR/\tan(\omega T / 2) \ .$$

The impedances of the capacitor and inductor are no longer directly or inversely proportional to the frequency ω, but are functions of $\tan(\omega T/2)$. The frequency response of these impedances and thus of the whole discrete-time filter is distorted according to (5.71). To sum up, we can conclude that the approximation of the analog first-order all-pass filter by a simple delay T has the same effect as applying the bilinear transform for the transition from the continuous-time to a discrete-time representation of the frequency-dependent impedances of the capacitor and inductor.

5.6.2.5 Observance of the Realisability Condition

Five of the seven port impedances in Fig. 5-58 are determined by the internal resistance of the source R_1, the terminating resistance of the network R_2 and the three elements of the ladder network C_1, C_2 and L. The choice of R_3 and R_4 seems to be arbitrary at first glance. It turns out, however, that they have to be determined in such a way that the filter becomes realisable. A look at the block diagrams of the three-port adapters (Fig. 5-55 and Fig. 5-56) shows that for each port there exists a direct path between input a_i and output b_i. If we directly connect two adapters, as is the case twice in our example, we find delayless loops in the overall filter structure. The problem can be solved by making at least one of the involved ports reflection-free. This means that no share of the incident wave a_i at port i reaches the output b_i. b_i must therefore be composed of shares of the incident waves a_j $(i \neq j)$ at the remaining ports only. A reflection-free port i manifests itself by a vanishing reflectance S_{ii} in the scattering matrix. With respect to the scattering matrix of the three-port parallel adapter (5.64), for instance, this means that for a reflection-free port 2 the parameter γ_2 becomes unity. The resulting scattering matrix (5.75) possesses only one free parameter γ_1.

$$S = \begin{pmatrix} \gamma_1 - 1 & 1 & 1 - \gamma_1 \\ \gamma_1 & 0 & 1 - \gamma_1 \\ \gamma_1 & 1 & -\gamma_1 \end{pmatrix} \tag{5.75}$$

The port impedance R_2 of the reflection-free port is no longer independent. It results from the condition $\gamma_2 = 1$ and is fixed by the other ports by the relation

$$R_2 = R_1 // R_3 = \frac{R_1 R_3}{R_1 + R_3} \ . \tag{5.76}$$

The adapter coefficient γ_1 is related to the independent coefficients R_1 and R_3 in the following way:

$$\gamma_1 = R_3/(R_1 + R_3) .\tag{5.77}$$

Fig. 5-61 shows symbol and block diagram of the corresponding three-port parallel adapter according to (5.75), with a reflection-free port 2. It is fully defined by the coefficient γ_1. The implementation requires one multiplier and four adders. If we track the signal paths in the block diagram, there is no longer a direct path between the input and output of port 2.

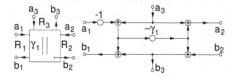

Fig. 5-61
Three-port parallel adapter with reflection-free port 2

Similar considerations can be made for the three-port serial adapter. If the parameter γ_2 of the scattering matrix (5.65) is unity, port 2 becomes reflection-free. The resulting scattering matrix (5.78) possesses only one free parameter γ_1.

$$S = \begin{pmatrix} 1-\gamma_1 & \gamma_1 & \gamma_1 \\ 1 & 0 & -1 \\ 1-\gamma_1 & \gamma_1-1 & \gamma_1 \end{pmatrix}\tag{5.78}$$

The port impedance R_2 of the reflection-free port is no longer independent. Because of the condition $\gamma_2 = 1$, it is fixed by the impedance of the other ports by the relation

$$R_2 = R_1 + R_3 .\tag{5.79}$$

The adapter coefficient γ_1 is related to the independent coefficients R_1 and R_3 in the following way:

$$\gamma_1 = R_1/(R_1 + R_3) .\tag{5.80}$$

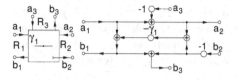

Fig. 5-62
Three-port serial adapter with reflection-free port 2

Fig. 5-62 shows symbol and block diagram of the corresponding three-port serial adapter according to (5.78) with a reflection-free port 2. It is fully defined by the coefficient γ_1. The implementation requires one multiplier and four adders. Again there is no longer a direct path between the input and output of port 2.

Example 5-5

Calculate the adapter coefficients for a third-order Chebyshev filter with 1.256 dB passband ripple and a cutoff frequency of $0.2 f_s$. The graph of the magnitude of this filter is shown in Fig. 5-60.

We take the following values for the capacitors and the inductor from the filter design table [57]:

$$C_1 = C_2 = 2{,}211315/(2\pi f_c R)$$
$$L = 0{,}947550\,R/2\pi f_c$$

R_1 and R_2 are chosen equal to the terminating resistors of the network. R_5, R_6, and R_7 are determined by the values of the capacitors, the inductor, and the cutoff frequency. R_3 and R_4 are chosen in such a way, that the respective ports of the parallel adapters become reflection-free.

$$R_1 = R_2 = R$$
$$R_5 = R_7 = T/2C_{1/2} = 0.284138\,R$$
$$R_6 = 2L/T = 1.508073R$$
$$R_3 = R_4 = R_1 \,//\, R_5 = R_2 \,//\, R_7 = 0.221267\,R \ .$$

In order to guarantee realisability, we choose parallel adapters with a reflection-free port. The serial adapter in the middle is independently configurable at all ports (Fig. 5-63).

$$\gamma_1 = \gamma_2 = R_5/(R_1 + R_5) = R_7/(R_2 + R_7) = 0.221267$$
$$\gamma_3 = \gamma_4 = 2R_3/(R_3 + R_4 + R_6) = 2R_4/(R_3 + R_4 + R_6) = 0.226870$$

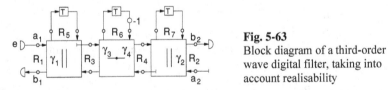

Fig. 5-63
Block diagram of a third-order wave digital filter, taking into account realisability

The block diagram shown in Fig. 5-63 can be further simplified. The serial adapter in the middle is symmetrical since $R_3 = R_4$ and hence $\gamma_3 = \gamma_4$. In this case, one coefficient γ is sufficient to completely describe its behaviour. Figure 5-64 depicts the structure of the symmetrical serial adapter. One multiplier and six adders are needed. Assuming $\gamma = \gamma_1 = \gamma_2$, we can derive the following scattering matrix from (5.65):

$$S = \begin{pmatrix} 1-\gamma & \gamma & \gamma \\ \gamma & 1-\gamma & -\gamma \\ 2-2\gamma & -2+2\gamma & -1+2\gamma \end{pmatrix} . \qquad\qquad (5.81)$$

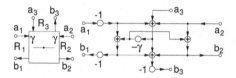

Fig. 5-64
Symmetrical three-port serial adapter

Figure 5-65 shows the block diagram of the corresponding symmetrical parallel adapter. One multiplier and six adders are required. Assuming again $\gamma = \gamma_1 = \gamma_2$, it follows from (5.64):

$$S = \begin{pmatrix} \gamma-1 & \gamma & 2-2\gamma \\ \gamma & \gamma-1 & 2-2\gamma \\ \gamma & \gamma & 1-2\gamma \end{pmatrix}.$$

(5.82)

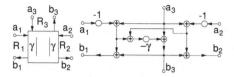

Fig. 5-65
Symmetrical three-port parallel adapter

In summary, the flowgraph of the whole filter is composed of two three-port parallel adapters with a reflection-free port, one symmetrical three-port serial adapter and three delays. The implementation requires three multipliers, fourteen adders and three delays. It is noticeable that wave digital filters in general require more adders than the corresponding direct form.

5.6.2.6 Second-Order Wave Digital Filter Block

The characteristics of filters that are realised as a parallel or series arrangement of second-order filter blocks essentially depend on the properties of the partial filters. Apart from direct-form (Fig. 5-22) and normal-form filters (Fig. 5-39), second-order filter blocks derived from wave digital filters are a further option. These introduce, at least partially, the good properties of this filter type into the overall filter. Compared to a full realisation as a wave digital filter, the advantage of low parameter sensitivity gets partially lost. The good stability properties, however, are fully kept.

In order to derive a second-order filter block based on wave digital filters, we choose a simple resonant circuit, as depicted in Fig. 5-66, as the analog reference filter. This circuit acts as a bandpass filter with the following transfer function:

$$H(p) = \frac{R_2 Cp}{1 + (R_1 + R_2)Cp + LCp^2}.$$

By means of two three-port serial adapters, this filter can be easily converted into a wave digital filter structure (Fig. 5-67). Using the scattering matrix (5.78)

Fig. 5-66
Second-order bandpass filter

we can derive the following relationships between the state variables z_1 and z_2, the signals v_1 and v_2 in the connecting branches and the input and output signals of the filter x and y.

$$z_1(n+1) = (1-\gamma_1)x(n) + (\gamma_1-1)v_2(n) - \gamma_1 z_1(n)$$
$$z_2(n+1) = (\gamma_2-1)v_1(n) + \gamma_2 z_2(n)$$
$$v_1(n) = x(n) + z_1(n) \qquad v_2(n) = -z_2(n)$$
$$y(n) = \gamma_2 v_1(n) + \gamma_2 z_2(n)$$

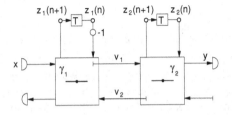

Fig. 5-67
Wave digital filter flowgraph of a second-order resonant circuit

By elimination of v_1 and v_2 we obtain a state-space representation of the filter that can be easily represented by a block diagram (Fig. 5-68).

$$z_1(n+1) = z_2(n) - \gamma_1\big(z_1(n) + z_2(n) + x(n)\big) + x(n)$$
$$z_2(n+1) = -z_1(n) + \gamma_2\big(z_1(n) + z_2(n) + x(n)\big) - x(n) \qquad (5.83)$$
$$y(n) = \gamma_2\big(z_1(n) + z_2(n) + x(n)\big)$$

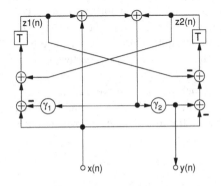

Fig. 5-68
Block diagram of a second-order band-pass filter based on wave digital filters

The following state-space coefficients can be derived from (5.83):

$$A = \begin{pmatrix} -\gamma_1 & 1-\gamma_1 \\ \gamma_2 - 1 & \gamma_2 \end{pmatrix} \qquad b = \begin{pmatrix} 1-\gamma_1 \\ \gamma_2 - 1 \end{pmatrix} \qquad c^{\mathrm{T}} = \begin{pmatrix} \gamma_2 & \gamma_2 \end{pmatrix} \qquad d = \gamma_2 .$$

Using (5.43) we can calculate the transfer function of this filter.

$$H(z) = \gamma_2 \frac{z^2 - 1}{z^2 + (\gamma_1 - \gamma_2)z + 1 - \gamma_1 - \gamma_2}$$

Figure 5-68 can be easily extended to cover the more general case of a second-order filter block that is characterised by the coefficients b_0, b_1 and b_2 in the numerator polynomial of the transfer function.

$$H(z) = \frac{b_0 z^2 + b_1 z + b_2}{z^2 + a_1 z + a_2}$$

Figure 5-69 shows the corresponding block diagram with the additional coefficients β_0, β_1 and β_2. The following relationships exist between the coefficients of the block diagram and those of the transfer function:

$$b_1 = \beta_1 - \beta_2 \qquad\qquad b_2 = \beta_0 - \beta_1 - \beta_2 \qquad\qquad b_0 = \beta_0$$
$$a_1 = \gamma_1 - \gamma_2 \qquad\qquad a_2 = 1 - \gamma_1 - \gamma_2 .$$

In Chap. 8 we will investigate in more detail the properties of this filter structure [53].

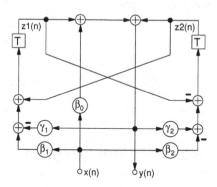

Fig. 5-69
A general second-order filter block based on wave digital filters

5.7 Allpass Filters

Allpass filters are useful building blocks in digital signal processing. An allpass system is a system whose magnitude response is constant for all frequencies, i.e.,

$$\left| H(e^{j\omega T}) \right| = c, \qquad\qquad \text{for all } \omega .$$

An obvious application of allpass networks is the equalisation of phase or group delay distortion without affecting the magnitude response. Selective IIR filters, in particular, feature a significant rise of the group delay in the vicinity of the cutoff frequencies. Equalisation means in this context to add extra delay in those frequency ranges where the given delay characteristic exhibits low values. In order to preserve the shape of a signal in the time domain, the resulting group delay response should be as flat as possible in the passband of the filter.

The implementation of selective filters with special properties is another wide field of applications for which allpass filters form the basis. Examples are notch filters, tunable filters, complementary filters and filter banks. Because of the special property of allpass networks that the gain is strictly limited to unity, filters can be designed which exhibit low sensitivity to coefficient imprecision [64].

Delays are a special kind of allpass systems. It is trivial to implement delays in the digital domain as long as the delay time is an integer multiple of the sampling period T. Merely a memory is required with read and write access simulating the behaviour of a shift register. In the analog domain, delays have been realised with devices such as magnetic tape loops or bucket-brigade circuits. Today's digital implementations, based on digital signal processors with A/D and D/A converters as peripherals, are superior over these analog solutions with respect to cost and performance parameters like bandwidth, linearity, and signal-to-noise ratio. Delays are used in a variety of applications in digital audio and video processing. Effects like echo, reverberation, or phasing, which must not be missing in any music production to enrich the sound, are based on delays.

Frequency, magnitude, phase, and group delay response of a delay by M sampling periods can be written as

$$H(e^{j\omega T}) = e^{-j\omega M T}$$
$$\left| H(e^{j\omega T}) \right| = 1$$
$$b(\omega) = \omega M T$$
$$\tau_g(\omega) = M T \quad .$$
(5.84)

If M is an integer, the delay can be realised by a z-transfer function of the form

$$H(z) = z^{-M} \quad .$$

A non-integer delay needs to make visible values of the signal occurring between the actual sampling instants. These do not really exist and have to be interpolated in an expedient manner. One possible way to achieve this is to use an IIR allpass filter which is designed to approximate an arbitrary, possibly non-integer delay according a given optimisation criterion.

Not only constant delay but also constant phase is of interest in digital signal processing. The ideal 90° phase shifter, also called the Hilbert transformer, is the basic building block to form so-called complex or analytic signals which consist of the original signal as the real part and the 90°-shifted signal, also called the quadrature signal, as the imaginary part. Applications that make use of analytic signal processing are among others: digital communication systems, modems, radar systems, antenna beamforming, and single sideband modulators.

5.7.1 Allpass Transfer Function

If we reverse the sequence of the coefficients of a polynomial, we obtain a so-called mirror-image polynomial. In Sect. 5.2.3, we demonstrated that mirror-image pairs of polynomials have identical magnitude responses but very different phase responses. So if we choose one of these polynomials as the denominator of a rational fractional function and the mirror-image counterpart as the according numerator polynomial, we obtain a function with unity magnitude but frequency-dependent phase. A filter with such a frequency characteristic is commonly referred to as an allpass filter. The transfer function of an allpass filter has the general form

$$H_A(z) = \frac{\sum_{i=0}^{N} a_i z^{-(N-i)}}{\sum_{i=0}^{N} a_i z^{-i}} \ .$$

By factoring out the term z^{-N} in the numerator, we obtain

$$H_A(z) = \frac{z^{-N} \sum_{i=0}^{N} a_i z^{i}}{\sum_{i=0}^{N} a_i z^{-i}} \ . \tag{5.85}$$

If we denote the denominator polynomial by $D(z)$ with

$$D(z) = \sum_{i=0}^{N} a_i z^{-i} \ ,$$

we can express the transfer function of an allpass filter as

$$H_A(z) = \frac{z^{-N} D(1/z)}{D(z)} \ . \tag{5.86}$$

In order to obtain a stable filter, all zeros of $D(z)$, i.e. the poles of $H(z)$, must lie within the unit circle in the z-plane. We showed in Sect. 5.2.3 that mirror-image polynomials have reciprocal conjugate zero pairs in the z-plane. Since all poles (zeros of the denominator) need to lie within the unit circle, all zeros of the numerator polynomial must lie outside the unit circle. Figure 5-70 shows possible locations of poles and zeros of an allpass filter.

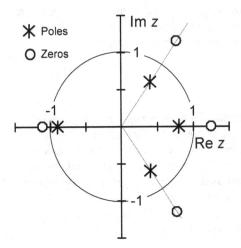

Fig. 5-70
Possible locations of poles and zeros of
allpass filters

Let us denote the phase response of the denominator by $b_D(\omega)$. Since the numerator polynomial has reciprocal conjugate zeros with respect to the denominator, the phase of the numerator $b_N(\omega)$ can be expressed according to (5.21) as

$$b_N(\omega) = N\omega T - b_D(\omega)$$

where N is the order of the polynomial. Since the phase response of the allpass filter $b_A(\omega)$ is

$$b_A(\omega) = b_N(\omega) - b_D(\omega) \; ,$$

$b_A(\omega)$ is related to the phase of the denominator as

$$b_A(\omega) = N\omega T - 2b_D(\omega) \; . \tag{5.87}$$

As the zeros of $D(z)$ all lie within the unit circle, the phase $b_D(\omega)$ will oscillate symmetrically about zero with a total phase shift of zero as the frequency ω goes from 0 to $2\pi/T$. For the phase response of the allpass filter, this means according to (5.87) that the overall phase shift is $2\pi N$ as ω goes from 0 to $2\pi/T$. If the phase shift of a Nth-order allpass filter is less than $2\pi N$, it can be concluded that there must be poles lying outside the unit circle. Such a filter would thus be unstable. Substitution of (5.5) in (5.87) yields a relation which allows the calculation of the phase response from the coefficients a_i of the filter.

$$b_A(\omega) = N\omega T - 2\arctan \frac{\displaystyle\sum_{i=0}^{N} a_i \sin i\omega T}{\displaystyle\sum_{i=0}^{N} a_i \cos i\omega T} \tag{5.88}$$

Differentiation of (5.87) with respect to the frequency ω yields the group delay of the allpass filter.

$$\tau_{gA}(\omega) = NT - 2\tau_{gD}(\omega) \tag{5.89}$$

Substitution of (3.31) in (5.89) results in an expression which allows the calculation of the group delay from the denominator polynomial of the allpass transfer function.

$$\tau_{gA}(\omega) = NT + 2T\,\mathrm{Re}\frac{zD'(z)}{D(z)}\bigg|_{z=e^{j\omega T}} \tag{5.90}$$

The first-order allpass filter possesses a pole/zero pair on the real z-axis. The denominator polynomial can be expressed as $D(z) = 1 - rz^{-1}$ where $|r| < 1$. Using (5.86), the transfer function assumes the form

$$H(z) = \frac{z^{-1}D(1/z)}{D(z)} = z^{-1}\frac{1-rz}{1-rz^{-1}} \ .$$

The frequency response is obtained by substituting $e^{j\omega T}$ for z.

$$H(e^{j\omega T}) = e^{-j\omega T}\frac{1-re^{j\omega T}}{1-re^{-j\omega T}} = e^{-j\omega T}\frac{1-r\cos\omega T - jr\sin\omega T}{1-r\cos\omega T + jr\sin\omega T} \tag{5.91}$$

The phase response can be expressed according to (5.87) as

$$b(\omega) = \omega T - 2b_D(\omega)$$

$$b(\omega) = \omega T + 2\arctan\frac{r\sin\omega T}{1-r\cos\omega T} \ . \tag{5.92}$$

Figure 5-71 shows the phase response of the first-order allpass filter for various values of the pole radius r.

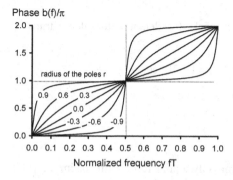

Fig. 5-71
Phase response of the first-order allpass filter with the pole radius r as a parameter

The group delay response is calculated using (5.90).

$$\tau_g(\omega) = T + 2T \left. \mathrm{Re}\frac{zD'(z)}{D(z)}\right|_{z=e^{j\omega T}} = T + 2T \left. \mathrm{Re}\frac{z(rz^{-2})}{1-rz^{-1}}\right|_{z=e^{j\omega T}}$$

$$\tau_g(\omega) = T \left. \mathrm{Re}\frac{1+rz^{-1}}{1-rz^{-1}}\right|_{z=e^{j\omega T}} = T \left. \mathrm{Re}\frac{(1+rz^{-1})(1-rz)}{(1-rz^{-1})(1-rz)}\right|_{z=e^{j\omega T}}$$

$$\tau_g(\omega) = T \left. \mathrm{Re}\frac{1-r^2+r(z^{-1}-z)}{1+r^2-r(z+z^{-1})}\right|_{z=e^{j\omega T}} = T\,\mathrm{Re}\frac{1-r^2-2jr\sin\omega T}{1+r^2-2r\cos\omega T}$$

$$\tau_g(\omega) = T\frac{1-r^2}{1+r^2-2r\cos\omega T} \tag{5.93}$$

Figure 5-72 shows the group delay response of the first-order allpass filter. For poles on the positive real z-axis, the group delay has a peak at $\omega = 0$, for poles on the negative real axis, the maximum occurs at $\omega = \pi/T$.

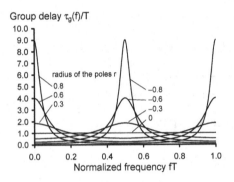

Group delay $\tau_g(f)/T$

radius of the poles r
0.8
0.6
0.3
−0.8
−0.6
−0.3
0

Normalized frequency fT

Fig. 5-72
Group delay response of the first-order allpass filter with the pole radius r as a parameter

The maximum value of the group delay is calculated as

$$\tau_{g\,max} = \frac{1+|r|}{1-|r|}\,.$$

For the second-order allpass building block, the denominator of the transfer function has the form

$$D(z) = (1 - r_\infty\,e^{j\varphi_\infty}\,z^{-1})(1 - r_\infty\,e^{-j\varphi_\infty}\,z^{-1}) = 1 - 2r_\infty\cos\varphi_\infty\,z^{-1} + r_\infty^2 z^{-2}\,.$$

Using (5.86), we get the transfer function

$$H(z) = \frac{z^{-2} - 2r_\infty\cos\varphi_\infty\,z^{-1} + r_\infty^2}{1 - 2r_\infty\cos\varphi_\infty\,z^{-1} + r_\infty^2 z^{-2}}$$

As expected, numerator and denominator polynomial form a mirror-image pair. For the calculation of phase and group delay, the frequency response is advantageously written in the form

$$H(e^{j\omega T}) = e^{-2j\omega T}\frac{(1 - r_\infty e^{j\varphi_\infty}e^{j\omega T})(1 - r_\infty e^{-j\varphi_\infty}e^{j\omega T})}{(1 - r_\infty e^{j\varphi_\infty}e^{-j\omega T})(1 - r_\infty e^{-j\varphi_\infty}e^{-j\omega T})}$$

$$H(e^{j\omega T}) = e^{-j\omega T}\frac{1 - r_\infty e^{+j(\omega T - \varphi_\infty)}}{1 - r_\infty e^{-j(\omega T - \varphi_\infty)}}\; e^{-j\omega T}\frac{1 - r_\infty e^{+j(\omega T + \varphi_\infty)}}{1 - r_\infty e^{-j(\omega T + \varphi_\infty)}}\;.$$

This equation consists of two terms which look quite similar as the corresponding expression for the first-order filter (5.91). Merely the normalised frequency variable ωT is shifted by $\pm\varphi_\infty$. So we can directly write down the corresponding equations for the calculation of the phase response and of the group delay response of the second-order filter.

$$b(\omega) = 2\omega T + 2\arctan\frac{r_\infty \sin(\omega T - \varphi_\infty)}{1 - r_\infty \cos(\omega T - \varphi_\infty)} + 2\arctan\frac{r_\infty \sin(\omega T + \varphi_\infty)}{1 - r_\infty \cos(\omega T + \varphi_\infty)}$$

$$(5.94)$$

Figure 5-73 shows graphs of the phase response for various values of the pole radius r_∞. The total phase shift amounts to 4π as the frequency goes from 0 to up to the sampling frequency $2\pi/T$. The curves oscillate symmetrically about a linear-phase straight line. The deviations from the linear shape are higher the closer the pole radius approaches unity. The slope of the phase curve is highest at the pole frequency φ_∞/T.

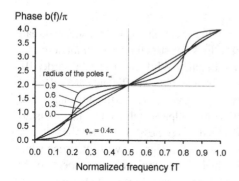

Fig. 5-73
Phase response of a second-order allpass filter with the pole radius r_∞ as a parameter, $\varphi_\infty = 0.4\pi$

The group delay response of the second-order allpass filter can be directly derived from (5.93). We simply have to sum up two versions of (5.93) with the normalised frequency axis shifted by $+\varphi_\infty$ and $-\varphi_\infty$ respectively.

$$\tau_g(\omega) = T\left[\frac{1 - r_\infty^2}{1 + r_\infty^2 - 2r_\infty \cos(\omega T - \varphi_\infty)} + \frac{1 - r_\infty^2}{1 + r_\infty^2 - 2r_\infty \cos(\omega T + \varphi_\infty)}\right] \quad (5.95)$$

Graphs of the group delay for various pole radii r_∞ are shown in Fig. 5-74. The maximum of the curves occurs at the pole frequency φ_∞/T.

Higher-order allpass filters can always be separated into first- and second-order sections. As long as the pole radius is less than unity, i.e. the filter section is stable, (5.93) and (5.95) are always positive. Since the group delay contributions sum up in a cascade arrangement, it can be generally stated that the group delay of a stable allpass filter is always positive. This is equivalent to the observation that the phase response curves of stable allpass filters are monotonously increasing.

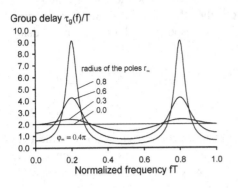

Group delay $\tau_g(f)/T$

Fig. 5-74
Group delay response of a second-order allpass filter with the pole radius r_∞ as a parameter, $\varphi_\infty = 0.4\pi$

5.7.2 Allpass Filter Structures

All filter structures, introduced in the previous sections, can be used to create allpass filters. But it is of special interest to look for structures which make use of the fact that the filter coefficients are equal in pairs. While a general Nth-order IIR filter is determined by $2N+1$ independent coefficients, N coefficients are sufficient to fully specify an Nth-order allpass filter. Such structures have the advantage that, regardless of the choice of the coefficients, allpass behaviour, i.e. constant magnitude response, is always guaranteed. Needless to say, the chosen coefficients must result in stable filters. In the following, we will introduce some of the common allpass filter structures which are canonical with respect to the number of (coefficient) multiplications. In some cases, these structures are also canonical regarding the number of delays equivalent to the memory requirements of the implementation.

5.7.2.1 Direct form

Figure 5-75 shows an Nth-order direct form I filter with allpass coefficients forming a mirror-image pair. If the recursive part of the filter is sketched upside down, it becomes obvious that in each case two signals are added which are multiplied by the same coefficient.

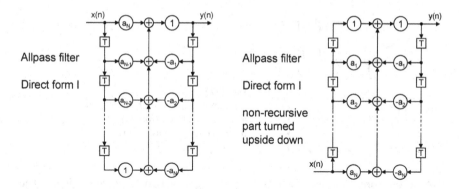

Fig. 5-75   Direct form allpass filter structure

We can now factor out the coefficients. This means that the two signals are first added and subsequently multiplied by the respective coefficient. Figure 5-76 shows the resulting structure which is canonical with respect to the number multiplications. The implementation requires N multipliers, $2N$ delays and $2N$ two-input adders.

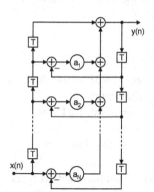

Fig. 5-76
Allpass filter structure with reduced number of multipliers

Higher order filters are often realised as a cascade arrangement of first- and second-order filter sections which are therefore of special interest. A first-order allpass filter is fully defined by one coefficient a_1. The corresponding transfer function can be expressed as

$$H(z) = \frac{a_1 + z^{-1}}{1 + a_1 z^{-1}} \, . \tag{5.96}$$

The block diagram is easily derived from Fig. 5-76 by removing all branches except for the one with the coefficient a_1. Fig 5-77a shows the resulting structure. Fig. 5-77b is simply a redrawn version of the same diagram. Fig. 5-77c shows the corresponding transposed structure.

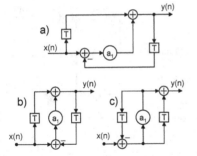

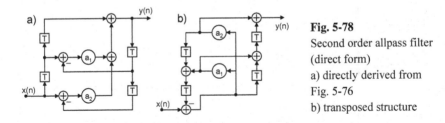

Fig. 5-77
First order allpass filter (direct form)
a) directly derived from Fig. 5-76
b) the same structure redrawn
c) transposed structure

A second order allpass filter is fully defined by two filter coefficients. The transfer function is of the form

$$H(z) = \frac{a_2 + a_1 z^{-1} + z^{-2}}{1 + a_1 z^{-1} + a_2 z^{-2}} \; .$$

(5.97)

The corresponding block diagram is derived from Fig. 5-76 by removing all branches except for those with the coefficients a_1 and a_2. Fig 5-78 shows the resulting structure and the according transposed version.

Fig. 5-78
Second order allpass filter
(direct form)
a) directly derived from
Fig. 5-76
b) transposed structure

The direct-form structures derived in this section are not canonical with respect to the required number of delays. The method presented in the following allows the synthesis of further allpass filters with one or two coefficients. Under these we will also find structures managing with the minimum number of delays.

5.7.2.2 Multiplier Extraction

We consider an Nth-order allpass filter network consisting of N coefficient multipliers and a number of adders and delays. The idea is to extract the coefficient multipliers from the block diagram as illustrated in Fig. 5-79. As a result, we obtain a block with $N+1$ input/output pairs. N of these pairs are used to externally connect the N coefficient multipliers, the remaining pair forms the input and output of the filter. Since the multipliers are extracted, the remaining circuit consists only of adders and delays.

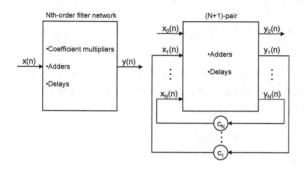

Fig. 5-79
Illustration of the multiplier
extraction approach

Since the $(N+1)$-pair network is linear, the output signals $y_j(n)$ are the sum of the reactions to the input signals $x_i(n)$. The transfer behaviour from input port i to output port j is described by the z-transfer function $h_{ji}(z)$. Equations (5.98) summarise the relations between input and output signals for a network with three input/output pairs.

$$Y_0(z) = X_0(z)h_{00}(z) + X_1(z)h_{01}(z) + X_2(z)h_{02}(z)$$
$$Y_1(z) = X_0(z)h_{10}(z) + X_1(z)h_{11}(z) + X_2(z)h_{12}(z) \tag{5.98}$$
$$Y_2(z) = X_0(z)h_{20}(z) + X_1(z)h_{21}(z) + X_2(z)h_{22}(z)$$

In the context of our problem, these transfer functions must possess the following properties:
- The $h_{ij}(z)$ are only made up of adders and delays.
- The structure between input port $x_0(n)$ and output port $y_0(n)$ must exhibit allpass behaviour.

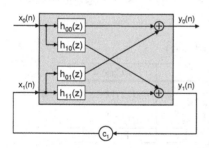

Fig. 5-80
General structure of a first-order allpass filter
with one coefficient multiplier

As an example, we will derive first-order allpass structures with one filter coefficient. We take Fig. 5-80 as the basis for the following derivation. The input ports X_0 and X_1 and the output ports Y_0 and Y_1 are related as

$$Y_0(z) = h_{00}(z)\,X_0(z) + h_{01}(z)\,X_1(z)$$
$$Y_1(z) = h_{10}(z)\,X_0(z) + h_{11}(z)\,X_1(z) \tag{5.99}$$
$$X_1(z) = c_1\,Y_1(z)\;.$$

By elimination of X_1 and Y_1 we obtain the transfer function of the filter structure.

$$Y_0(z) = \frac{h_{00}(z) + c_1\left[h_{10}(z)\,h_{01}(z) - h_{11}(z)\,h_{00}(z)\right]}{1 - c_1\,h_{11}(z)} X_0(z) \tag{5.100}$$

Comparison of (5.100) with the transfer function of a first-order allpass filter (5.96) yields relations to determine the transfer functions $h_{ij}(z)$ in such a way that Fig. 5-80 turns into an allpass filter.

$$H(z) = \frac{h_{00}(z) + c_1\left[h_{10}(z)\,h_{01}(z) - h_{11}(z)\,h_{00}(z)\right]}{1 - c_1\,h_{11}(z)} = \frac{z^{-1} + a_1}{1 + a_1\,z^{-1}}$$

Comparison of numerator and denominator leads to

$$h_{00}(z) = z^{-1} \tag{5.101a}$$

$$h_{11}(z) = -z^{-1} \tag{5.101b}$$

$$h_{10}(z)\,h_{01}(z) - h_{11}(z)\,h_{00}(z) = 1 \tag{5.101c}$$

$$c_1 = a_1$$

$h_{00}(z)$ and $h_{11}(z)$ are simple unit delays. Substitution of (5.101a) and (5.101b) in (5.101c) results in the following relationship between $h_{10}(z)$ and $h_{01}(z)$:

$$h_{10}(z)\,h_{01}(z) = 1 - z^{-2} = (1 - z^{-1})(1 + z^{-1}) \tag{5.102}$$

From Eq. (5.102), we can derive four possible combinations of transfer functions $h_{10}(z)$ and $h_{01}(z)$ which result in allpass transfer functions. Table 5-2 summarises these combinations and indicates the figures in which the according allpass filter block diagrams can be found.

Table 5-2　Four possible combinations of $h_{10}(z)$ and $h_{01}(z)$

Case 1	Case 2	Case 3	Case 4
$h_{10}(z) = 1$	$h_{10}(z) = 1 - z^{-2}$	$h_{10}(z) = 1 - z^{-1}$	$h_{10}(z) = 1 + z^{-1}$
$h_{01}(z) = 1 - z^{-2}$	$h_{01}(z) = 1$	$h_{01}(z) = 1 + z^{-1}$	$h_{01}(z) = 1 - z^{-1}$
Fig. 5-77c	Fig. 5-77b	Fig. 5-81	Fig. 5-82

Case 1 and Case 2 in Table 5-2 lead to the filter structures shown in Fig. 5-77 which we derived from the direct form. So we will have a closer look at Case 3 and Case 4 which lead to new structures.

For case 3, we have

$$h_{10}(z) = 1 - z^{-1}, \quad h_{01}(z) = 1 + z^{-1},$$
$$h_{00}(z) = z^{-1} \text{ and } h_{11}(z) = -z^{-1}.$$

Substitution in (5.99) yields

$$Y_0(z) = z^{-1}X_0(z) + (1 + z^{-1})X_1(z) = z^{-1}(X_0(z) + X_1(z)) + X_1(z)$$
$$Y_1(z) = (1 - z^{-1})X_0(z) - z^{-1}X_1(z) = -z^{-1}(X_0(z) + X_1(z)) + X_0(z)$$
$$X_1(z) = a_1\, Y_1(z) \ .$$

By introduction of a new internal signal $s(n)$, we obtain the following set of equations which can be easily mapped into a block diagram.

$$Y_0(z) = S(z) + X_1(z)$$
$$Y_1(z) = -S(z) + X_0(z)$$
$$S(z) = z^{-1}(X_0(z) + X_1(z))$$
$$X_1(z) = a_1\, Y_1(z) \ .$$

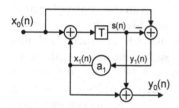

Fig. 5-81
Structure of a first-order canonical allpass filter

Fig. 5-81 shows the resulting structure which is made up of 3 two-input adders/subtractors, 1 delay and 1 coefficient multiplier. The filter is canonical with respect to both the number of delays and the number of coefficient multipliers.

Case 4 can be treated in the same way. As a result, we obtain the structure in Fig. 5-82 which is the transposed version of Fig. 5-81. Apart from the position of a minus sign, both block diagrams are identical

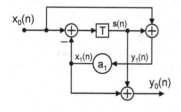

Fig. 5-82
Structure of the transposed first-order canonical allpass filter

Second-order allpass filters can be synthesised in the same way. Starting point is a structure with two extracted coefficient multipliers as shown in Fig. 5-83. The input and output ports are related as

$$Y_0(z) = h_{00}(z)\, X_0(z) + h_{01}(z)\, X_1(z) + h_{02}(z)\, X_2(z)$$
$$Y_1(z) = h_{10}(z)\, X_0(z) + h_{11}(z)\, X_1(z) + h_{12}(z)\, X_2(z)$$
$$Y_2(z) = h_{20}(z)\, X_0(z) + h_{21}(z)\, X_1(z) + h_{22}(z)\, X_2(z)$$
$$X_1(z) = c_1\, Y_1(z) \qquad\qquad X_2(z) = c_2\, Y_2(z) \ .$$

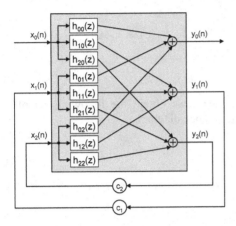

Fig. 5-83
General structure of a second-order allpass filter with two coefficient multipliers

It is shown in [54] that 24 different allpass structures with the transfer function (5.97) can be derived from this set of equations where twelve structures are the transposed versions of the remaining twelve. The direct-form structures shown in Fig. 5-78 are contained in this set. As a further result of the above method, we show in Fig. 5-84 the block diagram of a canonical filter only requiring 2 coefficient multipliers, 2 delays and 4 two-input adders. Besides the hardware requirements, the roundoff noise generated in the filter is another important criterion for the choice of the appropriate structure. [54] includes a thorough analysis of the output noise generated by the various realisations. The methods to determine quantisation noise generated in a filter network are treated in detail in Chap. 8 of this book.

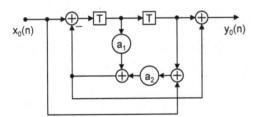

Fig. 5-84
Structure of a second-order canonical allpass filter

5.7.2.3 Wave Digital Filter

The implementation of allpass filters based on wave digital filter (WDF) structures is straightforward. Any one-port representing a lossless analog reference network is an allpass filter if we choose the incident wave port a as the input and the reflected wave port b as the output of the filter. Since the network is lossless, no energy is dissipated inside. The incident power must therefore be equal to the reflected power. With R denoting the port resistance of the one-port, we have

$$\frac{a^2}{R} = \frac{b^2}{R}$$

or

$$|a| = |b|.$$

So a and b can only differ with respect to the phase.

Any lossless electrical reference network consisting of inductors and capacitors turns into a WDF structure made up of adaptors and delays. In the following we will look at structures that implement first-order, second-order and higher-order allpass filters.

The simplest configuration consists of a two-port adaptor and a delay as shown in Fig. 5-85. In the analog reference domain, this corresponds to a capacitor connected to a voltage source with internal resistance R. The transfer function of the filter

$$H(z) = \frac{B_1(z)}{A_1(z)}$$

can be easily derived using the relations between incident and reflected waves characterising a two-port adaptor (5.61).

$$B_1 = -r A_1 + (1 + r) A_2$$
$$B_2 = (1 - r) A_1 + r A_2$$

The second port is terminated by a delay which can be expressed as

$$A_2 = B_2 z^{-1}.$$

By elimination of A_2 and B_2 we obtain

$$H(z) = \frac{B_1(z)}{A_1(z)} = -\frac{r - z^{-1}}{1 - rz^{-1}},$$

which is an allpass transfer function.

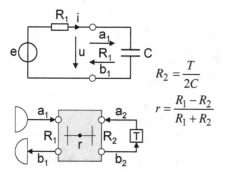

$$R_2 = \frac{T}{2C}$$

$$r = \frac{R_1 - R_2}{R_1 + R_2}$$

Fig. 5-85
First-order allpass filter based on the reflectance of a capacitor

Figure 5-86 shows the resulting block diagram in detail. The network consists of 1 coefficient multiplier, 1 delay and 3 two-input adders/subtractors. The realisation is canonical with respect to the number of multipliers and delays. In the same way, a first-order allpass filter can be realised based on the reflectance of an inductor. The only modification in the block diagram is an additional inverter in the delay path. The corresponding z-transfer function reads as

$$H(z) = -\frac{r+z^{-1}}{1+rz^{-1}} \ .$$

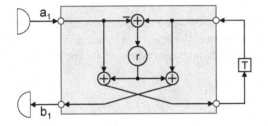

Fig. 5-86
WDF implementation of a first-order allpass filter in detail

The reflectance of lossless parallel and serial resonant circuits as shown in Fig. 5-87 is a second-order allpass transfer function. These circuits can therefore be used as the analog prototype networks for the derivation of block-diagrams of WDF second-order allpass filters. Serial and parallel connection of capacitor and inductor are realised in the wave quantity domain by means of three-port adaptors. We connect to port 1 the equivalent of a capacitor which is a delay and to port 2 the equivalent of an inductor which is a delay plus inverter. Incident and reflected waves at port 3 act as the input and output of the filter.

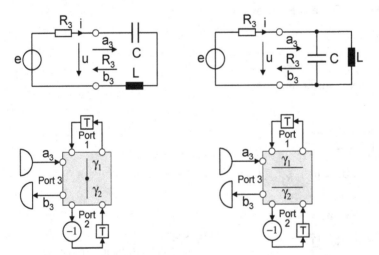

Fig. 5-87   Analog prototypes and discrete-time realisation of second-order allpass filters

According to (5.65a), incident and reflected waves of a serial three-port adapter are related as

$$B_1 = (1-\gamma_1)A_1 + \gamma_1 A_2 + \gamma_1 A_3$$
$$B_2 = \gamma_2 A_1 + (1-\gamma_2)A_2 - \gamma_2 A_3$$
$$B_3 = (2-\gamma_1-\gamma_2)A_1 - (2-\gamma_1-\gamma_2)A_2 - (1-\gamma_1-\gamma_2)A_3 .$$

Port 1 and port 2 are terminated by delays which can be expressed as

$$A_1 = z^{-1}B_1$$
$$A_2 = -z^{-1}B_2$$

By elimination of A_1, B_1, A_2, and B_2, we obtain the allpass transfer function of the implementation based on the serial adapter.

$$H_s(z) = \frac{B_3(z)}{A_3(z)} = \frac{(\gamma_1+\gamma_2-1)+(\gamma_1-\gamma_2)z^{-1}+z^{-2}}{1+(\gamma_1-\gamma_2)z^{-1}+(\gamma_1+\gamma_2-1)z^{-2}}$$

The circuit with the parallel three-port adapter leads to a similar result. Merely some signs are different.

$$H_p(z) = \frac{B_3(z)}{A_3(z)} = -\frac{(\gamma_1+\gamma_2-1)-(\gamma_1-\gamma_2)z^{-1}+z^{-2}}{1-(\gamma_1-\gamma_2)z^{-1}+(\gamma_1+\gamma_2-1)z^{-2}}$$

Another allpass structure, which can be easily extended towards higher filter orders, is based on the behaviour of transmission line sections used as the analog prototype of a wave digital filter implementation. It is well known from transmission line theory that lossless transmission lines with open or short circuit termination behave like capacitors or inductors depending on the frequency and the line length. The reflectance measured at the input of such a line section or of a serial arrangement of multiple lossless line segments must be an allpass transfer function as explained above.

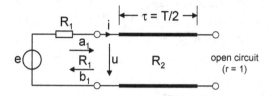

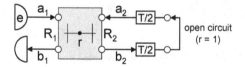

Fig. 5-88
Allpass filter derived from the reflectance of an open line

Figure 5-88 shows an open line with a one-way delay of $T/2$, connected to a voltage source with an internal resistance of R_1. The line impedance is chosen as R_2. It is easy to draw an equivalent wave flow diagram. The matching between R_1 and R_2 is realised with a two-port adaptor. Both waves a_2 and b_2 experience a delay of $T/2$ while traversing the line. At the open termination, the wave is completely reflected with equal sign. As a result we obtain the same WDF implementation as shown in Fig. 5-85 since the two delays of $T/2$ can be combined to one of T.

In the next step, we append another piece of line with impedance R_3 to the circuit above. The one-way delay of this line segment is again chosen as $T/2$. Fig. 5-89 shows the resulting analog circuit and the equivalent WDF structure. The two line segments are coupled in the wave quantity domain by another two-port adaptor which matches the impedances R2 and R3. Without changing the functionality of the circuit, we can combine again the delays of $T/2$ to one (Fig. 5-89).

In the following we derive the z-transfer function of this circuit which gives us the relationship between the two filter coefficients r_{12} and r_{34} and the coefficients of the transfer function. For the two-port adaptors, we can write down the following equations using (5.61).

$$B_1 = -r_{12} A_1 + (1 + r_{12}) A_2$$
$$B_2 = (1 - r_{12}) A_1 + r_{12} A_2$$
$$B_3 = -r_{34} A_3 + (1 + r_{34}) A_4$$
$$B_4 = (1 - r_{34}) A_3 + r_{34} A_4$$

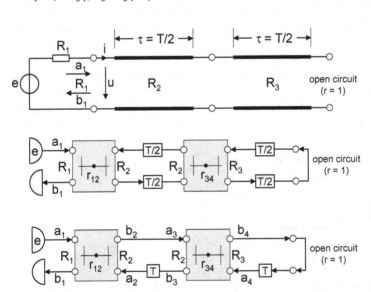

Fig. 5-89   Second-order allpass filter based on the reflectance of coupled line segments

The interconnection of ports yields further equations.

$$A_2 = B_3 \, z^{-1}$$
$$A_3 = B_2$$
$$A_4 = B_4 \, z^{-1}$$

By elimination of A_2, B_2, A_3, B_3, A_4, and B_4, we obtain the desired relationship.

$$H(z) = \frac{B_1}{A_1} = \frac{-r_{12} + r_{34}(r_{12} - 1)z^{-1} + z^{-2}}{1 + r_{34}(r_{12} - 1)z^{-1} - r_{12}z^{-2}} = \frac{a_2 + a_1 \, z^{-1} + z^{-2}}{1 + a_1 \, z^{-1} + a_2 \, z^{-2}}$$

The concept of coupled line segments can be easily extended to the realisation of allpass filter of arbitrary order.

The one port adaptor whose details are shown in Fig. 5-86 is not the only possible structure to implement allpass filters as a chain of adaptors and delays. Let us consider a general two-port with a scattering matrix given by

$$\begin{pmatrix} B_1 \\ B_2 \end{pmatrix} = \begin{pmatrix} S_{11} & S_{12} \\ S_{21} & S_{22} \end{pmatrix} \begin{pmatrix} A_1 \\ A_2 \end{pmatrix}$$

Port 2 of this two-port be terminated by a delay z^{-1} as depicted in Fig. 5-90

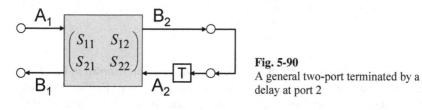

Fig. 5-90
A general two-port terminated by a delay at port 2

What are the constraints imposed on the scattering coefficients which finally result in an allpass transfer function $H(z) = B_1(z)/A_1(z)$? The following equations describe the behaviour of the two-port and of the terminating delay.

$$B_1 = S_{11}A_1 + S_{12}A_2$$
$$B_2 = S_{21}A_1 + S_{22}A_2$$
$$A_2 = B_2 z^{-1}$$

Elimination of A_2 and B_2 yields

$$H(z) = \frac{B_1(z)}{A_1(z)} = \frac{S_{11} + (S_{12}S_{21} - S_{11}S_{22})z^{-1}}{1 - S_{22}z^{-1}} \, . \tag{5.103}$$

An allpass transfer function requires that numerator and denominator polynomial form a mirror-image pair. For an allpass filter having a gain of +1 at $\omega = 0$ ($z = 1$), the mirror image condition calls for

$$S_{11} = -S_{22} \qquad \text{and}$$
$$S_{12}S_{21} - S_{11}S_{22} = 1.$$

(5.104)

With $k = S_{11}$ we have

$$S_{12}S_{21} = 1 - k^2.$$

(5.105)

Substitution of (5.104) and (5.105) in (5.103) yields the allpass transfer function

$$H(z) = \frac{k + z^{-1}}{1 + kz^{-1}}$$

So any two-port with a scattering matrix of the form

$$\begin{pmatrix} k & S_{12} \\ S_{21} & -k \end{pmatrix} \quad \text{with } S_{12}S_{21} = 1 - k^2$$

can be used to construct allpass filters by a cascade arrangement of two-ports and delays as in Fig. 5-89. Some possible combinations of the transmittance coefficients S_{12} and S_{21} are listed in the following.

1. $S_{12} = 1 - k^2$ $S_{21} = 1$
2. $S_{12} = 1$ $S_{21} = 1 - k^2$
3. $S_{12} = \sqrt{(1 - k^2)}$ $S_{21} = \sqrt{(1 - k^2)}$
4. $S_{12} = 1 - k$ $S_{21} = 1 + k$
5. $S_{12} = 1 + k$ $S_{21} = 1 - k$

The 5th option can be realised with the WDF two-port adaptor shown in Fig. 5-86 where k is related to the reflection coefficient r as $k = -r$. Further two-port implementations will be introduced in the context of the Gray and Markel Lattice/Ladder implementations in Sect. 5.7.3.1.

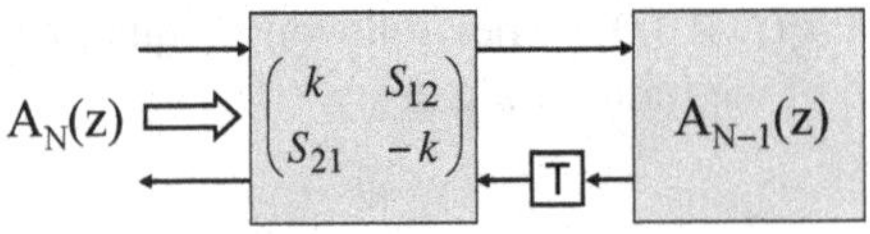

Fig. 5-91
Extraction of a first-order section from a Nth-order allpass filter

Fig. 5-91 shows a Nth-order allpass filter network which is realised with a two-port terminated by a delay arranged in series with an allpass filter of order $N-1$. The transfer function $A_N(z)$ of this circuit is calculated as

$$A_N(z) = \frac{k + A_{N-1}(z)z^{-1}}{1 + kA_{N-1}(z)z^{-1}}.$$

(5.106)

It is interesting to note that the limit of $A_N(z)$ equals k as z approaches infinity.

Let us now consider the inverse problem. A first-order section, consisting of a two-port and a delay, is to be separated from a Nth-order allpass transfer function.

The procedure consists of two steps. First, the parameter k of the separated two-port is determined by taking the limit of $A_N(z)$ for z approaching infinity. The transfer function of the remaining allpass filter of order $N-1$ is obtained by solving (5.106) for $A_{N-1}(z)$.

$$A_{N-1}(z) = \frac{z(A_N(z)-k)}{1-kA_N(z)} \qquad (5.107)$$

By recursive application of this algorithm, a given allpass transfer function can be completely decomposed into a chain of two-ports and delays as in Fig. 5-92.

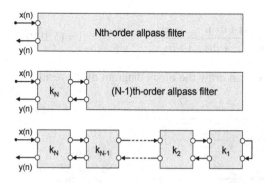

Fig. 5-92
Recursive method to design WDF-based allpass filters

Example 5-6

Given the transfer function of a second-order allpass filter. Determine the coefficients k_1 and k_2 of a lattice filter implementation. Derive the relationship between the coefficients of the transfer function a_1 and a_2 and the coefficients of the filter.

The transfer function of a second-order allpass filter can be written in the form

$$A_2(z) = \frac{a_2 + a_1 z^{-1} + z^{-2}}{1 + a_1 z^{-1} + a_2 z^{-2}} = \frac{a_2 z^2 + a_1 z + 1}{z^2 + a_1 z + a_2} \ .$$

The coefficient k_2 is calculated as the limit value of $A_2(z)$ as z approaches infinity.

$$k_2 = \lim_{z \to \infty} \frac{a_2 z^2 + a_1 z + 1}{z^2 + a_1 z + a_2} = a_2 \qquad (5.108)$$

We extract a two-port with the coefficient $k_2 = a_2$ from the second-order filter and calculate the transfer function of the remaining first-order block.

$$A_1(z) = \frac{z\left(A_2(z)-k_2\right)}{1-k_2 A_2(z)}$$

$$A_1(z) = \frac{z\left[\dfrac{a_2 z^2 + a_1 z + 1}{z^2 + a_1 z + a_2} - a2\right]}{1 - a_2 \dfrac{a_2 z^2 + a_1 z + 1}{z^2 + a_1 z + a_2}}$$

$$A_1(z) = \frac{a_1 z + (1 + a_2)}{(1 + a_2)z + a_1}$$

The two-port k_1 of this first-order block is again calculated as the limit of $A_1(z)$ as z approaches infinity.

$$k_1 = \lim_{z \to \infty} A_1(z) = \lim_{z \to \infty} \frac{a_1 z + (1 + a_2)}{(1 + a_2)z + a_1} = \frac{a_1}{1 + a_2} \tag{5.109}$$

With k_1 and k_2 determined, we can draw the block diagram of the resulting WDF allpass filter.

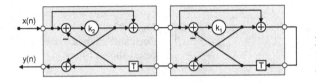

Fig. 5-93
Second-order WDF allpass filter

According to (5.108) and (5.109), the coefficients k_1 and k_2 of the filter and a_1 and a_2 of the transfer function are related as

$$k_2 = a_2 \quad \text{and}$$

$$k_1 = \frac{a_1}{1 + a_2}.$$

Using these coefficients, the transfer function of the filter can now be written in the form

$$A_2(z) = \frac{a_2 + a_1 z^{-1} + z^{-2}}{1 + a_1 z^{-1} + a_2 z^{-2}} = \frac{k_2 + k_1(1 + k_2)z^{-1} + z^{-2}}{1 + k_1(1 + k_2)z^{-1} + k_2 z^{-2}} \ .$$

5.7.3 Applications

At first glance, allpass filters, which are an exemplar of non-selective filters, do not seem to be appropriate candidates for realising selective filters. But there are numerous examples where allpass filters are advantageously used as building blocks in frequency-selective applications. Some of these are introduced in this section.

5.7.3.1 The Gray And Markel Filter Structure

Gray and Markel [28] proposed a filter realisation which is based on filter structures similar to the WDF allpass filters which are derived from the reflectance of coupled line segments (Fig. 5-89). These filters possess the same good properties as wave digital filters in terms of stability and low coefficient sensitivity. Figure 5-94 shows the general block diagram.

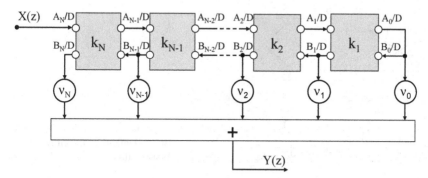

Fig. 5-94 The Markel and Gray filter structure

Each of the cascaded blocks contains a network which realises a scattering matrix of the form

$$\begin{pmatrix} k_m & S_{12m} \\ S_{21m} & -k_m \end{pmatrix} \quad \text{with } S_{12m}S_{21m} = 1 - k_m^2 \tag{5.110}$$

and a delay as shown in Fig. 5-95.

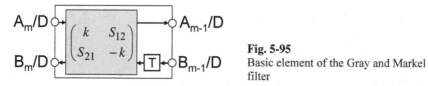

Fig. 5-95
Basic element of the Gray and Markel filter

The $A_m(z)$ and $B_m(z)$ are the numerator polynomials of the transfer functions between the input of the filter and the respective points within the filter structure. All these transfer function have a common denominator polynomial $D(z)$. The output of the filter is a linear combination of the signals which appear at the connecting points in the backward path of the structure.

$$Y(z) = \sum_{m=0}^{N} v_m \frac{B_m(z)}{D(z)} X(z) = \frac{\sum_{m=0}^{N} v_m B_m(z)}{D(z)} X(z) \tag{5.111}$$

Besides the two-port adaptor whose details are shown in Fig. 5-86, Gray and Markel introduced further basic structures whose scattering matrices have the desired form (5.110). These are shown in Fig. 5-96. The various implementations are commonly named after the number of multipliers used in the structure. Cascading the one-multiplier and two-multiplier implementations yields a lattice-type filter structure while the three-multiplier and four-multiplier versions result in a ladder structure.

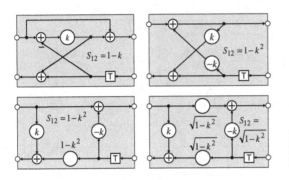

Fig. 5-96
Basic elements of the Gray and Markel filter.

The transfer function of the filter to be implemented with the structure presented in Fig. 5-94 is given as

$$H(z) = \frac{N(z)}{D(z)} = \sum_{i=0}^{N} b_i z^{-i} \bigg/ \sum_{i=0}^{N} a_i z^{-i}$$

A recursive algorithm will be used to determine the N k-parameters and the $N+1$ tap parameters v_m. The algorithm is initialised at the left port of block N in Fig 5-94. To the input, we apply the input signal $X(z)$ of the filter. We can write the input signal in the form

$$X(z) = X(z)\frac{D(z)}{D(z)} \ .$$

$A_N(z)$ can therefore be initialised as

$$A_N(z) = D(z) \ .$$

Due to the choice of the filter structure we know that, with respect to the left-side output port, the filter behaves as an allpass filter. So the output signal can be expressed as

$$X(z)\frac{z^{-N}D(1/z)}{D(z)} \ .$$

$B_N(z)$ can therefore be initialised as

$$B_N(z) = z^{-N}D(1/z) \ .$$

We choose the denominator polynomial of the allpass filter equal to the denominator polynomial of the target transfer function $H(z)$ because $D(z)$ appears as the denominator in (5.111).

The basic procedure of the algorithm is to calculate recursively the functions $A_m(z)$ and $B_m(z)$ for $m = N-1$ down to $m = 0$. The inputs and outputs of the basic element according to Fig. 5-95 are related by the scattering matrix as

$$\begin{pmatrix} B_m \\ A_{m-1} \end{pmatrix} = \begin{pmatrix} k_m & S_{12m}z^{-1} \\ S_{21m} & -k_m z^{-1} \end{pmatrix} \begin{pmatrix} A_m \\ B_{m-1} \end{pmatrix}$$

$$B_m = k_m A_m + S_{12m}z^{-1}B_{m-1}$$

$$A_{m-1} = S_{21m}A_m - k_m z^{-1}B_{m-1}$$

Solving these equations for $A_{m-1}(z)$ and $B_{m-1}(z)$ yields the following recurrence formulas

$$A_{m-1}(z) = \frac{A_m(z) - k_m B_m(z)}{S_{12m}}$$

$$B_{m-1}(z) = \frac{z(B_m(z) - k_m A_m(z))}{S_{12m}} \qquad (5.112)$$

$A_m(z)$ and $B_m(z)$ have the property that
- the degree of the polynomials is reduced by one with every iteration step, and
- both polynomials form a mirror image pair.

$A_m(z)$ and $B_m(z)$ are polynomials of the form

$$A_m(z) = \sum_{i=0}^{m} d_{mi} z^{-i} \quad \text{and} \quad B_m(z) = \sum_{i=0}^{m} d_{mi} z^{-(m-i)} .$$

$P_m(z)$ is an auxiliary polynomial which is required to calculate the tap coefficients v_m of the filter.

$$P_m(z) = \sum_{i=0}^{m} p_{mi} z^{-i}$$

The algorithm starts with the initialisation of the various polynomials:
$A_N(z) = D(z)$, $B_N(z) = z^{-N}D(1/z)$, $P_N(z) = N(z)$, $m = N$
The coefficients k_m and v_m are calculated recursively in the following way:
1. The parameter k_m is calculated as $k_m = d_{mm}/d_{m0}$.   (5.113)
2. The tap coefficient v_m is calculated as p_{mm}/d_{m0}.   (5.114)
3. Apply (5.112) to calculate $A_{m-1}(z)$ and $B_{m-1}(z)$.
4. Calculate $P_{m-1}(z) = P_m(z) - v_m B_m(z)$.   (5.115)
5. $m = m-1$, go to 1

Example 5-7

Use the one-multiplier basic element to realise a Gray and Markel lattice filter with the transfer function

$$H(z) = \frac{N(z)}{D(z)} = \frac{1 - 3z^{-1} + 3z^{-2} - z^{-3}}{1 - 0.1z^{-1} - 0.4z^{-2} + 0.45z^{-3}}.$$

The filter is determined by the three k-parameters k_3, k_2, and k_1 and the four tap coefficients v_3, v_2, v_1, and v_0

First we initialise the polynomials required for the algorithm.

$$A_3(z) = D(z) = 1 - 0.1z^{-1} - 0.4z^{-2} + 0.45z^{-3}$$

$$B_3(z) = z^{-3}D(1/z) = z^{-3} - 0.1z^{-2} - 0.4z^{-1} + 0.45$$

$$P_3(z) = N(z) = 1 - 3z^{-1} + 3z^{-2} - z^{-3}$$

k_3 and v_3 are calculated using (5.113) and (5.114).

$$k_3 = d_{33} / d_{30} = 0.45/1.0 = 0.45$$

$$v_3 = p_{33} / d_{30} = -1.0/1.0 = -1.0$$

$$S_{21,3} = 1 - k_3 = 0.55$$

For the next step of the algorithm, $A_2(z)$, $B_2(z)$ and $P_2(z)$ are calculated using (5.112) and (5.115).

$$A_2(z) = \frac{A_3(z) - k_3 B_3(z)}{S_{12,3}} = 1.45 + 0.145z^{-1} - 0.645z^{-2}$$

$$B_2(z) = z^{-2}A_2(1/z) = 1.45z^{-2} + 0.145z^{-1} - 0.645$$

$$P_2(z) = P_3(z) - v_3 B_3(z) = 1.45 - 3.4z^{-1} + 2.9z^{-2}$$

$$k_2 = d_{22} / d_{20} = -0.645/1.45 = -0.445$$

$$v_2 = p_{22} / d_{20} = 2.9/1.45 = 2.0$$

$$S_{21,2} = 1 - k_2 = 1.445$$

$$A_1(z) = \frac{A_2(z) - k_2 B_2(z)}{S_{12,2}} = 0.805 + 0.145z^{-1}$$

$$B_1(z) = z^{-1}A_1(1/z) = 0.805z^{-1} + 0.145$$

$$P_1(z) = P_2(z) - v_2 B_2(z) = 2.741 - 3.691z^{-1}$$

$$k_1 = d_{11} / d_{10} = 0.145/0.805 = 0.181$$

$$v_1 = p_{11} / d_{10} = -3.691/0.805 = -4.588$$

$$S_{21,1} = 1 - k_1 = 0.819$$

$$A_0(z) = \frac{A_1(z) - k_1 B_1(z)}{S_{12,1}} = 0.95$$

$$B_0(z) = z^{-0} A_0(1/z) = 0.95$$

$$P_0(z) = P_1(z) - v_1 B_1(z) = 3.408$$

$$v_0 = p_{00}/d_{00} = 3.408/0.95 = 3.588$$

It should be noted that the k-parameters for a given filter specification are independent of the choice of the structure of the basic element (Fig. 5-96). The tap coefficients, however, may be different because the distribution of gain is different among these structures.

5.7.3.2 The Digital Lattice Filter

One of the interesting properties of allpass filters is the possibility to find structurally-bounded implementations which guarantee unity gain for all frequencies in a wide range of filter coefficients. The coefficients of these structures only influence the frequency and the quality factor of the poles and zeros, but have no influence on the gain at all. In the context of wave digital filters, we showed that structurally-bounded filters exhibit low coefficient sensitivity in the passband. Allpass filters offer an alternative way to realise filters which can cope with short coefficient wordlengths.

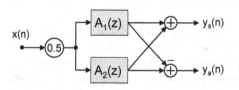

Fig. 5-97
Parallel connection of allpass filters

We consider the parallel connection of two allpass filters as shown in Fig. 5-97. The allpass filters are characterised by the transfer functions $A_1(z)$ and $A_2(z)$. The filter structure has two outputs which represent the sum and difference of the two allpass output signals. The respective transfer functions can be expressed as

$$H_s(z) = \frac{P(z)}{D(z)} = \frac{1}{2}\left(A_1(z) + A_2(z)\right) \quad \text{and}$$

$$\tag{5.116}$$

$$H_a(z) = \frac{Q(z)}{D(z)} = \frac{1}{2}\left(A_1(z) - A_2(z)\right) \ .$$

If $H_s(z)$ and $H_a(z)$ can be realised as shown in Fig. 5-97, the corresponding allpass transfer functions can be calculated as

$$A_1(z) = H_s(z) + H_a(z) = \left(P(z) + Q(z)\right)/D(z) \quad \text{and} \tag{5.117a}$$

$$A_2(z) = H_s(z) - H_a(z) = \left(P(z) - Q(z)\right)/D(z) \ . \tag{5.117b}$$

As long as the phase responses of both allpass filters are different, the overall magnitude response of the structure will be frequency-dependent. A frequency range, in which $A_1(z)$ and $A_2(z)$ are (more or less) in-phase, is a passband for $H_s(z)$ and a stopband for $H_a(z)$. We have the reverse situation when $A_1(z)$ and $A_2(z)$ are out-of-phase. The magnitude responses at the two outputs are thus complementary with respect to each other. Since the gain of the allpass filters is strictly unity for all frequencies, the magnitude response of the parallel structure in Fig. 5-97 can never exceed unity.

$$\frac{1}{2}\left|A_1(e^{j\omega T}) \pm A_2(e^{j\omega T})\right| \leq 1 \qquad \text{for all } \omega$$

So we have found a structure which fulfils all prerequisites for low coefficient sensitivity. There are, in principle, two possible proceedings to synthesise a transfer function as the parallel connection of two allpass sections.

In the first approach, the complete filter design including the allpass decomposition takes place in the analog domain. In this context, we make use of the properties of symmetric LC-networks which may be realised as a lattice and whose transfer functions can be expressed as the sum of two allpass transfer functions. In Sect. 2.8.5, we derived the conditions under which a transfer function possesses these properties: Given a transfer function $t(p) = u(p)/d(p)$. Determine a transfer function $\rho(p) = v(p)/d(p)$ such that

$$\left|t(j\omega)\right|^2 + \left|\rho(j\omega)\right|^2 = 1 \quad \text{or} \quad \left|u(j\omega)\right|^2 + \left|v(j\omega)\right|^2 = \left|d(j\omega)\right|^2 .$$

If the polynomial $u(p)$ is even and the polynomial $v(p)$ is odd or vice versa, $t(p)$ and $\rho(p)$ can be realised as the parallel connection of two allpass sections. The calculated transfer functions of the analog allpass filters are then converted to transfer functions in the variable z by using an appropriate algorithm such as the bilinear transform. The resulting digital allpass filters are realised with one of the structurally-bounded allpass structures as introduced in this chapter.

In the second approach, the allpass decomposition is performed in the digital domain. In the following, we will derive the conditions under which a given transfer function is realisable as the difference or sum of two allpass transfer functions in z [73].

According to (5.86), the transfer function of an Nth-order allpass filter can be generally expressed in the form

$$A(z) = z^{-N} \frac{d(1/z)}{d(z)} .$$

We calculate the sum and the difference of two allpass transfer functions

$$A_1(z) = z^{-N_1} \frac{d_1(1/z)}{d_1(z)} \quad \text{and} \quad A_2(z) = z^{-N_2} \frac{d_2(1/z)}{d_2(z)} , \qquad (5.118)$$

which results in

$$H_s(z) = \frac{1}{2}(A_1(z) + A_2(z)) = \frac{1}{2}\frac{z^{-N_1}d_1(1/z)d_2(z) + z^{-N_2}d_2(1/z)d_1(z)}{d_1(z)d_2(z)} \quad (5.119a)$$

$$H_a(z) = \frac{1}{2}(A_1(z) - A_2(z)) = \frac{1}{2}\frac{z^{-N_1}d_1(1/z)d_2(z) - z^{-N_2}d_2(1/z)d_1(z)}{d_1(z)d_2(z)} \quad (5.119b)$$

Analysis of (5.119) shows that the coefficients of the numerator polynomial

$$\sum_{i=0}^{M} b_i z^i$$

have certain symmetry properties. For the plus sign in (5.119a), the coefficients are symmetric, meaning that $b_i = b_{M-i}$. The minus sign in (5.119b) results in antisymmetric coefficients with $b_i = -b_{M-i}$. The numerator polynomials $P(z)$ and $Q(z)$ are thus linear-phase with an additional 90° phase shift in the case of $Q(z)$ (refer to Sect. 5.2.6). So we have found a first requirement which a transfer function has to fulfil to be realisable as the parallel connection of two allpass filters: The numerator polynomial must be linear-phase with symmetric or antisymmetric coefficients.

Making use of (5.119), it can be easily shown that

$$H_s(z)H_s(1/z) + H_a(z)H_a(1/z) = 1 \ . \quad (5.120)$$

For $z = e^{j\omega T}$, (5.120) assumes the form

$$H_s(e^{-j\omega T}) H_s(e^{j\omega T}) + H_a(e^{-j\omega T}) H_a(e^{j\omega T}) = \left|H_a(e^{j\omega T})\right|^2 + \left|H_s(e^{j\omega T})\right|^2 = 1$$

which again proves the complementary behaviour of $H_s(z)$ and $H_a(z)$. This relation is the discrete-time version of the Feldtkeller equation (2.78) which relates the transmittance and reflectance of resistively terminated lossless networks.

If we express $H_s(z)$ and $H_a(z)$ in terms of their numerator and denominator polynomials, (5.120) assumes the form

$$\frac{P(z)}{D(z)}\frac{P(1/z)}{D(1/z)} + \frac{Q(z)}{D(z)}\frac{Q(1/z)}{D(1/z)} = 1 \quad \text{or}$$

$$P(z)P(1/z) + Q(z)Q(1/z) = D(z)D(1/z) \ . \quad (5.121)$$

Thus $D(z)D(1/z)$ must be representable as the sum of the terms $P(z)P(1/z)$ and $Q(z)Q(1/z)$, where the coefficients of the polynomial $P(z)$ are symmetric and those of $Q(z)$ antisymmetric. If this condition is fulfilled, the transfer functions $P(z)/D(z)$ and $Q(z)/D(z)$ can be realised as the parallel connection of two allpass sections.

In order to check if this condition is fulfilled we have to calculate

- $Q(z)$ from $D(z)$ and $P(z)$ if the given transfer function has a numerator polynomial with symmetric coefficients, and
- $P(z)$ from $D(z)$ and $Q(z)$ if the given transfer function has a numerator polynomial with antisymmetric coefficients.

Multiplication of (5.117a) by (5.117b) yields a useful relation for this purpose.

$$\left(H_s(z) + H_a(z)\right)\left(H_s(z) - H_a(z)\right) = A_1(z)A_2(z)$$

By substitution of (5.118), we obtain

$$H_s^2(z) - H_a^2(z) = z^{-(N_1+N_2)} \frac{d_1(1/z)}{d_1(z)} \frac{d_2(1/z)}{d_2(z)} = z^{-N} \frac{D(1/z)}{D(z)}$$

$$\frac{P^2(z)}{D^2(z)} - \frac{Q^2(z)}{D^2(z)} = z^{-N} \frac{D(1/z)}{D(z)}$$

$$P^2(z) - Q^2(z) = z^{-N} D(z)D(1/z) \qquad (5.122)$$

(5.122) provides the square of the desired polynomial $P(z)$ or $Q(z)$, which is of the form

$$s_0 + s_1 z^{-1} + s_2 z^{-2} + \ldots + s_{2L-1} z^{-(2L-1)} + s_{2L} z^{-2L} \ .$$

Only if the allpass decomposition of the given transfer function is possible, this polynomial will be the square of another polynomial. In this case, the coefficients a_n of $P(z)$ or $Q(z)$ can be calculated recursively as follows:

$$a_0 = \sqrt{s_0} \ , \ a_1 = s_1/2a_0$$

$$a_n = \frac{s_n - \sum_{k=1}^{n-1} a_k a_{n-k}}{2a_0} \qquad \text{for } n = 2 \ldots L \qquad (5.123)$$

Example 5-8

Realise the transfer function

$$H(z) = \frac{(1+z^{-1})^3}{30.371 - 48.854z^{-1} + 38.761z^{-2} - 12.278z^{-3}}$$

of a third-order Chebyshev filter, if possible, as the parallel connection of two allpass filters.

The coefficients of the numerator polynomial are symmetric. We thus have:

$$P(z) = 1 + 3z^{-1} + 3z^{-2} + z^{-3} \quad \text{and}$$

$$D(z) = 30.371 - 48.854z^{-1} + 38.761z^{-2} - 12.278z^{-3}$$

We determine $Q^2(z)$ using (5.122).

$$P^2(z) = 1 + 6z^{-1} + 15z^{-2} + 20z^{-3} + 15z^{-4} + 6z^{-5} + z^{-6}$$

$$z^{-3}D(z)D(1/z) = -372.888 + 1777.051z^{-1} - 3853.347z^{-2}$$
$$+ 4962.370z^{-3} - 3853.347z^{-4} + 1777.051z^{-5} - 372.888z^{-6}$$

$$Q^2(z) = P^2(z) - z^{-3}D(z)D(1/z)$$

$$Q^2(z) = 373.888 - 1771.051z^{-1} + 3868.347z^{-2} - 4942.370z^{-3}$$
$$+ 3868.347z^{-4} - 1771.050z^{-5} + 373.888z^{-6}$$

Using (5.123), we obtain the coefficients of the Polynomial $Q(z)$.

$$q_0 = \sqrt{s_0} = \sqrt{373.888} = 19.336$$

$$q_1 = \frac{s_1}{2q_0} = \frac{-1.771.051}{38.672} = -45.796$$

$$q_2 = \frac{s_2 - q_1 q_1}{2q_0} = \frac{3868.347 - 45.796 \times 45.796}{38.672} = 45.796$$

$$q_3 = \frac{s_3 - 2q_1 q_2}{2q_0} = \frac{-4942.370 + 2 \times 45.796 \times 45.796}{38.672} = -19.336$$

$$Q(z) = -19.336z^{-3} + 45.796z^{-2} - 45.796z^{-1} + 19.336$$

$Q(z)$ is a polynomial with antisymmetric coefficients. $H(z)$ is thus realisable as the parallel connection of two allpass filters. The allpass transfer functions are calculated using (5.117).

$$A_1(z) = \frac{P(z) + Q(z)}{D(z)} = \frac{-18.336z^{-3} + 48.796z^{-2} - 42.796z^{-1} + 20.336}{30.371 - 48.854z^{-1} + 38.761z^{-2} - 12.278z^{-3}}$$

$$A_1(z) = 0.670 \frac{(1 - 0.604z^{-1})(1 - 1.501z^{-1} + 1.493z^{-2})}{(1 - 0.604z^{-1})(1 - 1.005z^{-1} + 0.670z^{-2})}$$

$$A_1(z) = \frac{0.670 - 1.005z^{-1} + z^{-2}}{1 - 1.005z^{-1} + 0.670z^{-2}}$$

$$A_2(z) = \frac{P(z) - Q(z)}{D(z)} = \frac{20.336z^{-3} - 42.796z^{-2} + 48.796z^{-1} - 18.336}{30.371 - 48.854z^{-1} + 38.761z^{-2} - 12.278z^{-3}}$$

$$A_2(z) = -0.604 \frac{(1 - 1.656z^{-1})(1 - 1.005z^{-1} + 0.670z^{-2})}{(1 - 0.604z^{-1})(1 - 1.005z^{-1} + 0.670z^{-2})}$$

$$A_2(z) = \frac{-0.604 + z^{-1}}{1 - 0.604z^{-1}}$$

The sum of $A_1(z)$ and $A_2(z)$ yields the desired transfer function.

An Lth-order allpass transfer function can only assume very specific values at $z = 1$ ($\omega = 0$) and $z = -1$ ($\omega = \pi/T$):

$$A(z) = z^{-L}\frac{d(1/z)}{d(z)} \quad , \text{thus}$$

$$A(z = 1) = 1 \quad \text{and} \quad A(z = -1) = (-1)^{L}.$$

This also imposes restrictions on the filter characteristics which can be realised with the parallel arrangement of allpass filters. Table 5-3 shows which values $H_s(z)$ and $H_a(z)$ can assume for $z = 1$ and $z = -1$ depending on the order of the transfer function to be realised.

Table 5-3   Properties of realisable transfer functions

Filter degree	$H_s(z) = A_1(z) + A_2(z)$		$H_a(z) = A_1(z) - A_2(z)$	
N	$z = 1$	$z = -1$	$z = 1$	$z = -1$
even	1	± 1	0	0
odd	1	0	0	± 1

Thus high-pass and low-pass filters can only be realised with odd filter orders whereas bandpass and bandstop filters require an even filter order.

In Chap. 2, we derived the conditions under which Butterworth, Chebyshev and Cauer filters can be realised as resistively terminated lossless lattice networks whose transfer functions can be represented as the sum or difference of two allpass transfer functions. The same conditions apply to digital implementations of these filters if the transfer function $H(z)$ is derived from the transfer function $H(p)$ of the analog implementation via the bilinear transform: odd-order low-pass and high-pass filters and two times odd-order bandpass and bandstop filters can be realised as the parallel connection of two digital allpass sections.

5.7.3.3 Notch Filter

The notch filter is a useful type of filter in numerous applications. It attenuates highly a particular frequency component while leaving nearby frequencies relatively unaffected. Examples for the use of notch filters are acoustic feedback control in public address systems and hearing aids, removal of fundamental and harmonic components of 50/60 Hz hum, suppression of the fundamental spectral component in harmonic distortion measurement, meter pulse suppression in telecommunication systems, and further applications in the fields of sonar and ultrasonic systems and motion detection.

The notch filter has two parameters which should be advantageously adjustable independently from each other, namely notch frequency and 3-dB rejection bandwidth. The desired filter characteristic can be realised with a special form of the structure we discussed in the previous section, the parallel connection of two allpass filters (Fig. 5-98). One branch is made up of a second-order allpass filter, the other simply consists of a straight-through connection with the transfer function $H(z) = 1$, which can be considered as the simplest form of an allpass network.

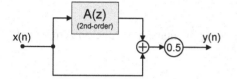

Fig. 5-98
Structure of the notch filter

Fig. 5-99 shows the phase characteristic of a second-order allpass filter. At $f = 0$ and $f = 0.5/T$, allpass and direct path in Fig. 5-98 are in phase resulting in unity gain at these frequencies. At a certain frequency ($f = 0.3/T$ in Fig. 5-99), allpass and direct path are out of phase which yields a perfect cancellation at the output adder. The width of the rejection band is determined by the slope of the phase response curve around the notch frequency. The closer the pole/zero pair of the allpass filter is located to the unit circle, the steeper is the graph of the phase characteristic and the narrower becomes the notch band. In the following, we will analyse the transfer function of the notch circuit and derive the dependencies of notch frequency and notch bandwidth on the filter coefficients.

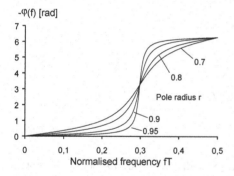

Fig. 5-99
Phase characteristic of a second-order allpass filter.

The transfer function of the structure in Fig. 5-98 can be written as

$$H(z) = \tfrac{1}{2}\left(1 + A(z)\right) .$$

Substitution of the transfer function of a second-order allpass section (5.97) yields

$$H(z) = \frac{1}{2}\left(1 + \frac{a_2 + a_1 z^{-1} + z^{-2}}{1 + a_1 z^{-1} + a_2 z^{-2}}\right) = \frac{\frac{1+a_2}{2} + a_1 z^{-1} + \frac{1+a_2}{2} z^{-2}}{1 + a_1 z^{-1} + a_2 z^{-2}} .$$

By replacing z with $e^{j\omega T}$ and after some algebra, we obtain the following expression for the square magnitude of the frequency response of the notch filter.

$$\left|H(e^{j\omega T})\right|^2 = \frac{\left(\frac{a_1}{1+a_2} + \cos\omega T\right)^2}{\left(\frac{1-a_2}{1+a_2}\right)^2 + \left(\frac{a_1}{1+a_2}\right)^2 + 2\frac{a_1}{1+a_2}\cos\omega T + \left(1 - \left(\frac{1-a_2}{1+a_2}\right)^2\right)\cos^2\omega T} \tag{5.124}$$

The notch frequency ω_0 is obtained by setting the numerator equal to zero, which yields

$$\cos\omega_0 T = \frac{-a_1}{1+a_2} \quad \text{or}$$

$$\omega_0 = \frac{1}{T}\arccos\left(\frac{-a_1}{1+a_2}\right) \tag{5.125}$$

By substitution of (5.125), (5.124) simplifies to

$$\left|H(e^{j\omega T})\right|^2 = \frac{(\cos\omega T - \cos\omega_0 T)^2}{\left(\frac{1-a_2}{1+a_2}\right)^2\sin^2\omega T + (\cos\omega T - \cos\omega_0 T)^2}$$

The frequencies ω_1 and ω_2, where the gain of the filter drops by 3 dB, can be determined by equating this equation with ½.

$$\frac{(\cos\omega_{1/2} T - \cos\omega_0 T)^2}{\left(\frac{1-a_2}{1+a_2}\right)^2\sin^2\omega_{1/2} T + (\cos\omega_{1/2} T - \cos\omega_0 T)^2} = \frac{1}{2}$$

$$(\cos\omega_{1/2} T - \cos\omega_0 T)^2 = \left(\frac{1-a_2}{1+a_2}\right)^2\sin^2\omega_{1/2} T$$

$$\cos\omega_{1/2} T - \cos\omega_0 T = \pm\frac{1-a_2}{1+a_2}\sin\omega_{1/2} T \tag{5.126}$$

With the plus sign in (5.126), we obtain the 3-dB frequency ω_1 below the notch frequency, with the minus sign the 3-dB point ω_2 above the notch frequency.

$$\cos\omega_1 T - \cos\omega_0 T = +\frac{1-a_2}{1+a_2}\sin\omega_1 T$$

$$\cos\omega_2 T - \cos\omega_0 T = -\frac{1-a_2}{1+a_2}\sin\omega_2 T$$

By taking the difference of these equations, we can eliminate the notch frequency.

$$\cos\omega_2 T - \cos\omega_1 T = -\frac{1-a_2}{1+a_2}(\sin\omega_2 T - \sin\omega_1 T)$$

Applying some addition rules of trigonometry and using

$$\omega_2 - \omega_1 = \Delta\omega_{3dB},$$

we finally obtain

$$\tan\frac{\Delta\omega_{3\mathrm{dB}}T}{2} = \frac{1-a_2}{1+a_2} \quad \text{or}$$

$$a_2 = \frac{1-\tan(\Delta\omega_{3\mathrm{dB}}T/2)}{1+\tan(\Delta\omega_{3\mathrm{dB}}T/2)}.$$

(5.127)

With (5.125) and (5.127), we have found the desired relationships between the filter characteristics, defined by ω_0 and $\Delta\omega_{3\mathrm{dB}}$, and the coefficients of the allpass transfer function. But we have not yet reached the goal that notch frequency and rejection bandwidth can be adjusted independently. This problem can be solved by choosing a lattice or wave digital filter allpass structure as shown in Fig. 5-89 or Fig. 5-93. For the structure shown in Fig. 5-93, we derived the relations between the filter coefficients k_1 and k_2 and the coefficients a_1 and a_2 of the transfer function in Example 5-6 as

$$k_2 = a_2$$

$$k_1 = \frac{a_1}{1+a_2}.$$

Substitution in (5.125) and (5.127) yields

$$k_1 = -\cos\omega_0 T$$

$$k_2 = \frac{1-\tan\Delta\omega_{3\mathrm{dB}}T/2}{1+\tan\Delta\omega_{3\mathrm{dB}}T/2}$$

(5.128)

The notch frequency ω_0 is solely determined by the filter coefficient k_1 and the rejection bandwidth $\Delta\omega_{3\mathrm{dB}}$ by k_2.

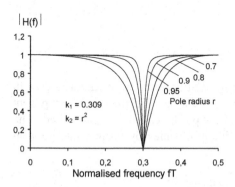

Fig. 5-100
Magnitude response of the notch filter

In Fig. 5-99, we showed the typical shape of the phase characteristic of second-order allpass filters for various pole radii. If we insert these allpass filters into the circuit of Fig. 5-98, we obtain notch filters with magnitude responses as depicted in Fig. 5-100.

Example 5-9

Determine the coefficient k_1 and the 3-dB rejection bandwidths of the notch characteristics shown in Fig. 5-100.

The notch frequency is $f_0 = 0.3/T$. According to (5.128), the coefficient k_1 calculates as

$$k_1 = -\cos \omega_0 T = -\cos(2\pi \times 0.3/T \times T) = -\cos 0.6\pi = 0.309 .$$

The coefficient a_2 of the second-order transfer function is the square of the pole radius. This can be easily shown by multiplying two complex-conjugate pole terms of the form

$$\frac{1}{(z - r_0 e^{j\varphi_0})} \frac{1}{(z - r_0 e^{-j\varphi_0})} = \frac{1}{z^2 - 2r_0 \cos \varphi_0 z + r_0^2} = \frac{1}{z^2 + a_1 z + a_2}$$

The 3-dB rejection bandwidth can be calculated using (5.127):

$$\tan \frac{\Delta\omega_{3dB}T}{2} = \frac{1 - a_2}{1 + a_2} = \frac{1 - r^2}{1 + r^2}$$

$$\Delta\omega_{3dB} = 2\pi \, \Delta f_{3dB} = \frac{2}{T} \arctan \frac{1 - r^2}{1 + r^2}$$

$$\Delta f_{3dB} = \frac{f_s}{\pi} \arctan \frac{1 - r^2}{1 + r^2}$$

Substituting the pole radii of Fig. 5-100, we obtain the following values for the rejection bandwidth:

$r = 0.7$ $\Delta f_{3dB} = 0.105 f_s$
$r = 0.8$ $\Delta f_{3dB} = 0.069 f_s$
$r = 0.9$ $\Delta f_{3dB} = 0.033 f_s$
$r = 0.95$ $\Delta f_{3dB} = 0.016 f_s$

By taking the difference in Fig. 5-98 instead of the sum, we obtain a filter with inverse characteristic which allows to isolate a specific frequency component from a given signal. A filter of this type is called a peak filter. The corresponding magnitude response is simply obtained by turning the graph in Fig. 5-100 upside down.

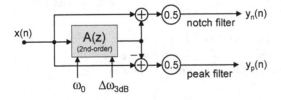

Fig. 5-101
Simultaneous realisation of a notch and peak filter

Figure 5-101 shows how notch and peak filter can be combined in one structure. The transfer functions of notch and peak filter are allpass and power complementary as they are realised as the sum and difference of two allpass transfer functions.

5.7.3.4 Equalisers

Notch and peak filter as introduced in the previous paragraph are examples of selective filters realised with digital allpass sections. A simple extension turns the structure of Fig. 5-101 into a peaking filter with the well-known bell-shaped magnitude response which cuts or boosts the spectrum of a signal by a certain amount around a given centre frequency. Multiple of these filter sections arranged in series make up a multi-band parametric equaliser.

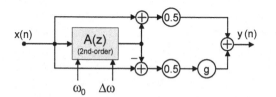

Fig. 5-102
Structure of a parametric equaliser

To realise the named shape of a magnitude response, an additional gain control is required being effective around the centre frequency. Forming the sum of the notch and peak filter outputs of Fig. 5-101, the latter weighted with a gain coefficient g as shown in Fig. 5-102, results in the desired filter structure. For $g = 1$, the circuit has exactly unity gain for all frequencies. Higher or lower values of g cut or boost the signal around the centre frequency. A gain of $g = 0$ results in a notch filter.

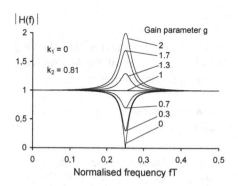

Fig. 5-103
Magnitude response of a parametric equaliser with the gain as a parameter

This structure offers three independently adjustable parameters which determine the magnitude characteristic of the equaliser section:
- The midband frequency ω_0 to control the frequency where the equaliser becomes effective.

- The width of the frequency band $\Delta\omega$ which is affected by the filter. This width is defined as the distance between the two frequencies where the gain assumes a value of $\sqrt{(0.5(1+g^2))}$. For $g = 0$ (notch characteristic), $\Delta\omega$ is the 3dB-bandwidth.
- The amount by which the spectrum of the signal is boosted or cut around ω_0.

Fig. 5-103 shows equalisation curves for various gain factors g.

Another type of equalisers is the shelving filter. Shelving filters cut or boost the high or low end of the audio frequency range. Figure 5-104 shows the typical magnitude response of a low-shelving filter. The high-frequency range is unaffected while the low-frequency range is cut or boosted depending on the gain factor g_l.

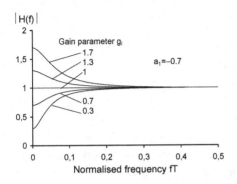

Fig. 5-104
Magnitude response of a low shelving filter with the gain as a parameter

The filter structure which realises this behaviour is similar to the structure of the peaking filter. The second-order allpass section is merely replaced by a first-order section which is characterised by the transfer function

$$H(z) = \frac{a_1 + z^{-1}}{1 + a_1 z^{-1}} \ .$$

Figure 5-105 shows a structure which realises both a low- and a high-shelving filter. The coefficient g_l determines the low-frequency gain of the low-shelving filter, g_h controls the high frequency gain of the high-shelving filter.

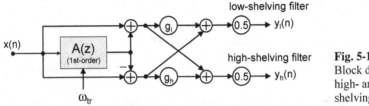

Fig. 5-105
Block diagram of a high- and low-shelving filter

The coefficient a_1 of the allpass transfer function determines the transition frequency ω_{tr} between the unaffected and the cut or boosted frequency ranges.

The transition frequency is defined as the frequency where the gain assumes a value of $\sqrt{(0.5(1+g^2))}$ which is the root mean square of the low- or high-end gains of the filter. For $g = 0$, the transition frequency equals the 3dB-cutoff frequency of the filter. Filter coefficient a_1 and transition frequency ω_{tr} are related as

$$\cos \omega_{tr} T = \frac{-2a_1}{1+a_1^2} .$$

Figure 5-106 shows the magnitude response of a high shelving filter for various values of the coefficient a_1.

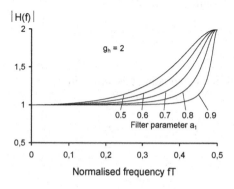

Fig. 5-106
Magnitude response of a high shelving filter with the filter coefficient a_1 as a parameter

Example 5-10

Establish the z-transfer function and calculate the magnitude response of the low-shelving filter. Derive the relationship between the coefficient a_1 of the allpass section and the 3-dB cutoff frequency for $g_1 = 0$.

From block diagram 5-105, we can derive the following relationship between the input and output of the low-shelving filter:

$$Y(z) = 0.5g_1(1+ A(z))X(z) + 0.5(1 - (A(z))X(z)$$

This results in the transfer function

$$H(z) = \frac{Y(z)}{X(z)} = 0.5[g_1(1+ A(z))+(1 - A(z))] .$$

Substitution of the allpass transfer function (5.96) yields

$$H(z) = 0.5\left[g_1\left(1+\frac{a_1 + z^{-1}}{1+a_1 z^{-1}}\right) + \left(1 - \frac{a_1 + z^{-1}}{1+a_1 z^{-1}}\right)\right] \quad \text{and}$$

$$H(z) = 0.5\frac{g_1(1+a_1)+(1-a_1)+[g_1(1+a_1)-(1-a_1)]z^{-1}}{1+a_1 z^{-1}} .$$

The frequency response is obtained by substituting $e^{j\omega T}$ or $\cos\omega T + j\sin\omega T$ for z.

$$H(e^{j\omega T}) = \frac{1}{2}\frac{g_1(1+a_1)+(1-a_1)+[g_1(1+a_1)-(1-a_1)](\cos\omega T - j\sin\omega T)}{1+a_1(\cos\omega T - j\sin\omega T)}$$

The desired magnitude response is the absolute value of the complex expression above. After some lengthy algebra, we finally arrive at

$$\left|H(e^{j\omega T})\right|^2 = \frac{1}{2}\frac{g_1^2(1+a_1)^2+(1-a_1)^2+[g_1^2(1+a_1)^2-(1-a_1)^2]\cos\omega T}{1+a_1^2+2a_1\cos\omega T}.$$

For $g_1 = 0$, the squared magnitude assumes the form

$$\left|H(e^{j\omega T})\right|^2 = \frac{1}{2}\frac{(1-a_1)^2(1-\cos\omega T)}{1+a_1^2+2a_1\cos\omega T}.$$

The 3-dB cutoff frequency ω_c is found by equating this term with ½.

$$\frac{1}{2}\frac{(1-a_1)^2(1-\cos\omega_c T)}{1+a_1^2+2a_1\cos\omega_c T} = \frac{1}{2}$$

$$\cos\omega_c T = \frac{-2a_1}{1+a_1^2}$$

$$\omega_c T = \arccos\frac{-2a_1}{1+a_1^2}$$

At the frequency defined in this way, the magnitude response assumes a value of

$$\left|H(e^{j\omega_c T})\right| = \sqrt{\frac{1+g_1^2}{2}}$$

for any other value of g_1.

5.7.3.5 Approximately Linear-Phase Filters

There are numerous applications where group delay distortion in the context of filtering cannot be tolerated. If filtering takes place in the digital domain, FIR filters are the first choice since this structure allows the realisation of exactly linear-phase frequency responses. The main drawbacks of FIR filters are the high computational effort compared to recursive minimum-phase realisations and the relatively high delay. An allpass-based structure as shown in Fig. 5-107 can be a compromise by offering approximately linear-phase behaviour with the computational effort of IIR implementations.

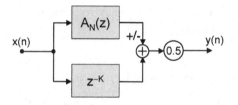

Fig. 5-107
Structure to implement approximately
linear-phase filters

This structure basically calculates the sum or difference of the outputs of a
K-sample delay and an Nth-order allpass filter where K and N are in the same
order of magnitude. In order to realise a frequency-selective overall response, the
phase response of the allpass filter must be designed in such a way that the phase
difference between the two paths assumes values as close as possible to $2k\pi$ in the
passband and $(2k+1)\pi$ in the stopband where k is integer. So the two signals sum
up in the one case and cancel out each other more or less perfectly in the other.

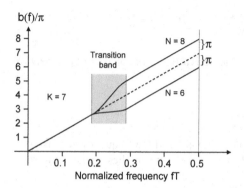

Fig. 5-108
Delay and allpass phase responses for a
low-pass/high-pass filter design

Fig. 5-108 shows the linear-phase response of a delay line (dotted line) and two
possible phase responses of an allpass filter which is appropriately designed to
realise a low-pass or high-pass filter. We take the sum of both branches in
Fig. 5-107 to realise a low-pass filter because the signals sum up at low
frequencies and cancel out each other at high frequencies due to the phase
difference of π. By taking the difference accordingly, we realise a high-pass filter.
The basic principle of this filter implementation is to imitate the linear-phase
behaviour of the delay in the passband and in the stopband by an allpass filter. So,
with the exception of the transition band of the filter, also the sum and the
difference of the responses in both branches will be linear-phase. In fact, it can be
shown that the group delay response of an allpass sum or allpass difference is the
average of the group delay responses of both individual allpass filters. Let

$$H_{A1}(\omega) = |H_{A1}(\omega)| e^{j\varphi_1(\omega)} = e^{-jb_1(\omega)}$$

and

$$H_{A2}(\omega) = |H_{A2}(\omega)| e^{j\varphi_2(\omega)} = e^{-jb_2(\omega)}$$

be the frequency responses of two allpass filters. The frequency response of the allpass sum and of the allpass difference can be written as

$$H_{A\pm}(\omega) = e^{-jb_1(\omega)} \pm e^{-jb_2(\omega)}$$
$$H_{A\pm}(\omega) = \cos b_1(\omega) \pm \cos b_2(\omega) - j(\sin b_1(\omega) \pm \sin b_2(\omega))$$

(5.129)

The phase response of the sum and difference is calculated as the negative argument of (5.129).

$$b_\pm(\omega) = -\arctan\left(-\frac{\sin b_1(\omega) \pm \sin b_2(\omega)}{\cos b_1(\omega) \pm \cos b_2(\omega)}\right)$$

Using addition theorems of trigonometric functions results in

$$b_+(\omega) = \arctan \tan \frac{b_1(\omega)+b_2(\omega)}{2} = \frac{b_1(\omega)+b_2(\omega)}{2}$$
$$b_-(\omega) = -\arctan \cot \frac{b_1(\omega)+b_2(\omega)}{2} = \frac{b_1(\omega)+b_2(\omega)}{2} - \frac{\pi}{2}$$

The group delay is finally calculated as the derivative of the phase with respect to the frequency ω.

$$\tau_{g\pm} = \frac{\tau_{g1}+\tau_{g2}}{2}$$

The overall phase shift of an N-sample delay is $N\pi$ if the frequency goes from 0 to half the sampling frequency π/T (5.84). The same applies to an Nth-order allpass filter(5.88).

$$b(\omega) = N\omega T - 2\arctan \frac{\sum_{i=0}^{N} a_i \sin i\omega T}{\sum_{i=0}^{N} a_i \cos i\omega T}$$

The first term represents an N-sample delay while the second term is a periodic function which becomes zero at $\omega=0$ and $\omega=\pi/T$. The coefficients a_i thus determine how the phase response of the allpass filter $b(\omega)$ oscillates around the linear phase shift $N\omega T$.

For the implementation of a low-pass or high-pass filter, the length of the delay K and the order of the allpass filter N must differ by one in order to realise a phase difference of π at half the sampling frequency. Fig. 5-109 illustrates the shape of the required phase response for bandpass or bandstop filters. We have now three regions where the allpass filter must approximate a linear-phase response. The phase difference at $\omega=\pi/T$ must be 0 or 2π. With one of these 4 possible curves implemented, taking the difference of both branches results in a bandpass filter characteristic, taking the sum leads to a bandstop response.

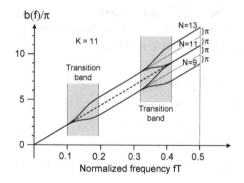

Fig. 5-109
Delay and allpass phase responses for a
bandpass/bandstop filter design

If we have found a set of coefficients a_i by an appropriate design procedure,
which realises the desired approximation of the linear phase in the passband and
stopband, it is not guaranteed, however, that the resulting allpass filter is stable.
Stable poles with $r_\infty < 1$ always result in a positive group delay response as can be
derived from (5.93)and (5.95). This is equivalent to a monotonously increasing
phase response since phase is calculated as the integral over the group delay. Also
slopes of zero or close to zero are hard to realise considering that the group delay
contributions of the pole/zero pairs always sum up into the direction of positive
values. Figure 5-110 shows an example of a design specification which is not
realisable with a stable allpass filter. The same low-pass/high-pass filter as in
Fig. 5-108 is to be realised with a lower filter order. The lower curve must now
have a negative slope in the transition band in order to reach the lower linear-
phase region within the given transition band. A third-order allpass filter is thus
not able to meet the given design specification whereas the upper curve may be
approximated by a stable fifth-order allpass filter. But even if the target phase
characteristic is monotonously increasing, the approximating phase function may
still be unstable. Filter design algorithms often yield approximations which
oscillate around the target phase so that negative slopes may occur anyhow around
some frequencies. If a design yields poles outside the unit circle, the length of the
delay and the order of the allpass filter need to be increased as necessary.

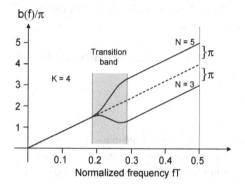

Fig. 5-110
Example of a desired phase response
which leads to an unstable allpass filter

For a given filter design problem, the allpass coefficients a_i are to be determined in such a way that the error $\varepsilon(\omega)$ between the desired phase response $b_d(\omega)$ and the actual phase response of the allpass filter is minimised.

$$\varepsilon(\omega) = b_{Allp}(\omega) - b_d(\omega) = N\omega T - 2\arctan\frac{\displaystyle\sum_{i=0}^{N} a_i \sin i\omega T}{\displaystyle\sum_{i=0}^{N} a_i \cos i\omega T} - b_d(\omega)$$

For a low-pass filter design as illustrated in Fig. 5-108, for instance, the desired phase response would be stated as

$$b_d(\omega) = \begin{cases} K\omega T & \text{for } 0 \le \omega \le 0.4\pi/T \\ K\omega T + \pi & \text{for } 0.6\pi/T \le \omega \le \pi/T \end{cases} .$$

The transition band remains unspecified and is commonly not taken into account in the optimisation process. So no filter resources are wasted to enforce a certain shape of the phase response in the transition band considering that it is only essential to achieve the desired linear phase in the passband and in the stopband. The most common criteria to obtain an optimum set of filter coefficients a_i are:

- Minimisation the mean square error:
  $$\min_{a_i} \int_0^{\pi/T} \varepsilon^2(\omega)\, d\omega$$

- Minimisation of the maximum error:
  $$\min_{a_i} \max |\varepsilon(\omega)| \quad \text{in the frequency range } 0 \le \omega \le \pi/T$$
  This criterion leads to an equiripple error function $\varepsilon(\omega)$.

- Minimisation of the error by a maximally flat approximation of the desired phase response in a limited number of points. This is achieved by setting the error and a number of derivatives of the error with respect to the frequency to zero in these points.

- Design of the N coefficients a_i of the allpass filter in such a way that the phase error becomes zero at N specified frequency points.

More details about the calculation of allpass filter coefficients based on desired phase characteristics can be found in Chap. 6.

5.7.3.6 Filter Bank Structure

The concept of approximately linear-phase recursive high-pass and low-pass filters as introduced in the previous section can be easily extended to construct bandpass filter banks. The structure shown in Fig. 5-107 can already be considered as the simplest form of a filter bank. It splits the spectrum of the input signal into a high-frequency and a low-frequency portion. By addition of a further allpass section as shown in Fig. 5-111, we obtain a 3-band filter bank which splits the input signal into a low-band, mid-band, and high-band signal.

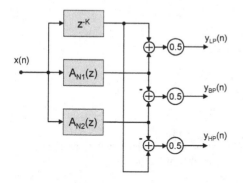

Fig. 5-111
Filter structure of a 3-band filter bank

The required phase responses of the allpass filters are shown in Fig. 5-112. Allpass section 1 realises a linear phase response with a phase jump of π taking place at ω_{c1}. For allpass filter 2 , this jump occurs at ω_{c2} with $\omega_{c1} < \omega_{c2}$.

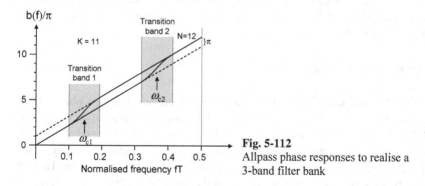

Fig. 5-112
Allpass phase responses to realise a 3-band filter bank

Taking the sum of the outputs of the delay and of allpass filter 1 yields a low-pass filter response, taking the difference of the outputs of the delay and of allpass filter 2 results in a high-pass filter response as described in the previous section. By taking the difference of the two allpass filter outputs, we obtain a filter with a bandpass characteristic. In the frequency band between the two cutoff frequencies ω_{c1} and ω_{c2} where the two phase responses differ by π, the signals sum up. Below ω_{c1} and above ω_{c2}, the output signals of the allpass filters cancel out each other. The structure of Fig. 5-111 can be easily extended to any arbitrary number of bands b as shown in Fig. 5-113. The advantages of this kind of filter bank realisation are:
- approximately linear phase in the passbands,
- lower delay compared to strictly linear-phase FIR filter implementations,
- simple reconstruction of the original signal from the filter bank outputs.

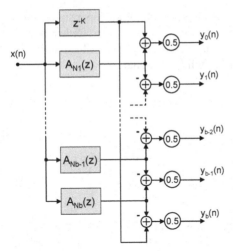

Fig. 5-113
Bandpass filter bank with an arbitrary number of frequency bands

The latter aspect is of importance if an application requires the synthesis of the original signal from the bandpass signals of the filter bank. It can be easily verified that the summation of all filter bank outputs exactly yields the input signal which is merely delayed by K sampling periods. Filter banks based on conventional FIR or IIR bandpass filters require much more filtering effort to reconstruct the original signal.

5.7.3.7 90-Degree Phase Shift

The Hilbert transform, if applied to a time-domain signal, shifts the phase of the signal by 90° or $\pi/2$. The Hilbert transform has numerous applications in digital signal processing, for instance in the fields of image processing, speech processing, medical signal processing, communications, geophysics, radar, high-energy physics, or electron microscopy. Adding the Hilbert transform H of a signal $x(t)$ as the imaginary part to $x(t)$ yields the so-called analytic signal $\hat{x}(t)$.

$$\hat{x}(t) = x(t) + jH x(t)$$

The analytic signal provides important information about the temporal distribution of energy and the instantaneous frequency and phase of a signal.

The discrete-time Hilbert transformer is a filter with unity magnitude and a constant phase shift of $\pi/2$ in the frequency interval $0 < \omega < \pi/T$ and of $-\pi/2$ in the interval $-\pi/T < \omega < 0$. This definition leads to an odd phase response function of ω which is a necessary property of real systems (real filter coefficients, real impulse response).

$$b_{\text{Hilbert}}(\omega) = \begin{cases} \pi/2 & \text{for } 0 < \omega < \pi/T \\ -\pi/2 & \text{for } -\pi/T < \omega < 0 \end{cases}$$

$$H(\omega) = |H(\omega)| e^{-jb(\omega)}$$

$$H_{\text{Hilbert}}(\omega) = \begin{cases} 1\, e^{-j\pi/2} = -j & \text{for } 0 < \omega < \pi/T \\ 1\, e^{j\pi/2} = j & \text{for } -\pi/T < \omega < 0 \end{cases} \qquad (5.130)$$

The ideal impulse response of the Hilbert transformer is of the form

$$h_{\text{Hilbert}}(n) = \frac{1 - \cos \pi n}{\pi n} = \begin{cases} 0 & \text{for } n \text{ even} \\ \dfrac{2}{\pi n} & \text{for } n \text{ odd} \end{cases} . \qquad (5.131)$$

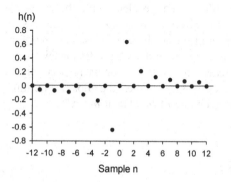

Fig. 5-114
Impulse response of the Hilbert
transformer according to (5.109)

Figure 5-114 shows the graph of $h_{\text{Hilbert}}(n)$. The impulse response is not causal and cannot be realised directly. Any approximation of the Hilbert transform with an IIR or FIR allpass filter must therefore truncate the sequence at some negative $n = -n_0$ and shift the resulting sequence by n_0 sampling periods to the right to make the filter causal. The latter means a delay by n_0 sampling periods. In order to compensate this, also the original signal needs to be delayed by the same amount as sketched in Fig. 5-115.

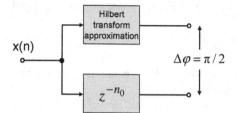

Fig. 5-115
Delay compensation of the Hilbert
transform

So the target phase response to be approximated is a constant phase shift of $\pi/2$ plus a delay of n_0 samples as shown in Fig. 5-116. The larger the chosen delay is, the higher is the order of the allpass filter and the better can the desired phase response be approximated.

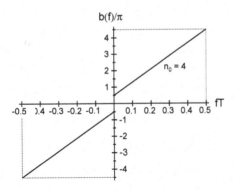

Fig. 5-116
Phase response of the ideal Hilbert
transformer with a 4-sample delay

For the realisations with IIR allpass filters, the problem with the curve in Fig. 5-116 is that the phase of the filter is always zero at $\omega = 0$ and always an integer multiple of π at $\omega = \pi/T$. The desired linear-phase response can therefore not be achieved at low frequencies near $\omega = 0$ and at high frequencies near $\omega = \pi/T$ where the phase must finally reach values of zero or $n\pi$ respectively as illustrated in Fig. 5-117. If the delay is chosen as n_0 samples, the allpass filter must be of the order n_0+1 to finally end up with a phase of $(n_0+1)\pi$ at $\omega = \pi/T$.

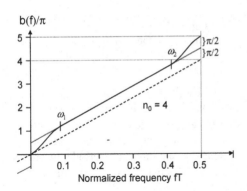

Fig. 5-117
Approximation of the ideal Hilbert
transformer with a 5th-order allpass
filter

For the design of an appropriate allpass filter, the target phase response is stated as

$$b(\omega) = n_0\omega T + \pi/2 \qquad \text{for} \quad \omega_1 \leq \omega \leq \omega_2 .$$

The frequency ranges below ω_1 and above ω_2 (Fig. 5-117) remain unspecified and are not taken into account in the optimisation process.

Example 5-11
Derive the impulse response of the Hilbert transformer from the frequency response given by (5.130).
The impulse response is calculated as the inverse Fourier transform of the frequency response (3.14).

$$h(n) = \frac{1}{2\pi} \int_{-\pi}^{+\pi} H(e^{j\omega T}) e^{jn\omega T} d\omega T$$

Substitution of the frequency response (5.130) yields:

$$h_{\text{Hilbert}}(n) = \frac{1}{2\pi} \int_{-\pi}^{0} j e^{j\omega nT} d\omega T + \frac{1}{2\pi} \int_{0}^{\pi} (-j) e^{j\omega nT} d\omega T$$

$$= \frac{j}{2\pi} \frac{e^{jn\omega T}}{jn} \bigg|_{\omega T=-\pi}^{0} - \frac{j}{2\pi} \frac{e^{jn\omega T}}{jn} \bigg|_{\omega T=0}^{\pi}$$

$$= \frac{1}{2\pi n}\left(1 - e^{-j\pi n} - e^{-j\pi n} + 1\right) = \frac{1 - e^{-j\pi n}}{\pi n}$$

$$= \frac{1 - \cos(-\pi n) - j\sin(-\pi n)}{\pi n}$$

$$h_{\text{Hilbert}}(n) = \frac{1 - \cos \pi n}{\pi n}$$

5.7.3.8 Fractional Delay

There are applications in digital signal processing, for instance in the fields of communications, audio and music processing, speech coding and synthesis, and time delay estimation, where it is required to delay a signal by fractions of the sampling period. Delaying signals by integer multiples of the sampling period is an easy task. A simple shift register structure is sufficient for this purpose. For the realisation of fractional delays it is required to get access to intermediate values between two consecutive samples of the signal sequence. These must be obtained by an appropriate interpolation procedure.

. A straightforward solution to this problem would be to convert the signal, which needs to be delayed, back into the analog domain and to resample the analog signal at time instants which realise the desired non-integer delay with respect to the original sequence of samples.

The ideal reconstruction of the analog signal from the given samples $f(n)$ can be achieved using the sinc-function (4.12) as derived in Sect. 4.4.2.4.

$$f_a(t) = \sum_{n=-\infty}^{+\infty} f(n) \operatorname{sinc} \pi(t/T - n)$$

We delay $f_a(t)$ by an arbitrary time interval Δ

$$f_a(t - \Delta) = \sum_{n=-\infty}^{+\infty} f(n) \operatorname{sinc} \pi[(t - \Delta)/T - n]$$

and resample the delayed analog signal at a rate of $1/T$.

$$f_a(nT - \Delta) = \sum_{m=-\infty}^{+\infty} f(m)\operatorname{sinc}\pi[(nT - \Delta)/T - m]$$

$$f_\Delta(n) = f_a(nT - \Delta) = \sum_{m=-\infty}^{+\infty} f(m)\operatorname{sinc}\pi(n - m - \Delta/T)$$

The new delayed sequence $f_\Delta(n)$ is thus obtained by convolution of the original sequence $f(n)$ with the function $\operatorname{sinc}\pi(n-\Delta/T)$.

$$f_\Delta(n) = f(n) * \operatorname{sinc}\pi(n - \Delta/T)$$

So a non-integer delay Δ can be realised directly in the digital domain. A filter is required with the impulse response

$$h(n) = \operatorname{sinc}\pi(n - \Delta/T). \qquad (5.132)$$

This impulse response is not causal and of infinite length. So we are facing the same problem as in the case of the Hilbert transformer. For practical use, for example in a FIR filter implementation, $h(n)$ must be trimmed to a finite length and shifted such that $h(n) = 0$ for $n < 0$. The longer $h(n)$ is chosen, the better is the approximation of a constant delay and of a unit gain. The window method as introduced in Sect. 7.4.2 may be used to alleviate the consequences of the truncation of the impulse response.

Example 5-12
Calculate and sketch the ideal impulse response of a filter which delays a sequence by half a sampling period.
The target filter has to shift the input sequence by $T/2$:

$$h(n) = \operatorname{sinc}\pi(n - 1/2)$$

$$h(n) = \frac{\sin\pi(n-1/2)}{\pi(n-1/2)} = \frac{\sin(n\pi - \pi/2)}{\pi(n-1/2)} = \frac{-\cos n\pi}{\pi(n-1/2)}$$

$$h(n) = \frac{2}{\pi}\frac{(-1)^{n+1}}{2n-1}$$

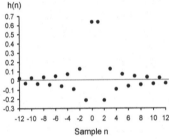

A simple circuit which was introduced in Example 5-1 (Sect. 5.2.5) may be useful in applications where the frequency range of interest is much lower than half the sampling frequency. Only one coefficient is necessary to control the delay Δ.

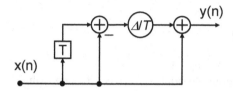

Fig 5-118
Simple first-order interpolator

The calculation of group delay response and magnitude response of this circuit is performed in detail in Example 5-1. The desired delay Δ directly appears as a coefficient in the filter structure. So this realisation may be advantageously utilised in applications where a variable delay is needed. In contrast, for the previous FIR filter approach, a complete filter design procedure must be performed each time a new delay has to be configured, and typically a large number of coefficients must be loaded into the filter algorithm.

IIR allpass filters may also be used to design fractional-delay filters. The desired constant group delay is $n_0T + \Delta$. The according phase response can be expressed as

$$b(\omega) = \omega(n_0T + \Delta) .$$

At $\omega = \pi/T$, the ideal phase shift would be $n_0\pi + \pi\Delta/T$. But allpass filters can only realise phase values of integer multiples of π at half the sampling frequency. The desired phase response can therefore only be approximated up to a certain frequency ω_1 as shown in Fig. 5-119.

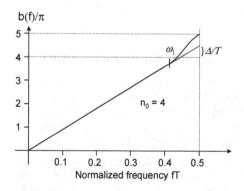

Fig. 5-119
Phase response of a 5th-order fractional-delay allpass filter

If we let the phase unspecified in the frequency range $\omega_1 < \omega < \pi/T$, the phase response that we get as a result of an allpass filter design will automatically tend to an integer multiple of π at $\omega = \pi/T$. In the example of Fig 5-119, n_0 is 4, and the order of the allpass filter is chosen as $N = 5$. So the phase will tend to a value of 5π at $\omega = \pi/T$. The higher we choose n_0 and N, the higher is the accuracy of the linear-phase characteristic of the fractional-delay filter in the frequency range $0 \le \omega \le \omega_1$.

The design target for a fractional-delay filter is a phase response of the form

$$b(\omega) = \omega(n_0 T + \Delta) \quad \text{for } 0 \le \omega \le \omega_1$$
$$\text{unspecified} \qquad \text{for } \omega_1 < \omega \le \pi/T \;.$$

The order of the allpass filter is chosen as $N = n_0 + 1$.

The algorithms to determine the appropriate allpass filter coefficients, which approximate this target phase response, will be introduced in Chap. 6.

The explanations in this section show that an additional delay is required for the interpolation of sample values. This holds for FIR and IIR interpolation filters. This extra delay is in both cases proportional to the order of the filter and thus to the number of available coefficients to approximate the desired behaviour. So there is a trade-off between the accuracy of the interpolation and the delay of the filter. If the application requires less delay between input and output, the only possibility is to delay the input signal, too. This leads to an overall delay of input and output signal, but the extra delay of the interpolator with respect to the input may be compensated if required as demonstrated in Fig. 5-120.

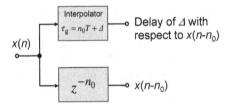

Fig. 5-120
Compensation of the delay of the interpolation filter

5.8 Software Implementation of Digital Filter Algorithms

The filter algorithms that we derived in this chapter are represented in the form of block diagrams or flowgraphs. For the implementation of these algorithms in hardware or software, the graphical form must be converted into a sequence of instructions which are usually executed periodically or in a loop. For the design of DSP algorithms in silicon, VHDL is the common language to describe the behaviour of algorithms. The code for general purpose processors or digital signal processors is usually written in C or assembler. The main advantages of C code are faster development cycles and greater portability while assembler is superior with respect to efficiency. Assembler allows a better access to the specific resources of a given signal processor than C. An established method is to combine the advantages of both languages. Only time-critical parts of the code are implemented in assembler while the rest of the code is still written in C. With recent advances in compiler technology, high-level languages such as C or C++ can make more efficient use of the special architectures and instruction sets of DSPs which makes assembler programming increasingly obsolete.

There are two basic types of digital signal processors on the market, fixed-point and floating-point. Floating point processors are easier to program and provide a higher computational precision which may be advantageous for certain applications. Fixed-point processors, on the other hand, have cost advantages and typically consume less power. The differences in cost and ease of use, however, become increasingly less pronounced. In the audio and video communication world, fixed-point implementations are still predominant since widely used codec algorithms such as MPEG, JPEG, H.264, G.711, or G.729 were designed to be performed in fixed-point. These algorithms are, in general, bit exact so that the greater precision and dynamic range of floating point cannot be used at all. These mass applications implicate high production volumes of fixed-point processors which is one of the reasons for the cost advantage of this processor type. So it is still a frequent task of DSP programmers to write fixed-point code, be it for economical reasons, for power consumption reasons, or because the main application calls for a fixed-point DSP.

Regardless of the target processor platform, the development of DSP algorithms often starts with a floating-point implementation. Effects due to finite computing precision or overflow are avoided at this stage. Also scaling which will become important in integer arithmetic implementations is not needed. This approach widely avoids initial coding errors. The main purpose of this first step is to verify that the algorithm is doing what it is supposed to do according to the specification. Starting from this first implementation, the code can be step-by-step optimised and adapted to the target processor. The initial floating-point implementation is always available as a reference and can be used to detect errors after each modification of the code. In this section, we will have a closer look at this very first step where a block diagram is converted to a floating point implementation. A full introduction to DSP programming would go beyond the scope of this book.

5.8.1 State-Space Structures

We will describe a robust procedure to derive program code from a given state-space representation of a filter which can be regarded as the most general case of a digital filter structure. State-space structures which we introduced in Sect. 5.4 can be characterised as follows (Fig. 5-35):

1. Each input value of a delay element is calculated as the linear combination of the outputs of the delay elements and of the current input signal $x(n)$ of the filter.

2. The current output sample $y(n)$ of the filter is calculated as a linear combination of the outputs of the delay elements and of the input signal $x(n)$.
3. When these calculations are completed, the inputs of the delay elements are transferred to the output of the delay elements.
4. When the next input sample is available, we continue with step 1.

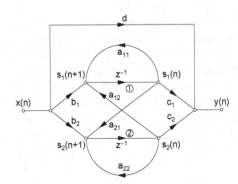

Fig. 5-121
General second-order state-space structure

The general second-order state-space filter (Fig. 5-121) may serve as an example. The following system of linear equations can be established:

$$y(n) = dx(n) + c_1 s_1(n) + c_2 s_2(n)$$
$$s_1(n+1) = b_1 x(n) + a_{11} s_1(n) + a_{12} s_2(n) \qquad (5.133)$$
$$s_2(n+1) = b_2 x(n) + a_{21} s_1(n) + a_{22} s_2(n) \ .$$

We define an array s[i] which holds the state variables s_1 and s_2 and the input sample $x(n)$.

$$s[0] = x(n), \quad s[1] = s_1(n), \quad s[2] = s_2(n)$$

A second array t[i] holds the temporary intermediate result of the calculation (5.133).

$$t[0] = y(n), \quad t[1] = s_1(n+1), \quad t[2] = s_2(n+1)$$

So we can express (5.133) in matrix form as

$$t = \begin{pmatrix} d & c_1 & c_2 \\ b_1 & a_{11} & a_{12} \\ b_2 & a_{21} & a_{22} \end{pmatrix} s \ . \qquad (5.134)$$

When all values of the array t are calculated according to (5.134), the elements of t are copied to the array s.

$$s = t$$

This corresponds to the transfer of the input values of the delay elements to the respective outputs.

The element s[0] which contained the input sample $x(n)$ at the beginning of the filter routine contains the according output sample $y(n)$ in the end. For convenience, we store the $N*N$ coefficients of the state-space filter row by row in a linear array coeff[i].

$$\text{coeff}[0] = d, \quad \text{coeff}[1] = c_1, \quad \text{coeff}[2] = c_2, \quad \text{coeff}[3] = b_1, \quad \text{coeff}[4] = a_{11}, \quad \ldots$$

So a first version of a C routine could look as follows:

```
void State_Space(double s[], double coeff[]) {
    /* General second-order state-space filter.        */
    /* s[1] and s[2] are the state variables.          */
    /* s[0] is used to pass the input sample x(n)      */
    /* to the routine. Upon return, s[0] contains      */
    /* the calculated output sample y(n). coeff[]      */
    /* is filled with the coefficients of the state-   */
    /* space coefficient matrix arranged row by row .*/
    double t[3];
    t[0] = coeff[0]*s[0]+coeff[1]*s[1]+coeff[2]*s[2];
    t[1] = coeff[3]*s[0]+coeff[4]*s[1]+coeff[5]*s[2];
    t[2] = coeff[6]*s[0]+coeff[7]*s[1]+coeff[8]*s[2];
    s[0] = t[0];
    s[1] = t[1];
    s[2] = t[2];
    return;}
```

(5.134) can be readily generalised to state-space filters of arbitrary order N as expressed by (5.135).

$$\begin{pmatrix} y(n) \\ s_1(n+1) \\ s_2(n+1) \\ \vdots \\ s_N(n+1) \end{pmatrix} = \begin{pmatrix} d & c_1 & c_2 & \cdots & c_N \\ b_1 & a_{11} & a_{12} & \cdots & a_{1N} \\ b_2 & a_{21} & a_{22} & \cdots & a_{2N} \\ \vdots & \vdots & \vdots & \ddots & \vdots \\ b_N & b_{N1} & b_{N2} & \cdots & b_{NN} \end{pmatrix} \begin{pmatrix} x(n) \\ s_1(n) \\ s_2(n) \\ \vdots \\ s_N(n) \end{pmatrix} \qquad (5.135)$$

Our filter subroutine must be modified in order to cope with a variable filter order. The order N is passed as an additional parameter to the routine. The temporary buffer t[] must be created with a dynamically assigned size. The actual calculations are performed in "for" or "while" loops where the limits of the indices are determined by the order N of the filter.

The following code listing tries to follow the style of an assembler program which makes effective use of the available DSP resources. The program avoids "for" loops and uses "while" loops instead. The indices in these loops are decremented down to zero because it is more effective in assembler to test for zero than to compare a value with a given constant in order to detect the end of the loop.

```
void State_Space(double s[], double coeff[], int N)    {
   /* **  General state-space filter of order N  ** */
   /* Array s[] contains the state variables.        */
   /* s[0] is used to pass the input sample x(n)     */
   /* to the routine. Upon return, s[0] contains     */
   /* the calculated output sample y(n). coeff[]      */
   /* is filled with the coefficients of the state-  */
   /* space coefficient matrix arranged row by row. */
   int i, j, k;
   double acc, *t;
   t = malloc(++N*sizeof(double));
   k = N;
   i = N*N;
   while (k != 0){
      acc=0;
      j=N;
      while (j != 0) acc += coeff[--i]*s[--j]; /* MAC operation */
      t[--k]=acc;}
   k = N;
   while (k != 0) s[k]=t[--k];
   free(t);
   return;}
```

The heart of the algorithm is the bordered statement in the listing. Such a statement can be found in almost every DSP application. It gives the following commands to the CPU:

- Decrement the index i and load the ith value of the coefficient array coeff[] from memory.
- Decrement the index j and load the jth value of the state variables array s[] from memory.
- Multiply both values.
  Add the result to the contents of a special register, the so-called accumulator.

Modern DSPs do this all for you within one processor cycle. Such a combined sequence of instructions is commonly called a MAC (Multiply and ACcumulate). In order to be able to perform so many operations in such a short period, special processor architectures are required. General purpose CPUs commonly have a von Neumann architecture which features a common memory and a common bus for data and instructions. In order to avoid this bottleneck, DSPs have separate buses and memory pages for data and instructions which is called a Harvard architecture. In order to fetch two operands simultaneously, e.g. for a multiplication as described above, the buses are often duplicated.

5.8.2 Program Flow

Filter routines as described in the previous section are usually executed in a loop where the samples of an input sequence are processed and transformed into a sequence of output samples. This loop may look quite different depending on the application.

Before the filter subroutine is called for the first time, the state variables s[] have to be initialised to zero. This avoids initial transient oscillations of the filter which are superimposed to the output signal. These have nothing to do with the input signal but are caused by the fact the filter is not in the quiescent state at the beginning.

We consider in the following some examples demonstrating how the filter routine may be embedded in a loop where data are loaded, processed, and stored back. If the input sequence input[i] is available in an array within the program, for instance for testing or in a simulation, the filter routine may appear in a "for" loop.

```
. . . . . . . . . .
for (i=0, i<Nsamples, i++) {
   s[0] = input[i];
   State_Space(s,coeff,N);
   output[i] = s[0]; }
. . . . . . . . . .
```

For off-line filtering of digital signals which are stored in a file on hard disk, the loop may take the form of a "while" loop where the condition in the while statement continuously checks if the end of the file is reached.

```
. . . . . . . . . .
while(!feof(Infile)) {
   fread(s,sizeof(short),1,Infile);
   State_Space(s,coeff,N);
   fwrite(s,sizeof(short),1,Outfile); }
. . . . . . . . . .
```

In case of real-time applications which usually run on digital signal processors, the call of the filter routine must be synchronised with the transfer rate of signal data via hardware I/O ports or with the conversion rate of A/D and D/A converters. It has to be further considered that often a number of concurrent applications share the resources of the DSP.

A frequently applied solution to this problem is to separate the tasks of the actual signal treatment and of serving the hardware registers of ports or converters. A timer in the DSP generates a clock signal which is used to synchronise the port clocks or to control the conversion rate of A/D and D/A converters. This timer also generates periodical HW interrupts which trigger the execution of an interrupt service routine. This routine reads HW input registers and stores the data into an input buffer in memory. In the same way, data are transferred from an output buffer in memory to the respective HW output register. Furthermore, a flag is set which indicates to the associated application that new input data are available and recent output data are transferred to the output register. The following pseudo code fragment demonstrates the principle.

```
void interrupt i_o_handler() {
    /*  Service routine of the   */
    /*  timer interrupt          */
    if (flagn == 1) errormessage(DSP_OVERLOAD);
    InBuf = input_register;
    output_register = OutBuf;
    flagn = 1;
    return; }
```

The associated application which processes the input sample clears the flag when the new output sample is calculated and made available in the output buffer. If the interrupt routine detects that the flag is still set, this is an indication that the DSP is overloaded because it was not able to process the input sample in the interval between two interrupts. The routine which processes the input sample, in our example a state-space filter, is embedded in a loop in which the flags of all interrupt routines are cyclically checked. If an active flag is detected, the corresponding application is executed.

```
. . . . . . . . . . .
while (!Stop) {
    if (flag1) application1();
    if (flag2) application2();
    . . . . . . . . . . . . . .
    if (flagn) {
        s[0] = InBuf
        State_Space(s,coeff);
        OutBuf = s[0];
        flagn = 0; }
    . . . . . . . . . . . . . .
    if (flagm) applicationm(); }
. . . . . . . . . . . .
```

5.8.3 Second-Order Filter Blocks

With increasing filter order, direct-form filters run into problems with respect to coefficient sensitivity and stability. This issue will be discussed in more detail in Chap. 8. One possibility to avoid these problems is to fractionise higher-order filters into sections of first and second order. The series connection of such low-order filters is therefore a frequent approach in filter design.

In this chapter, we introduced a variety of possibilities to realise second-order filter sections. Examples are
- direct form,
- normal form, and
- WDF based structures.

Also allpass filters can be realised in different ways using for example the
- direct form or a
- WDF or Gray and Markel implementation.

In the following, we will derive the mathematical algorithms for some of these basic structures.

Direct-form II

We start with the common direct-form II implementation whose block diagram is shown in Fig. 5-122. The two delay elements are numbered as ① and ②. For the derivation of the algorithm, we apply the approach introduced further up in this section.

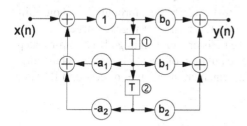

Fig. 5-122
Direct-form II filter of second-order

First we calculate the value of the output sample y of the filter and the temporary values t_1 and t_2 that are applied to the inputs of the delay elements. From Fig. 5-122 we can derive

$$t_1 = x - a_1 s_1 - a_2 s_2 \qquad (1)$$
$$y = b_0 t_1 + b_1 s_1 + b_2 s_2 \qquad (2)$$
$$t_2 = s_1 \qquad (3).$$

Then we transfer the input values as the new state variables s_1 and s_2 to the output of the delay elements.

$$s_2 = t_2 \qquad (4)$$
$$s_1 = t_1 \qquad (5)$$

With some modifications and simplifications, we make this set of equations ready for implementation in a C subroutine. In (1) we substitute s_0 for x since we pass the input value in the array element s[0] to the subroutine. Since s_0 is not needed later on, we can directly assign the calculated output value y to s_0 because we use the array element s[0] also to pass the calculated output sample back to the calling program. (3) and (4) can be directly combined since s_2 is not needed in later calculations. Note that the temporary variable t_2 is not needed anymore. This results in the following optimised set of equations.

$$t_1 = s_0 - a_1 s_1 - a_2 s_2 \qquad \text{(i)}$$
$$s_0 = b_0 t_1 + b_1 s_1 + b_2 s_2 \qquad \text{(ii)}$$
$$s_2 = s_1 \qquad \text{(iii)}$$
$$s_1 = t_1 \qquad \text{(iv)}$$

It is recommended to always start with the foolproof approach: Calculate the temporary input values of the delays and the output sample of the filter, then shift the input values of the delays to the respective outputs. After that, try to simplify the set of equations by dropping unnecessary copy statements and removing temporary variables that are not needed. These modifications must be made with care. As an example, equations (4) and (5) may be reversed in order without any problems. This is not the case for equations (iii) and (iv). The two delay elements form a shift register where, after the shift operation, input1 appears as the output1 and the former input2 appears as the output2. If we reverse the order of (iii) and (iv), this does not work anymore. The input1 directly falls through the whole shift register to the output2. So the general rule for such shift register structures is to start the copy statements at the end of the register and then to go back step-by-step to the beginning. This guarantees proper emulation of the shift register behaviour.

The following C listing is an implementation of the direct-form II filter discussed above.

```
void Direct_FormII(double s[], double coeff[]) {
    /* ** Second-order direct form filter type II ** */
    /* (2nd canonic form)                           */
    /* s[1] and s[2] are the state variables.       */
    /* s[0] is used to pass the input sample x(n)   */
    /* to the routine. Upon return, s[0] contains   */
    /* the calculated output sample y(n). coeff[]    */
    /* is filled with the filter coefficients in    */
    /* the order a1, a2, b0, b1, b2.                 */
    double t1;
    t1 = s[0]-coeff[0]*s[1]-coeff[1]*s[2];
    s[0]=coeff[2]*t+coeff[3]*s[1]+coeff[4]*s[2];
    s[2]=s[1];
    s[1]=t1;
    return;}
```

Example 5-13
Derive the DSP algorithm for the transposed direct-form structure as shown in Fig. 5-123.

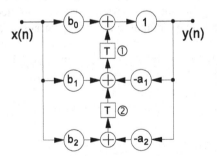

Fig. 5-123
Transposed direct-form filter of second order

The following set of equations can be derived from Fig. 5-123.

$$y = s_1 \qquad + b_0 x \quad (1)$$
$$t_1 = s_2 - a_1 y + b_1 x \quad (2)$$
$$t_2 = \quad -a_2 y + b_2 x \quad (3)$$
$$s_1 = t_1 \qquad\qquad (4)$$
$$s_2 = t_2 \qquad\qquad (5)$$

(2) can be substituted in (4) because s_1 is not used in (5). In the same way, (3) can be directly substituted in (5).

$$y = b_0 x + s_1$$
$$s_1 = b_1 x - a_1 y + s_2$$
$$s_2 = b_2 x - a_2 y$$

Finally taking into account that we use s_0 for passing the input and output sample values between calling program and subroutine, we get the following set of equations:

$$y = b_0 s_0 + s_1$$
$$s_1 = b_1 s_0 - a_1 y + s_2$$
$$s_2 = b_2 s_0 - a_2 y$$
$$s_0 = y$$

The translation of these equations into a C subroutine could look like this

```
void Direct_FormIItrans(double s[], double coeff[]) {
    /* Second-order transposed direct-formII filter */
    /* s[1] and s[2] are the state variables.       */
    /* s[0] is used to pass the input sample x(n)    */
    /* to the routine. Upon return, s[0] contains    */
    /* the calculated output sample y(n). coeff[]     */
    /* is filled with the filter coefficients in the */
    /* order b0, b1, a1, b2, a2.                      */
    double y;
    y    = coeff[0]*s[0]              +s[1];
    s[1] = coeff[1]*s[0]-coeff[2]*y+s[2];
    s[2] = coeff[3]*s[0]-coeff[4]*y;
    s[0] = y;
    return; }
```

Normal-form filter

The second-order normal-form filter can be considered as a special form of the general state-space filter shown in Fig. 5-121 which realises the general second-order transfer function

$$H(z) = \frac{g_0 + g_1 z^{-1} + g_2 z^{-2}}{1 + a_1 z^{-1} + a_2 z^{-2}} .$$

The normal form is characterised by the following set of state-space coefficients which determine the denominator polynomial of the second-order transfer function.

$$a_{11} = \alpha$$
$$a_{12} = -\beta \quad \text{with} \quad \alpha = -a_1/2$$
$$a_{21} = \beta \qquad \beta = \sqrt{a_2 - a_1^2/4}$$
$$a_{22} = \alpha$$

The remaining state-space coefficients d, b_1, b_2, c_1, and c_2 only influence the numerator polynomial of the transfer function. Since 5 parameters are available to realise the 3 numerator coefficients g_0, g_1, and g_2, there are numerous alternatives available to implement a given transfer function. One possible choice is to locate all calculations at the input of the filter. In this case, we have $c_1 = 1$ and $c_2 = 0$. The remaining parameters are calculated as

$$d = g_0$$
$$b_1 = 2\alpha g_0 + g_1$$
$$b_2 = [(\beta^2 - \alpha^2)g_0 - \alpha g_1 - g_2]/\beta.$$

So if we feed the parameter set (5.136) into the state-space subroutine, we obtain the desired normal-form filter.

$\text{coeff}[0] = g_0$ $\qquad\qquad\qquad\qquad$ $\text{coeff}[1] = 1.0$ $\quad$ $\text{coeff}[2] = 0.0$

$\text{coeff}[3] = 2\alpha g_0 + g_1$ $\qquad\qquad\qquad$ $\text{coeff}[4] = \alpha$ $\quad$ $\text{coeff}[5] = -\beta$ $\quad$ (5.136)

$\text{coeff}[6] = [(\beta^2 - \alpha^2)g_0 - \alpha g_1 - g_2]/\beta$ $\quad$ $\text{coeff}[7] = \beta$ $\quad$ $\text{coeff}[8] = \alpha$

If we intend to calculate the non-recursive part of the filter at the output of the normal-form structure, we have to apply the following set of parameters.

$\text{coeff}[0] = g_0$ $\quad$ $\text{coeff}[1] = 2\alpha g_0 + g_1$ $\quad$ $\text{coeff}[2] = -[(\beta^2 - \alpha^2)g_0 - \alpha g_1 - g_2]/\beta$

$\text{coeff}[3] = 1.0$ $\quad$ $\text{coeff}[4] = \alpha$ $\qquad\qquad$ $\text{coeff}[5] = -\beta$

$\text{coeff}[6] = 0.0$ $\quad$ $\text{coeff}[7] = \beta$ $\qquad\qquad$ $\text{coeff}[8] = \alpha$

It might also be advantageous to distribute the calculations which are related to the numerator polynomial to the input coefficients b_1 and b_2 and to the output coefficients c_1 and c_2. For floating-point implementations, this choice is of little relevance. In the fixed-point case, however, this degree of freedom of the state-space structure can be used to optimise the performance of the filter with respect to noise and overflow behaviour.

WDF-based second-order filter section

The block diagram of the WDF-based second-order filter section is shown in Fig. 5-124. The following set of equations can be derived.

$$y = s_1 + s_2 + \beta_0 x$$
$$t_1 = s_2 - \gamma_1 y + \beta_1 x$$
$$t_2 = -s_1 + \gamma_2 y - \beta_2 x$$
$$s_1 = t_1$$
$$s_2 = t_2$$

The intermediate storage in t_2 is not required. The array element s_0 is used to pass the input sample to the subroutine and to pass back the output sample.

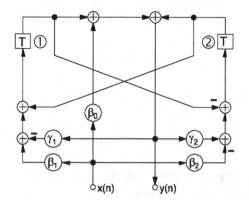

Fig. 5-124
WDF-based second-order filter section

$$y = s_1 + s_2 + \beta_0 s_0$$
$$t_1 = s_2 - \gamma_1 y + \beta_1 s_0$$
$$s_2 = -s_1 + \gamma_2 y - \beta_2 s_0$$
$$s_1 = t_1$$
$$s_0 = y$$

From this set of equations, we can derive the following C-code.

```
void WDF(double s[], double coeff[]) {
    /* Second-order WDF-based structure           */
    /* s[1] and s[2] are the state variables.     */
    /* s[0] is used to pass the input sample x(n) */
    /* to the routine. Upon return, s[0] contains */
    /* the calculated output sample y(n). coeff[] */
    /* is filled with the filter coefficients in the */
    /* order beta0, gamma1, beta1, gamma2, beta2.  */
    double y,t1;
    y   =  s[1]+s[2]       +coeff[0]*s[0];
    t1  =  s[2]-coeff[1]*y+coeff[2]*s[0];
    s[2] = -s[1]+coeff[3]*y-coeff[4]*s[0];
    s[1] = t1;
    s[0] = y;
    return; }
```

Second-order direct-form allpass section

The block diagram of the direct-form second-order allpass section is shown in Fig. 5-125. The following set of equations can be derived.

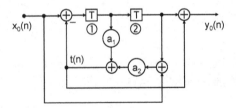

Fig. 5-125
Second-order direct-form allpass section

$$t = a_1 s_1 + a_2 (s_2 + x)$$
$$y = s_2 + t$$
$$t_1 = x - t$$
$$t_2 = s_1$$
$$s_2 = t_2$$
$$s_1 = t_1$$

The intermediate storage in t_2 is not required. The array element s_0 is used to pass the input sample to the subroutine and to pass back the output sample.

$$t = a_1 s_1 + a_2 (s_2 + s_0)$$
$$t_1 = s_0 - t$$
$$s_0 = s_2 + t$$
$$s_2 = s_1$$
$$s_1 = t_1$$

From this set of equations, we can derive the following C-code.

```
void Allpass_DF(double s[], double coeff[]) {
    /* Second-order direct-form allpass filter       */
    /* s[1] and s[2] are the state variables.        */
    /* s[0] is used to pass the input sample x(n)    */
    /* to the routine. Upon return, s[0] contains    */
    /* the calculated output sample y(n). coeff[]    */
    /* is filled with the filter coefficients in the */
    /* order a1, a2.                                  */
    double t,t1;
    t    = coeff[0]*s[1]+coeff[1]*(s[2]+s[0]);
    t1   = s[0]-t;
    s[0] = s[2]+t;
    s[2] = s[1];
    s[1] = t1;
    return; }
```

Gray and Markel second-order allpass section

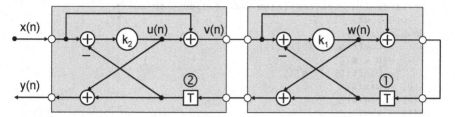

Fig. 5-126   Gray and Markel second-order allpass section

Fig. 5-126 shows the block diagram of a second-order allpass filter based on the one-multiplier Gray and Markel structure. We denote three internal nodes of the block diagram as u, v, and w. The reason is that the intermediate results stored in these variables are used twice in the course of the further calculations so that we can save arithmetic operations. We can establish the following set of operations.

$$u = k_2(x - s_2)$$
$$v = u + x$$
$$w = k_1(v - s_1)$$
$$y = u + s_2$$
$$t_1 = v + w$$
$$t_2 = w + s_1$$
$$s_2 = t_2$$
$$s_1 = t_1$$

The temporary variables t_1 and t_2 can be eliminated. s_0 is again used as the input and output buffer of the subroutine.

$$u = k_2(s_0 - s_2)$$
$$v = u + s_0$$
$$w = k_1(v - s_1)$$
$$s_0 = u + s_2$$
$$s_2 = w + s_1$$
$$s_1 = w + v$$

From this set of equations, we can derive the following C-code.

```
void Allpass_GM(double s[], double coeff[]) {
    /* Second-order allpass filter using the Gray     */
    /* and Markel structure                           */
    /* s[1] and s[2] are the state variables.         */
    /* s[0] is used to pass the input sample x(n)     */
```

```
/* to the routine. Upon return, s[0] contains    */
/* the calculated output sample y(n). coeff[]     */
/* is filled with the filter coefficients in the */
/* order k2, k1.                                   */
double u,v,w;
u     = coeff[0]*(s[0]-s[2]);
v     = u + s[0];
w     = coeff[1]*(v-s[1]);
s[0]  = u+s[2];
s[2]  = w+s[1];
s[1]  = w+v;
return; }
```

Cascading low-order sections to obtain higher order filters is easily accomplished. The filter subroutine is called several times in a row, each time with a different set of coefficients and with a different set of state variables. The output value of the preceding filter section is copied to the input of the next section. The following listing shows the example of a 6th-order direct-form filter.

```
while (!stop) {
    s1[0] = GetInputSample;
    Direct_FormII(s1,coeff1);
    s2[0] = s1[0];
    Direct_FormII(s2,coeff2);
    s3[0] = s2[0];
    Direct_FormII(s3,coeff3);
    PutOutputSample(s3[0]); }
```

It is important to note that each partial filter requires its own set of state variables. Each of the second-order sections is an independent filter which must save its state variables until the next sample is processed.

5.8.4 FIR Filters

Fig. 5-127 shows the block diagram of the FIR filter which we will convert into a software implementation. We start again with calculating the value of the output sample $y(n)$ of the filter and the temporary values that are applied to the inputs of the delay elements.

$$y = b_0 x + \sum_{i=1}^{M} b_i s_i$$

We substitute s_0 for x since we pass the input value x in the array element s_0 to the subroutine.

$$y = \sum_{i=0}^{M} b_i s_i$$

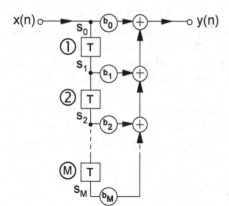

Fig. 5-127
Block diagram of the FIR filter

The delay elements are arranged in series forming a shift register. For the temporary values t_i we therefore have

$$t_i = s_{i-1} \quad \text{for } i = 1 \ldots M.$$

Finally we transfer the input values t_i as the new state variables s_i to the outputs of the delay elements.

$$s_i = t_{i-1} \quad \text{for } i = 1 \ldots M$$

The latter relations describe the behaviour of the shift register which can be combined to the simplified form

$$s_i = s_{i-1} \quad \text{for } i = M \ldots 1.$$

As explained further above, the copy operations must start at the end of the shift register to achieve the desired behaviour. The temporary variables t_i are not needed in this case. The following C subroutine implements the FIR filter algorithm.

```
void FIR_filter(double s[], double coeff[], int M)        {
    /* Mth order FIR filter. The array s[] contains the */
    /* state variables. s[0] is used to pass the input  */
    /* sample x(n) to the routine. Upon return, s[0]    */
    /* contains the calculated output sample y(n).      */
    /* coeff[] is filled with the filter coefficients   */
    /* in the order b0 ... bM.                          */
    int i;
    double acc = 0.0;
    for (i=M; i >= 0; i--) acc += coeff[i]*s[i];
    for (i=M; i > 0; i--) s[i]=s[i-1];
    s[0]=acc;
    return; }
```

The filter algorithm mainly consists of two for loops. The first one performs a sequence of MAC operations to calculate the output sample. The second for loop shifts the input signal by one position through the chain of delay elements. In view of the fact that the MAC operations typically require only one processor cycle, the shift operation causes a considerable part of the processor load. The latter can be avoided if the shift register is realised in the form of a cyclic buffer. The whole shift operation then collapses to decrementing an index register which points to the beginning of the shift register.

If an index register in the DSP is configured to control a cyclic buffer of length *N*, the hardware of the DSP takes care that the index value always remains within the range 0 ... *N*−1 when the index register is incremented or decremented. Fig. 5-128 demonstrates this behaviour.

..., N-2, N-1, 0, 1, 2, ..., N-3, N-2, N-1, 0, 1, 2, ..., N-3, N-2, N-1, 0, 1, ...

◄─── decrement   increment ───►

Fig. 5-128 Principle of a cyclic index register

The following C routine makes use of a cyclic buffer to perform the simplified shift operation.

```
void FIR_filter(double *io,double s[],double coeff[],int *p,int M) {
    /* Mth order FIR filter. The array s[] contains    */
    /* the state variables. coeff[] is filled with the */
    /* filter coefficients in the order b0 ... bM.      */
    /* p is the pointer to the beginning of the shift   */
    /* register in the circular buffer. This pointer    */
    /* must be saved until the next call of the         */
    /* subroutine, like the state variables. The input  */
    /* sample x(n) is passed via i0 to the subroutine.  */
    /* Upon return, the variable to which io points     */
    /* contains the output value y(n).                  */
    int i;
    double acc = 0.0;
    s[*p]=*io;
    i=M+1;
    while (i != 0) acc += coeff[--i]*s[cycdec(p,M)];
    cycdec(p,M);                    /* shift operation */
    *io=acc;
    return; }
```

The function cycdec performs a cyclic decrement of the index p. If the index is decremented when zero, it is set to M.

```
int cycdec(int *p, int M) {
    /* cyclic decrement of *p */
    if ((*p)-- == 0) *p=M;
    return *p; }
```

Note that, in assembler, this function needs not to be explicitly written. The hardware of the DSP automatically controls the number range of the index register.

Passing parameters to a subroutine requires some additional processor cycles. Parameters are copied onto the stack before the subroutine is called and removed from the stack upon return. Also the local variables of a subroutine are stored on the stack. These stack operations can be avoided if global variables are used instead. In computer software development, it is commonly recommended to use local variables to make code more universal and readable. But in DSP applications it might be more efficient to use global variables. This increases speed of code but reduces modularity.

5.8.5 Fixed-Point Implementation

Limited number range or overflow are not an issue if we perform our calculations in floating-point arithmetic. The exponent of a floating-point number automatically shifts the binary point (the equivalent of the decimal point in decimal arithmetic) to a position which allows the best possible utilisation of the dynamic range of the mantissa. The strategy is to minimise the number of leading zeros in the number representation. In the case of fixed-point processors, however, we must map the signal and coefficient values to integers which are typically represented by 16, 24, or 32 bits. The FIR filter algorithm may serve as an example to demonstrate the peculiarities of fixed-point implementations.

According to (5.1), the output sample of the FIR filter is calculated as

$$y(n) = \sum_{i=0}^{M} b_i x(n-i) \tag{5.137}$$

which is basically a sequence of MAC operations. For a fixed-point implementation, all signal and coefficient values must be represented by integers. For the signal representation, this is not a problem. We assume, e.g. for 16-bit processing in 2th-complement arithmetic, that the allowed signal values lie in the range from -2^{15} to $+2^{15}-1$.

Filter coefficients are typically in the order of one or below. In order to prepare (5.137) for integer calculation, we multiply all coefficients b_i by an integer power of two (2^s) such that the coefficient with the largest magnitude is just representable with the available 16 bits. The fractional digits of these scaled coefficients are dropped by rounding to the nearest integer. This rounding results in filter coefficients which deviate more or less from the ideal values as obtained from filter design. Depending on the sensitivity of the filter to coefficient variations, the actual filter characteristic will deviate at a certain degree from the specification.

The multiplication of 16-bit signal values by 16-bit coefficients yields results of 32-bit length. Since the MAC operation adds the result of the multiplication to the contents of the accumulator, the latter must have a length of 32 bits plus a number of guard bits which avoid possible overflow. Overflow occurs if the calculated sum in a sequence of MAC operations requires more bits than available in the accumulator.

In the next step, the multiplication of the coefficients by 2^s is compensated by dividing the sum in the accumulator by 2^s. The fractional digits obtained by this division are dropped by rounding to the nearest integer. Such a division by 2^s is commonly realised by shifting the contents of the accumulator by s bits to the right. This rounding introduces another type of error into the filter algorithm which is commonly referred to as the quantisation error. Quantisation manifests itself as a noiselike signal which is superimposed to the useful signal.

In the last step, the result in the accumulator is stored back to memory with the original wordlength of 16 bits. If the result of the described calculations exceeds the number range of a 16-bit integer, we are in a overflow situation which must be handled appropriately. A common action is saturation where the actual value is replaced by the maximum value that can be expressed with 16 bits.

The described procedures to handle FIR filter calculations in a fixed-point environment can be mathematically summarised as

$$y(n) = \text{overflow}\left(\text{round}\left(2^{-s} \sum_{i=0}^{M} \text{round}\left(2^{s} b_i \right) x(n-i) \right) \right).$$

Comparison of this expression with (5.137) immediately reveals that fixed-point implementations require a lot more consideration. Issues like scaling to handle non-integer coefficients, coefficient imprecision, quantisation noise, and overflow handling are much less pronounced or even not present in floating-point implementations, where equations like (5.137) can be implemented in a straight-forward manner.

The consequences of quantisation, overflow, and finite coefficient wordlength are analysed in detail in Chap. 8.

6 Design of IIR Filters

6.1 Introduction

The different mathematical expressions for the transfer functions of FIR and IIR filters also result in very different design methods for both filter types. While a given frequency response is approximated in case of an IIR filter by a rational fractional transfer function of z with poles and zeros (5.30), the FIR filter is simply characterised by zeros and a polynomial of z (5.2).

In the case of IIR filters we can take advantage of their relationship to continuous-time filters made up of networks of discrete elements. Both have impulse (or unit-sample) responses of infinite length and are described by rational fractional transfer functions in the frequency domain. We can start with the actual filter design procedure completely in the continuous-time domain by making use of the whole range of classical approximation methods, such as Chebyshev, Butterworth or Bessel, for instance, and of numerous design tables available in the literature. The transformation of the obtained coefficients of the analog reference filters into those of the target discrete-time filters could be performed, in principle, using a pocket calculator. The kernel of commercial digital filter design software is in general an analog filter design program with an appendage to calculate the digital filter coefficients.

It seems confusing at first glance that various methods are available for the transition from the continuous-time to the discrete-time domain as a look at the headings of this chapter shows. Why does not exist one universal method to transfer all characteristics of the analog filter in the time and frequency domain to corresponding discrete-time realisations? The first clue to this problem was the fact that the frequency response of the continuous-time filter is a rational fractional function of $j\omega$ while the frequency response of the discrete-time system is periodic and a function of $e^{j\omega T}$. Thus we cannot expect to obtain identical frequency responses within the frequency range $-\pi/T \leq \omega \leq +\pi/T$ for both cases. Also in the time domain, discrete-time and continuous-time systems show marked differences concerning the mathematical description of the system behaviour. The relations between the input and output signals of the system are expressed in the one case by an integral, the convolution integral (1.1), in the other case by a sum, the convolution sum (3.3). It can be generally stated that the characteristics of continuous-time and discrete-time systems converge more the smaller the chosen sampling period T. This statement is true for both the time and the frequency domain. In the limiting case when T approaches zero, the exponential $e^{j\omega T}$ can be replaced by its argument and the sum converges to an integral.

D. Schlichthärle, *Digital Filters*, DOI: 10.1007/978-3-642-14325-0_6,
© Springer-Verlag Berlin Heidelberg 2011

As a rule, it is desirable to choose the sampling frequencies of digital systems to be not much higher than is required for processing the respective signals in accordance with the sampling theorem. Sampling rate, and thus processing speed, has an important influence on hardware costs. Examples for the increase of costs are the need for higher signal processing speed, faster digital logic, higher A/D and D/A conversion rate and lower memory access times. Another effect can be observed in the context of non-real-time applications where data have to be stored. As the amount of sampled data increases with the sampling rate, larger storage capacities, in terms of RAM, hard disk or CD for instance, are required, which is a further cost factor.

In the following section we will show that, under certain circumstances, convolution sum and convolution integral yield comparable system behaviour. The limiting conditions, however, to reach this result are hard to satisfy in practice.

6.2 Preservation Theorem of the Convolution

The behaviour of continuous-time LTI systems is described by the convolution integral according to (1.1).

$$y_a(t) = \int_{-\infty}^{+\infty} x_a(\tau)\, h_a(t-\tau)\, d\tau \tag{6.1}$$

Assuming that the input signal $x_a(t)$ is band-limited and meets the requirements of the sampling theorem, we can derive a relation which permits the numerical calculation of the output signal $y_a(t)$ at discrete instants of time nT. Figure 6-1 illustrates the problem.

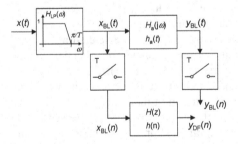

Fig. 6-1
Illustration of the preservation theorem of the convolution

A band-limited signal $x_{BL}(t)$ is applied to the input of a continuous-time filter which is characterised by its transfer function $H_a(j\omega)$ or alternatively by its impulse response $h_a(t)$. The output signal $y_{BL}(t)$ of the filter is also band-limited provided that the filter is linear and time-invariant. Both signals, $x_{BL}(t)$ and $y_{BL}(t)$ are sampled at a rate of $1/T$ resulting in the discrete-time sequences $x_{BL}(n)$ and

$y_{BL}(n)$. The question now is how a discrete-time filter, characterised by the transfer function $H(z)$ and the unit-sample response $h(n)$, needs to look like in order to produce the same output signal, i.e. $y_{DF}(n) = y_{BL}(n)$. The two paths differ in the order of sampling and filtering. In the one case, the signal is filtered in the continuous-time domain and then sampled. In the other, the signal is first sampled and then filtered in the discrete-time domain.

Input and output signal of the continuous-time filter are related as

$$Y_{BL}(j\omega) = H_a(j\omega)X_{BL}(j\omega) \; .$$

The output signal $y_{BL}(t)$ can be found by inverse Fourier transform.

$$y_{BL}(t) = \frac{1}{2\pi} \int_{-\infty}^{+\infty} H_a(j\omega)X_{BL}(j\omega)e^{j\omega t} d\omega$$

The samples of $y_{BL}(t)$ are obtained by the substitution $t = nT$.

$$y_{BL}(nT) = y_{BL}(n) = \frac{1}{2\pi} \int_{-\infty}^{+\infty} H_a(j\omega)X_{BL}(j\omega)e^{jn\omega T} d\omega \qquad (6.2)$$

Since $X_{BL}(j\omega)$ is band-limited, the spectrum is fully defined by the samples $x_{BL}(n)$. According to (4.4b), both are related as

$$X_{BL}(j\omega) = \begin{cases} T\displaystyle\sum_{k=-\infty}^{+\infty} x_{BL}(k)e^{-jk\omega T} & \text{for } |\omega| < \pi/T \\[2mm] 0 & \text{for } |\omega| \geq \pi/T \end{cases} \; .$$

Substitution in (6.2) yields

$$y_{BL}(n) = \frac{T}{2\pi} \int_{-\pi/T}^{+\pi/T} H_a(j\omega) \sum_{k=-\infty}^{+\infty} x_{BL}(k)e^{-jk\omega T}e^{jn\omega T} \; .$$

Interchanging integration and summation gives

$$y_{BL}(n) = T \sum_{k=-\infty}^{+\infty} x_{BL}(k) \frac{1}{2\pi} \int_{-\pi/T}^{+\pi/T} H_a(j\omega)e^{-j(n-k)\omega T} \; .$$

The integral term represents an inverse Fourier transform of $H_a(j\omega)$ which calculates the samples of the impulse response $h_a(t)$ taken at the instants $t = (n-k)T$. Since the integration interval is limited to the frequency range $|\omega| < \pi/T$, the calculation of the samples only makes use of the spectral components up to half the sampling frequency. The integral thus yields the samples of a band-limited version of the impulse response of the analog filter which we denote in the following by the index BL.

$$y_{BL}(n) = T \sum_{k=-\infty}^{+\infty} x_{BL}(k) h_{aBL}(n-k)$$

The output signal of the discrete-time filter is obtained by the convolution

$$y_{DF}(n) = \sum_{k=-\infty}^{+\infty} x_{BL}(k) h(n-k) \tag{6.3}$$

where $h(n)$ is the unit-sample response of the filter. So both paths in Fig. 6-1 provide the same output signal if the impulse response of the continuous-time filter and the unit-sample response of the discrete-time filter are related as

$$h(n) = T h_{aBL}(n) .$$

In summary, convolution integral (6.1) and convolution sum (6.3) provide the same output samples if the following conditions are satisfied:
1. The spectrum of the input signal must be limited to the Nyquist frequency.
2. The spectrum of the impulse response of the analog filter must also be limited to the Nyquist frequency before it is sampled to obtain the unit-sample response of the discrete-time filter to be realised.

The second requirement avoids a possible aliasing and thus distortion of the frequency response in the baseband. Such aliasing occurs, as shown in Chap. 4, by the periodic continuation of the spectrum if the impulse response is sampled and the sampling theorem is not fulfilled.

The preservation theorem of the convolution provides, in principle, the desired method to numerically imitate the convolution integral in the discrete-time domain. At the same time, we obtain a transfer function in the frequency domain that is identical to the one of the corresponding analog filter in the frequency range $|\omega| < \pi/T$. We merely have to design a digital filter that has a unit-sample response corresponding to the samples of the low-pass limited impulse response of the analog reference filter. In most practical applications, however, this second requirement of the preservation theorem can only hardly be met or cannot be fulfilled at all.

The fewest problems can be expected if the analog reference filter is a low-pass filter whose frequency response has reached such a high attenuation at $f_s/2$ that aliasing does not cause noticeable distortion of the frequency response in the baseband. This condition is approximately fulfilled by higher-order low-pass and bandpass filters whose cutoff frequencies lie well below the Nyquist frequency.

In all other cases, an additional sharp low-pass filter is required to limit the frequency response of the analog filter strictly to $f_s/2$ before the impulse response is sampled. Unfortunately, this low-pass filter, whose characteristic is contained in the sampled impulse response, has to be implemented in the discrete-time domain, too, which may increase the complexity of the digital filter considerably. The filter order of the discrete-time system would exceed the order of the original analog reference filter by far. Even if this extra complexity is tolerated, the described procedure is not an ideal solution since filters with sharp transition bands cause considerable phase distortions which would destroy the exact reproduction of the system behaviour in the time domain. A further drawback is the fact that the additional low-pass filter has a transition band with finite width which cannot be used any more. This is especially annoying for the design of high-pass and bandstop filters which, in the ideal case, have no attenuation at the Nyquist frequency.

The discussion so far makes clear that it is difficult to exactly reproduce the frequency response of an analog filter by a discrete-time realisation. Experience with practical applications shows, however, that such an exact reproduction is seldom needed. In the following, we will show how rational fractional transfer functions of z can be derived for all filter types that preserve the order of the analog reference filter and approximate the characteristics of this filter with an accuracy sufficient for most practical cases.

6.3 Approximation in the Time Domain

We start with a group of design methods which exactly reproduce the response of continuous-time filters to standard input signals such as pulses, steps, or ramps in the discrete-time domain. These methods are therefore summarised under the term "approximation in the time domain". Preserving the order of the analog reference filter is only possible if we tolerate the violation of the sampling theorem in the transition from the continuous-time to the discrete-time domain. The named violation of the sampling theorem has the consequence that these methods are not universally valid. They only yield exact results in the time domain for special classes of input signals. Since we are dealing with linear time- (shift-) invariant systems, the exact reproduction in the time domain also applies to signals which are composed of sequences of pulses, steps, or kinks, provided that these occur in the sampling time grid. The deviations between the frequency responses of continuous-time reference filter and resulting discrete-time filters will depend on the amount of aliasing that occurs when we sample impulse, unit step, and ramp response.

6.3.1 Continuous-Time Filter Responses

For the continuous-time reference filter, we assume a general transfer function of the form

$$H(p) = \frac{\sum_{j=0}^{M} b_j p^j}{p^N + \sum_{i=0}^{N-1} a_i p^i} . \tag{6.4}$$

The filter coefficient a_N is normalised to unity. If the degrees of numerator and denominator are equal, we apply long division in order to reduce the degree of the numerator polynomial by one.

$$H(p) = b_N + \frac{\sum\limits_{j=0}^{N-1}(b_j - a_j b_N)p^j}{p^N + \sum\limits_{i=0}^{N-1} a_i p^i}. \tag{6.5}$$

Since the resulting rational fractional transfer function is strictly proper, it can be expanded in partial fractions.

$$H(p) = b_N + \sum_{k=1}^{N} \frac{c_k}{p - p_{\infty k}} \tag{6.6}$$

$p_{\infty k}$ is the kth pole of the filter. c_k is the kth partial-fraction coefficient. Evaluation of (6.4) and (6.6) for $p = 0$ reveals the relation

$$\frac{b_0}{a_0} = b_N + \sum_{k=1}^{N} \frac{c_k}{-p_{\infty k}} \tag{6.7}$$

which will be useful in the following calculations. The constant term in (6.6) can be realised as a bypass with gain b_N. The summation term represents a low-pass filter since the transfer function

$$H(p) = \sum_{k=1}^{N} \frac{c_k}{p - p_{\infty k}} \tag{6.8}$$

approaches zero as p goes to infinity. (6.8) will be used in the following as the analog reference filter for the design of a related discrete-time filter with similar characteristics in the time domain. The constant b_N will be later added to the obtained z-transfer function in order to realise the bypass with frequency-independent gain in the discrete-time domain.

In the first step, we calculate the impulse response $h(t)$ by inverse Laplace transform of (6.8).

$$h(t) = \sum_{k=1}^{N} c_k\, e^{p_{\infty k} t}\, u(t) \tag{6.9}$$

The terms in the summation formula represent the exponentially decaying characteristic oscillations of the filter. The unit step function takes into account that the filter is causal, and therefore $h(t) = 0$ for $t < 0$.

The Laplace transform of the unit step function $u(t)$ is

$$U(p) = 1/p .$$

The unit step response $A(p)$ of the continuous-time filter is therefore obtained in the frequency domain as

$$A(p) = U(p)H(p) = \frac{1}{p}H(p) .$$

Since division by p in the frequency domain is equivalent to integration in the time domain, the unit step response $a(t)$ is obtained as

$$a(t) = \int_0^t \left(\sum_{k=1}^{N} c_k\, e^{p_{\infty k}\tau}\, u(\tau) \right) d\tau$$

$$a(t) = \sum_{k=1}^{N} \frac{c_k}{p_{\infty k}} e^{p_{\infty k}\tau} \bigg|_0^t\, u(t)$$

$$a(t) = \sum_{k=1}^{N} \frac{c_k}{p_{\infty k}} (e^{p_{\infty k}t} - 1)\, u(t) \tag{6.10}$$

The Laplace transform of the ramp function

$$r(t) = \frac{t}{T} u(t) \quad \text{is}$$

$$R(p) = \frac{1}{p^2 T}$$

The ramp response $M(p)$ of the continuous-time filter is therefore obtained in the frequency domain as

$$M(p) = \frac{1}{p^2 T} H(p) = \frac{1}{pT} A(p) .$$

So $a(t)$ has to be integrated again to yield the ramp response $m(t)$ in the time domain.

$$m(t) = \frac{1}{T} \int_0^t \left(\sum_{k=1}^{N} \frac{c_k}{p_{\infty k}} (e^{p_{\infty k}\tau} - 1) u(\tau) \right) d\tau$$

$$m(t) = \sum_{k=1}^{N} \frac{c_k}{p_{\infty k}^2 T} e^{p_{\infty k}\tau} \bigg|_0^t\, u(t) - \sum_{k=1}^{N} \frac{c_k}{p_{\infty k}} \frac{\tau}{T} \bigg|_0^t\, u(t)$$

$$m(t) = \sum_{k=1}^{N} \frac{c_k}{p_{\infty k}^2 T} (e^{p_{\infty k}t} - 1) u(t) - \sum_{k=1}^{N} \frac{c_k}{p_{\infty k}} \frac{t}{T} u(t) \tag{6.11}$$

6.3.2 Sampling the Analog Filter Responses

The transition to the discrete-time domain is accomplished by sampling $h(t)$, $a(t)$, and $m(t)$. The continuous time variable t is replaced by the discrete time variable nT. But in doing so, we encounter a problem. The named filter responses contain jumps at $t = 0$ for which it is difficult to derive meaningful sample values. The step function is defined just before and after switching but not at the time instant of the discontinuity. So the question is which value to choose for $h(0)$, $a(0)$, or $m(0)$. A promising approach is to limit the bandwidth of the signals which avoids the occurrence of discontinuities. As a side effect, aliasing would be reduced resulting in a better reproduction of the frequency response of the analog reference filter. With respect to the latter, the ideal case would be to limit the bandwidth of $h(t)$, $a(t)$, or $m(t)$ to half the sampling frequency. These named responses include the following types of functions:
- Unit step function
- Decaying exponential
- Ramp function

For each of these functions, we will analyse in the following the consequences of limiting the bandwidth to half the sampling frequency.

The unit step function $u(t)$ has the spectrum

$$U(j\omega) = \pi\,\delta(\omega) + \frac{1}{j\omega} \ .$$

The band-limited unit step function is obtained by inverse Fourier transform with the frequency limits $\pm\pi/T$.

$$u_{\mathrm{BL}}(t) = \frac{1}{2\pi} \int_{-\pi/T}^{+\pi/T} \left(\pi\delta(\omega) + 1/j\omega\right) e^{j\omega t}\,\mathrm{d}\omega$$

$$u_{\mathrm{BL}}(t) = \frac{1}{2\pi} \int_{-\pi/T}^{+\pi/T} \pi\delta(\omega) e^{j\omega t}\,\mathrm{d}\omega + \frac{1}{2\pi} \int_{-\pi/T}^{+\pi/T} \frac{e^{j\omega t}}{j\omega}\,\mathrm{d}\omega$$

$$u_{\mathrm{BL}}(t) = \frac{1}{2} + \frac{1}{2\pi} \int_{-\pi/T}^{+\pi/T} \frac{\cos\omega t + j\sin\omega t}{j\omega}\,\mathrm{d}\omega = \frac{1}{2} + \frac{1}{2\pi} \int_{-\pi/T}^{+\pi/T} \frac{\sin\omega t}{\omega}\,\mathrm{d}\omega$$

$$u_{\mathrm{BL}}(t) = \frac{1}{2} + \frac{1}{2\pi} \int_{-\pi t/T}^{+\pi t/T} \frac{\sin\omega t}{\omega t}\,\mathrm{d}\omega t \qquad (6.12)$$

Closely related to the integral in the above equation is the sine integral function $\mathrm{Si}(x)$ which is defined as

$$\mathrm{Si}(x) = \int_{0}^{x} \frac{\sin t}{t}\,\mathrm{d}t \ .$$

Substitution in (6.8) yields the expression

$$u_{\mathrm{BL}}(t) = \frac{1}{2} + \frac{1}{2\pi} 2\,\mathrm{Si}(\pi t/T) = \frac{1}{2} + \frac{1}{\pi}\mathrm{Si}(\pi t/T)\ .$$

Band-limited unit step function

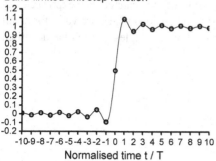

Normalised time t / T

Fig. 6-2
Band-limited unit step function $u_{\mathrm{BL}}(t)$

The Si-function cannot be expressed by elementary functions. Values may be found in tables or calculated using known series expansions. Figure 6-2 shows the graph of the band-limited unit step function which oscillates about zero for $t < 0$ and about one for $t > 0$. For $t = 0$, the function assumes a value of ½. Unfortunately, the sampling instants do no coincide with the points where the function assumes the values 0 or 1. But still a good approximation to the sampled values shown in Fig. 6-2 would be the sequence ... 0, 0, ½, 1, 1 ... . In contrast to the discrete-time unit step sequence $u(n)$ that we introduced in Chap. 3, the sequence derived from the band-limited step function $u_{\mathrm{BL}}(t)$ assumes a value of ½ for $n = 0$. Formally we can express this sequence as

$$u_{\mathrm{BL}}(n) \approx u(n) - 0.5\delta(n)\ . \qquad (6.13)$$

A frequent expression occurring in the time response of filters is the complex exponential of the form

$$f(t) = e^{p_\infty t}\, u(t) \qquad (6.14)$$

which has a jump at $t = 0$. This signal has the spectrum

$$F(\mathrm{j}\omega) = \frac{1}{\mathrm{j}\omega - p_\infty}\ .$$

The band-limited version of $f(t)$ is obtained by the inverse Fourier transform

$$f_{\mathrm{BL}}(t) = \frac{1}{2\pi} \int\limits_{-\pi/T}^{+\pi/T} \frac{e^{\mathrm{j}\omega t}}{\mathrm{j}\omega - p_\infty}\, \mathrm{d}\omega\ .$$

$$f_{\text{BL}}(t) = \frac{-j}{2\pi} e^{p_\infty t} \text{Ei}(j\omega t - p_\infty t) \Big|_{-\pi/T}^{+\pi/T}$$

$$f_{\text{BL}}(t) = \frac{-j}{2\pi} e^{p_\infty t} \left(\text{Ei}(j\pi t/T - p_\infty t) - \text{Ei}(-j\pi t/T - p_\infty t) \right) \qquad (6.15)$$

Ei(x) is the so-called exponential integral which cannot be expressed by elementary functions. There only exist series expansions which are valid for small or for large arguments. Figures 6-3 and 6-4 show examples of such band-limited responses.

Band-limited decaying exponential

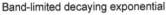
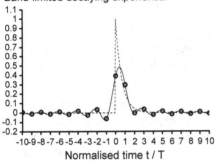

Fig. 6-3
Example of a band-limited decaying exponential
Dashed line: unlimited
Solid line: band-limited

Band-limited decaying sinusoid

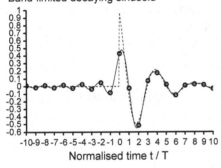

Fig. 6-4
Example of a band-limited decaying sinusoid
Dashed line: unlimited
Solid line: band-limited

As in the case of the unit step function, the largest deviation between original and band-limited function occurs in the vicinity of the discontinuity. The value of the band-limited function at $t = 0$ is not in the middle of the step. The exact value depends on the complex pole frequency p_∞. Evaluation of (6.15) around $t = 0$ yields

$$f_{\text{BL}}(0) = \frac{1}{2} - \frac{1}{\pi} \arctan \frac{-2 \, \text{Re} \, p_\infty T/\pi}{1 - \left| p_\infty T/\pi \right|^2} - \frac{j}{4\pi} \ln \frac{1 + \left| p_\infty T/\pi \right|^2 - 2 \, \text{Im} \, p_\infty T/\pi}{1 + \left| p_\infty T/\pi \right|^2 + 2 \, \text{Im} \, p_\infty T/\pi}$$

$$f_{BL}(0) = \frac{1}{2} - d(p_\infty)$$ (6.16)

For poles close to the origin, $f_{BL}(0)$ approaches a value of ½. Apart from $t = 0$, original and band-limited function match quite well. So we choose the samples of (6.14) as an approximation of the band-limited signal and use (6.16) as the sample value for $n = 0$.

$$f_{BL}(n) \approx e^{p_\infty T n} u(n) - \left(\frac{1}{2} + d(p_\infty)\right)\delta(n)$$

The ramp function is defined as

$$r(t) = \frac{t}{T} u(t) \ .$$

The band-limited version of $r(t)$ can be calculated as

$$r_{BL}(t) = \frac{t}{T}\left(\frac{1}{2} + \frac{1}{\pi}\text{Si}(\pi t/T)\right) + \frac{1}{\pi^2}\cos \pi t/T \ .$$

Figure 6-5 shows the graph of this function. The ramp function is continuous in $t = 0$. The band-limited function deviates only little from the original function $r(t)$. We use, therefore, the sequence

$$r_{BL}(n) = r(n) = \frac{nT}{T}u(nT) = nu(n)$$

as the discrete-time equivalent of $r(t)$.

Band-limited ramp function

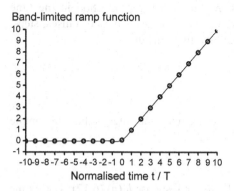

Normalised time t / T

Fig. 6-5
Band-limited ramp function $r_{BL}(t)$

Using the sequences of the basic functions that we derived in this section, we can define the following band-limited discrete-time approximations of the impulse response $h(t)$ (6.6), of the unit step response $a(t)$ (6.7), and of the ramp response $m(t)$ (6.8) of the analog reference filter.

$$h_a(n) = \sum_{k=1}^{N} c_k \left(e^{p_{\infty k} T n} u(n) - \frac{1}{2} \delta(n) \right) - \frac{1}{T} D_h \delta(n)$$

$$\text{with } D_h = \sum_{k=1}^{N} c_k T d(p_{\infty k})$$

(6.17)

$$a_a(n) = \sum_{k=1}^{N} \frac{c_k}{p_{\infty k}} (e^{p_{\infty k} T n} - 1) u(n) - D_a \delta(n)$$

$$\text{with } D_a = \sum_{k=1}^{N} \frac{c_k}{p_{\infty k}} d(p_{\infty k})$$

(6.18)

$$m_a(n) = \sum_{k=1}^{N} \frac{c_k}{-p_{\infty k}} n u(n) + \sum_{k=1}^{N} \frac{c_k}{p_{\infty k}^2 T} (e^{p_{\infty k} T n} - 1) u(n) - D_m \delta(n)$$

$$\text{with } D_m = \sum_{k=1}^{N} \frac{c_k}{p_{\infty k}^2 T} d(p_{\infty k})$$

(6.19)

In the next sections, we will derive the z-transfer functions of discrete-time filters which exactly reproduce the above responses of the analog reference filter. We will discuss three cases:

- the unit-sample sequence $\delta(n)$ at the input results in $h_a(n)$ at the output,
- the unit-step sequence $u(n)$ at the input results in $a_a(n)$ at the output,
- the ramp sequence $r(n)$ at the input results in $m_a(n)$ at the output.

The related discrete-time filter design procedures are referred to as the impulse-invariant, step-invariant, and ramp-invariant design methods. Since we are dealing with linear time- (shift-) invariant systems, the exact reproduction in the time domain also applies to signals which are composed of sequences of pulses, steps, or kinks, provided that these occur on the sampling time grid.

6.3.3 Impulse-Invariant Design

In this case, the stimulus signal $x(n)$ of the discrete-time filter is the unit sample impulse $\delta(n)$. The z-transform of this signal is

$$X(z) = 1 \ .$$

The desired output signal is the sampled impulse response $h_a(n)$ (6.17). Taking the z-transform of $h_a(n)$ yields

$$H_a(z) = \sum_{k=1}^{N} c_k \left(\frac{z}{z - e^{p_{\infty k} T}} - \frac{1}{2} \right) - \frac{1}{T} D_h$$

The transfer function of the impulse-invariant discrete-time filter is obtained as

$$H(z) = \frac{H_a(z)}{X(z)} = \frac{\sum\limits_{k=1}^{N} c_k \left(\dfrac{z}{z - e^{p_{\infty k}T}} - \dfrac{1}{2} \right) - \dfrac{1}{T} D_h}{1}$$

$$H(z) = \sum_{k=1}^{N} c_k \left(\frac{z}{z - e^{p_{\infty k}T}} - \frac{1}{2} \right) - \frac{1}{T} D_h \qquad (6.20)$$

(6.20) is the z-transfer function of the impulse-invariant discrete-time filter. When a unit sample sequence is applied to the input, the sampled impulse response of the analog reference filter appears at the output.

The frequency response of the impulse-invariant filter is the quotient of the Fourier transforms of the output sequence $y(n) = h_a(n)$ and of the input sequence $x(n) = \delta(n)$.

$$H(e^{j\omega T}) = \frac{H_a(e^{j\omega T})}{1}$$

According to (4.3), the discrete-time Fourier transform (DTFT) of the sequence $h_a(n)$ can be expressed by the spectrum of the original analog impulse response $h_a(t)$ as

$$H_a(e^{j\omega T}) = \frac{1}{T} \sum_{k=-\infty}^{+\infty} H_a[j(\omega + k\omega_s)] \ .$$

The frequency responses of the impulse-invariant discrete-time and of the analog reference filter are therefore related as

$$H(e^{j\omega T}) = \frac{1}{T} \sum_{k=-\infty}^{+\infty} H_a[j(\omega + k\omega_s)] = \frac{1}{T} \sum_{k=-\infty}^{+\infty} H_a(j\omega + jk2\pi/T) \ . \qquad (6.21)$$

The frequency response of the discrete-time filter is obtained by periodic continuation of the frequency response of the analog filter with the period $2\pi/T$. Strictly band-limited analog filters are not realisable. So the spectrum of $h(t)$ will always extend beyond the Nyquist frequency π/T. The frequency response of the discrete-time filter will therefore be more or less distorted depending on the degree of aliasing. Bad attenuation in the stopband and deviations of the DC or overall gain will be consequence. By proper choice of the constant D_h, these effects can be remedied to a certain degree.

Equation (6.21) shows that the frequency response of the discrete-time filter is $1/T$ times the frequency response of the analog reference filter. In order to take this factor into account, we multiply the transfer function (6.20) by T. With this modification, the discrete-time filter has exactly the periodically continued frequency response of the analog reference filter. In the final step, we have to add the constant b_N to realise the bypass signal path in the discrete-time filter.

$$H(z) = b_N + \sum_{k=1}^{N} c_k T \left(\frac{z}{z - e^{p_{\infty k} T}} - \frac{1}{2} \right) - D_h \qquad (6.22)$$

The constant D_h which only influences the sample $h(0)$ of the impulse response can be used in several ways to optimise the frequency response of the filter:

1. We can calculate D_h exactly using (6.16) and (6.17). This minimises the overall aliasing error between the frequency response of the analog filter and the impulse-invariant discrete-time filter.

2. We can set D_h to zero which is equivalent to the assumption that $h(0)$ assumes a value corresponding to half the step size of the analog impulse response at $t = 0$. This simplifies the design procedure. The deviation with respect to the exact calculation of D_h is tolerable in most practical cases.

3. We can determine D_h such that the magnitude of the analog reference filter and of the impulse-invariant discrete-time filter match at $\omega = 0$ $(z = 1)$.

4. We can determine D_h such that the magnitude of the analog reference filter and of the impulse-invariant discrete-time filter match at an arbitrary frequency.

Once again, all these variants only differ by the value of the sample $h(0)$ which is in any way undetermined when we attempt to replicate an analog impulse response showing a jump at $t = 0$. All other samples of $h(n)$ exactly match the analog impulse response. For case 2., the transfer function can be expressed as

$$H(z) = b_N + \sum_{k=1}^{N} \frac{c_k T}{2} \frac{1 + e^{p_{\infty k} T} z^{-1}}{1 - e^{p_{\infty k} T} z^{-1}} \; .$$

Matching the gain of analog and discrete-time filter at $\omega = 0$ $(z = 1)$ results in the expression

$$H(z) = \frac{b_0}{a_0} + \sum_{k=1}^{N} \frac{c_k T}{1 - e^{-p_{\infty k} T}} \frac{1 - z^{-1}}{1 - e^{p_{\infty k} T} z^{-1}} \; . \qquad (6.23)$$

Equation (6.23) is a partial-fraction representation of the z-transfer function which can be readily converted to the standard rational fractional form. The impulse-invariant design method guarantees stability since poles of the analog filter in the left half of the p-plane (Re $p_{\infty k} < 0$) result in poles of the discrete-time filter with magnitude less than unity, as can be verified by inspection of Equ. (6.23).

Example 6-1

A resonant circuit with the transfer function

$$H(P) = \frac{P}{P^2 + 0.1P + 1} \qquad f_r = 0.2/T \qquad (\omega_r = 0.4\pi/T)$$

is to be realised as an impulse-invariant discrete-time filter. The Q-factor of the circuit amounts to 10, the resonant frequency is $0.2\,f_s$. The poles of the filter are the zeros of the denominator polynomial.

$$P^2 + 0.1P + 1 = 0$$

$$P_{\infty 1} \approx -0.05 + j \qquad P_{\infty 2} \approx -0.05 - j$$

The transfer function can therefore be expressed as

$$H(P) = \frac{P}{(P + 0.05 - j)(P + 0.05 + j)} \; . \tag{6.24}$$

For the transition to the partial fraction representation, we have to determine the partial fraction coefficients A_1 and A_2:

$$H(P) = \frac{A_1}{P + 0.05 - j} + \frac{A_2}{P + 0.05 + j} \; .$$

Multiplying out and comparison of the coefficients with (6.24) yields

$$A_1 = 0.5 + j0.025 \qquad A_2 = 0.5 - j0.025$$

$$H(P) = \frac{0.5 + j0.025}{P + 0.05 - j} + \frac{0.5 - j0.025}{P + 0.05 + j} \; .$$

The relation

$$P = \frac{p}{\omega_r} = \frac{p}{0.4\pi/T}$$

leads to the unnormalised form

$$H(p) = \frac{(0.628 + j0.031)/T}{p - (-0.063 + j1.255)/T} + \frac{(0.628 - j0.031)/T}{p - (-0.063 - j1.255)/T} \; .$$

For the transition to the impulse-invariant discrete-time system, the following coefficients can be taken from this transfer function:

$$p_{\infty 1} = (-0.063 + j1.255)/T \qquad p_{\infty 2} = (-0.063 - j1.255)/T$$
$$c_1 = (0.628 + j0.031)/T \qquad c_2 = (0.628 - j0.031)/T$$

Substitution of these coefficients in (6.23) gives to the desired transfer function

$$H(z) = 0 + \frac{(0.628 + j0.031)}{1 - e^{0.063 - j1.255}} \frac{1 - z^{-1}}{1 - e^{-0.063 + j1.255} z^{-1}}$$

$$+ \frac{(0.628 - j0.031)}{1 - e^{0.063 + j1.255}} \frac{1 - z^{-1}}{1 - e^{-0.063 - j1.255} z^{-1}}$$

or, in rational fractional representation,

$$H(z) = \frac{0.628 - 0.056z^{-1} - 0.554z^{-2}}{1 - 0.583z^{-1} + 0.882z^{-2}} .$$

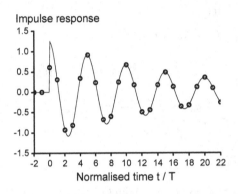

Fig. 6-6
Impulse response of the continuous-time (solid line) and discrete-time realisation (circles) of an impulse invariant resonant circuit

Figure 6-6 demonstrates that the unit sample (impulse) response of the discrete-time filter is made up of samples of the impulse response of the continuous-time reference filter. At the discontinuity $t = 0$, the exact value of the sample is chosen such that the frequency response is optimised at $\omega = 0$. Figure 6-7 depicts the frequency response of the discrete-time filter in comparison to the frequency response of the analog reference filter.

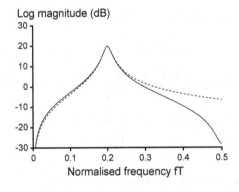

Fig. 6-7
Frequency response of the continuous-time (dashed line) and discrete-time (solid line) realisation of an impulse invariant resonant circuit

Due to the choice of the design parameter D_h, the magnitude of the discrete-time implementation is exactly zero at $\omega = 0$. At low frequencies, both curves match very well. At the resonance frequency, both filters have exactly the desired gain of 20 dB. While approaching the Nyquist frequency, the effect of aliasing becomes increasingly noticeable. The magnitude of the discrete-time filter drops stronger than that of the reference filter.

Figure 6-8 shows a further example. In this case we approximate a second-order low-pass filter. The pole frequency is located at $\omega = 0.6\pi/T$ where the magnitude response shows a peak. The curves match well at low frequencies and at the resonant frequency. Near the Nyquist frequency, the effect of aliasing becomes increasingly noticeable.

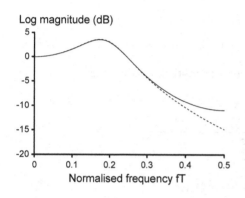

Fig. 6-8
Dashed line:

$$H(P) = \frac{1}{1 + 0.707P + P^2}, f_c = 0.2/T$$

Solid line:

$$H(z) = \frac{0.315 + 1.811z^{-1} + 0.261z^{-2}}{1 - 0.494z^{-1} + 0.411z^{-2}}$$

The example of a simple first-order low-pass filter is shown in Fig. 6-9. Merely at very low frequencies, there is a good match between the frequency responses of the discrete-time filter and the continuous-time reference filter. The effect of aliasing is especially pronounced since a first order low-pass filter does not greatly contribute to the lowering of high frequencies before sampling.

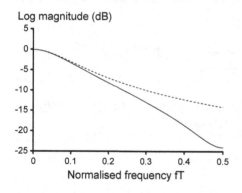

Fig. 6-9
Dashed line:

$$H(P) = \frac{1}{1 + P}, f_c = 0.1/T$$

Solid line:

$$H(z) = \frac{0.281 + 0.185z^{-1}}{1 - 0.533z^{-1}}$$

6.3.4 Step-Invariant Design

For the step-invariant filter design, the stimulus signal $x(n)$ of the discrete-time filter is the unit step sequence $u(n)$. The z-transform of this signal is

$$X(z) = \frac{z}{z-1} \ .$$

The desired output signal is the sampled step response $a_a(n)$ (6.18). Taking the z-transform of $a_a(n)$ yields

$$A_a(z) = \sum_{k=1}^{N} \frac{c_k}{P_{\infty k}} \frac{z}{z - e^{P_{\infty k}}} - \sum_{k=1}^{N} \frac{c_k}{P_{\infty k}} \frac{z}{z-1} - D_a \ .$$

The transfer function of the step-invariant discrete-time filter is obtained as

$$H(z) = \frac{A_a(z)}{X(z)} = \frac{\displaystyle\sum_{k=1}^{N} \frac{c_k}{P_{\infty k}} \frac{z}{z - e^{P_{\infty k}}} - \sum_{k=1}^{N} \frac{c_k}{P_{\infty k}} \frac{z}{z-1} - D_a}{\dfrac{z}{z-1}}$$

$$H(z) = \sum_{k=1}^{N} \frac{c_k}{-P_{\infty k}} + \sum_{k=1}^{N} \frac{c_k}{P_{\infty k}} \frac{1-z^{-1}}{1-e^{P_{\infty k}} z^{-1}} - D_a(1-z^{-1}) \ . \tag{6.25}$$

(6.25) is the z-transfer function of the step-invariant discrete-time filter. When a unit step sequence is applied to the input, the sampled step response of the analog reference filter appears at the output.

The frequency response of the step-invariant filter is the quotient of the Discrete-Time Fourier Transforms (DTFTs) of the output sequence $y(n) = a_a(n)$ and the input sequence $x(n) = u(n)$.

$$H(e^{j\omega T}) = \frac{A_a(e^{j\omega T})}{\dfrac{1}{1-e^{-j\omega T}} + \pi \sum_{n=-\infty}^{+\infty} \delta(\omega T - n2\pi)} = A_a(e^{j\omega T})(1-e^{-j\omega T}) \tag{6.26}$$

According to (4.3), the DTFT of the sequence $a_a(n)$ can be expressed by the spectrum of the original analog step response $a_a(t)$ as

$$A_a(e^{j\omega T}) = \frac{1}{T} \sum_{n=-\infty}^{+\infty} A_a(j\omega + jn2\pi/T) \ . \tag{6.27}$$

Since the continuous-time step response is the integral of the impulse response, we have the following relationship of the Fourier transforms in the frequency domain:

$$A_a(j\omega) = \frac{H_a(j\omega)}{j\omega} + \pi H_a(0)\delta(\omega) \ . \tag{6.28}$$

Substitution of (6.32) and (6.28) in (6.26) yields

$$H(e^{j\omega T}) = \sum_{n=-\infty}^{+\infty} \frac{1-e^{-j\omega T}}{j(\omega T+n2\pi)} H_a(j\omega+jn2\pi/T)$$

and after some algebra

$$H(e^{j\omega T}) = \sum_{n=-\infty}^{+\infty} H_a(j\omega+jn2\pi/T)\frac{\sin(\omega T/2+n\pi)}{\omega T/2+n\pi}e^{-j(\omega T/2+n\pi)}. \qquad (6.29)$$

The frequency response of the step-invariant filter is obtained by periodic continuation of the frequency response of the analog reference filter which is weighted by a sinc function in this case. This weighting has some interesting consequences:

- The gain of the discrete-time filter equals the gain of the analog reference filter at $\omega = 0$. The point $\omega = 0$ is alias-free.
- Apart from the aliasing errors, this weighting leads to additional systematic distortion (sinc distortion) which amounts to about 4 dB at the Nyquist frequency.

In the final step, we have to add the constant b_N to the z-transfer function (6.25) to realise the bypass signal path in the discrete-time filter.

$$H(z) = b_N + \sum_{k=1}^{N} \frac{c_k}{-p_{\infty k}} + \sum_{k=1}^{N} \frac{c_k}{p_{\infty k}}\frac{1-z^{-1}}{1-e^{p_{\infty k}}z^{-1}} - D_a(1-z^{-1})$$

By making use of (6.7), we obtain the transfer function of the step-invariant filter in the form

$$H(z) = \frac{b_0}{a_0} + \sum_{k=1}^{N} \frac{c_k}{p_{\infty k}}\frac{1-z^{-1}}{1-e^{p_{\infty k}}z^{-1}} - D_a(1-z^{-1}) \qquad (6.30)$$

The constant D_a which only influences the sample $a(0)$ of the step response can be used in several ways to optimise the frequency response of the filter:

1. We can calculate D_a exactly using (6.16) and (6.18). This optimises the overall aliasing error between the frequency response of the analog filter and the step-invariant discrete-time filter.
2. We can set D_a to zero which is equivalent to the assumption that the exponential sequence in $a(n)$ assumes for $n = 0$ a value corresponding to half the step size of the corresponding analog function at $t = 0$. This simplifies the design procedure. The deviation with respect to the exact calculation of D_a is tolerable in most practical cases.
3. We can determine D_a such that the magnitude of the analog reference filter and of the step-invariant discrete-time filter match at an arbitrary frequency.

For $z = 1$ ($\omega = 0$), (6.30) assumes a value of b_0/a_0 which is exactly the magnitude of the analog reference filter at $\omega = 0$. So D_a offers a degree of freedom to match the magnitude of the step-invariant filter with the reference filter at a further frequency. If we disregard D_a in (6.30), the z-transfer function simplifies to

$$H(z) = \frac{b_0}{a_0} + \sum_{k=1}^{N} \frac{c_k}{p_{\infty k}} \frac{1 - z^{-1}}{1 - e^{p_{\infty k}} z^{-1}} \; . \qquad (6.31)$$

Equation (6.31) is a partial-fraction representation of the z-transfer function which can readily be converted to the standard rational fractional form. Stability is guaranteed since poles $p_{\infty k}$ of the analog reference filter in the left half of the p-plane result in poles of the discrete-time filter with magnitude less than unity as can be verified by inspection of Equ. (6.31).

Example 6-2

The resonant circuit that we already used as an example in the previous section is to be realised as a step-invariant discrete-time filter. We can make use of the poles and partial fraction coefficients $p_{\infty 1}$, $p_{\infty 2}$, c_1 and c_2 calculated there. Substitution of these coefficients in (6.31) leads to the following transfer function in partial-fraction form

$$H(z) = 0 + \frac{(0.628 + j0.031)}{(-0.063 + j1.255)} \frac{(1 - z^{-1})}{(1 - e^{-0.063 + j1.255} z^{-1})}$$
$$+ \frac{(0.628 - j0.031)}{(-0.063 - j1.255)} \frac{(1 - z^{-1})}{(1 - e^{-0.063 - j1.255} z^{-1})} \; ,$$

$$H(z) = 0 + \frac{-0.5\,j(1 - z^{-1})}{1 - e^{-0.063 + j1.255} z^{-1}} + \frac{0.5\,j(1 - z^{-1})}{1 - e^{-0.063 - j1.255} z^{-1}} \; ,$$

or, by multiplying out in the standard rational fractional form,

$$H(z) = \frac{0.894 z^{-1} - 0.894 z^{-2}}{1 - 0.583 z^{-1} + 0.882 z^{-2}} \; .$$

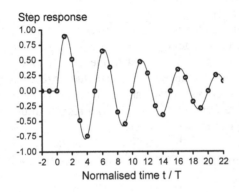

Step response

Normalised time t / T

Fig. 6-10
Step response of the continuous-time (solid line) and discrete-time realisation (circles) of a step-invariant resonant circuit

Figure 6-10 shows that the samples of the step response of the discrete-time filter coincide with the step response of the continuous-time reference filter. Concerning the magnitude response, there is good correspondence at low frequencies between discrete-time and continuous-time filter as can be seen from Fig. 6-11. The resonance peak of the discrete-time filter is slightly lower than the peak of the analog reference filter. This is a consequence of the weighting with the sinc function. Near the Nyquist frequency, aliasing becomes increasingly effective.

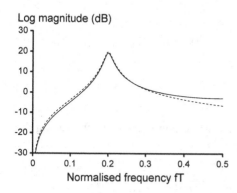

Fig. 6-11
Frequency response of the continuous-time (dashed line) and discrete-time (solid line) realisation of a step-invariant resonant circuit

The example of a second-order low-pass filter as depicted in Fig. 6-12 shows similar behaviour. The approximation at low frequencies is good while larger deviations can be observed as the frequency approaches $\omega = \pi/T$.

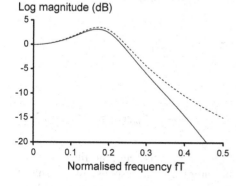

Fig. 6-12
Dashed line:

$$H(P) = \frac{1}{1 + 0.707P + P^2}, f_c = 0.2/T$$

Solid line:

$$H(z) = \frac{0.529z^{-1} + 0.388z^{-2}}{1 - 0.494z^{-1} + 0.411z^{-2}}$$

Figure 6-13 shows the example of a first-order high-pass filter. In this case, aliasing and sinc-weighting obviously causes an increase of the overall gain.

6.3.5 Ramp-Invariant Design

For the ramp-invariant filter design, the stimulus signal $x(n)$ of the discrete-time filter is the ramp sequence $r(n) = n\,u(n)$. The z-transform of this signal is

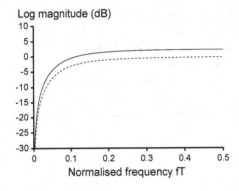

Log magnitude (dB)

Normalised frequency fT

Fig. 6-13
Dashed line:

$$H(P) = \frac{P}{1+P}, f_c = 0.1/T$$

Solid line:

$$H(z) = \frac{1 - z^{-1}}{1 - 0.533 z^{-1}}$$

$$X(z) = \frac{z}{(z-1)^2}$$

The desired output signal is the sampled ramp response $m_a(n)$ (6.19). Taking the z-transform of $m_a(n)$ yields

$$M_a(z) = \sum_{k=1}^{N} \frac{c_k}{-P_{\infty k}} \frac{z}{(z-1)^2} + \sum_{k=1}^{N} \frac{c_k}{p_{\infty k}^2 T}\left(\frac{z}{z - e^{P_{\infty k}T}} - \frac{z}{z-1}\right) - D_m$$

$$M_a(z) = \sum_{k=1}^{N} \frac{c_k}{-P_{\infty k}} \frac{z}{(z-1)^2} + \sum_{k=1}^{N} \frac{c_k}{p_{\infty k}^2 T} \frac{z(e^{P_{\infty k}T} - 1)}{(z-1)(z - e^{P_{\infty k}T})} - D_m$$

The transfer function of the ramp-invariant filter is obtained as

$$H(z) = \frac{Ma(z)}{X(z)} = \frac{\displaystyle\sum_{k=1}^{N} \frac{c_k}{-P_{\infty k}} \frac{z}{(z-1)^2} + \sum_{k=1}^{N} \frac{c_k}{p_{\infty k}^2 T} \frac{z(e^{P_{\infty k}T} - 1)}{(z-1)(z - e^{P_{\infty k}T})} - D_m}{\dfrac{z}{(z-1)^2}}$$

$$H(z) = \sum_{k=1}^{N} \frac{c_k}{-P_{\infty k}} + \sum_{k=1}^{N} \frac{c_k(e^{P_{\infty k}T} - 1)}{p_{\infty k}^2 T} \frac{(1 - z^{-1})}{(1 - e^{P_{\infty k}T} z^{-1})} - D_m z(1 - z^{-1})^2 \qquad (6.32)$$

(6.32) is the z-transfer function of the ramp-invariant discrete-time filter. When a ramp sequence is applied to the input, the sampled ramp response of the analog reference filter appears at the output. What remains to be done is to add the constant b_N to the z-transfer function to realise the bypass signal path in the discrete-time filter.

$$H(z) = b_N + \sum_{k=1}^{N} \frac{c_k}{-P_{\infty k}} + \sum_{k=1}^{N} \frac{c_k(e^{P_{\infty k}T} - 1)}{p_{\infty k}^2 T} \frac{(1 - z^{-1})}{(1 - e^{P_{\infty k}T} z^{-1})} - D_m z(1 - z^{-1})^2$$

By making use of (6.7), we obtain the transfer function of the ramp-invariant filter in the form

$$H(z) = \frac{b_0}{a_0} + \sum_{k=1}^{N} \frac{c_k (e^{P_{\infty k} T} - 1)}{p_{\infty k}^2 T} \frac{1 - z^{-1}}{1 - e^{P_{\infty k} T} z^{-1}} - D_m z (1 - z^{-1})^2 \qquad (6.33)$$

The last term in (6.33) is not causal due to the factor z. So if there is a need to optimise the frequency response of the ramp-invariant filter in some way using the constant D_m, the whole transfer function needs to be multiplied by z^{-1} which is equivalent to the introduction of a delay by one sample period. The DC gains of ramp-invariant filter and analog reference filter are identical since

$$H(z = 1) = H(p = 0) = b_0 / a_0 .$$

The frequency response of the ramp-invariant discrete-time filter is related to the frequency response of the analog reference filter as

$$H(e^{j\omega T}) = \sum_{n=-\infty}^{+\infty} H_a(j\omega + jn2\pi/T) \left(\frac{\sin(\omega T/2 + n\pi)}{\omega T/2 + n\pi} \right)^2$$

The frequency response of the step-invariant filter is obtained by periodic continuation of the frequency response of the analog reference filter which is weighted by the square of a sinc function. This weighting has some interesting consequences:

- The gain of the discrete-time filter equals the gain of the analog reference filter at $\omega = 0$. The point $\omega = 0$ is alias-free.
- Apart from the aliasing errors, this weighting leads to additional systematic distortion (sinc distortion) which amounts to a loss of about 8 dB at the Nyquist frequency.

Example 6-3

A ramp-invariant first-order high-pass filter is to be realised based on the continuous-time reference filter

$$H(P) = \frac{P}{P+1} , \quad f_c = 0.2/T .$$

Since numerator and denominator polynomial are of the same degree, a constant needs to be extracted in the first step.

$$H(P) = 1 - \frac{1}{P+1}$$

The relation

$$P = \frac{p}{\omega_c} = \frac{p}{2\pi f_c} = \frac{pT}{0.4\pi}$$

leads to the unnormalised form

$$H(p) = 1 - \frac{0.4\pi/T}{p + 0.4\pi/T} \; .$$

Pole frequency and partial fraction coefficient can be derived from the above transfer function as

$$p_\infty = -0.4\pi/T \qquad c = -0.4\pi/T \; .$$

The desired z-transfer function is obtained by substitution of these coefficients in (6.33).

$$H(z) = \frac{-0.4\pi(e^{-0.4\pi} - 1)}{(0.4\pi)^2} \frac{1 - z^{-1}}{1 - e^{-0.4\pi} z^{-1}}$$

$$H(z) = \frac{0.569 - 0.569 z^{-1}}{1 - 0.285 z^{-1}}$$

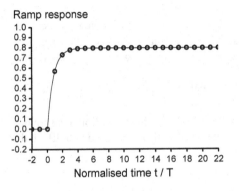

Ramp response

Normalised time t / T

Fig. 6-14
Ramp response of the continuous-time (solid line) and discrete-time realisation (circles) of a first-order high-pass filter

Figure 6-14 shows that the samples of the ramp response of the discrete-time high-pass filter coincide with the ramp response of the continuous-time reference filter. The magnitude responses of both filters as depicted in Fig. 6-15 show good agreement over the whole frequency range.

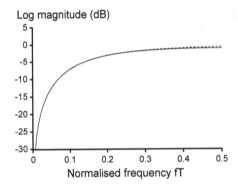

Log magnitude (dB)

Normalised frequency fT

Fig. 6-15
Dashed line:

$$H(P) = \frac{P}{1 + P}, f_c = 0.2/T$$

Solid line:

$$H(z) = \frac{0.569(1 - z^{-1})}{1 - 0.285 z^{-1}}$$

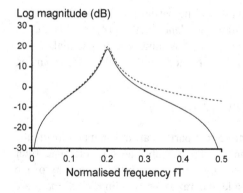

Fig. 6-16
Dashed line:
$$H(P) = \frac{P}{1 + 0.1P + P^2}, f_c = 0.2/T$$

Solid line:
$$H(z) = \frac{0.528 - 0.023z^{-1} - 0.505z^{-2}}{1 - 0.583z^{-1} + 0.882z^{-2}}$$

The magnitude responses of discrete-time and continuous-time resonant circuit show good correspondence at low frequencies as can be seen from Fig. 6-16. The resonance peak of the discrete-time filter is lower than the peak of the analog reference filter. This is a consequence of the weighting with the square of the sinc function. Near the Nyquist frequency, aliasing becomes increasingly effective. The same tendency can be observed for the example of a second-order low-pass filter (Fig. 6-17). The drop of the magnitude response with increasing frequency is doubled compared to the step-invariant method due to the weighting with the square of the sinc function

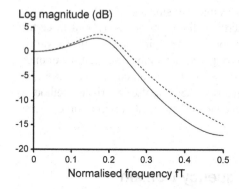

Fig. 6-17
Dashed line:
$$H(P) = \frac{1}{1 + 0.707P + P^2}, f_c = 0.2/T$$

Solid line:
$$H(z) = \frac{0.199 + 0.593z^{-1} + 0.126z^{-2}}{1 - 0.494z^{-1} + 0.411z^{-2}}$$

6.3.6 Summary

The z-transfer functions that we derived for the design of impulse-invariant, step-invariant, and ramp-invariant discrete-time filters all have one similar basic structure:

$$H(z) = H(\omega = 0) + \sum_{k=1}^{N} C_k \frac{1 - z^{-1}}{1 - e^{P_{\infty k}T} z^{-1}} \ .$$

The three types only differ in the calculation of the partial-fraction coefficients C_k. For a given continuous-time reference filter, the denominator polynomials are the same for all design methods. The difference is in the numerator polynomials.

The poles of the z-transfer function are related to the poles of the analog reference filter as

$$z_{\infty k} = e^{p_{\infty k}T} = e^{\text{Re } p_{\infty k}T}\left(\cos(\text{Im } p_{\infty k}T) + j\sin(\text{Im } p_{\infty k}T)\right) .$$

So a pole in the left-half p-plane (negative real part) is always mapped inside the unit circle in the z-plane. This guarantees that a stable analog reference filter always results in a related stable discrete-time filter. Due to the periodicity of the sine and cosine functions, poles outside the range $-\pi/T < \text{Im} p_{\infty k}T < +\pi/T$ are mirrored into the Nyquist band. This leads to filter characteristics which have little to do with the desired characteristics of the analog reference filter. It makes therefore no sense to apply the proposed design methods to continuous-time transfer functions which have poles outside the Nyquist band. This problem can always be solved by choosing the sampling period T of the discrete-time filter small enough.

If the approximation in the time domain is used to realise a given magnitude response, the impulse-invariant method is superior because, apart from the unavoidable aliasing errors, no further distortion of the characteristics of the analog reference filter takes place. The magnitude responses of step-invariant and ramp-invariant filters, on the contrary, are subject to sinc distortion.

We recall that the linear-phase property of filters is closely related to certain symmetry properties of the impulse response. Applying the impulse-invariant design method will pass on these symmetry properties to the discrete-time domain. So it can be expected that the phase response of the designed discrete-time filter will have similar linear-phase properties. The impulse-invariant method is therefore one possibility to realise Bessel characteristics with discrete-time filters.

6.4 Approximation in the Frequency Domain

According to Sect. 3.3.1, the frequency response of a discrete-time system is derived from the transfer function $H(z)$ by substitution of (6.34).

$$z = e^{j\omega T} \tag{6.34}$$

z is replaced by the term $e^{j\omega T}$. In the design process of the digital filter we have the inverse problem. Given a frequency response $H(j\omega)$. How can we find the transfer function $H(z)$ of a realisable and stable discrete-time filter? The basic idea is to solve (6.34) for $j\omega$ in order to get a transformation that provides a transfer function of z derived from a frequency response $H(j\omega)$.

$$j\omega = \ln z / T \tag{6.35a}$$

We have no problem if the given frequency response is a polynomial or a rational fractional function of $e^{j\omega T}$, since with

$$H(e^{j\omega T}) \quad \rightarrow \quad H(e^{\ln z / T \times T}) = H(z)$$

we obtain a polynomial or a rational fractional function of z that can be realised with the filter structures introduced in Chap. 5. It is our intention, however, to use analog filters as a reference for the design of digital filters. The frequency response of analog filters is a rational fractional function of $j\omega$. Application of (6.35a) would yield rational fractional transfer functions of $\ln z / T$ which, unfortunately, are not realisable with any of the known structures.

$$H(j\omega) \quad \rightarrow \quad H(\ln z / T)$$

We can generalise (6.35a) to the transformation

$$p = \ln z / T , \tag{6.35b}$$

which would allow us to start from the transfer function $H(p)$ of the analog reference filter.

$$H(p) \quad \rightarrow \quad H(\ln z / T)$$

How can (6.35b), which is a transformation from the p-plane of the continuous-time domain into the z-plane of the discrete-time domain, be modified in such a way that
1. the desired frequency response is maintained and
2. the resulting transfer function in z is stable and realisable?

Series expansion of the logarithm in (6.35b) and an appropriate truncation of the series is a possibility to achieve rational fractional functions in z. We investigate in the following two candidates of series expansions which result in transformations with different properties.

6.4.1 Difference Method

A possible series expansion of the logarithm has the following form:

$$\ln z = \frac{(z-1)}{z} + \frac{(z-1)^2}{2z^2} + \frac{(z-1)^3}{3z^3} + \frac{(z-1)^4}{4z^4} + \dots$$

with $\mathrm{Re}\, z > 1/2$.

Truncating this series after the first term yields the following transformation from the p-plane into the z-plane:

$$p = \frac{z-1}{zT} = \frac{1-z^{-1}}{T} . \tag{6.36}$$

This transformation corresponds to the method which is often applied for the numerical solution of differential equations, where derivatives are replaced by differences. Figure 6-18 illustrates the situation. The derivative (tangent) at $t = nT$ is approximated by the secant through the points $x(nT)$ and $x((n-1)T)$. The error of this approximation depends heavily on the shape of the function in the considered interval.

$$x'(t)\Big|_{t=nT} \approx \frac{x(t)-x(t-T)}{T} = \frac{x(n)-x(n-1)}{T}\Big|_{t=nT}$$

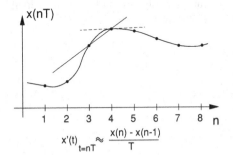

$$x'(t)\Big|_{t=nT} \approx \frac{x(n)-x(n-1)}{T}$$

Fig. 6-18
Approximation of the first derivative by a difference

Laplace transform of this relation yields

$$p\, X(p) \approx \frac{X(p)-X(p)\mathrm{e}^{-pT}}{T}$$

$$p \approx \frac{1-\mathrm{e}^{-pT}}{T} \ .$$

Since $\mathrm{e}^{pT} = z$, the frequency variable p can be approximated by the rational term in z (6.36):

$$p \approx \frac{1-z^{-1}}{T} = \frac{1}{T}\frac{z-1}{z} \ .$$

According to the rules of the z-transform, this relation means calculation of the difference between the current and the past sample (backward difference) and division by the sampling period T, which is in fact the slope of the secant and thus an approximation of the derivative. With decreasing T, the approximation improves, since the secant more closely approaches the tangent.

We consider now the stability of discrete-time systems whose transfer functions are obtained from analog reference filters using (6.36). As the poles p_∞ of stable analog filters lie in the left half of the p-plane, the transformation (6.36) results in the following inequality which must be fulfilled by poles z_∞ in the z-domain:

$$\mathrm{Re}\, p_\infty = \mathrm{Re}\frac{1-z_\infty^{-1}}{T} < 0 \ .$$

From this inequality we can derive a condition for the location of stable poles in the z-plane:

$$(\mathrm{Re}\, z_\infty - 0.5)^2 + (\mathrm{Im}\, z_\infty)^2 < 0.25 \ .$$

The poles of the analog filter are mapped into a circle with the centre $z = 0.5$ and a radius of 0.5 (Fig. 6-19). It becomes clear that stability of the discrete-time filter is guaranteed since all stable poles in the left half of the p-plane are transformed inside the unit circle of the z-plane. Unfortunately, with transformation (6.36) we can reach only a very small area of the unit circle. We cannot map any pole at all, wherever its stable reference pole may lie in the p-plane, into the left half of the unit circle, which represents the frequency range $f_s/4 \dots f_s/2$. In the right half of the unit circle, the possibility of locating poles close to the unit circle is very limited. That means that poles with a high Q-factor, as required for sharp filters, can only be realised far below the sampling frequency in the vicinity of $\omega = 0$ or $z = 1$. If we want to approximate the frequency response of an analog filter with sufficient precision, we must choose a sampling frequency much higher (by a factor of ten or more) than the highest pole frequency of the filter. In practice, this means in most cases that the sampling frequency is also much higher than required to satisfy the sampling theorem for the signals to be processed.

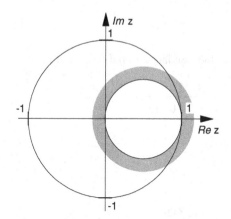

Fig. 6-19
Mapping of the left half of the p-plane into the z-plane in the case of the difference method

Example 6-4
Apply the difference method to the second-order low-pass filter specified below:

$$H(P) = \frac{1}{P^2 + 0.1P + 1}, f_c = 0.3/T \qquad (\omega_c = 0.6\pi/T) \qquad (6.37)$$

$$P = \frac{p}{\omega_c} = \frac{p}{2\pi f_c} = \frac{p}{0.6\pi/T} \ .$$

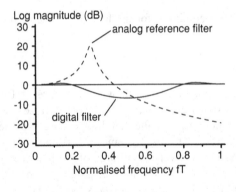

Fig. 6-20
Design of a second-order low-pass filter using the difference method

Using design rule (6.36) results in the following expression for the normalised frequency P:

$$P = \frac{(1-z^{-1})/T}{0.6\pi/T} = \frac{1-z^{-1}}{0.6\pi} .$$

Substitution in the transfer function (6.37) yields the transfer function in z of the discrete-time filter:

$$H(z) = \frac{1}{\left(\dfrac{1-z^{-1}}{0.6\pi}\right)^2 + 0.1\left(\dfrac{1-z^{-1}}{0.6\pi}\right) + 1} .$$

Multiplying out the brackets leads to the standard rational fractional form

$$H(z) = \frac{0.75z^2}{z^2 - 0.46z + 0.21} = \frac{0.75}{1 - 0.46z^{-1} + 0.21z^{-2}} .$$

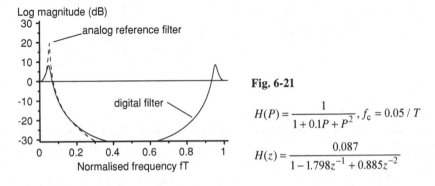

Fig. 6-21

$$H(P) = \frac{1}{1 + 0.1P + P^2}, f_c = 0.05/T$$

$$H(z) = \frac{0.087}{1 - 1.798z^{-1} + 0.885z^{-2}}$$

Figure 6-20 illustrates that the behaviour of a pole pair with Q-factor 10 cannot be recognised any more. A far better approximation can be reached when the pole is shifted to lower frequencies, as shown in Fig. 6-21. The cutoff frequency is now

$f_c = 0.005/T$. However, the Q-factor is still too low, because the transformation (6.36) is not able to locate the pole pair close enough to the unit circle.

6.4.2 Bilinear Transform

A further series expansion of the logarithm is of the following form:

$$\ln z = 2\left(\frac{z-1}{z+1} + \frac{1}{3}\left(\frac{z-1}{z+1}\right)^3 + \frac{1}{5}\left(\frac{z-1}{z+1}\right)^5 \cdots\right)$$

with $\mathrm{Re}\,z > 0$.

Truncation of the series after the first term leads to the following transformation from the p-plane into the z-plane:

$$p = \frac{2}{T}\frac{z-1}{z+1} \tag{6.38}$$

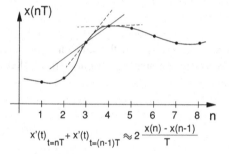

$$x'(t)\Big|_{t=nT} + x'(t)\Big|_{t=(n-1)T} \approx 2\,\frac{x(n) - x(n-1)}{T}$$

Fig. 6-22
Approximation of the first derivative in the case of the bilinear transform

This transformation, commonly referred to as the bilinear transform, is also based on the approximation of the derivative by a secant through two points. The assumption made here is that the secant through two points is a good approximation of the average of the slopes of the curve at these two points (Fig. 6-22).

$$\frac{x'(t)+x'(t-T)}{2}\Bigg|_{t=nT} \approx \frac{x(t)-x(t-T)}{T}\Bigg|_{t=nT} = \frac{x(n)-x(n-1)}{T} \tag{6.39}$$

Laplace transform of relation (6.39) yields

$$\frac{pX(p)+pX(p)\mathrm{e}^{-pT}}{2} \approx \frac{X(p)-X(p)\mathrm{e}^{-pT}}{T}$$

$$\frac{p\,(1+\mathrm{e}^{-pT})}{2} \approx \frac{1-\mathrm{e}^{-pT}}{T}\,.$$

Since $z = e^{pT}$, we have

$$\frac{p\,(1+z^{-1})}{2} \approx \frac{1-z^{-1}}{T} \;,$$

and, after solving for p,

$$p \approx \frac{2}{T}\frac{z-1}{z+1} \;.$$

The transformation is only usable if it leads to stable discrete-time filters. As the poles p_∞ of stable analog filters lie in the left half of the p-plane, the bilinear transform (6.38) results in the following inequality which must be fulfilled by the poles z_∞ in the z-domain.

$$\text{Re } p_\infty = \text{Re}\,\frac{2}{T}\frac{z_\infty-1}{z_\infty+1} < 0$$

From this inequality, we can derive a condition for the mapping of stable analog poles into the z plane.

$$(\text{Re } z_\infty)^2 + (\text{Im } z_\infty)^2 < 1$$

The left half of the p-plane is mapped into the interior of the unit circle in the z-plane, as depicted in Fig. 6-23. This means that a stable continuous-time filter always results in a stable discrete-time filter.

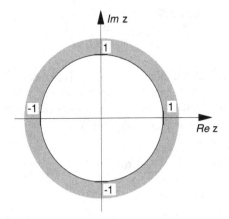

Fig. 6-23
Mapping of the left half of the p-plane into the unit circle of the z-plane using the bilinear transform

The application of the difference method that we introduced in the previous section results in a poor approximation of the frequency response of the analog reference filter. In the following, we investigate the corresponding behaviour of the bilinear transform which converts the transfer function $H(p)$ of the analog filter into the transfer function in z of a related discrete-time filter.

$$H(p) \quad \xrightarrow{\text{bilinear transform}} \quad H\left(\frac{2}{T}\frac{z-1}{z+1}\right)$$

We compare the frequency response of the analog filter, which is obtained by the substitution $p = j\omega$, with that of the discrete-time filter which results from the substitution $z = e^{j\omega T}$.

$$H(j\omega) \quad \xrightarrow{\text{bilinear transform}} \quad H\left(\frac{2}{T}\frac{e^{j\omega T}-1}{e^{j\omega T}+1}\right)$$

With

$$\frac{2}{T}\frac{e^{j\omega T}-1}{e^{j\omega T}+1} = \frac{2}{T}\frac{e^{j\omega T/2}-e^{-j\omega T/2}}{e^{j\omega T/2}+e^{-j\omega T/2}} = \frac{2}{T}\tanh(j\omega T/2) = j\frac{2}{T}\tan(\omega T/2)$$

we obtain the following comparison between the frequency responses of analog and discrete-time system:

$$\underset{H(j\omega)}{\text{analog system}} \quad \xrightarrow{\text{bilinear transform}} \quad \underset{H\left(j\frac{2}{T}\tan(\omega T/2)\right)}{\text{discrete - time system}} \;\cdot$$

It is obvious that the frequency response of the discrete-time system can be derived from the frequency response of the analog reference filter by a simple distortion of the frequency axis. ω is substituted by ω' which can be expressed as

$$\omega' = \frac{2}{T}\tan(\omega T/2) \; . \tag{6.40}$$

If ω progresses in the range from 0 up to half the sampling frequency π/T, ω' progresses from 0 to ∞. The bilinear transform thus compresses the frequency response of the analog filter in the range $-\infty \le \omega \le +\infty$ to the frequency range $-\pi/T \le \omega \le +\pi/T$. Due to the tangent function, this compressed range is then periodically continued with the period $2\pi/T$. The described behaviour has the following consequences to be considered for the design of digital filters:

Aliasing
Since the frequency response of the analog filter is compressed prior to the periodic continuation, overlapping of spectra (aliasing), which caused problems in the case of the impulse- and step-invariant designs, is ruled out when the bilinear transform is applied. As a result, passband ripple and stopband attenuation of the analog reference filter are exactly preserved.

Filter Order

Due to the compression of the frequency axis of the analog reference filter, the sharpness of the transition band(s) increases compared to the original frequency response. This means that, if the bilinear transform is applied, a given tolerance scheme can be satisfied with a lower filter order than we needed for the corresponding analog filter. The relations to calculate the required filter order which we introduced in Chap. 2 are nevertheless still valid. Instead of the original stopband and passband edge frequencies, we have to use the corresponding frequencies obtained by the frequency warping (6.40). The closer the transition band is located to the Nyquist frequency, the higher the saving in terms of filter order. In the next section we will calculate an example of a filter design which will demonstrate the reduction of the filter order by using the bilinear transform.

Choice of the Cutoff Frequencies

The most important characteristic frequencies of filters are the edge frequencies of the tolerance scheme or the 3-dB cutoff frequency. These should be maintained in any case if the transfer function of the analog reference filter is converted into the transfer function of a corresponding discrete-time system. Due to the compression of the frequency axis caused by the bilinear transform, all characteristic frequencies are shifted towards lower frequencies, which is demonstrated in Fig. 6-24 by means of the cutoff frequency of a second-order low-pass filter.

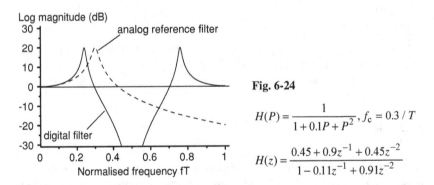

Fig. 6-24

$$H(P) = \frac{1}{1 + 0.1P + P^2}, f_c = 0.3 / T$$

$$H(z) = \frac{0.45 + 0.9z^{-1} + 0.45z^{-2}}{1 - 0.11z^{-1} + 0.91z^{-2}}$$

This behaviour can be compensated for by shifting all characteristic frequencies to higher frequencies using (6.40). The analog filter design is then performed using these corrected values. Application of the bilinear transform to obtain a discrete-time equivalent of the analog reference filter shifts all characteristic frequencies back to the desired position. Figure 6-25 illustrates this approach.

Phase Response

The imitation of the flat group-delay responses of Bessel filters is not possible if the bilinear transform is applied. Bessel filters exhibit an almost linearly increasing phase in the passband. Due to the nonlinear compression of the frequency axis, the slope of the phase curve increases more and more when approaching the Nyquist frequency. This behaviour is equivalent to an ever-increasing group delay. We come back to the solution of this problem in Sect. 6.4.4.

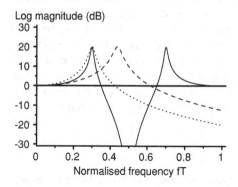

Log magnitude (dB)

Fig. 6-25
Design of a second-order low-pass filter using the bilinear transform
<u>dotted line:</u> continuous-time reference filter
<u>dashed line:</u> shift of the cutoff frequency using (6.40)
<u>solid line:</u> resulting discrete-time filter after bilinear transform

Example 6-5

We reconsider the second-order low-pass filter of Example 6-4:

$$H(P) = \frac{1}{P^2 + 0.1P + 1}, f_c = 0.3/T \qquad (\omega_c = 0.6\pi/T) \quad (6.41)$$

$$P = \frac{p}{\omega_c} = \frac{p}{2\pi f_c} = \frac{p}{0.6\pi/T} \ .$$

Derive the z-transfer function of a discrete-time filter using the bilinear transform.

Using design rule (6.38) results in the following expression for the normalised frequency P:

$$P = \frac{2/T}{0.6\pi/T} \frac{z-1}{z+1} = \frac{1}{0.3\pi} \frac{z-1}{z+1} \ .$$

Substitution in the continuous-time transfer function (6.41) yields the z-transfer function of the discrete-time filter:

$$H(z) = \frac{1}{\left(\dfrac{1}{0.3\pi} \dfrac{z-1}{z+1}\right)^2 + 0.1\left(\dfrac{1}{0.3\pi} \dfrac{z-1}{z+1}\right) + 1} \ .$$

Multiplying out the brackets leads to the standard rational fractional form

$$H(z) = \frac{0.45z^2 + 0.9z + 0.45}{z^2 - 0.11z + 0.91} = \frac{0.45 + 0.9z^{-1} + 0.45z^{-2}}{1 - 0.11z^{-1} + 0.91z^{-2}} \ .$$

Figure 6-24 depicts the corresponding magnitude response. The effect of the compression of the frequency axis is clearly visible. The gain at $\omega = 0$ and in the maximum, however, is preserved exactly.

6.4.3 Practical Filter Design Using the Bilinear Transform

For reasons that we will consider in more detail in Chap. 8, the cascade arrangement of second-order filter blocks as shown in Fig. 5-30 is of special importance. It is therefore a frequent task in practice to convert analog second-order filter blocks to corresponding discrete-time realisations using the bilinear transform. All first- and second-order filter blocks that can occur in the context of Butterworth, Chebyshev, Cauer and Bessel filters, also taking into account the possible filter transformations leading to high-pass, bandpass and bandstop filter variants, have normalised transfer functions of the following general form:

$$H(P) = \frac{B_0 + B_1 P + B_2 P^2}{1 + A_1 P + A_2 P^2} \ . \tag{6.42}$$

In order to obtain the unnormalised representation, we have to substitute $P = p/\omega_c$. Application of the bilinear transform results in

$$P = \frac{p}{\omega_c} = \frac{2}{\omega_c T} \frac{z-1}{z+1} \ . \tag{6.43}$$

Substitution of (6.43) in (6.42) yields the desired transfer function in z.

$$H(z) = \frac{\dfrac{4B_2}{(\omega_c T)^2}\left(\dfrac{z-1}{z+1}\right)^2 + \dfrac{2B_1}{\omega_c T}\left(\dfrac{z-1}{z+1}\right) + B_0}{\dfrac{4A_2}{(\omega_c T)^2}\left(\dfrac{z-1}{z+1}\right)^2 + \dfrac{2A_1}{\omega_c T}\left(\dfrac{z-1}{z+1}\right) + 1} \tag{6.44}$$

(6.44) can be readily rearranged to the normal form (6.45).

$$H(z) = V \frac{1 + b_1 z^{-1} + b_2 z^{-2}}{1 + a_1 z^{-1} + a_2 z^{-2}} \tag{6.45}$$

With the introduction of the notations

$$
\begin{aligned}
&C_1 = 2A_1/(\omega_c T) && D_1 = 2B_1/(\omega_c T)\\
&C_2 = 4A_2/(\omega_c T)^2 && D_2 = 4B_2/(\omega_c T)^2
\end{aligned}
\tag{6.46}
$$

the coefficients of (6.45) can be calculated using the following relations.

Second-Order Filter Blocks

$$
\begin{aligned}
V &= (B_0 + D_1 + D_2)/(1 + C_1 + C_2)\\
b_1 &= 2(B_0 - D_2)/(B_0 + D_1 + D_2)\\
b_2 &= (B_0 - D_1 + D_2)/(B_0 + D_1 + D_2)\\
a_1 &= 2(1 - C_2)/(1 + C_1 + C_2)\\
a_2 &= (1 - C_1 + C_2)/(1 + C_1 + C_2)
\end{aligned}
\tag{6.47}
$$

First-order Filter Blocks

$$V = (B_0 + D_1)/(1 + C_1)$$
$$b_1 = (B_0 - D_1)/(B_0 + D_1)$$
$$b_2 = 0 \qquad\qquad (6.48)$$
$$a_1 = (1 - C_1)/(1 + C_1)$$
$$a_2 = 0$$

In the case of low-pass filters, the coefficients A_1, A_2 and B_2 can be directly taken from the filter tables in the Appendix. For other filter types like high-pass, bandpass or bandstop, these coefficients are calculated using the relations for the respective filter transformation derived in Sect. 2.7.

Example 6-6

Apply the relations derived above to determine the filter coefficients of a filter which is specified by the following low-pass tolerance scheme:

passband ripple:	1 dB	$\rightarrow$	$v_p = 0.891$	$\rightarrow$	$v_{pnorm} = 1.96$
stopband attenuation:	30 dB	$\rightarrow$	$v_s = 0.032$	$\rightarrow$	$v_{snorm} = 0.032$
passband edge frequency:			$f_p = 0.35/T$		
stopband edge frequency:			$f_s = 0.40/T$.		

The filter characteristic is Butterworth. First we determine the design parameters N and ω_c of the analog filter without pre-distortion of the frequency axis so that we will be able to compare these results later with the parameters of the analog reference filter for the design of the discrete-time filter. The necessary order of the filter is determined using (2.14a).

$$N = \frac{\log\dfrac{1.96}{0.032}}{\log\dfrac{0.40}{0.35}} = 30.8 \qquad \text{or rounded up to} \quad N = 31$$

The 3-dB cutoff frequency is obtained by using relation (2.14b).

$$\omega_c = \omega_p \times 1.96^{1/31}$$
$$\omega_c = 2\pi \times 0.35/T \times 1.022$$
$$\omega_c = 0.715\pi/T = 2.25/T \qquad\qquad f_c = 0.3577/T$$

An analog filter satisfying the above tolerance scheme is thus determined by the two design parameters $N = 31$ and $\omega_c = 2.25/T$.

For the design of the analog reference filter as the basis for a discrete-time filter design, the edge frequencies of the tolerance scheme must be first transformed using relation (6.40).

$$\omega_p' = 2/T \tan(2\pi \times 0.35/T \times T/2) = 3.93/T$$
$$\omega_s' = 2/T \tan(2\pi \times 0.40/T \times T/2) = 6.16/T$$

The filter order of the discrete-time filter is calculated as

$$N = \frac{\log \dfrac{1.96}{0.032}}{\log \dfrac{6.16}{3.93}} = 9.16 \qquad \text{or rounded up to} \quad N = 10 \ .$$

Equation (2.14b) is used again to calculate the 3-dB cutoff frequency.

$$\omega_c = \omega_p \times 1.96^{1/10} = 3.93 / T \times 1.070$$

$$\omega_c = 4.205 / T \qquad\qquad f_c = 0.669 / T$$

As already predicted, the order of the reference filter for the digital filter design is considerably lower than the one of an analog filter satisfying the same tolerance scheme. The 3-dB cutoff frequency is shifted to higher values, which finally results in the correct position when the bilinear transform is applied.

The starting point for the design of the discrete-time filter are the normalised coefficients of the analog reference filter. For a tenth-order Butterworth filter, which consists of five second-order filter blocks, we find the following coefficients in the corresponding table in the Appendix.

$$A_{11} = 0.312869 \qquad\qquad A_{21} = 1.0$$
$$A_{12} = 0.907981 \qquad\qquad A_{22} = 1.0$$
$$A_{13} = 1.414214 \qquad\qquad A_{23} = 1.0$$
$$A_{14} = 1.782013 \qquad\qquad A_{24} = 1.0$$
$$A_{15} = 1.975377 \qquad\qquad A_{25} = 1.0$$
$$B_{0v} = 1 \qquad\qquad\qquad B_{1v} = B_{2v} = 0$$
$$V_{tot} = 1$$

Relations (6.46) and (6.47) are now used to calculate the coefficients of the five partial discrete-time filters, which, arranged in cascade, have the desired frequency response.

	Filter 1	Filter 2	Filter 3	Filter 4	Filter 5
V	0.462	0.482	0.527	0.603	0.727
a_1	0.715	0.746	0.815	0.934	1.126
a_2	0.132	0.182	0.291	0.479	0.783
b_1	2	2	2	2	2
b_2	1	1	1	1	1

Figure 6-26 shows the magnitude responses of the five partial filters, which all have a common asymptote with a slope of 12 dB/octave. While filter 4 and filter 5 show resonant peaks in the range of the cutoff frequency, the remaining second-order blocks exhibit a monotonous rise of the attenuation.

The cascade arrangement of the five partial filters yields the magnitude response shown in Fig. 6-27a and Fig. 6-27b. As a result we obtain the expected flat response of a Butterworth filter which is fully compliant with the prescribed tolerance limits.

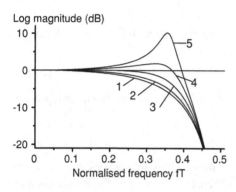

Fig. 6-26
Magnitude responses of the five partial filters of a tenth-order Butterworth filter

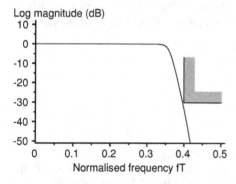

Fig. 6-27a
Magnitude of the designed Butterworth filter (detail of the stopband edge)

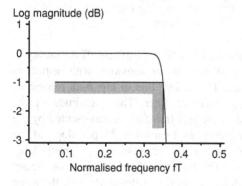

Fig. 6-27b
Magnitude of the designed Butterworth filter (detail of the passband edge)

6.4.4 IIR Filters with Bessel Characteristics

Bessel filters exhibit an approximately constant group delay or a linear phase in a wide range of frequencies. If we apply the bilinear transform to the transfer function, this property gets lost, unfortunately. We start by considering the influence of the bilinear transform on the ideal phase response of a linear phase filter, which can be expressed as

$$b(\omega) = \omega t_0 \ .$$

We obtain the corresponding phase response after bilinear transform by substitution of (6.40).

$$b(\omega) = (2t_0 / T)\tan(\omega T / 2) \tag{6.49}$$

Differentiation of (6.49) with respect to ω yields the corresponding group-delay characteristic.

$$\tau(\omega) = t_0(1 + \tan^2(\omega T / 2))$$

The group delay is approximately constant for very low frequencies only. Close to the Nyquist frequency $f_s/2$, the impact of the compression of the frequency axis becomes increasingly noticeable. The slope of the phase increases, which is equivalent with an ever-increasing group delay which finally approaches infinity.

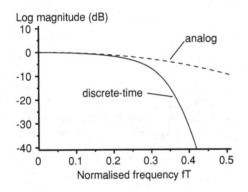

Fig. 6-28
Magnitude response of an analog fifth-order Bessel filter and of a related discrete-time filter obtained by bilinear transform

We choose the example of a fifth-order Bessel filter (3-dB cutoff frequency at $0.3 f_s$) to demonstrate the consequences of the bilinear transform with respect to the magnitude and group-delay response. The dashed line in Fig. 6-28 represents the magnitude response of the analog reference filter. The magnitude of the discrete-time transfer function obtained by bilinear transform is represented by the solid line. The compression of the frequency axis causes a sharper slope of the curve in the stopband. Figure 6-29 shows the corresponding group-delay responses. The analog filter exhibits the flat response that is typical for Bessel approximations. The curve for the related discrete-time filter starts with the same value at $\omega = 0$ but rises rapidly towards the Nyquist frequency.

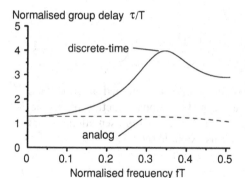

Fig. 6-29
Group delay response of an analog
fifth-order Bessel filter and of a related
discrete-time filter obtained by bilinear
transform

A way out of this problem is to predistort the phase response of the analog
reference filter in such a way that the bilinear transform yields a linear phase
again. An appropriate target phase response for the analog reference filter is the
inverse tangent:

$$b(\omega) = (2t_0 / T)\arctan(\omega T / 2) \ .$$

If the bilinear transform is applied to a transfer function with the above phase, we
obtain a strictly linear phase, as can be verified by substitution of (6.40). In the
following, we normalise the frequency ω to the delay t_0 ($\Omega = \omega t_0$) :

$$b(\Omega) = (2t_0 / T)\arctan(\Omega T / 2t_0) \ .$$

With the notation $\mu = 2t_0/T$ we finally arrive at

$$b(\Omega) = \mu \arctan(\Omega / \mu) \ . \tag{6.50}$$

According to (2.55), the phase of a transfer function of the form

$$H(P) = \frac{1}{1 + P + A_2 P^2 + A_3 P^3 + A_4 P^4 + ...}$$

can be expressed as

$$b(\Omega) = \arctan \frac{\Omega - A_3 \Omega^3 + A_5 \Omega^5 - ...}{1 - A_2 \Omega^2 + A_4 \Omega^4 - ...} \ . \tag{6.51}$$

In the case of an analog filter design, (6.51) has to approximate a linear phase
of the form $b(\Omega) = \Omega$ in the best possible way. Our analog reference filter for the
digital filter design, however, has to approximate the predistorted phase response
(6.50)

$$\mu \arctan(\Omega / \mu) = \arctan \frac{\Omega - A_3 \Omega^3 + A_5 \Omega^5 - ...}{1 - A_2 \Omega^2 + A_4 \Omega^4 - ...}$$

or

$$f(\Omega) = \tan(\mu \arctan(\Omega/\mu)) = \frac{\Omega - A_3\Omega^3 + A_5\Omega^5 - \ldots}{1 - A_2\Omega^2 + A_4\Omega^4 - \ldots}. \tag{6.52}$$

According to [22], $f(\Omega)$ can be expanded into a continued fraction that, if truncated after the Nth term, leads to a rational fractional function of degree N. (6.53) shows the systematic of this continued-fraction representation, which is truncated after the fifth partial quotient in this example ($N = 5$):

$$f(\Omega) = \tan(\mu \arctan(\Omega/\mu)) \approx \cfrac{1}{\cfrac{1}{\Omega} - \cfrac{1-1/\mu^2}{\cfrac{3}{\Omega} - \cfrac{1-4/\mu^2}{\cfrac{5}{\Omega} - \cfrac{1-9/\mu^2}{\cfrac{7}{\Omega} - \cfrac{1-16/\mu^2}{\cfrac{9}{\Omega} - \ldots}}}}}. \tag{6.53}$$

For $N = 1 \ldots 3$ we obtain the following rational fractional approximations of $f(\Omega)$:

$N = 1$ $\qquad$ $f(\Omega) \approx \Omega$

$N = 2$ $\qquad$ $f(\Omega) \approx \dfrac{3\Omega}{3 - (1-1/\mu^2)\Omega^2}$

$N = 3$ $\qquad$ $f(\Omega) \approx \dfrac{15\Omega - (1-4/\mu^2)\Omega^3}{15 - (6-9/\mu^2)\Omega^2}.$

These rational fractional approximations of $f(\Omega)$ directly provide the desired filter coefficients A_i, as a comparison with (6.52) shows.

$N = 1$ $\qquad$ $H(P) = \dfrac{1}{1+P}$

$N = 2$ $\qquad$ $H(P) = \dfrac{3}{3 + 3P + (1-1/\mu^2)P^2}$ $\qquad$ (6.54)

$N = 3$ $\qquad$ $H(P) = \dfrac{15}{15 + 15P + (6-9/\mu^2)P^2 + (1-4/\mu^2)P^3}$

The parameter μ has to satisfy a certain condition in order to guarantee stability of the filter. From the theory of Hurwitz polynomials (polynomials with poles exclusively in the left half of the P-plane) it is known that all coefficients of the continued-fraction expansion must be positive. This leads to the condition

$$1 - (N-1)^2/\mu^2 > 0$$

or

$$\mu = 2t_0 / T > N - 1 . \tag{6.55}$$

The delay t_0 of the discrete-time Bessel filter must therefore exceed a certain minimum value in order to obtain stable filters with all poles located in the left half of the P-plane.

$$t_0 > (N-1)T/2$$

The primary drawback of discrete-time filters designed in this way is the fact that the allowable range of the 3-dB cutoff frequencies is strongly limited, especially for higher filter orders. Filters with cutoff frequencies above the limits indicated in Fig. 6-30 are unstable.

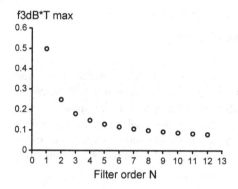

Fig. 6-30
Maximum achievable 3-dB cutoff frequency of the discrete-time Bessel filter against the filter order

Due to these limitations, the usability of the predistorted Bessel filter is highly restricted. Alternatives for the design of recursive filters with approximately linear phase in the passband are the impulse-invariant filters which preserve the temporal structure of the impulse response and the approximately linear-phase structure based on the parallel connection of a delay and an allpass filter that we introduced in Chap. 5.

6.5 Allpass Filter Design

There are numerous DSP applications which are based on the use of digital allpass filters. Some examples were introduced in Chap. 5 as for instance fractional delay filters, Hilbert transformers, or approximately linear-phase recursive filters. The approximation of a desired phase response $b_d(\omega)$ by the phase response $b_A(\omega)$ of an allpass filter of order N is therefore a frequent task in digital signal processing. There are various approximation criteria which aim at minimising the error $e(\omega)$ between the desired and the actual phase response which can be expressed as

$$e(\omega) = b_d(\omega) - b_A(\omega) . \tag{6.56}$$

According to (5.88), the phase response of a stable allpass filter is calculated as

$$b_A(\omega) = N\omega T - 2\arctan\frac{\sum\limits_{r=0}^{N} a_r \sin r\omega T}{\sum\limits_{r=0}^{N} a_r \cos r\omega T} \ . \tag{6.57}$$

The coefficient a_0 can always be normalised to unity so that the allpass phase response is fully determined by N coefficients a_i with $1 \le i \le N$.

$$b_A(\omega) = N\omega T - 2\arctan\frac{\sum\limits_{r=1}^{N} a_r \sin r\omega T}{1 + \sum\limits_{r=1}^{N} a_r \cos r\omega T}$$

$b_A(\omega)$ is composed of the linear-phase term $N\omega T$ and an odd periodic function which passes through zero at all integer multiples of the frequency π/T. Since the group delay $\tau_g(\omega)$ of a stable allpass filter is not negative, $b_A(\omega)$ must increase monotonously. In order to arrive at a stable filter as the result of an allpass design, the desired phase response $b_d(\omega)$ must match these general characteristics of an allpass filter. Once the order N of the filter is decided, the desired phase must fulfil the following requirements:

- $b_d(0) = 0$.
- $b_d(\pi/T) = N\pi$.
- $b_d(\omega)$ increases monotonously.

However, we cannot guarantee the stability of allpass digital filters when the phase error of the approximation is large. But if the filter order is selected reasonably so that the phase error is small enough, the designed allpass filter will become stable. Higher accuracy of the approximation requires more coefficients and thus a higher filter order to reproduce the desired phase. A better approximation is therefore always coupled with a higher delay of the filter due to the linear-phase term of the allpass phase response.

By substitution of (6.57) in (6.56), the design error e_i at a given frequency ω_i can be expressed as

$$e(\omega_i) = b_d(\omega_i) - N\omega_i T + 2\arctan\frac{\sum\limits_{r=0}^{N} a_r \sin r\omega_i T}{\sum\limits_{r=0}^{N} a_r \cos r\omega_i T} \tag{6.58}$$

The desired phase of the allpass filter may be specified in the whole frequency range $0 < \omega < \pi/T$ (full-band approximation) or in one or more frequency bands.

The entity of these bands is commonly referred to as the frequency range of interest. Only this range is taken into account when the coefficients a_r are determined to minimise the error $e(\omega)$ in some sense.

Most approximation methods result in a phase response which will oscillate about the desired phase curve. The peak values of the design error are in the same order of magnitude for all frequencies of interest. In the case of a Chebyshev approximation, the error even has an equiripple characteristic. There are applications, however, which require a much lower error in one band while it is acceptable to have higher phase deviations in another band. A good example is the approximately linear-phase filter that we introduced in Sect. 5.7.3.5. In order to achieve high attenuation in the stopband, only small deviations from the linear-phase characteristic are allowed. In contrast, commonly accepted passband ripples in the order of 0.1 dB to 1 dB can cope with less accurate approximations of the desired phase. The solution to this problem is the introduction of a frequency-dependent weighting function $W(\omega)$ by which the error function (6.58) is multiplied.

$$e(\omega_i) = W(\omega_i)\left[b_d(\omega_i) - N\omega_i T + 2\arctan\frac{\displaystyle\sum_{r=0}^{N} a_r \sin r\omega_i T}{\displaystyle\sum_{r=0}^{N} a_r \cos r\omega_i T}\right] \tag{6.59}$$

Frequency bands to which is assigned a high weighting in the approximation process will show lower error values, those with a low weighting will show higher error values than in the unweighted case.

Relation (6.59) is highly nonlinear in the coefficients a_r. Nonlinear optimisation methods are computationally expensive and often require good initial solutions for convergence to the desired result. Optimisation of the error in the one or another way directly based on (6.59) is therefore mathematically an ambitious task. It would be desirable to convert the nonlinear phase design problem into a linear approximation problem. Unfortunately, there is no equivalent transformation of (6.59) available which yields a linear relationship between the phase error and the filter coefficients. But this relation may be expressed in a form which allows a quasilinear treatment of the approximation problem. With the substitution

$$\beta_d(\omega) = \frac{b_d(\omega) - N\omega T}{2}$$

we obtain

$$\frac{e(\omega_i)}{2W(\omega_i)} = \beta_d(\omega_i) + \arctan\frac{\displaystyle\sum_{r=0}^{N} a_r \sin r\omega_i T}{\displaystyle\sum_{r=0}^{N} a_r \cos r\omega_i T}$$

$$\sin\left(\frac{e(\omega_i)}{2W(\omega_i)}\right) = \sin\left(\beta_d(\omega_i) + \arctan\frac{\sum\limits_{r=0}^{N} a_r \sin r\omega_i T}{\sum\limits_{r=0}^{N} a_r \cos r\omega_i T}\right). \qquad (6.60)$$

The sine summation in (6.60) is the negative imaginary part, the cosine summation the real part of the frequency response of the denominator polynomial $D(z)$ of the allpass transfer function.

$$D(e^{j\omega T}) = \sum_{r=0}^{N} a_r e^{-jr\omega T} = \sum_{r=0}^{N} a_r \cos r\omega T - j\sum_{r=0}^{N} a_r \sin r\omega T$$
$$= D_{\mathrm{Re}}(\omega) + j D_{\mathrm{Im}}(\omega)$$

So (6.60) can be rewritten in the form

$$\sin\left(\frac{e(\omega_i)}{2W(\omega_i)}\right) = \sin\left(\beta_d(\omega_i) - \arctan\frac{D_{\mathrm{Im}}(\omega_i)}{D_{\mathrm{Re}}(\omega_i)}\right)$$

$$\sin\left(\frac{e(\omega_i)}{2W(\omega_i)}\right) = \sin\beta_d(\omega_i)\cos\arctan\frac{D_{\mathrm{Im}}(\omega_i)}{D_{\mathrm{Re}}(\omega_i)} - \cos\beta_d(\omega_i)\sin\arctan\frac{D_{\mathrm{Im}}(\omega_i)}{D_{\mathrm{Re}}(\omega_i)}$$

Since

$$\cos(\arg D(e^{j\omega T})) = \cos\left(\arctan\frac{D_{\mathrm{Im}}(\omega_i)}{D_{\mathrm{Re}}(\omega_i)}\right) = \frac{D_{\mathrm{Re}}(\omega)}{|D(e^{j\omega T})|} \quad \text{and}$$

$$\sin(\arg D(e^{j\omega T})) = \sin\left(\arctan\frac{D_{\mathrm{Im}}(\omega_i)}{D_{\mathrm{Re}}(\omega_i)}\right) = \frac{D_{\mathrm{Im}}(\omega)}{|D(e^{j\omega T})|}$$

we finally yield

$$\sin\left(\frac{e(\omega_i)}{2W(\omega_i)}\right) = \sin\beta_d(\omega_i)\frac{D_{\mathrm{Re}}(\omega_i)}{|D(e^{j\omega T})|} - \cos\beta_d(\omega_i)\frac{D_{\mathrm{Im}}(\omega_i)}{|D(e^{j\omega T})|}$$

$$\sin\left(\frac{e(\omega_i)}{2W(\omega_i)}\right)|D(e^{j\omega T})| = \sum_{r=0}^{N} a_r(\cos r\omega_i T \sin\beta_d(\omega_i) + \sin r\omega_i T \cos\beta_d(\omega_i))$$

$$= \sum_{r=0}^{N} a_r \sin(r\omega_i T + \beta_d(\omega_i))$$

If the phase error $e(\omega_i)$ is small enough, which should be the case in realistic design scenarios, the sine function can be replaced by its argument which leads to the following expression:

$$e(\omega_i) = \frac{\sum_{r=0}^{N} a_r \sin\left(r\omega_i T + \beta_{\mathrm{d}}(\omega_i)\right)}{\left|D(\mathrm{e}^{\mathrm{j}\omega T})\right| \big/ 2W(\omega_i)} \tag{6.61}$$

At first glance, (6.61) does not seem to be very helpful. The numerator is the desired linear expression in the filter coefficients a_r, but $|D(\omega)|$ in the denominator is also dependent on the filter coefficients. So we still have nonlinear dependencies. In an iterative procedure, however, we can use the coefficients of the previous iteration step to determine $|D(\omega)|$ for the new iteration. What then remains is the solution of a linear optimisation problem in each iteration step. It can be observed that $|D(\omega)|$ commonly has rather smooth graphs whose shape is not dramatically sensitive to the variations of the filter coefficients that occur in the course of the iteration. Allpass approximations based on this concept therefore show good convergence behaviour. In [44] and [34], for instance, the error measure (6.61) or a similar expression is used as the starting point for a least mean square or Chebyshev allpass design. As the initial condition for the first iteration step it is sufficient to make a simple assumption such as $|D(\omega)| = 1$.

As mentioned above, the phase response of the allpass filter is fully determined by N coefficients. So one coefficient can be chosen arbitrarily. For some applications it is advantageous to set a_0 equal to unity which results in expression (6.62).

$$e(\omega_i) = \frac{\sum_{r=1}^{N} a_r \sin\left(r\omega_i T + \beta_{\mathrm{d}}(\omega_i)\right)}{\left|D(\mathrm{e}^{\mathrm{j}\omega T})\right| \big/ 2W(\omega_i)} + \frac{\sin \beta_{\mathrm{d}}(\omega_i)}{\left|D(\mathrm{e}^{\mathrm{j}\omega T})\right| \big/ 2W(\omega_i)} \tag{6.62}$$

We will consider in the following various criteria for a good approximation of a given phase response:

- Interpolation, where the error $e(\omega)$ becomes zero at N frequency points.
- Minimisation of the mean square error between the desired and the designed phase response.
- Minimisation of the maximum error of $e(\omega)$ which leads to an equiripple error characteristic.
- MAXFLAT design, where the error $e(\omega)$ and some of its derivatives becomes zero at a limited number of frequencies.

6.5.1 Interpolation

The most straightforward approach for the design of allpass filters is interpolation. The desired phase response is specified at N distinct frequencies. This gives us N conditional equations to determine the N unknown filter coefficients a_r. The phase response of the allpass filter designed in this way interpolates the N desired phase values.

The phase response of the allpass filter equals the desired phase values at N frequencies ω_i with $i = 1 \ldots N$. This means that the error $e(\omega)$ vanishes at these frequencies.

$$e(\omega_i) = 0 \quad \text{for} \quad i = 1 \ldots N$$

Substitution in (6.62) yields

$$\sum_{r=1}^{N} a_r \sin\left(r\omega_i T + \beta_d(\omega_i)\right) + \sin\beta_d(\omega_i) = 0$$

$$\sum_{r=1}^{N} a_r \sin\left(r\omega_i T + \beta_d(\omega_i)\right) = -\sin\beta_d(\omega_i) \qquad (6.63)$$

where

$$\beta_d(\omega_i) = \frac{b_d(\omega_i) - N\omega_i T}{2} \quad \text{for} \quad i = 1 \ldots N \qquad (6.64)$$

is determined by the desired phase values. (6.63) can be written in vector form.

$$\mathbf{Sa = b} \qquad\qquad (6.65)$$

with

$$s_{ir} = \sin\left(r\omega_i T + \beta_d(\omega_i)\right)$$
$$b_i = -\sin\beta_d(\omega_i) \qquad\qquad \begin{aligned} i &= 1 \ldots N \\ r &= 1 \ldots N \end{aligned}$$

$\mathbf{a}$ is the vector of the unknown coefficients a_r which can be found by solving the system of linear equations (6.65).

$$\mathbf{a = S^{-1}b}$$

As pointed out in [44], matrix equation (6.65) may not always be numerically well-behaved. This is especially true when the phase is approximated in narrow frequency bands which leads to clustered poles and zeros. Numerically well-behaving algorithms should therefore be used for the matrix inversion like QR decomposition or single value decomposition [60].

We present in the following some examples for the allpass design based on interpolation.

The desired phase response of the Hilbert transformer was derived in Sect. 5.7.3.7 as

$$b_d(\omega) = n_0 \omega T + \pi/2 \quad \text{for} \quad \omega_1 \le \omega \le \omega_2$$

The band between ω_1 and ω_2 is the frequency range of interest. Below ω_1, $b(\omega)$ must approach zero as ω goes to zero. Above ω_2, $b(\omega)$ may approach $(n_0 + 1)\pi$ as ω goes to π/T as depicted in Fig. 5-117. This requires a filter order of

$N = n_0 + 1$. As an alternative, $b(\omega)$ may approach an angle of $n_0\pi$ which requires a filter order of $N = n_0$. We start with considering the first case in more detail. As an input to (6.64), we calculate the desired phase values at N equidistant frequency points ω_i:

$$b_{\mathrm{d}}(\omega_i) = n_0\omega_i T + \pi/2 = (N-1)\omega_i T + \pi/2 \qquad (6.66)$$

with

$$\omega_i = \omega_1 + \frac{\omega_2 - \omega_1}{N-1} i \quad \text{for} \quad i = 0 \ldots N-1 \ .$$

Substitution in (6.64) yields

$$\beta_{\mathrm{d}}(\omega_i) = \frac{(N-1)\omega_i T + \pi/2 - N\omega_i T}{2}$$
$$\beta_{\mathrm{d}}(\omega_i) = \pi/4 - \omega_i T/2 \ . \qquad (6.67)$$

(6.67) can be used as an input for the calculation of the matrix and vector elements of (6.65). The filter coefficients are obtained by solving this system of linear equations. As an example, Fig. 6-31 shows the phase response of an 8th-order Hilbert transformer with $\omega_1 = 0.1\pi/T$ and $\omega_2 = 0.9\pi/T$ which interpolates eight given phase values.

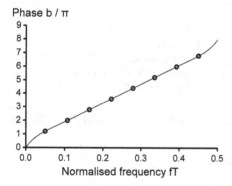

Phase b / π

<table>
<tr><td>Normalised frequency fT</td></tr>
</table>

Fig. 6-31
Hilbert transformer with $\omega_1 = 0.1\pi/T$, $\omega_2 = 0.9\pi/T$, $N = n_0 + 1 = 8$

We get a better insight into the details of the approximation when we subtract the linear-phase part $n_0\omega T$ from the obtained phase response. Fig. 6-32 shows that the deviation from the ideal phase of $\pi/2$ is not uniform in the band of interest. It is larger in the vicinity of the band edges and smaller in the middle. The zeros of the error are equally spaced because we defined the desired phase values at equally spaced frequency points. The error near the band edges may be reduced by increasing the density of prescribed phase values in this region. This method is referred to as frequency weighting.

Alternatively we can design a Hilbert transformer in such a way that $N = n_0$. Fig. 6-30 shows the resulting phase deviations from the ideal 90° phase shift. It

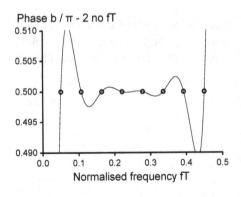

Fig. 6-32
Phase response of the Hilbert transformer less the linear-phase component $n_0 \omega T$

turns out that such a design commonly results in much larger errors than in the case before. Interestingly, a design with odd filter order yields a real pole/zero pair on the unit circle which cancels out in the allpass transfer function. So assuming an odd filter order results in an allpass filter with even order. This does not happen in the case $N = n_0 + 1$.

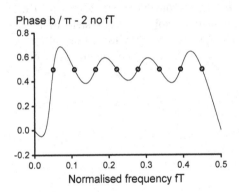

Fig. 6-33
Phase response of the Hilbert transformer less the linear-phase component $n_0 \omega T$

According to Sect. 5.7.3.8, the desired phase response of a fractional delay filter is of the form

$$b_d(\omega) = \omega(n_0 T + \Delta) \quad \text{for} \quad 0 \le \omega \le \omega_1$$
$$\text{unspecified} \qquad\qquad \text{for} \quad \omega_1 < \omega \le \pi / T$$

$n_0 T$ is the integer part of the desired delay, Δ is the fractional part. The band between 0 and ω_1 is the frequency range of interest. Above ω_1, $b(\omega)$ will approach $N\pi$ as ω goes to π / T. As an input to (6.64), we calculate the desired phase values at N equidistant frequency points ω_i:

$$b_d(\omega_i) = (n_0 T + \Delta)\omega_i$$

with

$$\omega_i = \frac{\omega_1}{N-1} i \quad \text{for} \quad i = 0 \ldots N-1 \ .$$

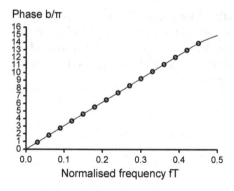

Fig. 6-34
Phase response of a 15th-order fractional delay filter with $n_0 = 15$, $\Delta = 0.5T$, and $\omega_1 = 0.9\pi/T$

Substitution in (6.64) yields

$$\beta_{\mathrm{d}}(\omega_i) = \frac{(n_0 T - NT + \Delta)\omega_1}{2(N-1)}i \quad.$$ (6.68)

We first consider the case $N = n_0$. Equation (6.68) then simplifies to

$$\beta_{\mathrm{d}}(\omega_i) = \frac{\Delta\omega_1}{2(N-1)}i \quad.$$

By solving the linear system of equations (6.65) we obtain the desired allpass filter coefficients. Figure 6-34 shows the resulting phase response for $N = n_0 = 15$ and $\Delta = 0.5$. The frequency range of interest extends from 0 to $0.9\pi/T$.

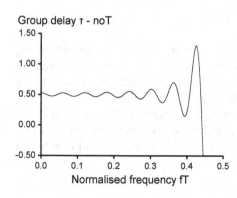

Fig. 6-35
Group delay of a 15th-order fractional delay filter with $n_0 = 15$, $\Delta = 0.5T$, and $\omega_1 = 0.9\pi/T$.

   More details concerning the quality of the approximation become visible if we show the graph of the group delay response (Fig. 6-35). At low frequencies, there is a rather good approximation of the desired delay of $\Delta = 0.5T$. But as the frequency approaches the upper edge of the band of interest, the ripple of the group delay response becomes increasingly large. In cases where such deviations from the ideal response are not tolerable, the order of the filter $N = n_0$ needs to be increased which will reduce the ripple.

Interestingly, we get much better results if we choose $N = n_0 + 1$. In this case, the desired delay $n_0 T + \Delta$ is smaller than the linear-phase delay NT of the allpass filter. With the same filter order, much less ripple can be achieved as demonstrated in Fig. 6-36. Again, the order of the allpass filter is 15. But the integer part n_0 of the desired delay is chosen as 14.

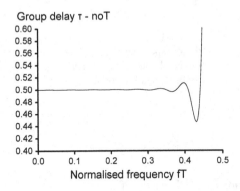

Group delay τ - noT

Fig. 6-36
Group delay of a 15th-order fractional delay filter with $n_0 = 14$, $\Delta = 0.5T$, and $\omega_1 = 0.9\pi/T$.

In order to realise an almost linear-phase low-pass filter, the sum is taken of the phase of a delay of length KT and the phase of an appropriately designed allpass filter of order $N = K + 1$ (refer to Sect. 5.7.3.5). The desired phase response of the allpass filter can be expressed as

$$b_{\mathrm{d}}(\omega) = \begin{cases} K\omega T & \text{for } 0 \le \omega \le \omega_1 \\ \text{undefined} & \text{for } \omega_1 < \omega < \omega_2 \\ K\omega T + \pi & \text{for } \omega_2 \le \omega \le \pi/T \ . \end{cases}$$

Below ω_1, delay and allpass filter are in phase. So both signals sum up. Above ω_2, delay and allpass filter are antiphase. So both signals cancel out. Altogether, we obtain a low-pass magnitude characteristic. By taking the difference instead of the sum of both paths, we obtain the complementary high-pass filter.

N support points are needed to determine the N unknown filter coefficients. Of these, N_1 points are placed in the passband and N_2 points in the stopband. As a first choice, the support points may be distributed evenly over the bands of interest. So we approximately have

$$\frac{N_1}{N_2} \approx \frac{\omega_1}{\pi/T - \omega_2}$$

The desired support points are calculated as

$$b_{\mathrm{d}}(\omega_i) = \frac{K\omega_1 T}{N_1} i \qquad\qquad\qquad\text{for } i = 1 \ldots N_1$$

$$b_{\mathrm{d}}(\omega_i) = \pi + K\omega_2 T + \frac{K(\pi - \omega_2 T)}{N_2}(i - N_1 - 1) \quad \text{for } i = N_1 + 1 \ldots N$$

As an example, we consider a low-pass filter designed with the following set of parameters:

$$N_1 = 8 \qquad \omega_1 = 0.6\pi / T$$
$$N_2 = 4 \qquad \omega_2 = 0.7\pi / T$$

Figure 6-37 shows the 12 support points and the interpolating phase response of the allpass filter.

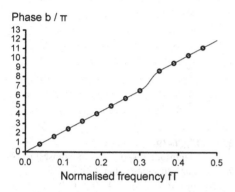

Phase b / π

Fig. 6-37
Allpass phase response to realise an almost linear-phase low-pass filter with $N_1 = 8$, $\omega_1 = 0.6\pi/T$, $N_2 = 4$, $\omega_2 = 0.7\pi/T$.

The phase difference $\Delta b(\omega)$ between delay path and allpass path is represented in Fig. 6-38. As desired, $\Delta b(\omega)$ approximates zero in the lower frequency band and a value of π in the upper band.

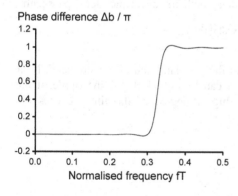

Phase difference Δb / π

Fig. 6-38
Phase difference between delay and allpass filter to realise an almost linear-phase low-pass filter with $N_1 = 8$, $\omega_1 = 0.6\pi/T$, $N_2 = 4$, $\omega_2 = 0.7\pi/T$.

According to Fig. 5-107, the transfer function of the almost linear-phase low-pass filter can be expressed as the sum of a delay term and an allpass term.

$$H(j\omega) = 0.5(e^{-jb_D(\omega)} + e^{-jb_A(\omega)})$$
$$H(j\omega) = 0.5\left(\cos b_D(\omega) - j\sin b_D(\omega) + \cos b_A(\omega) - j\sin b_A(\omega)\right)$$
$$|H(j\omega)| = \sqrt{\left(\cos b_D(\omega) + \cos b_A(\omega)\right)^2 + \left(\sin b_D(\omega) + \sin b_A(\omega)\right)^2}$$

The magnitude of the transfer function is obtained after some trigonometric manipulations as

$$|H(j\omega)| = \left|\cos\frac{b_D(\omega)-b_A(\omega)}{2}\right| = \left|\cos\frac{\Delta b(\omega)}{2}\right| \qquad (6.69)$$

Using (6.69), we can calculate the magnitude response of our example filter from the phase difference represented in Fig. 6-38. The result is shown in Fig. 6-39 on a logarithmic scale.

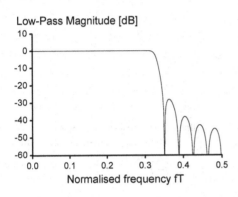

Fig. 6-39
Magnitude response of an almost linear-phase low-pass filter with $N_1 = 8$, $\omega_1 = 0.6\pi/T$, $N_2 = 4$, $\omega_2 = 0.7\pi/T$.

We finally check how good the approximation of a constant group delay is in the passband and in the stopband. Figure 6-40 shows the group delay response of the low-pass filter. As derived in Sect. 5.7.3.5, this group delay response is calculated as the average of the group delay of the delay path and of the parallel allpass path.

$$\tau_g(\omega) = \frac{KT + \tau_{Allp}(\omega)}{2} = \frac{(N-1)T + \tau_{Allp}(\omega)}{2}$$

The group delay is almost constant in the passband and in the stopband. Near the band edges to the transition band, it can be observed that the ripple of the delay curve is increasing. Choosing a higher degree of the filter reduces the amplitude of the ripple.

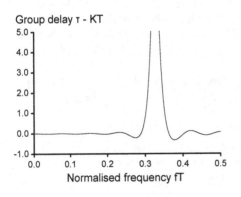

Fig. 6-40
Group delay of an almost linear-phase low-pass filter with $N_1 = 8$, $\omega_1 = 0.6\pi/T$, $N_2 = 4$, $\omega_2 = 0.7\pi/T$.

The achievable stopband attenuation of our example filter is about 30 dB as can be seen from Fig. 6-39. The stopband performance can be improved by assigning more support points to the stopband. Fig. 6-41 shows the logarithmic magnitude of a 12th-order filter as before, but here 6 support points are equally assigned to the passband and to the stopband. The stopband attenuation is now about 55 dB. It should be mentioned, however, that the delay ripple in the passband is now higher due to the reduced number of available support points in the passband. The concept of varying the density and distribution of support points in order to manipulate the filter characteristics in a desired manner is referred to as frequency weighting.

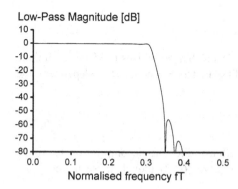

Fig. 6-41
Magnitude response of an almost linear-phase low-pass filter with $N_1 = 6$, $\omega_1 = 0.6\pi/T$, $N_2 = 6$, $\omega_2 = 0.7\pi/T$.

6.5.2 Least Mean Square Design

The goal of the least mean square design is to determine the allpass filter coefficients a_r such that the mean square error

$$\sqrt{\int_0^{\pi/T} e^2(\omega)\,d\omega}$$

is minimised. For a numerical treatment of the approximation problem, we calculate the error $e(\omega)$ on a dense grid of frequencies and replace the integral with a summation.

$$\sqrt{\sum_{i=0}^{gridsize-1} e^2(\omega_i)} \quad \text{with} \quad \omega_i = \frac{\pi}{T}\frac{i}{gridsize-1} \tag{6.70}$$

The grid size is commonly chosen about one order of magnitude higher than the order N of the allpass filter. Using (6.62), the error values $e(\omega_i)$ can be expressed in the following vector form.

$$e = Sa - b \tag{6.71}$$

with

$$e_i = \left(e(\omega_0), \ldots, e(\omega_{gridsize-1})\right)$$

$$a_r = (a_1, \ldots, a_N) \quad , \text{the unknown coefficients}$$

$$s_{ir} = \frac{\sin\left(r\omega_i T + \beta_d(\omega_i)\right)}{\left|D_{\text{prev}}(e^{j\omega_i T})\right|/2W(\omega_i)}$$

$$b_i = -\frac{\sin \beta_d(\omega_i)}{\left|D_{\text{prev}}(e^{j\omega_i T})\right|/2W(\omega_i)} \qquad i = 0 \ldots gridsize-1, \quad r = 1 \ldots N$$

$$\beta_d(\omega_i) = \frac{b_d(\omega_i) - N\omega_i T}{2}$$

The problem of finding a set of coefficients **a** that minimizes (6.70) has a unique solution. Provided that the N columns of the matrix **S** are linearly independent, the optimum vector **a** can be calculated as

$$a = \left[S^T S\right]^{-1} S^T b \; . \tag{6.72}$$

Again, numerically well-behaving algorithms like QR decomposition or single value decomposition should be used to solve (6.72) though these methods are computationally expensive.

In order to compare the performance of the least mean square algorithm with the performance of interpolation, we will use the same examples as in the previous section. In the following examples we set the weighting function $W(\omega)$ to unity in the whole frequency range of interest.

The desired phase of the Hilbert transformer can be expressed as

$$b_d(\omega) = n_0 \omega T + \pi/2 \quad \text{for} \quad \omega_1 \leq \omega \leq \omega_2 \; .$$

We consider the case $N = n_0 + 1$. As an input to (6.71), the desired phase is evaluated on a dense grid of frequencies.

$$b_d(\omega_i) = n_0 \omega_i T + \pi/2 = (N-1)\omega_i T + \pi/2$$

with

$$\omega_i = \omega_1 + \frac{\omega_2 - \omega_1}{gridsize-1} i \quad \text{for} \quad i = 0 \ldots gridsize-1$$

$$\beta_d(\omega_i) = \pi/2 - \omega_i T/2$$

Figure 6-42 shows the phase response of an 8th-order allpass filter with $\omega_1 = 0.1\pi/T$ and $\omega_2 = 0.9\pi/T$ which minimises the mean square deviation between desired and actual phase. In the frequency range $\omega_1 < \omega < \omega_2$, the phase response is linear and shows an additional phase shift of $\pi/2$. Outside the frequency range of interest, the value of the phase is forced to 0 and 8π respectively.

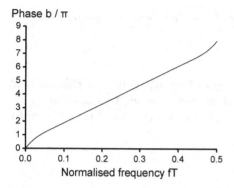

Fig. 6-42
Phase of a Hilbert transformer with
$\omega_1 = 0.1\pi/T$, $\omega_2 = 0.9\pi/T$, $N = n_0 + 1 = 8$

Removal of the linear-phase part $n_0\omega T$ of the phase response reveals more details concerning the quality of the approximation (Fig. 6-43). A comparison with the results of the interpolation method (Fig. 6-32) shows that the overshot of the phase near the edges of the band of interest is notably reduced. In return, we observe an increase of the ripple in the centre of the band. On the whole, the amplitudes of the ripple are much more uniform compared to the interpolation method.

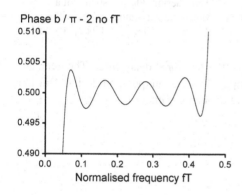

Fig. 6-43
Phase less the linear-phase part of a
Hilbert transformer with $\omega_1 = 0.1\pi/T$,
$\omega_2 = 0.9\pi/T$, $N = n_0 + 1 = 8$

The desired phase of a fractional delay filter is of the form

$$b_d(\omega) = \omega(n_0 T + \Delta) \quad \text{for} \quad 0 \leq \omega \leq \omega_1 .$$

$n_0 T$ is the integer part of the desired delay, Δ is the fractional part. We consider the case $N = n_0 + 1$. As an input to (6.71), the desired phase is evaluated on a dense grid of frequencies.

$$b_d(\omega_i) = (n_0 T + \Delta)\omega_i$$

with

$$\omega_i = \frac{\omega_1}{gridsize - 1}i \quad \text{for} \quad i = 0 \ldots gridsize - 1$$

$$\beta_d(\omega_i) = \frac{(\Delta - T)\omega_1/2}{gridsize - 1}i$$

For the calculation of an example, we choose the following specification of a fractional delay filter:

$$N = n_0 + 1 = 15, \qquad \omega_1 = 0.9\pi/T, \qquad \Delta = 0.5T.$$

Fig. 6-44 shows the resulting phase response of the allpass filter. Within the band of interest, the phase is linear with a slope corresponding to a delay of $14.5T$. Above ω_1, the phase approaches a value of 15π as it is required for a 15th-order allpass filter.

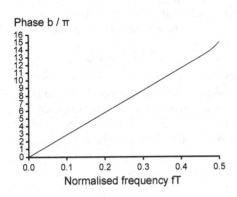

Fig. 6-44
LMS design: Phase response of a fractional delay allpass filter with $N = n_0 + 1 = 15$, $\omega_1 = 0.9\pi/T$, $\Delta = 0.5T$

Figure 6-45 represents the group delay of the fractional delay filter. The overshot near the band edge is about halved compared to the interpolation design. But the ripple amplitude is increased in the rest of the band of interest.

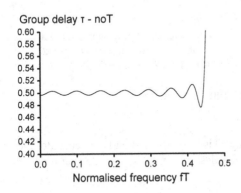

Fig. 6-45
LMS design: Group delay response of a fractional delay allpass filter with $N = n_0 + 1 = 15$, $\omega_1 = 0.9\pi/T$, $\Delta = 0.5T$

The desired phase of an almost linear-phase low-pass filter can be expressed in the form

$$b_d(\omega) = \begin{cases} K\omega T & \text{for } 0 \leq \omega \leq \omega_1 \\ \text{undefined} & \text{for } \omega_1 < \omega < \omega_2 \\ K\omega T + \pi & \text{for } \omega_2 \leq \omega \leq \pi/T. \end{cases}$$

The desired phase characteristic is evaluated on a dense grid of frequencies. The frequency range of interest is the concatenation of the passband and the stopband of the filter. The grid frequencies are evenly distributed over this frequency range.

Again we choose the same design data as in the previous section in order to demonstrate the performance of the LMS design method.

$$\omega_1 = 0.6\pi/T, \qquad \omega_2 = 0.7\pi/T, \qquad N = 12$$

Figure 6-46 shows the resulting phase response of the allpass filter. The slope of the phase characteristic corresponds to a group delay of $11T$. Taking into account the additional phase shift of π introduced in the transition band, the phase ends up at a value of 12π at half the sampling frequency π/T.

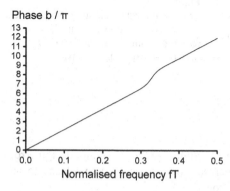

Fig. 6-46
LMS Design: Phase response of the allpass filter of an almost linear-phase low-pass filter with $N = 12$, $\omega_1 = 0.6\pi/T$, $\omega_2 = 0.7\pi/T$

The magnitude of the low-pass filter is depicted in Fig. 6-47. The stopband attenuation is increased by about 6 dB compared to the interpolation method. The ripple amplitudes are more uniform.

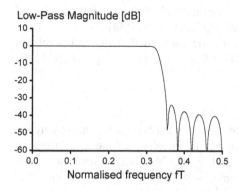

Fig. 6-47
LMS Design: Magnitude response of the almost linear-phase low-pass filter with $N = 12$, $\omega_1 = 0.6\pi/T$, $\omega_2 = 0.7\pi/T$

In the previous examples, the weighting function $W(\omega)$ was set to unity. In the following, we will demonstrate how weighting can be used to improve the filter performance in a desired manner. The 12th-order low-pass filter of the previous example exhibits a stopband attenuation of about 34 dB (highest peak in the stopband). We perform now the same filter design, but with $W(\omega) = 1$ in the passband and $W(\omega) = 10$ in the stopband, which reduces the design error in the stopband resulting in a higher attenuation as can be seen from Fig. 6-48. The filter achieves a stopband attenuation of about 47 dB.

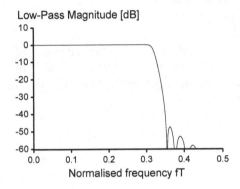

Fig. 6-48
LMS design with weighting: Magnitude response of the almost linear-phase low-pass filter with $N = 12$, $\omega_1 = 0.6\pi/T$, $\omega_2 = 0.7\pi/T$, $W_1 = 1$, $W_2 = 10$

6.5.3 Quasi-Equiripple Design

In some applications, it might be desirable to minimize the maximum design error occurring in the frequency range of interest. This design objective leads to an equiripple characteristic of the error function $e(\omega)$.

In the last example, we demonstrated that weighting can be used to influence the error function in a desired manner. The basic idea, here, is to assign to the frequency ranges where $e(\omega)$ is large a higher weighting $W(\omega)$ than to the frequencies where the error is small in order to make the ripple amplitudes more uniform. The envelope $\text{env}(\omega)$ of the magnitude of $e(\omega)$ contains the desired information to create an appropriate weighting function [69] [48].

We start with $W_0(\omega) = 1$ and determine the filter coefficients recursively as described for the LMS design using (6.71). After each iteration step, we update the weighting function as follows:

$$W_{k+1}(\omega) = W_k(\omega)\,\text{env}_k(\omega) \ .$$

So the weighting function for the next iteration step is obtained by multiplying the weighting function of the previous step by the envelope of the error function resulting from the previous iteration step. The algorithm is terminated when the variation of the ripple amplitudes becomes smaller than a given number. The magnitude of the denominator polynomial $|D(\omega)|$ in (6.71) can be omitted (set to unity) which reduces computational complexity and speeds up the convergence. Note that $|D(\omega)|$ is just another weighting function. With or without this additional weighting, the algorithm will finally converge to an equiripple error characteristic.

The following figures demonstrate that this modified least mean square algorithm results in equiripple designs. Figure 6-49 shows the phase response of the Hilbert transformer, Fig. 6-50 the magnitude of the almost linear-phase recursive filter which already served as examples earlier in this chapter. It turns out that the equiripple design is superior to the least squares design near the band edges whereas the least squares design has much smaller ripple magnitude elsewhere.

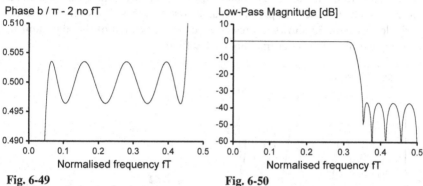

Fig. 6-49
Quasi-equiripple design: Hilbert transformer, $N = n_0 + 1 = 8$ $\omega_1 = 0.1\pi/T$, $\omega_2 = 0.9\pi/T$

Fig. 6-50
Quasi-equiripple design: almost linear-phase low-pass filter, $N = 12$, $\omega_1 = 0.6\pi/T$, $\omega_2 = 0.7\pi/T$

Figure 6-51 shows the group delay response of the exemplary fractional delay filter from Fig. 6-45. The adaptive weighting method to achieve equiripple behaviour seems to fail in this case. The ripple amplitude is still increasing as the frequency approaches the band edge. But we must keep in mind that we made the phase response equiripple which does not automatically mean that the group delay is equiripple, too. The group delay is the derivative of the phase response with respect to frequency. So the peak values of the group delay correspond to the maximum slopes of the phase curve between the peaks and valleys which are not equidistant.

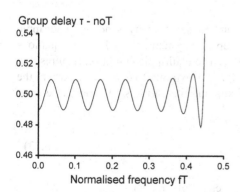

Fig. 6-51
Quasi-equiripple phase design: fractional delay allpass filter, $N = n_0 + 1 = 15$, $\omega_1 = 0.9\pi/T$, $\Delta = 0.5T$. Group delay is not equiripple.

We can overcome this problem by choosing a different update strategy for the weighting function. Intuitively, instead of using the envelope of the phase error, we calculate the envelope of the group delay error which is defined as

$$e_k(\omega) = \tau_d(\omega) - \tau_{Ak}(\omega),$$

where $\tau_d(\omega)$ is the desired group delay characteristic and $\tau_{Ak}(\omega)$ is the group delay of the allpass filter in the kth iteration step. Now the weighting function $W_k(\omega)$ converges such that the group delay characteristic of the allpass filter becomes equiripple. Figure 6-52 shows the result of this modification of the algorithm for the example of the 15th-order fractional delay filter.

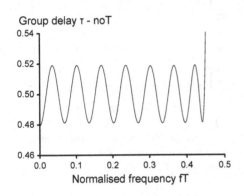

Group delay τ - noT

Fig. 6-52
Group delay characteristic of the 15th-order fractional delay filter. The delay error is used to update the weighting function $W_k(\omega)$.

In the literature, the method described in this section is referred to as a quasi-equiripple design because the equiripple condition is not explicitly stated in the algorithm. It has been shown, however, that the results obtained come very close to or even equal the outcome of true equiripple algorithms, but general convergence of the adaptive LMS algorithm to the optimum result has never been proved analytically [48].

6.5.4 True Equiripple Design

Remez-type algorithms are proven means to solve many kinds of equiripple design problems. Starting point is the setup of a system of $N + 1$ linear equations in the $N + 1$ unknown filter coefficients a_r. The equiripple condition requires that the error $e(\omega)$ oscillates around zero with $N + 1$ extreme frequencies ω_i where the error assumes values with equal magnitude but alternating sign. The phase error can therefore be expressed as

$$e(\omega_i) = (-1)^i \delta \quad \text{with } i = 0 \ldots N \tag{6.73}$$
$$\text{and } \omega_{i+1} > \omega_i > \omega_{i-1}.$$

Insertion of (6.61) in (6.73) yields the desired set of equations.

$$\frac{\sum_{r=0}^{N} a_r \sin(r\omega_i T + \beta_d(\omega_i))}{\left|D(e^{j\omega_i T})\right| / 2W(\omega_i)} = (-1)^i \delta \quad \text{for} \quad i = 0 \dots N$$

$$\sum_{r=0}^{N} a_r \sin(r\omega_i T + \beta_d(\omega_i)) = (-1)^i \delta \left|D(e^{j\omega_i T})\right| / 2W(\omega_i)$$

The extreme frequencies ω_i and the magnitude of the denominator $|D(\omega)|$ are initially unknown. So an iterative procedure is required with a similar approach as used for the LMS design: The extreme frequencies ω_i and the denominator polynomial $D(\omega)$ are calculated based on the coefficient set a_r obtained in the previous iteration step. In order to find the extreme frequencies, a search algorithm is applied to find the $N + 1$ extremas of the error $e(\omega)$ on a dense grid of frequencies.

$$\sum_{r=0}^{N} a_r \sin(r\omega_{\text{prev}i} T + \beta_d(\omega_{\text{prev}i})) = (-1)^i \delta \left|D_{\text{prev}}(e^{j\omega_{\text{prev}i} T})\right| / 2W(\omega_{\text{prev}i}) \tag{6.74}$$

$$\text{with} \quad i = 0 \dots N$$

For the first iteration step, the magnitude of the denominator polynomial is initialised as $|D(\omega)| = 1$. The $N + 1$ extreme frequencies ω_i are initially evenly distributed over the frequency range of interest. The value of δ can, in principle, be chosen arbitrarily. The magnitude of the coefficients a_r is proportional to δ. In the end, the coefficients can be normalised such that $a_0 = 1$. In order to guarantee that the right side of (6.74) remains in a reasonable order of magnitude in the course of the iteration, δ is chosen as

$$\delta = \frac{1}{N+1} \sum_{i=0}^{N} 2W(\omega_{\text{prev}i}) / D_{\text{prev}}(\omega_{\text{prev}i}). \tag{6.75}$$

(6.74) can be written in vector form as

$$\mathbf{Sa = b}$$

with

$a_r = (a_1, \dots, a_N)$, the unknown coefficients

$$s_{ir} = \sin(r\omega_{\text{prev}i} T + \beta_d(\omega_{\text{prev}i})) \quad \text{with} \quad i = 0 \dots N, \quad r = 0 \dots N \tag{6.76}$$

$$b_i = (-1)^i \delta \left|D_{\text{prev}}(e^{j\omega_{\text{prev}i} T})\right| / 2W(\omega_{\text{prev}i})$$

$$\beta_d(\omega_{\text{priv}i}) = \frac{b_d(\omega_{\text{priv}i}) - N\omega_{\text{priv}i} T}{2}$$

$\mathbf{a}$ is the vector of the unknown coefficients a_r which can be found by solving the system of linear equations (6.76).

$$\mathbf{a = S^{-1}b}$$

The Remez-type algorithm outlined above can be summarized as follows:

1. Determine an initial set of extreme frequencies ω_i which are equally spaced in the frequency range of interest.
2. Set $|D(\omega_i)| = 1$.
3. Determine δ as the average of $2W(\omega_i)/|D(\omega_i)|$ for $i = 0 ... N$ (6.75).
4. Setup the system of linear equations (6.76).
5. Solve system of equations for coefficient set a_r.
6. Calculate error $e(\omega)$ on a dense grid of frequencies using the coefficients a_r from 5.
7. Apply search algorithm to find new set of $N + 1$ extreme frequencies ω_i in $e(\omega)$.
8. Exit if ripple is sufficiently even.
9. Calculate $|D(\omega_i)|$ using the coefficients a_r from 5.
10. goto 3.

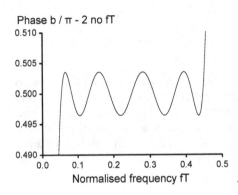

Phase b / π - 2 no fT

Fig. 6-53
Equiripple design: Hilbert transformer, $N = n_0 + 1 = 8$, $\omega_1 = 0.1\pi/T$, $\omega_2 = 0.9\pi/T$

For the Hilbert transformer as specified in Fig. 6-53, $N + 1 = 9$ extreme frequencies ω_i need to be equidistantly distributed over the frequency range of interest $0.1\pi/T ... 0.9\pi/T$ in order to initialise the algorithm.

$$\omega_i = 0.1(i+1)\pi/T \quad \text{for} \quad i = 0...8$$

The weighting $W(\omega_i)$ is assumed to be constant in the frequency range of interest which results in an equal ripple in that whole frequency range.

$$W(\omega_i) = 1.0 \quad \text{for} \quad i = 0...8$$

The magnitude of the denominator polynomial $|D(\omega_i)|$ is initialised as

$$|D(\omega_i)| = 1.0 \quad \text{for} \quad i = 0...8$$

The starting value of δ is 2 according to (6.75).

$$\delta = 2.0$$

The desired phase of the Hilbert transformer is

$$b_d = n_0 \omega_i T + \pi/2 = (N-1)\omega_i T + \pi/2 \quad \text{for} \quad i = 0...8$$

$$\beta_d = \frac{b_d(\omega_i) - N\omega_i T}{2} = \pi/4 - \omega_i T/2$$

With these initial data, the algorithm can be started. A stable equiripple error curve is obtained after about 4 to 5 iterations. Fig. 6-53 shows that the resulting phase characteristic is almost indistinguishable from the result of the quasi-equiripple design (Fig. 6-49).

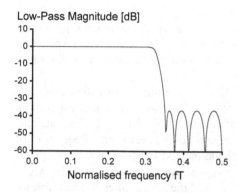

Low-Pass Magnitude [dB]

Normalised frequency fT

Fig. 6-54
Equiripple design: almost linear-phase low-pass filter, $N = 12$, $\omega_1 = 0.6\pi/T$, $\omega_2 = 0.7\pi/T$

For the almost linear-phase low-pass filter according to Fig. 6-54, the $N + 1 = 13$ extreme frequencies are equidistantly distributed over the concatenated passband $(0 ... 0.6\pi/T)$ and stopband $(0.7\pi/T ... \pi/T)$ where no extremas are allocated to the boundaries $\omega = 0$ and $\omega = \pi/T$. Since the desired phase assumes the fixed values 0 and $N\pi$ at these frequencies, the error curve cannot assume extreme values at $\omega = 0$ and $\omega = \pi/T$.

$$\omega_i = \frac{0.9(i+1)}{14}\pi/T \quad \text{for} \quad i = 0...8$$

$$\omega_i = \frac{0.9(i+1)+1.4}{14}\pi/T \quad \text{for} \quad i = 9...12$$

The ripple in the passband is chosen equal to the ripple in the stopband which requires a constant weighting function.

$$W(\omega_i) = 1.0 \quad \text{for} \quad i = 0...12$$

The magnitude of the denominator polynomial $|D(\omega)|$ is again initialised as

$$|D(\omega_i)| = 1.0 \quad \text{for} \quad i = 0...8$$

The starting value of δ is 2 according to (6.75).

$$\delta = 2.0$$

The desired phase can be expressed as

$$b_d(\omega_i) = n_0 \omega_i T = (N-1)\omega_i T \qquad \text{for} \quad \omega_i = 0 \ldots 0.6\pi/T$$
$$b_d(\omega_i) = n_0 \omega_i T + \pi = (N-1)\omega_i T + \pi \quad \text{for} \quad \omega_i = 0.7\pi/T \ldots \pi/T$$

and

$$\beta_d = \frac{b_d(\omega_i) - N\omega_i T}{2} = -\frac{\omega_i T}{2} \qquad \text{for} \quad \omega_i = 0 \ldots 0.6\pi/T$$
$$\beta_d = \frac{b_d(\omega_i) - N\omega_i T}{2} = \frac{\pi - \omega_i T}{2} \quad \text{for} \quad \omega_i = 0.7\pi/T \ldots \pi/T$$

With these initial data, the algorithm can be started. A stable equiripple error curve is obtained after about 6 to 7 iterations. Again, the magnitude response of the almost linear-phase filter Fig. 6-54 is almost indistinguishable from the result of the quasi-equiripple design (Fig. 6-50).

A comparison of the computational effort for the quasi-equiripple design and the Remez algorithm shows that the latter is superior. The Remez algorithm requires the solution of a linear system of equations of order $N + 1$. Computations with high frequency resolution are only required to find the extreme frequencies of the error function $e(\omega)$. The quasi-equiripple method is based on a least squares design which requires the solution of a system of equations with a grid size which is an order of magnitude higher than the order of the allpass filter N. The effort for finding the extreme frequencies in the one case is comparable to the effort of calculating the envelope of the error function in the other.

6.5.5 MAXFLAT Design

It is the goal of this design method to make the phase characteristic $b_A(\omega)$ of the allpass filter maximally flat in the vicinity of a limited number of frequencies ω_i. Maximally flat in this sense means that $b_A(\omega)$ assumes a desired phase value

$$b_A(\omega_i) = b_d(\omega_i)$$

and a desired group delay

$$b'_A(\omega_i) = \tau_{gA}(\omega_i) = \tau_{gd}(\omega_i)$$

at ω_i and that the higher derivatives up to the Kth order are zero.

$$b^{(\nu)}(\omega_i) = 0 \quad \text{for} \quad \nu = 2 \ldots K_i$$

The more higher derivatives are zero the closer the approximation of the tangent at ω_i. The parameter K_i is called the degree of tangency of the phase response at ω_i. This means for the design error of the phase that $e(\omega)$ and all its derivatives up to the Kth order are zero.

$$e^{(\nu)}(\omega_i) = 0 \quad \text{for} \quad \nu = 0 \ldots K_i \tag{6.77}$$

Unfortunately, for the system of equations expressed by (6.77) there does not exist a closed form solution in the general case. But the coefficients of allpass filters with maximally flat phase at $\omega = 0$ and $\omega = \pi/T$ can be easily calculated. In [66] and [70], the coefficients of all-pole filters of the form

$$H(z) = \frac{b_0}{D(z)} = \frac{b_0}{\displaystyle\sum_{i=0}^{N} a_i z^{-i}}$$

are derived for a maximally flat delay $\tau(\omega)$ at $\omega = 0$ and $\omega = \pi/T$. An allpass filter with the same coefficients

$$H_A(z) = \frac{z^{-N} D(z^{-1})}{D(z)} = \frac{\displaystyle\sum_{i=0}^{N} a_{N-i} z^{-i}}{\displaystyle\sum_{i=0}^{N} a_i z^{-i}}$$

has a group delay of

$$\tau_A(\omega) = 2\tau(\omega) + NT$$

which is likewise maximally flat at $\omega = 0$ and $\omega = \pi/T$. The design formulas for the all-pole filter can still be used. Merely the group delay parameter must be substituted as

$$\tau = \frac{\tau_d - NT}{2} \tag{6.78}$$

in order to have the desired group delay of the allpass filter $\tau_d(\omega)$ as the design parameter.

Two things are special at $\omega = 0$ and $\omega = \pi/T$:
- The phase is fixed ($b(0) = 0$, $b(\pi/T) = N\pi$) and cannot be arbitrarily chosen.
- Since the phase response is an odd function of ω, all even order derivatives are zero at $\omega = 0$ and $\omega = \pi/T$.

The phase response, therefore, needs to satisfy the following constraints:

$$b'(0) = \tau_d \quad \text{(if } K > 0\text{)}.$$
$$b^{(2v+1)}(0) = 0 \quad \text{for} \quad v = 1 \ldots K - 1.$$
$$b'(\pi/T) = \tau_d \quad \text{(if } L > 0\text{)}.$$
$$b^{(2v+1)}(\pi/T) = 0 \quad \text{for} \quad v = 1 \ldots L - 1.$$

K and L are the degrees of tangency at $\omega = 0$ and $\omega = \pi/T$ respectively. τ_d is the desired group delay which is assumed to be equal at $\omega = 0$ and $\omega = \pi/T$. In order to satisfy these $K + L$ constraints, an allpass filter of order $N = K + L$ is required which has N free coefficients. These coefficients can be calculated as [66]

$$a_n = \frac{(-1)^n}{(\tau_d / T + 1)_n} \binom{N}{n} \sum_{i=0}^{L} (-4)^i$$

$$\times \binom{L}{i} \frac{(\tau_d / T - N)_i (n - i + 1)_i (\tau_d / T - N + 2i)_{n-i}}{2(N + 1 - i)_i} \quad \text{for} \quad n = 0...N$$

(6.79)

where $\binom{m}{n}$ denotes the nth binominal coefficient and $(x)_n$ the rising factorial

$(x)_n = (x)(x+1)(x+2)...(x+n-1)$. (6.79) looks quite cumbersome but can be efficiently calculated using the recursive relationship

$$\frac{c_{i,n+1}}{c_{i,n}} = -\frac{N-n}{n-i+1} \frac{\tau_d / T - N + n + i}{\tau_d / T + 1 + n}$$

and

(6.80)

$$a_n = \sum_{i=0}^{L} c_{i,n}$$

Stability of the allpass filter is guaranteed if

$$\tau_d > (N-1)T \quad \text{for} \quad K = 0 \quad \text{or} \quad L = 0$$
$$\tau_d > (N-2)T \quad \text{for} \quad K \neq 0 \quad \text{and} \quad L \neq 0$$

The main applications for MAXFLAT allpass filters are fractional delay filters and low-pass or high-pass filters with maximally flat group delay characteristic.

The MAXFLAT filter has three independent design parameters,
- the tangency K at $\omega = 0$,
- the tangency L at $\omega = \pi/T$,
- and the desired group delay τ_d at $\omega = 0$ and/or $\omega = \pi/T$,

where the order of the allpass filter N is the sum of K and L.

The fractional delay filter makes use of the fact that τ_d can assume non-integer values. The parameter L is set to zero ($K = N$). So all filter resources are used to make the phase and group delay characteristic maximally flat in the vicinity of $\omega = 0$. With $L = 0$, (6.79) simplifies to

$$a_n = (-1)^n \binom{N}{n} \frac{(\tau_d / T - N)_n}{(\tau_d / T + 1)_n}.$$

The special case $L = 0$ has been described in various forms in the literature. Thiran [70], for example, calculates the filter coefficients a_n as

$$a_n = (-1)^n \binom{N}{n} \prod_{i=0}^{N} \frac{\tau_\mathrm{d}/T - N + i}{\tau_\mathrm{d}/T - N + n + i} \; .$$

In the original formulas from [66] and [70], the delay parameter is substituted according to (6.78) in order to have the desired delay of the allpass filter as a design parameter. Again a recursive relationship exists which simplifies the calculation of the coefficients.

$$\frac{a_{n+1}}{a_n} = -\frac{N-n}{n+1} \frac{\tau_\mathrm{d}/T - N + n}{\tau_\mathrm{d}/T + 1 + n} \; .$$

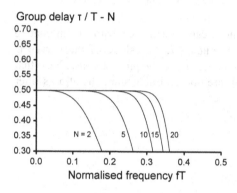

Group delay τ / T - N

Fig. 6-55
Group delay of the maximally flat fractional delay filter with the order of the allpass filter as a parameter,
$\tau_\mathrm{d} = (N + 0.5)\,T$

With increasing filter order, phase and group delay become flat over a wider frequency range. Fig. 6-55 demonstrates this behaviour for allpass filters with a desired maximally flat group delay of $\tau_\mathrm{d} = (N + 0.5)\,T$ at $\omega = 0$.

The filter structure to realise approximately linear-phase recursive filters is shown in Fig. 5-107. This structure basically forms the sum and difference of the responses of a delay and of an allpass filter. Assuming that the delay is of length $n_0 T$, a low-pass or high-pass magnitude response can be realised if the allpass filter approximates a phase of

$$b(\omega) = n_0 \omega T \qquad \text{around } \omega = 0, \text{ and}$$
$$b(\omega) = n_0 \omega T \pm \pi \quad \text{around } \omega = \pi/T \; .$$

The desired behaviour can be achieved if delay length n_0 and filter order N are related as

$$n_0 = N \pm 1 \; .$$

Fig. 6-56 illustrates the magnitude response of an approximately linear-phase, maximally flat low-pass filter of order $N = 8$ with $n_0 = N + 1$. The choice of the parameter L determines the cutoff frequency of the filter. The relation between the cutoff frequencies, e.g. the frequencies where the gain is 0.5, and the parameter L is not linear, unfortunately.

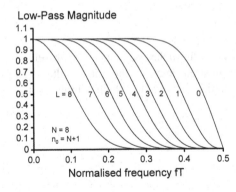

Fig. 6-56
Magnitude response of an approximately linear-phase, maximally flat low-pass filter of order $N = 8$ with $n_0 = N+1$ and L as a parameter.

A significant better magnitude performance can be achieved with the alternative realisation $n_0 = N - 1$ as demonstrated in Fig. 6-57. The slope of the curves is obviously steeper in the transition region compared to the previous case. Note that the cases $L = 0$ ($K = 8$) and $L = 8$ ($K = 0$) are not realisable since the allpass filter is not stable as stated earlier.

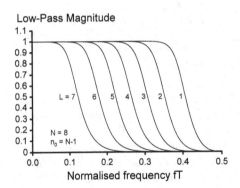

Fig. 6-57
Magnitude response of an approximately linear-phase, maximally flat low-pass filter of order $N = 8$ with $n_0 = N-1$ and L as a parameter.

Especially for lower filter orders, the range of available discrete cutoff frequencies is rather limited. An interpolation method is introduced in [66] which allows a continuous variation of the cutoff frequency.

7 Design of FIR Filters

7.1 Introduction

For FIR filters, no design techniques exist in the continuous time domain that could be used as a basis for the design of related digital filters as in the case of IIR filters. There are no design tables available with normalised coefficients of prototype filters that could be easily adapted to special design specifications. So each filter design has to pass, in principle, a complete mathematical design procedure. According to the chosen approximation method, a number of more or less complex algorithms are available, iterative or analytic, that in most cases require computer support.

It is the aim of all design methods to approximate a given frequency response as "close" as possible by a finite number of FIR filter coefficients. The starting point of all these methods is the assumption of idealised frequency responses or tolerance specifications in the passband and stopband. The finite length of the unit-sample response and hence the finite number of filter coefficients limits the degree of approximation to the given specification. Low variation of the magnitude (ripple) in the passband, high attenuation in the stopband and sharp cutoff are competing design parameters in this context.

There are a number of possible criteria for an optimum choice of coefficients that lead to "close" approximations of the design target. Four such criteria are considered in more detail in this chapter. Two of them aim at minimising an error measure defined with respect to the deviation of the target magnitude response from the actually realised response, whereas the two remaining criteria aim at matching the characteristics of the desired and realised magnitude responses at a limited number of frequencies:

- The filter coefficients are to be determined in such a way that the mean square deviation between desired and actually realised magnitude responses is minimised.
- The filter coefficients are to be determined in such a way that the magnitude of the maximum deviation between desired and actually realised amplitude responses is minimised.
- The filter coefficients have to be determined in such a way that desired and actually realised amplitude response coincide in few points with respect to the gain and a number of derivatives (in case of a low-pass design, for instance, at two points: $\omega = 0$ and $\omega = \pi/T$).
- n filter coefficients are to be determined in such a way that desired and actually realised magnitude responses coincide at n points. A system of n equations has to be solved to determine the coefficients in this case.

D. Schlichthärle, *Digital Filters*, DOI: 10.1007/978-3-642-14325-0_7,
© Springer-Verlag Berlin Heidelberg 2011

7.2 Linear-Phase Filters

The main advantage of the FIR filter structure is the realisability of exactly linear-phase frequency responses. That is why almost all design methods described in the literature deal with filters with this property. Since the phase response of linear-phase filters is known, the design procedures are reduced to real-valued approximation problems, where the coefficients have to be optimised with respect to the magnitude response only. In the following, we derive some useful relations in the context of linear-phase filters in the time and frequency domain.

An Nth-order FIR filter is determined by $N + 1$ coefficients and has, according to (5.3), the following frequency response:

$$H(e^{j\omega T}) = \sum_{r=0}^{N} b_r e^{-j\omega rT} \quad . \tag{7.1}$$

Extracting an exponential term out of the sum that corresponds to a delay of $NT/2$ (half the length of the shift register of the FIR filter) yields

$$H(e^{j\omega T}) = e^{-j\omega NT/2} \sum_{r=0}^{N} b_r e^{-j\omega(r-N/2)T} \quad . \tag{7.2}$$

For the coefficients b_r there exist two possible symmetry conditions that lead to the situation that, apart from the extracted linear-phase delay term, there are no further frequency-dependent phase variations contributing to the overall frequency response. These are:

$$b_r = b_{N-r} \tag{7.3a}$$

$$b_r = -b_{N-r} \quad . \tag{7.3b}$$

Condition (7.3a) leads to symmetrical unit-sample responses as shown in Fig. 7-1a and Fig. 7-1b. Condition (7.3b) yields antisymmetrical unit-sample responses as

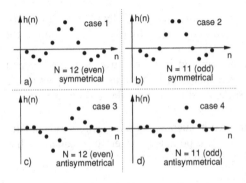

Fig. 7-1
Symmetry cases for linear-phase FIR filters

depicted in Fig. 7-1c and Fig. 7-1d. Taking the symmetry conditions (7.3) into account, the complex-valued sum in (7.2) can by expressed by real sine or cosine terms respectively.

$$H(e^{j\omega T}) = e^{-j\omega NT/2} \sum_{r=0}^{N} b_r \cos(r - N/2)\omega T$$

(7.4a)

$$\text{for } b_r = b_{N-r}$$

and

$$H(e^{j\omega T}) = je^{-j\omega NT/2} \sum_{r=0}^{N} -b_r \sin(r - N/2)\omega T$$

(7.4b)

$$\text{for } b_r = -b_{N-r}$$

The summations in (7.4a) and (7.4b) are purely real and thus represent a zero-phase filter with the frequency response $H_0(\omega)$. The zero-phase frequency response must not be confused with the amplitude (magnitude) response since $H_0(\omega)$ may also assume negative values. The following relationships exist between frequency response and zero-phase frequency response.

$$H(e^{j\omega T}) = e^{-j\omega NT/2} H_0(\omega) \qquad \text{for } b_r = b_{N-r}$$

(7.5a)

and

$$H(e^{j\omega T}) = je^{-j\omega NT/2} H_0(\omega) \qquad \text{for } b_r = -b_{N-r}$$

(7.5b)

Relations (7.4b) and (7.5b) contain, beside the linear-phase delay term, the additional factor j that causes a frequency-independent phase shift of 90°. Filters with antisymmetrical unit-sample responses are therefore well suited for FIR filter realisations of differentiators, integrators, Hilbert transformers and quadrature filters.

The relations (7.4) can be further simplified since half of the terms in the respective sums are identical in pairs because of the symmetry condition. The cases of even and odd filter orders must be treated separately so that we can distinguish between four cases in total (refer to Fig. 7-1) which feature certain symmetry properties in the frequency domain.

Case 1   **Symmetrical unit-sample response, N even (Fig. 7-1a)**

$$H_0(\omega) = \sum_{n=0}^{Nt} B(n)\cos n\omega T$$

with $Nt = N/2$, $B(0) = b(Nt)$

(7.6a)

and $B(n) = 2b(Nt - n) = 2b(Nt + n)$ for $n = 1...Nt$

symmetry condition : $H_0(\omega) = H_0(2\pi/T - \omega)$

Case 2   Symmetrical unit-sample response, N odd (Fig. 7-1b)

$$H_0(\omega) = \sum_{n=1}^{Nt} B(n)\cos(n-1/2)\omega T$$

with $Nt = (N+1)/2$

and $B(n) = 2b(Nt-n) = 2b(Nt+n-1)$ for $n = 1...Nt$ (7.6b)

symmetry conditions : $H_0(\omega) = -H_0(2\pi/T - \omega)$

$$H_0(\pi/T) = 0$$

Case 3   Antisymmetrical unit-sample response, N even (Fig. 7-1c)

$$H_0(\omega) = \sum_{n=0}^{Nt} B(n)\sin n\omega T$$

with $Nt = N/2, \ \ B(0) = b(Nt) = 0$

and $B(n) = 2b(Nt-n) = -2b(Nt+n)$ for $n = 1...Nt$ (7.6c)

symmetry conditions : $H_0(\omega) = -H_0(2\pi/T - \omega)$

$$H_0(0) = 0 \text{ and } H_0(\pi/T) = 0$$

Case 4   Antisymmetrical unit-sample response, N odd (Fig. 7-1d)

$$H_0(\omega) = \sum_{n=1}^{Nt} B(n)\sin(n-1/2)\omega T$$

with $Nt = (N+1)/2$

and $B(n) = 2b(Nt-n) = -2b(Nt+n-1)$ for $n = 1...Nt$ (7.6d)

symmetry conditions : $H_0(\omega) = H_0(2\pi/T - \omega)$

$$H_0(0) = 0$$

The frequency response of a zero-phase filter can be therefore expressed as the sum of Nt sine or cosine terms. The coefficients $B(n)$ thus have the meaning of the coefficients of a Fourier's decomposition of the periodical frequency response of the FIR filter. It has to be finally noted that the four considered symmetry cases exhibit different properties with respect to the realisable filter characteristics:

Case 1:   no restriction,

Case 2:   low-pass and bandpass filters only   since $H_0(\pi/T) = 0$,

Case 3:   bandpass filter only   since $H_0(0) = 0$ and $H_0(\pi/T) = 0$,

Case 4:   high-pass and bandpass filters only   since $H_0(0) = 0$.

7.3 Frequency Sampling

In case of the frequency sampling method, the $N+1$ filter coefficients of a Nth-order FIR filter are determined such that the frequency response coincides at $N+1$ distinct frequencies ω_i with the values of the desired frequency response $H_D(\omega_i)$. Using (7.1) we can establish a set of $N+1$ equations that allows the determination of the $N+1$ filter coefficients b_n.

$$H_D(\omega_i) = \sum_{n=0}^{N} b_n e^{-j\omega_i nT}$$

$$i = 0 \dots N$$

(7.7)

In order to obtain the filter coefficients, the set of equations (7.7) has to be solved with respect to the coefficients b_n. A particularly effective method of solving this set of equations can be derived if the frequencies ω_i are equally spaced over the range $\omega = 0 \dots 2\pi/T$.

$$\omega_i = i \frac{2\pi}{(N+1)T}$$

$$i = 0 \dots N$$

(7.8)

Substitution of (7.8) in (7.7) yields a mathematical relationship between the filter coefficients and the samples of the desired frequency response that is commonly referred to as the "Discrete Fourier Transform" (DFT).

$$H_D(\omega_i) = \sum_{n=0}^{N} b_n e^{-j2\pi in/(N+1)}$$

$$i = 0 \dots N, \quad \omega_i = i2\pi/[(N+1)T]$$

(7.9)

Relationship (7.9) represents a $(N+1)$ point DFT, which allows us to calculate $N+1$ samples of the frequency response from $N+1$ filter coefficients which, in case of the FIR filter, are identical with the unit-sample response. For our problem we have to derive an inverse DFT (IDFT) in order to calculate conversely the coefficients b_n from the samples of the frequency response $H_D(\omega_i)$. We start with multiplying both sides of (7.9) by the term $e^{j2\pi im/(N+1)}$ and sum both sides with respect to the index i. After exchanging the sums on the right side we obtain

$$\sum_{i=0}^{N} H_D(\omega_i) e^{j2\pi im/(N+1)} = \sum_{n=0}^{N} b_n \sum_{i=0}^{N} e^{-j2\pi i(n-m)/(N+1)}$$

$$\omega_i = i2\pi/[(N+1)T] \ .$$

(7.10)

Because of the periodicity and the symmetry properties of the complex exponential function, it can be readily shown that the right sum over i vanishes, with the exception of the case $m = n$ where the exponential term becomes unity. This simplifies (7.10) to the form

$$\sum_{i=0}^{N} H_D(\omega_i) e^{j2\pi i m/(N+1)} = b_m(N+1)$$

$$m = 0 \ldots N \ ,$$

or

$$b_m = \frac{1}{N+1} \sum_{i=0}^{N} H_D(\omega_i) e^{j2\pi i m/(N+1)} \tag{7.11}$$

$$\omega_i = i2\pi/[(N+1)T], m = 0 \ldots N$$

Relation (7.11) enables the calculation of the filter coefficients b_m from $N+1$ equally spaced samples of the desired frequency response $H_D(\omega)$. For the practical filter design it would be advantageous if we could prescribe the real-valued zero-phase instead of the complex-valued linear-phase frequency response. By substitution of (7.5), relation (7.11) can be modified in this direction.

$$b_m = \frac{1}{N+1} \sum_{i=0}^{N} H_{0D}(\omega_i) e^{j2\pi i (m-N/2)/(N+1)} \qquad \text{for } b_m = b_{N-m}$$

$$b_m = \frac{1}{N+1} \sum_{i=0}^{N} H_{0D}(\omega_i) j e^{j2\pi i (m-N/2)/(N+1)} \qquad \text{for } b_m = -b_{N-m}$$

$$\omega_i = i2\pi/[(N+1)T], \ m = 0 \ldots N$$

Because of the properties of the zero-phase frequency response for the various symmetry cases of the filter coefficients (7.6a–d), these relations can be further simplified, resulting in purely real expressions.

Case 1

$$b_m = b_{N-m} = \frac{1}{N+1}\left[H_{0D}(0) + 2\sum_{i=1}^{N/2} H_{0D}(\omega_i)\cos\big((m-N/2)\omega_i T\big) \right] \tag{7.12a}$$

$$m = 0 \ldots N/2, \ \omega_i = i2\pi/[(N+1)T]$$

Case 2

$$b_m = b_{N-m} = \frac{1}{N+1}\left[H_{0D}(0) + 2\sum_{i=1}^{(N-1)/2} H_{0D}(\omega_i)\cos\big((m-N/2)\omega_i T\big) \right] \tag{7.12b}$$

$$m = 0 \ldots (N-1)/2, \ \omega_i = i2\pi/[(N+1)T]$$

Case 3

$$b_m = -b_{N-m} = \frac{1}{N+1}\left[2\sum_{i=1}^{N/2}H_{0D}(\omega_i)\sin\big((m-N/2)\omega_i T\big)\right]$$

$$m = 0 \ldots N/2, \quad \omega_i = i2\pi/[(N+1)T]$$

(7.12c)

Case 4

$$b_m = -b_{N-m} = \frac{1}{N+1}\left[H_{0D}(\pi/T)\,\sin\big((m-N/2)\pi\big)\right.$$

$$\left. +2\sum_{i=1}^{(N-1)/2}H_{0D}(\omega_i)\sin\big((m-N/2)\omega_i T\big)\right]$$

$$m = 0 \ldots (N-1)/2, \quad \omega_i = i2\pi/[(N+1)T]$$

(7.12d)

Relation (7.12) enables the calculation of the filter coefficients for the various possible symmetry cases from equally spaced samples of the zero-phase frequency response. By means of three examples we will demonstrate the power and the drawbacks of the frequency sampling method.

Example 7-1

A 32nd-order low-pass filter with a cutoff frequency at $f_c T = 0.25$ or $\omega_c = (\pi/T)/2$ is to be designed by frequency sampling. The corresponding samples of the zero-phase frequency response are:

$$H_{0D}(\omega_i) = \begin{cases} 1 & \text{for } \omega_i < (\pi/T)/2 \\ 0 & \text{for } \omega_i > (\pi/T)/2 \end{cases} \qquad \text{or}$$

$$H_{0D}(i) = 1 \qquad \text{for} \qquad i = 0 \ldots 8 \ ,$$
$$H_{0D}(i) = 0 \qquad \text{for} \qquad i = 9 \ldots 16 \ .$$

(7.12a) yields the following filter coefficients:

$b(0)$ =	$b(32)$ =	0.0209	$b(9)$ =	$b(23)$ =	−0.0463
$b(1)$ =	$b(31)$ =	−0.0231	$b(10)$ =	$b(22)$ =	−0.0158
$b(2)$ =	$b(30)$ =	−0.0193	$b(11)$ =	$b(21)$ =	0.0643
$b(3)$ =	$b(29)$ =	0.0261	$b(12)$ =	$b(20)$ =	0.0154
$b(4)$ =	$b(28)$ =	0.0180	$b(13)$ =	$b(19)$ =	−0.1065
$b(5)$ =	$b(27)$ =	−0.0303	$b(14)$ =	$b(18)$ =	−0.0152
$b(6)$ =	$b(26)$ =	−0.0170	$b(15)$ =	$b(17)$ =	0.3184
$b(7)$ =	$b(25)$ =	0.0365	$b(16)$ =		0.5152
$b(8)$ =	$b(24)$ =	0.0163			

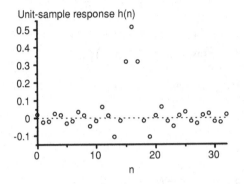

Fig. 7-2a
Unit-sample response of a
32nd-order low-pass filter
designed by frequency sampling

Fig. 7-2a shows a graphical representation of the 33 filter coefficients and hence of the unit-sample response of the filter. Using (7.6a), we calculate the corresponding magnitude response, which is depicted in Fig. 7-2b.

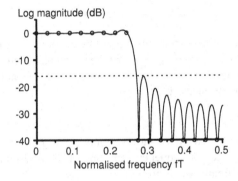

Fig. 7-2b
Magnitude response of a
32nd-order low-pass filter
designed by frequency sampling

It attracts attention that the achieved attenuation in the stopband of about 16 dB is rather poor, considering the relatively high filter order of $N = 32$. It must not be forgotten, however, that the specification of the filter leads to an extremely sharp transition band. The transition from the passband to the stopband happens between two neighbouring samples of the desired frequency response. We already pointed out earlier that high attenuation in the stopband and sharp cutoff are competing design targets.

In order to achieve a better stopband attenuation, we replace the abrupt transition between passband and stopband by a certain transition area in which the gain drops linearly. If we choose one intermediate point within the transition band, we obtain a sample sequence of the frequency response of the form ..., 1, 1, 0.5, 0, 0, ..., and in case of two intermediate points we have the sequence ..., 1, 1, 0.67, 0.33, 0, 0, ... accordingly. Figure 7-3 shows the resulting magnitude responses, again for a filter order of $N = 32$ and a cutoff frequency at $f_c T = 0.25$.

The stopband attenuation is noticeably improved. An optimum solution, however, is obviously not yet found, particularly since two interpolating values yield a worse result than only one. Lowering of the first lobe in the stopband would considerably improve the result. This can be achieved by varying the intermediate values until the optimum curves are achieved. The analytic method of applying the IDFT to the samples of the desired frequency response is thus supplemented by an iterative procedure which is applied to a limited number of values in the transition band. The optimum values in our example are 0.3908 for one and 0.5871 / 0.1048 for two intermediate values. Figure 7-4 depicts the corresponding magnitude responses. Two intermediate points in the transition band now yield a stopband attenuation more than 20 dB higher than only one.

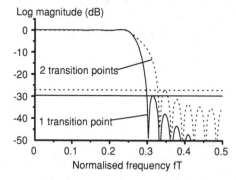

Fig. 7-3
Magnitude responses of FIR filters designed by frequency sampling with linear transition between passband and stopband

The frequency sampling method is apparently not well suited for the design of selective filters with sharp transition bands. The calculation of the coefficients is easy using relationship (7.12), but additional iteration is necessary to arrive at acceptable results. Furthermore, there are restrictions concerning the choice of the edge frequencies since these have to coincide with the sampling grid of the frequency response. In the following two examples, we discuss zero-phase frequency responses that exhibit smooth curves.

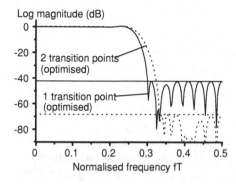

Fig. 7-4
Magnitude responses of FIR filters designed by frequency sampling with optimised intermediate values

Example 7-2

A differentiator is to be approximated by a tenth-order FIR filter. The corresponding transfer function features a constant phase shift of $\pi/2$ and a linearly rising magnitude. According to the considerations in Sect. 7.2, a case 4 filter is the right choice because the differentiator is a high-pass filter, in principle, with $H(0) = 0$. Since case 4 filters require an odd filter order, we have to choose $N = 11$. We define a linear rise of the zero-phase frequency response of the form

$$H_{0D}(\omega) = \frac{\omega}{\pi / T}$$

which has unity gain at the Nyquist frequency. The following gain values are used as an input to (7.12d) to calculate the filter coefficients b_n.

$$H_{0D}(\omega_i) = \frac{\omega_i}{\pi / T} = \frac{i 2\pi / \left((N+1)T\right)}{\pi / T} = \frac{2i}{N+1} \; , \quad i = 1 \ldots (N+1)/2$$

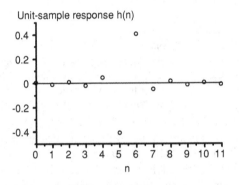

Fig. 7-5a
Unit-sample response of an eleventh-order differentiator designed by frequency sampling

Fig. 7-5a shows a graphical representation of the obtained 12 filter coefficients. These coefficients are equal to the samples of the unit-sample response of the differentiator.

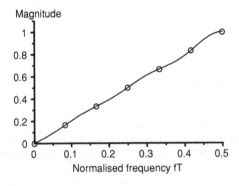

Fig. 7-5b
Magnitude response of an eleventh-order differentiator designed by frequency sampling

Fig. 7-5b depicts the sampling points, which lie on a straight line, and the plot of the approximated magnitude response of the FIR filter. In this case, the frequency sampling method yields an excellent result. The largest deviation with respect to the ideal curve can be observed in the vicinity of the Nyquist frequency. Since the magnitude curve is symmetrical about the frequency $\omega = \pi/T$, the slope becomes zero in this point. Because of the antisymmetry of the unit-sample response, the phase shift is exactly $\pi/2$.

Example 7-3

The sinc-distortion of a D/A converter with $\tau = T$ (Fig. 4-7) is to be compensated in the discrete-time domain by means of a 6th-order FIR filter. Since sinc-distortion is linear-phase according to Sect. 4.4, a linear-phase FIR filter is the appropriate choice for our problem.

The target frequency response is the reciprocal of

$$H(\omega) = \text{sinc}(\omega T/2) = \frac{\sin(\omega T/2)}{\omega T/2}$$

Four gain values are required as an input to Eq. (7.12a):

$$H(0) = 1.0$$

$$H\left(\frac{2}{7}\pi/T\right) = 1.034$$

$$H\left(\frac{4}{7}\pi/T\right) = 1.148$$

$$H\left(\frac{6}{7}\pi/T\right) = 1.381$$

Figure 7-6 shows the resulting magnitude response compared to the target curve and the sample values that have been used for the design. Up to a frequency of $\omega = 0.9\pi/T$, the magnitude of the designed filter deviates by less than 0.25 dB from the target. Because of the symmetry condition for $H_0(\omega)$, the slope has to be zero at $\omega = \pi/T$, which causes the deviation to be mainly concentrated in this frequency range.

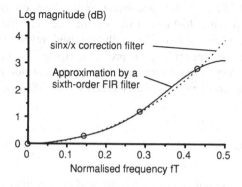

Fig. 7-6
Approximation of a sinx/x-equaliser characteristic by means of a sixth-order FIR filter

The previous two examples show that frequency sampling is a simple analytic method to approximate more or less smooth frequency responses by means of FIR filters. Selective filters designed with this method exhibit a bad attenuation behaviour in the stopband, which can only be improved by iterative variation of sample values in a defined transition band. The choice of the edge frequencies is restricted by the sampling grid of the zero-phase frequency response, which leads to severe restrictions, especially in the case of low filter orders. A filter specification in form of a tolerance scheme is not possible with this method.

We have seen, in the first example of a low-pass filter, that idealised frequency responses lead to bad results when frequency sampling is applied. The more realistic assumption of a finite slope in the transition band improves the results significantly. The approach in [47] goes one step further. On the basis of Chebyshev polynomials, it is possible to construct even more realistic frequency responses for which passband edge, stopband edge, passband ripple and stopband attenuation can be explicitly specified. Sampling of such frequency responses often leads to almost optimum results with regular ripple in the passband and stopband. The length of the filter and the degree of the involved Chebyshev polynomials can only be determined approximately from the filter specification. Some iteration steps are therefore required to find an optimum combination of these parameters.

We have included in the Appendix two C subroutines that the reader may easily integrate in his own filter design programs. The Procedure CoeffIDFT is an implementation of (7.12), which allows the calculation of the filter coefficients from the equally spaced samples of the given zero-phase frequency response. A special form of the inverse discrete Fourier transform (IDFT) [55] is used that takes into account the special symmetry conditions in the time and frequency domain and is optimised to handle real sequences in both domains. The procedure ZeroPhaseResp enables the calculation of the zero-phase frequency response of FIR filters from the filter coefficients b_n with arbitrary resolution.

7.4 Minimisation of the Mean-Square Error

The starting point for the FIR filter design according to this method is the requirement that the filter coefficients $B(n)$ have to be determined such that the mean square deviation between desired and actually realised responses is minimised which is a least-squares approximation problem. $H_D(\omega)$ is given in the form of a zero-phase frequency response which removes any consideration of the phase from our design problem. The least-squares error E_{LS} can be expressed as

$$E_{LS} = \frac{T}{\pi} \int_0^{\pi/T} (H_0(\omega) - H_{0D}(\omega))^2 \, d\omega . \tag{7.13}$$

Depending on the symmetry of the unit-sample response and on the filter order (even or odd), we substitute one of the equations (7.6a–d) in (7.13). In order to determine the minimum of E_{LS} we equate the derivatives of (7.13) with respect to the filter coefficients $B(n)$ with zero. Solving the resulting equations for $B(n)$ yields the desired set of coefficients.

We consider case 1 filters in more detail and substitute (7.6a) in (7.13).

$$E_{LS}[B(0), ..., B(Nt)] = \frac{T}{\pi} \int_0^{\pi/T} \left(\sum_{n=0}^{Nt} B(n) \cos n\omega T - H_{0D}(\omega) \right)^2 d\omega$$

$$\text{for } N \text{ even}, Nt = N/2$$

We differentiate this equation with respect to the νth filter coefficient $B(\nu)$ and set the resulting term to zero.

$$\frac{dE_{LS}}{dB(\nu)} = \int_0^{\pi/T} 2 \left(\sum_{n=0}^{Nt} B(n) \cos n\omega T - H_{0D}(\omega) \right) \cos \nu\omega T d\omega = 0$$

$$\sum_{n=0}^{Nt} B(n) \int_0^{\pi/T} \cos n\omega T \cos \nu\omega T d\omega = \int_0^{\pi/T} H_{0D}(\omega) \cos \nu\omega T d\omega \qquad (7.14)$$

On the left side of (7.14), only the case $n = \nu$ gives a nonzero contribution to the summation. All other terms in the sum vanish since we integrate over integer periods of cosine functions.

$$B(\nu) \int_0^{\pi/T} \cos^2 \nu\omega T d\omega = \int_0^{\pi/T} H_{0D}(\omega) \cos \nu\omega T d\omega$$

The coefficients $B(\nu)$ of a case 1 filter are thus calculated as

$$B(0) = \frac{T}{\pi} \int_0^{\pi/T} H_{0D}(\omega) d\omega$$

$$B(\nu) = \frac{2T}{\pi} \int_0^{\pi/T} H_{0D}(\omega) \cos \nu\omega T d\omega \qquad (7.15a)$$

$$\nu = 1 ... Nt \ .$$

Applying the same procedure to case 2 filters yields the following expression for the calculation of the coefficients $B(\nu)$.

$$B(\nu) = \frac{2T}{\pi} \int_0^{\pi/T} H_{0D}(\omega) \cos(\nu - 1/2) \omega T d\omega \qquad (7.15b)$$

$$\nu = 1 ... Nt$$

Using the relations between the filter coefficients $b(n)$ and the Fourier coefficients $B(n)$ given in (7.6a) and (7.6b), a single equation can be derived for even and odd filter orders to determine the filter coefficients $b(n)$.

$$b(v) = h(v) = \frac{T}{\pi} \int_0^{\pi/T} H_{0D}(\omega)\cos(v - N/2)\omega T \mathrm{d}\omega \qquad (7.16)$$

$$v = 0 \dots N$$

We consider a simple low-pass filter as an example, which is specified as follows:

$$H_{0D}(\omega) = \begin{cases} 1 & \text{for } 0 \le \omega < \omega_c \\ 0 & \text{for } \omega_c < \omega \le \pi/T \end{cases} \qquad (7.17)$$

Using (7.16) we obtain the following filter coefficients.

$$b(v) = h(v) = \frac{T}{\pi} \int_0^{\omega_c} \cos(v - N/2)\omega T \mathrm{d}\omega$$

$$b(v) = h(v) = \frac{\sin \omega_c T(v - N/2)}{\pi(v - N/2)} = \frac{\omega_c T}{\pi} \mathrm{sinc}(\omega_c T(v - N/2)) \qquad (7.18)$$

$$v = 0 \dots N$$

Depending on the filter order N, the sinc function (7.18) is a more-or-less good approximation of the infinite unit-sample response of an ideal low-pass filter as specified by (7.17). The truncation of the ideal unit-sample response to the $N + 1$ samples of the realised filter is called windowing.

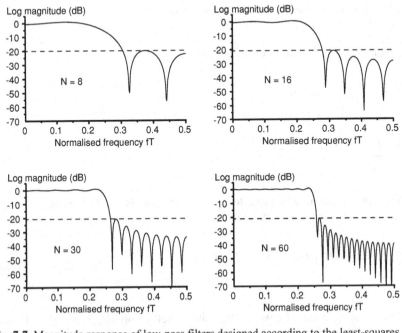

Fig. 7-7 Magnitude response of low-pass filters designed according to the least-squares error criterion ($f_c T = 0.25$)

Figure 7-7 depicts the magnitude responses of low-pass filters for various filter orders N, which have been designed using equation (7.18). It appears that the sharpness of the transition band goes up as the filter order increases. The stopband attenuation, however, characterised by the highest maximum in the stopband, is absolutely independent of the filter order. The attenuation of the first lobe in the stopband amounts to about 20 dB. This poor behaviour in the stopband is characteristic for the kind of windowing that we performed to truncate the length of the unit-sample response to $N + 1$ samples resulting in a FIR filter of order N. We have applied a so-called rectangular window that sharply cuts away the rest of the ideal unit-sample response and equally weights the samples within the window.

As in the case of the frequency sampling method, the magnitude response of the obtained filter oscillates around the ideal values in the passband and stopband. The largest deviation occurs in the area of the abrupt transition between passband and stopband. In the following, we will discuss two methods that lead to better results with respect to passband ripple and stopband attenuation.

Direct solution

The first method is based on a further development of equation (7.14) and takes into account that, for a realisable filter, the graph of the magnitude must have a finite slope in the transition area between passband and stopband. The method assumes a certain width of the transition band, but ignores this frequency interval in the mean-square error expression (7.13). As a consequence of this approach, the relatively simple computation of the coefficients $B(n)$ using (7.18) is no longer possible. Rather, we have to solve a system of equations of the form

$$C\,b = h$$

in order to obtain the coefficient vector b.

The window method

The infinite unit-sample response of the ideal prototype filter is multiplied by special window functions which result in better filter characteristics than we have obtained by the simple rectangular windowing as described above. These windows have the property that they taper smoothly to zero at each end, with the desired effect that the height of the side lobes in the stopband is diminished.

7.4.1 Direct Solution

The average of the mean-square deviation taken over the whole frequency interval $\omega = 0 \ldots \pi/T$, as expressed in (7.13), is to be minimised by an appropriate choice of the coefficients $B(n)$. Since the ideal sharp transition between passband and stopband as specified in (7.17) lies in this interval, the filter resources are largely used to approximate this abrupt drop in the gain. These resources are accordingly not available to optimise the filter characteristics with respect to passband ripple and stopband attenuation. An attenuation of the first side lobe in the stopband higher than about 21 dB is not achievable (refer to Fig. 7-7). A more realistic assumption about the magnitude curve in the transition band would certainly help. We have shown in the context of the frequency sampling method, however, that it is difficult to predict an optimum transition curve between passband and stopband. It is therefore an obvious approach to leave the transition band completely out of consideration when we write down the least-squares error expression (7.13) which is to be minimised.

$$E_{LS} = \frac{T}{\pi} \int_A \left(H_0(\omega) - H_{0D}(\omega) \right)^2 d\omega$$

$$A = \text{interval } 0 \ldots \pi/T \text{ excluding the transition band}$$

We thus optimise the characteristics in the passband and stopband according to the least-squares error criterion, but we do not make any effort to enforce a certain curve in the transition band which remains unspecified. In consequence, the integral is no longer taken over complete periods of the cosine as in (7.14) so that the relatively simple equations to calculate the coefficients $B(n)$ do not survive. Equation (7.14) rather takes on the more complex form

$$\sum_{n=0}^{Nt} B(n) \left(\int_0^{\omega_p} \cos n\omega T \cos v\omega T \, d\omega + \int_{\omega_s}^{\pi/T} \cos n\omega T \cos v\omega T \, d\omega \right) =$$

$$\int_0^{\omega_p} H_{0D}(\omega) \cos v\omega T \, d\omega + \int_{\omega_s}^{\pi/T} H_{0D}(\omega) \cos v\omega T \, d\omega \qquad (7.19)$$

$$n = 0 \ldots Nt \; .$$

Relation (7.19) can be easily written in the form of a matrix equation.

$$\boldsymbol{C}\,\boldsymbol{b} = \boldsymbol{h} \qquad (7.20)$$

The elements of the matrix $\boldsymbol{C}$ are calculated as

$$C_{vn} = \int_0^{\omega_p} \cos n\omega T \cos v\omega T \, d\omega + \int_{\omega_s}^{\pi/T} \cos n\omega T \cos v\omega T \, d\omega \qquad (7.21a)$$

$$v = 0 \ldots Nt, \; n = 0 \ldots Nt \; .$$

The matrix $\boldsymbol{C}$ is independent of the transfer function to be approximated. It only contains the information about the frequency ranges that are taken into account for the minimisation of the mean-square deviation between desired and realised magnitude responses. The vector $\boldsymbol{h}$ on the right side of (7.20) is determined as follows:

$$h_v = \int_0^{\omega_p} H_{0D}(\omega)\cos v\omega T\,d\omega + \int_{\omega_s}^{\pi/T} H_{0D}(\omega)\cos v\omega T\,d\omega \qquad (7.21b)$$

$$v = 0 \dots Nt \ .$$

In the case of low-pass filters, only the left integral contributes to the result since $H_{0D}(\omega)$ vanishes in the stopband $\omega = \omega_s \dots \pi/T$.

$$h_v = \int_0^{\omega_p} \cos v\omega T\,d\omega \qquad\qquad v = 0 \dots Nt \qquad\qquad (7.21c)$$

For high-pass filters we obtain, accordingly,

$$h_v = \int_{\omega_p}^{\pi/T} \cos v\omega T\,d\omega \qquad\qquad v = 0 \dots Nt \ . \qquad\qquad (7.21d)$$

Solving (7.20) with respect to b yields the desired filter coefficients in vector form

$$b = \left(B_0, B_1, B_2, \ \dots \ , B_{Nt}\right)^{\mathrm{T}} \ .$$

The evaluation of the integrals (7.21a–d) is elementary. Care has to be taken of the special cases $v = n$ and $v = n = 0$, which have to be treated separately.

$$C_{vn} = \begin{cases} 0.5\left(\dfrac{\sin(n-v)\omega_p T - \sin(n-v)\omega_s T}{(n-v)T} + \dfrac{\sin(n+v)\omega_p T - \sin(n+v)\omega_s T}{(n+v)T}\right) & n \ne v \\[4mm] 0.5\left(\omega_p + \pi/T - \omega_s + \dfrac{\sin(n+v)\omega_p T - \sin(n+v)\omega_s T}{(n+v)T}\right) & \begin{array}{l} n = v \\ n \ne 0, v \ne 0 \end{array} \\[4mm] \omega_p + \pi/T - \omega_s & n = 0, v = 0 \end{cases}$$

$$\qquad\qquad\qquad (7.22a)$$

$$v = 0 \dots Nt, \ n = 0 \dots Nt$$

$$h_v = \begin{cases} \dfrac{\sin v\omega_p T}{vT} & \text{for } v \ne 0 \\[3mm] \omega_p & \text{for } v = 0 \end{cases} \qquad \text{for low - pass filters} \qquad (7.22b)$$

$$v = 0 \dots Nt$$

$$h_v = \begin{cases} -\dfrac{\sin v\omega_p T}{vT} & \text{for } v \neq 0 \\ \pi/T - \omega_p & \text{for } v = 0 \end{cases} \quad \text{for high - pass filters} \tag{7.22c}$$

$$v = 0 \dots Nt$$

The starting point for the treatment of case 2 filters, which cover odd filter orders, is the substitution of (7.6b) in (7.13). The set of equations to determine the filter coefficients $B(n)$ assumes the form

$$\sum_{n=1}^{Nt} B(n) \left(\int_0^{\omega_p} \cos(n-1/2)\omega T \cos(v-1/2)\omega T \, d\omega \right.$$

$$\left. + \int_{\omega_s}^{\pi/T} \cos(n-1/2)\omega T \cos(v-1/2)\omega T \, d\omega \right) = \tag{7.23}$$

$$\int_0^{\omega_p} H_{0D}(\omega)\cos(v-1/2)\omega T \, d\omega + \int_{\omega_s}^{\pi/T} H_{0D}(\omega)\cos(v-1/2)\omega T \, d\omega$$

$$v = 0 \dots Nt \; .$$

Evaluation of the integrals yields the following relations for the calculation of the matrix C and the vector h:

$$C_{vn} = \begin{cases} 0.5\left(\dfrac{\sin(n-v)\omega_p T - \sin(n-v)\omega_s T}{(n-v)T} \right. \\ \quad \left. + \dfrac{\sin(n+v-1)\omega_p T - \sin(n+v-1)\omega_s T}{n+v-1)T} \right) & n \neq v \\ 0.5\left(\omega_p + \pi/T - \omega_s + \dfrac{\sin(n+v-1)\omega_p T - \sin(n+v-1)\omega_s T}{(n+v-1)T} \right) & n = v \end{cases} \tag{7.24a}$$

$$v = 0 \dots Nt, \; n = 0 \dots Nt$$

$$h_v = \frac{\sin\left((v-1/2)\omega_p T\right)}{(v-1/2)T} \quad \text{for low - pass filters} \tag{7.24b}$$

$$v = 0 \dots Nt$$

$$h_v = \frac{(-1)^{v+1} - \sin\left((v-1/2)\omega_p T\right)}{(v-1/2)T} \quad \text{for high - pass filters} \tag{7.24c}$$

$$v = 0 \dots Nt \; .$$

The following example is intended to illustrate the consequences of disregarding the transition band in the approximation process. We consider a 30th-order FIR filter with a cutoff frequency at $f_c = 0.25/T$. We will show the magnitude responses for normalised transition bandwidths of $\Delta = 0$, $\Delta = 0.4$ and $\Delta = 0.8$. The input parameters for the calculation of the elements of the matrix C and of the vector h using (7.22) are as follows:

① $\Delta = 0$ $f_p T = 0.25$ $f_s T = 0.25$ $Nt = 15$
② $\Delta = 0.4$ $f_p T = 0.23$ $f_s T = 0.27$ $Nt = 15$
③ $\Delta = 0.8$ $f_p T = 0.21$ $f_s T = 0.29$ $Nt = 15$.

The filter coefficients can only be determined with computer support. In the first place we calculate the elements of C and h using (7.22). Note that C consists of 256 matrix elements. Afterwards we solve the linear set of equations (7.20) using one of the well known methods such as the Gaussian elimination algorithm. Figure 7-8 shows the resulting magnitude responses.

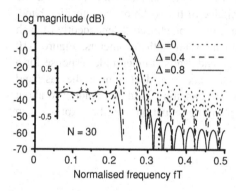

Fig. 7-8
Magnitude responses of FIR filters for various widths of the transition band Δ

The case $\Delta = 0$ corresponds to the magnitude response for $N = 30$ shown in Fig. 7-7, which only has a stopband attenuation of about 20 dB. With increasing width of the unspecified transition band we observe an increase of the attainable stopband attenuation and a reduction of the ripple in the passband. The price to pay for these improvements is a reduced selectivity of the filter.

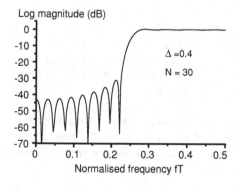

Fig. 7-9
Magnitude response of a FIR high-pass filter with the specification
$N = 30$
$f_s T = 0.23$
$f_p T = 0.27$

Figure 7-9 depicts the magnitude response of a high-pass filter which is specified as follows:

Filter order: $N = 30$
Cutoff frequency $f_c\,T = 0.25$
Normalised width of the transition band: $\varDelta = 0.4$.

These data yield the following input parameters for the determination of the matrix C (7.22a) and the vector h (7.22c):

$f_s\,T = 0.23$ $f_p\,T = 0.27$ $Nt = 15$.

The presented method, which minimises the mean-square deviation between desired and realised magnitude responses, offers some degree of freedom for the filter design which goes beyond the possibilities of frequency sampling. Besides the filter order and the cutoff frequency, which can be arbitrarily chosen, we can additionally influence the sharpness of the cutoff, the passband ripple and the stopband attenuation by an appropriate choice of the width of the transition band. Passband ripple and stopband attenuation cannot be chosen independently of each other, since there is a fixed relationship between both parameters. Figure 7-10 shows the deviation $\delta(\omega)$ between desired and realised magnitude response for a 30th-order low-pass filter with $\varDelta = 0.4$. The maximum deviations in the passband and stopband are equal in the vicinity of the transition band and amount to about $\delta_{max} = 0.073$. Passband ripple and stopband attenuation can be expressed in logarithmic scale as

$$D_p = 20\left|\log\left(1 + \delta_{max}\right)\right| = 0.7\text{dB}$$

$$D_s = -20\log\left|\delta_{max}\right| = 22.7\text{dB} \quad .$$

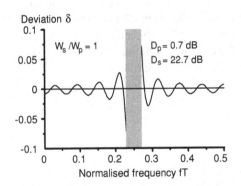

Fig. 7-10
Deviation from the ideal magnitude response for a 30th-order low-pass filter with $\varDelta = 0.4$

It would be certainly desirable if, for a given filter degree N, the stopband attenuation could be increased at the expense of a higher ripple in the passband or vice versa. By introduction of a weighting function $W(\omega)$ into the error measure (7.13), it is possible to influence the design in this direction.

$$E_{\mathrm{LS}} = \frac{T}{\pi} \int_A \left(W(\omega)[H_0(\omega) - H_{0\mathrm{D}}(\omega)] \right)^2 \mathrm{d}\omega$$

(7.25)

$$A = \text{interval} \, 0 \ldots \pi / T, \text{ excluding the transition band}$$

$W(\omega)$ determines at which degree the various frequency ranges are taken into account for the minimisation of the mean-square error. If $W(\omega)$ assumes higher values in the passband than in the stopband, for instance, the passband ripple will be reduced and the stopband attenuation becomes worse. If we choose a higher $W(\omega)$ in the stopband, we have the inverse behaviour. Only the ratio between the values of $W(\omega)$ in the passband and stopband is of importance; the absolute values can be chosen arbitrarily. Figure 7-11 shows again the error curve of a 30th-order low-pass filter with $\Delta = 0.4$, but the stopband is weighted 50 times higher than the passband. The maximum deviation in the stopband is now reduced to $\delta_{\mathrm{max}} = 0.022$, corresponding to an attenuation of 33.2 dB. The ripple in the passband increases to 2.1 dB.

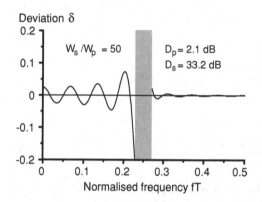

Fig. 7-11
Deviation from the ideal frequency response for a 30th-order low-pass filter with $\Delta = 0.4$ and different weighting of passband and stopband

There is no mathematical relation available in the literature that allows the calculation of the necessary filter order N, the width of the transition band Δ and the weighting function $W(\omega)$ from the data of a prescribed tolerance scheme. So some iteration steps are required, which could be best performed with computer support.

Finally, it must be noted that the described method can, in principle, also be used for bandpass and bandstop filters. Instead of two frequency intervals we have to optimise three intervals with respect to the least-squares error criterion: stopband–passband–stopband for bandpass filters or passband–stopband–passband for bandstop filters respectively. The transition bands, two in this case, are left out of consideration again.

7.4.2 The Window Method

We demonstrated in Sect. 7.4 how the frequency response is influenced by truncation (windowing) of the infinite unit-sample response of the ideal low-pass filter. A similar behaviour can be observed if the infinite unit-sample responses of ideal high-pass, bandpass and bandstop filters are truncated in the same way to obtain $N + 1$ samples ($n = 0 \ldots N$). These filter types and the already considered low-pass filter have the following truncated unit-sample responses.

low-pass

$$h(n) = \frac{\sin \omega_c T(n - N/2)}{\pi(n - N/2)} \qquad n = 0 \ldots N \tag{7.26a}$$

high-pass

$$h(n) = 1 - \frac{\sin \omega_c T(n - N/2)}{\pi(n - N/2)} \qquad n = 0 \ldots N \tag{7.26b}$$

bandpass

$$h(n) = \frac{\sin \omega_u T(n - N/2) - \sin \omega_l T(n - N/2)}{\pi(n - N/2)} \qquad n = 0 \ldots N \tag{7.26c}$$

bandstop

$$h(n) = 1 - \frac{\sin \omega_u T(n - N/2) - \sin \omega_l T(n - N/2)}{\pi(n - N/2)} \qquad n = 0 \ldots N \tag{7.26d}$$

ω_c is the cutoff frequency of the low-pass or high-pass filter. ω_l and ω_u are the lower and upper cutoff frequencies of the bandpass or bandstop filter.

Figure 7-12 shows the magnitude responses of FIR filters with unit-sample responses according to (7.26a–d). The filter order is $N = 32$ in all cases. The first lobe in the stopband is about 20 dB below the maximum for all filter types. A better stopband attenuation than 20 dB is apparently not realisable if the $N + 1$ samples are simply cut out of the infinite unit-step response of the ideal filter. This behaviour is characteristic for the kind of windowing that we have performed, i.e. a rectangular window in which all samples are equally weighted.

A better filter performance can be achieved in the stopband, however, if the samples are progressively reduced in magnitude as n approaches 0 or N from the centre of the window. It can be observed that the stopband attenuation may increase considerably but at the expense of a less rapid transition between passband and stopband. A multitude of windows with this property have been extensively studied in the literature, which differ in terms of the sharpness of the

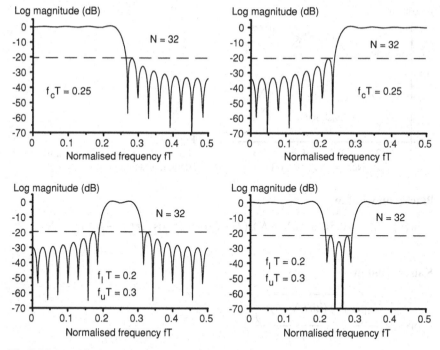

Fig. 7-12 Magnitude responses of low-pass, high-pass, bandpass and bandstop filter with rectangular window and $N = 32$

cutoff, the attainable stopband attenuation and the evenness of the ripple. Figure 7-13 depicts three examples of curves of such windows. In the following we give the mathematical expressions for the most important windows [62] with $0 \leq n \leq N$ in all cases.

Rectangular window

$$w(n) = 1$$

Triangular window

$$w(n) = 1 - |(2n - N)/(N + 2)|$$

Hamming window

$$w(n) = 0.54 + 0.46 \cos(\pi(2n - N)/N)$$

Hanning window

$$w(n) = 0.5 + 0.5 \cos(\pi(2n - N)/(N + 2))$$

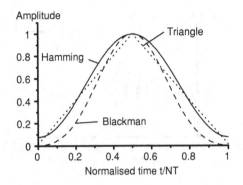

Fig. 7-13
Graphs of the triangular, Hamming and Blackman windows

Blackman window

$$w(n) = 0.42 + 0.5\cos\left(\pi(2n-N)/(N+2)\right) + 0.08\cos\left(2\pi(2n-N)/(N+2)\right)$$

Kaiser window

$$w(n) = \frac{I_0\left(\alpha\sqrt{1-(2n/N-1)^2}\right)}{I_0(\alpha)}$$

$\alpha = 0.1102\,(ATT-8.7)$	$ATT > 50$ dB
$\alpha = 0.5842\,(ATT-21)^{0.4} + 0.07886\,(ATT-21)$	21 dB $\leq ATT \leq 50$ dB
$\alpha = 0$	$ATT < 21$ dB

ATT is the desired stopband attenuation in dB. I_0 is the zeroth-order modified Bessel function of first kind.

Chebyshev window

$$W(f) = \delta_p T_{N/2}(\alpha\cos(2\pi fT) + \beta)$$
$$\alpha = (X_0+1)/2 \qquad\qquad \beta = (X_0-1)/2$$
$$X_0 = (3-\cos(2\pi\Delta F))/(1+\cos(2\pi\Delta F))$$
$$W(f) = \delta_p T_{N/2}(\alpha\cos(2\pi fT) + \beta) \tag{7.27}$$

The Chebyshev window is specified in the frequency domain. $T_{N/2}(x)$ is the $N/2$th-order Chebyshev polynomial according to (2.17). Figure 7-14 shows the shape of the spectrum of the Chebyshev window and the parameters that specify the window. Filter order N, ripple δ_p and width of the main lobe ΔF cannot be chosen arbitrarily; these parameters have to satisfy equation (7.27). In order to obtain the window in the time domain, $W(f)$ is sampled at $N+1$ equidistant frequency points. These samples are transformed into the discrete-time domain by inverse DFT using, for instance, the procedure CoeffIDFT in the Appendix.

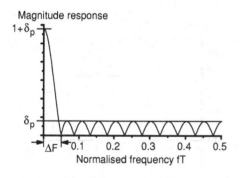

Fig. 7-14
Spectrum of a Chebyshev window

The first step in the practical filter design is the calculation of the truncated unit-sample response of the ideal filter, choosing one of the relations (7.26a–d), depending on the filter type. These calculated samples are afterwards multiplied by one of the window functions, which yields the unit-sample response and hence the coefficients of the desired FIR filter. Each of the window functions results in characteristic curves of the magnitude response. Figure 7-15 depicts the magnitude responses of 50th-order low-pass filters with $f_cT = 0.25$ that have been designed using the various window functions introduced in this section. For the calculation of the magnitude response from the filter coefficients we can again make use of the procedure **ZeroPhaseResp**.

The appropriate window is chosen according to the given specification with respect to the sharpness of the cutoff and the stopband attenuation of the filter under design. Five of the presented windows have fixed values for the stopband attenuation and are thus not very flexible. The width of the transition band is determined by the filter order. Table 7-1 summarises the main characteristics of the various windows. The triangular window has the special property that the magnitude response of the resulting filter has no ripple. The gain decreases more or less monotonously (refer to Fig. 7-15). The Hamming window leads to an almost even ripple which comes very close to the ideal case. Of all the windows with the transition bandwidth $4/NT$, the Hamming window possesses the highest stopband attenuation. The Blackman window has the highest stopband attenuation of all fixed windows but at the expense of a relatively wide transition band of $6/NT$, which is 3 times the width of the rectangular and 1.5 times the width of the Hamming window.

Window type	Transition width	Stopband attenuation
rectangular	2/NT	21 dB
triangular	4/NT	25 dB
Hamming	4/NT	53 dB
Hanning	4/NT	44 dB
Blackman	6/NT	74 dB
Kaiser	variable	variable
Chebyshev	variable	variable

Table 7-1
Characteristics of the window functions for the FIR filter design

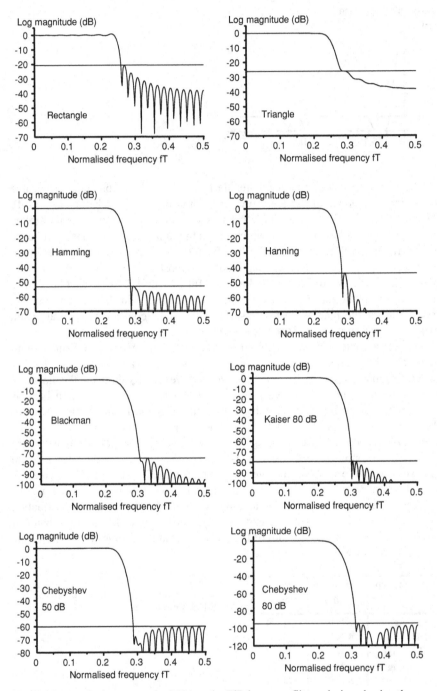

Fig. 7-15 Magnitude responses of 50th-order FIR low-pass filters, designed using the windows introduced in this section

The Kaiser window allows the explicit specification of the stopband attenuation. The width of the transition band Δ between passband and stopband can be adjusted through the filter order N. According to [39], the necessary filter order for a Kaiser window design can be approximately calculated as

$$N \approx \frac{ATT - 8}{14.36\,\Delta} \;.$$

(7.28)

The Chebyshev window enables the design of filters with relatively even ripple. The double width of the main lobe ΔF in Fig. 7-14 corresponds to about the width Δ of the transition band between passband and stopband of the realised filter. The ripple of the window is in general smaller than the ripple of the resulting transfer function so that we cannot exactly predict the achievable stopband attenuation. Relation (7.27) allows an approximate calculation of the required filter order. Choosing ΔF as half the width of the transition band and δ_p as the desired ripple δ of the filter is more likely to yield a too-pessimistic estimate of the filter order.

7.4.3 Arbitrary Magnitude Responses

If the desired magnitude $H_{0D}(\omega)$ is of arbitrary form and cannot be expressed by elementary functions, the definition of the least mean square error measure (7.25) by an integral is inappropriate. When assuming piecewise constant magnitude responses as in the previous section, the involved integrals can easily be solved. This is not case if the magnitude is defined by a mathematically complex function or if the desired magnitude response is the result of a measurement which exists in the form of a sequence of measured values. For the treatment of these cases, it is advantageous to evaluate the weighted error $\varepsilon(\omega)$ on a dense grid of frequencies ω_i

$$\varepsilon(\omega_i) = W(\omega_i)\big(H_0(\omega_i) - H_{0D}(\omega_i)\big) \qquad \text{with } i = 0\ldots gridsize - 1 \qquad (7.29)$$

and to minimise the sum

$$\sum_{i=0}^{gridsize-1} \varepsilon^2(\omega_i)$$

(7.30)

which is a numerical approximation of the integral (7.25). The gridsize is commonly chosen about one order of magnitude higher than the order N of the filter. For a fullband approximation problem, the frequencies ω_i are evenly distributed over the whole range $0 \ldots \pi/T$. If the desired magnitude is to be optimised in a number of bands, the ω_i are distributed over these bands while the transition bands are disregarded.

We start with considering case 1 filters and substitute (7.6a) in (7.29).

$$\varepsilon(\omega_i) = W(\omega_i)\left[\sum_{n=0}^{Nt} B(n)\cos n\omega_i T - H_{0D}(\omega_i)\right] \qquad \text{for } i = 0 \ldots gridsize - 1$$

This set of equations can be expressed in the following vector form.

$$e = Cb - h \qquad\qquad\qquad (7.31)$$

with

$$e_i = \left(\varepsilon(\omega_0), \ldots, \varepsilon(\omega_{gridsize-1})\right)$$
$$b_n = \left(B(0), \ldots, B(Nt)\right) \quad \text{, the unknown coefficients}$$
$$c_{in} = W(\omega_i)\cos n\omega_i T$$
$$h_i = W(\omega_i)H_{0D}(\omega_i) \qquad i = 0 \ldots gridsize - 1, \quad n = 0 \ldots Nt$$

The problem of finding a set of coefficients that minimises (7.30) has a unique solution. Provided that the $Nt + 1$ columns of the matrix C are linearly independent, the optimum coefficient vector b can be calculated as

$$b = [C^T C]^{-1} C^T h . \qquad\qquad\qquad (7.32)$$

Die FIR filter coefficients $b(n)$ are related to the obtained coefficients $B(n)$ as

$$b(Nt - n) = b(Nt + n) = B(n)/2 \quad \text{for } n = 1 \ldots Nt, \quad Nt = N/2$$
$$b(Nt) = B(0)$$

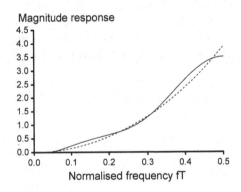

Magnitude response

Fig. 7-16
Magnitude response of a 6th-order sinc equaliser

Figure 7-16 shows the magnitude response of the 6th-order sinc equaliser from Example 7-3. Compared to the results of the frequency sampling method, the deviations from the desired magnitude response are larger, especially at low frequencies. The reason is that we try to minimize the error also in the vicinity of $\omega = \pi/T$ where the target magnitude response shows an ever-increasing slope while the magnitude of the FIR filter must have zero slope due to the symmetry condition of case 1 filters. In order to moderate the consequences of this contradiction, the least mean square criterion is only applied up to a certain frequency just below the Nyquist frequency. So no filter resources are wasted to improve the approximation in a small frequency region around the Nyquist frequency at the cost of the accuracy in the rest of the band of interest. Figure 7-17 shows the obtained magnitude response if we limit the application of the LMS criterion to the frequency range $0 \ldots 0.9\pi/T$. The error at $\omega = \pi/T$ is larger now, but the overall accuracy of the approximation is improved. Needless to say that further improvement can be achieved by increasing the order of the FIR filter.

Magnitude response

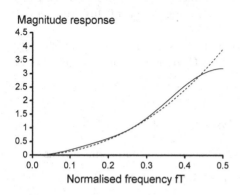

Fig. 7-17
Magnitude response of a 6th-order sinc equaliser with limitation of the approximation range to $0 \ldots 0.9\pi/T$

Figure 7-18 shows the result of a further design example. A low-pass filter is specified as follows:

Order of the FIR filter: $N = 30$ ($Nt = 15$)
Passband $0 \ldots 0.46\pi/T$: $H_{0D}(\omega) = 1$, $W(\omega) = 1$
Stopband $0.54\ \pi/T \ldots \pi/T$: $H_{0D}(\omega) = 0$, $W(\omega) = 10$
Transition band $0.46\ \pi/T \ldots 0.54\pi/T$: unspecified

Magnitude response

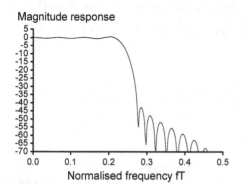

Fig. 7-18
Magnitude response of a 30th-order low-pass filter, designed using the least mean square error criterion

Case 1 filters are the most versatile type of linear-phase FIR filters. There are no restrictions concerning the magnitude at the band edges $\omega = 0$ and $\omega = \pi/T$. If an odd filter order is required for some reason, a case 2 filter must be realised which has in any case a magnitude of zero at $\omega = \pi/T$. The according matrix C and the vector h to solve (7.32) for the case 2 are calculated as

$$c_{in} = W(\omega_i)\cos(n-1/2)\omega_i T$$
$$h_i = W(\omega_i)H_{0D}(\omega_i) \qquad i = 0\ldots gridsize-1, \quad n = 1\ldots Nt \quad .$$

Die FIR filter coefficients $b(n)$ are related to the obtained coefficients $B(n)$ as

$$b(Nt-n) = b(Nt+n-1) = B(n)/2 \quad \text{for} \quad n = 1\ldots Nt, \quad Nt = (N+1)/2 \ .$$

Case 3 filters with even order and case 4 filters with odd order are used if a constant phase shift of $\pi/2$ is required, e.g. for the design of differentiators or Hilbert transformers.

The magnitude of case 3 filters is zero at $\omega = 0$ and $\omega = \pi/T$ which strongly limits the applicability. The according matrix C and the vector h to solve (7.32) for the case 3 are calculated as

$$c_{in} = W(\omega_i)\sin n\omega_i T$$
$$h_i = W(\omega_i)H_{0D}(\omega_i) \qquad i = 0\ldots gridsize-1, \quad n = 1\ldots Nt \quad .$$

Die FIR filter coefficients $b(n)$ are related to the obtained coefficients $B(n)$ as

$$b(Nt-n) = b(Nt+n) = B(n)/2 \quad \text{for} \quad n = 1\ldots Nt, \quad Nt = N/2$$
$$b(Nt) = 0 \qquad .$$

The magnitude of case 4 filters is zero at $\omega = 0$. There is no constraint at $\omega = \pi/T$. The matrix C and the vector h to solve (7.32) for the case 4 are calculated as

$$c_{in} = W(\omega_i)\sin(n-1/2)\omega_i T$$
$$h_i = W(\omega_i)H_{0D}(\omega_i) \qquad i = 0\ldots gridsize-1, \quad n = 1\ldots Nt \quad .$$

Die FIR filter coefficients $b(n)$ are related to the obtained coefficients $B(n)$ as

$$b(Nt-n) = -b(Nt+n-1) = B(n)/2 \quad \text{for} \quad n = 1\ldots Nt, \quad Nt = (N+1)/2 \ .$$

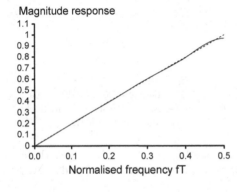

Magnitude response

Normalised frequency fT

Fig. 7-19
Magnitude response of an 11th-order differentiator
dashed line: ideal desired magnitude
solid line: 11th-order FIR filter

Figure 7-19 shows the magnitude response of the 11th-order differentiator from Example 7-2. The magnitude response of the designed filter FIR filter shows good correspondence with the desired linearly increasing gain of the ideal differentiator. A major deviation only occurs at $\omega = \pi/T$ where the magnitude curve must have zero slope. The matrix C and the vector h which were used to determine the coefficients $B(n)$ in (7.32) were calculated as

$$\omega_i = \frac{i\pi/T}{gridsize - 1} \quad \text{for } i = 0 \dots gridsize - 1$$

$$c_{in} = \sin(n - 1/2)\omega_i T$$

$$h_i = \frac{i}{gridsize - 1} \quad \text{for } i = 0 \dots gridsize - 1, \quad n = 1 \dots Nt, \quad Nt = (N+1)/2$$

7.5 Chebyshev Approximation

Inspection of Table 7-1 shows that, of all the windows with a transition width of $4/NT$, the Hamming window exhibits the highest stopband attenuation and thus the best performance for a given filter order N. A special property of the Hamming window is, as mentioned earlier, the almost even ripple in the stopband. Both properties – high stopband attenuation and even ripple – are related to each other. The stopband attenuation is defined as the level of the lobe with the largest amplitude in the stopband with the gain in the passband normalised to 0 dB (refer to Fig. 7-15). All other lobes with lower amplitude waste filter resources, in principle, since they lead in wide areas to a higher attenuation than required to realise the stopband attenuation as defined above in the whole stopband. The optimum utilisation of filter resources can apparently be achieved if all ripples in the stopband have equal amplitude. The same considerations also apply to the ripple in the passband. The achievement of magnitude responses with the named property is a Chebyshev approximation problem.

7.5.1 The Remez Exchange Algorithm

The starting point for an equiripple FIR filter design is the requirement that, by an appropriate choice of the coefficients $B(n)$, a frequency response $H(\omega)$ is achieved that minimises the maximum of the deviation $E(\omega)$ from a desired response $H_D(\omega)$. The target filter characteristic is specified in the form of a zero-phase frequency response $H_{0D}(\omega)$, which avoids further consideration of the phase. In order to avoid any filter resources being wasted for optimisations in the transition band, we only include passband and stopband in the approximation process as we have done previously in the context of the least-squares error approach. The frequency range $\omega = 0 \ldots \pi/T$, excluding the transition band, is denoted as A in the following. The error measure, the so-called Chebyshev error E_C, that has to be minimised in the approximation process can thus be mathematically expressed as

$$E_C = \max_{\omega \in A} |H_0(\omega) - H_{0D}(\omega)| = \max_{\omega \in A} |E(\omega)| . \tag{7.33}$$

For case 1 filters, which we consider for the time being, we obtain by substitution of (7.6a)

$$E_C[B(0), B(1), \ldots, B(Nt)] = \max_{\omega \in A} \left| \sum_{n=0}^{Nt} B(n) \cos n\omega T - H_{0D}(\omega) \right| . \tag{7.34}$$

The minimisation of E_C cannot be performed analytically, since the right-hand side of (7.34) is not a differentiable function. The determination of the optimum coefficients by setting the derivative of E_C with respect to the coefficients $B(n)$ to zero is therefore not possible. In order to solve (7.34), a well-known property of the Chebyshev approximation problem may be used. The alternation theorem defines properties of the maxima and minima of the error function $E(\omega)$ [61]:

Alternation theorem: If $H_0(\omega)$ is a linear combination of r cosine functions of the form

$$H_0(\omega) = \sum_{n=0}^{r-1} B(n) \cos n\omega T , \tag{7.35}$$

then a necessary and sufficient condition that $H_0(\omega)$ be the unique, best Chebyshev approximation to a continuous function $H_{0D}(\omega)$ on A, is that the error function $E(\omega)$ exhibits at least $r + 1$ extremal frequencies ω_i in A with the following properties: All extreme values have the same magnitude. The signs of the extreme values alternate:

$$E(\omega_i) = -E(\omega_{i+1}) \qquad \text{with } i = 1 \ldots r .$$

The magnitude of the extreme values $|E(\omega_i)|$ is denoted as δ in the following considerations.

For case 1 filters we have $r - 1 = Nt$. The number of alternating maxima and minima amounts to at least $N/2 + 2$ or $Nt + 2$. With this knowledge we can establish a set of $Nt + 2$ equations.

$$E(\omega_i) = H_0(\omega_i) - H_{0D}(\omega_i) = -(-1)^i \delta \qquad \text{or}$$

$$H_0(\omega_i) = H_{0D}(\omega_i) - (-1)^i \delta \qquad i = 0 \ldots Nt + 1 \tag{7.36}$$

Relation (7.36) expresses that $Nt + 2$ frequencies exist in A where the deviation between desired and realised frequency responses assumes the maximum value δ.

The signs of the extrema alternate. Substitution of (7.6a) in (7.36) yields a set of equations that can be solved with respect to the desired filter coefficients.

$$\sum_{n=0}^{Nt} B(n)\cos n\omega_i T = H_{0D}(\omega_i) - (-1)^i \delta \qquad i = 0 \dots Nt+1 \qquad (7.37)$$

By means of these $Nt + 2$ equations, we can determine the $Nt + 1$ unknown filter coefficients $B(n)$ as well as the maximum deviation δ. This only works, of course, if the extremal frequencies ω_i are known. For want of these values we start with a first rough estimation of the location of the extremal frequencies. A possible approach is to distribute the $Nt + 2$ frequencies evenly over the frequency interval A. Figure 7-20 illustrates that the transition band between the edge frequencies ω_p and ω_s remains unconsidered for the allocation of the frequencies ω_i.

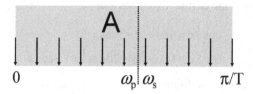

Fig. 7-20
First assumption concerning the location of the extremal frequencies

With these initial values as the first assumption, we solve the set of equations (7.37) with respect to the filter coefficients $B(n)$. These are used to calculate the zero-phase frequency response $H_0(\omega)$ of the filter and the error function $E(\omega) = H_0(\omega) - H_{0D}(\omega)$ with high resolution, which concludes the first iteration step. We have found a first approximation of the desired frequency response.

For the next iteration step, we search for the frequencies at which maxima and minima occur in the error curve $E(\omega)$. In the event that more than $Nt + 2$ extrema are found in A, we choose those with the largest peaks. The extremal frequencies determined in this way are used to solve the set of equations (7.37). With the obtained coefficient $B(n)$, we again calculate the zero-phase frequency response $H_0(\omega)$ of the filter and the error function $E(\omega)$ with high resolution, which concludes the second iteration step. The next step starts again with the search of extremal frequencies in $E(\omega)$. The described iteration procedure to find the position of the extremal frequencies in a Chebyshev approximation problem is the basic Remez exchange algorithm.

The iteration loop can be terminated if the extremal frequencies do not significantly change compared to the results of the previous iteration step. A good termination criterion is also provided by the quantity δ, which is a result of each iteration step. δ increases with each step and converges to the actual maximum deviation of the frequency response from the desired characteristic. If the relative change of δ compared to the previous step is smaller than a given limit or if it even decreases due to calculation inaccuracies, then the iteration is stopped.

The computational effort for the described algorithm is relatively high since we have to solve a set of equations in each iteration step. The modification of the algorithm that we describe in the following section makes use of the fact that the knowledge of the filter coefficients is not necessarily required during the iteration process. What we need is the frequency response and derived from that the error function $E(\omega)$ in high resolution in order to determine the extremal frequencies for the next iteration step. These curves can be calculated in a different way, as we will see in the next section.

7.5.2 Polynomial Interpolation

The simplification of the algorithm is based on the possibility of expressing the zero-phase frequency response according to (7.35) by a polynomial. We can show this by means of trigonometric identities which express the cosine of integer multiples of the argument by powers of the cosine. Examples are:

$$\cos 2x = 2\cos^2 x - 1 \qquad\qquad \cos 3x = 4\cos^3 x - 3\cos x$$

$$\cos 4x = 8\cos^4 x - 8\cos^2 x + 1 \qquad \cos 5x = 16\cos^5 x - 20\cos^3 x + 5\cos x$$

Equation (7.35) can therefore be written in the form

$$H_0(\omega) = \sum_{n=0}^{Nt} C(n)(\cos \omega T)^n \ ,$$

where the actual relationship between the $C(n)$ and $B(n)$ is not of direct interest. If we finally substitute $x = \cos \omega T$, we have the desired form of a polynomial.

$$H_0(\omega) = \sum_{n=0}^{Nt} C(n)x^n \qquad\qquad x = \cos \omega T \qquad\qquad (7.38)$$

The polynomial $H_0(\omega)$ of degree Nt can on the one hand be characterised by the $Nt + 1$ coefficients $C(0) \ldots C(Nt)$ as in (7.38), and also by $Nt + 1$ distinct points through which the graph of $H_0(\omega)$ passes. The Lagrange interpolation formula, here in the barycentric form [29, 61], allows a complete reconstruction of the polynomial and hence of the frequency response from these $Nt + 1$ points.

$$H_0(\omega) = \frac{\displaystyle\sum_{k=0}^{Nt} \left(\frac{\beta_k}{x - x_k} \right) H_0(\omega_k)}{\displaystyle\sum_{k=0}^{Nt} \left(\frac{\beta_k}{x - x_k} \right)} \qquad \text{with } x = \cos \omega T \text{ and } x_k = \cos \omega_k T$$

The $H_0(\omega_k)$ are the $Nt + 1$ points which are interpolated by the Lagrange polynomial. The coefficients β_k are calculated as

$$\beta_k = \prod_{\substack{i=0 \\ i \neq k}}^{Nt} \frac{1}{x_k - x_i} \ .$$

The interpolation can be done with arbitrary resolution. As the interpolation points we choose the alternating extremal values as expressed in (7.36), which are derived from the alternation theorem. $Nt + 1$ of the $Nt + 2$ extremal values are used to characterise the polynomial of Ntth degree.

$$H_0(\omega) = \frac{\displaystyle\sum_{k=0}^{Nt} \left(\frac{\beta_k}{x - x_k}\right)\left(H_{0D}(\omega_k) - (-1)^k \delta\right)}{\displaystyle\sum_{k=0}^{Nt} \left(\frac{\beta_k}{x - x_k}\right)} \tag{7.39}$$

with $x = \cos \omega T$ and $x_k = \cos \omega_k T$

The remaining extremal value is used to determine the ripple δ. The requirement that the polynomial (7.39) also passes through this last point yields a relation with δ as the only unknown.

$$H_{0D}(\omega_{Nt+1}) - (-1)^{Nt+1}\delta = \frac{\displaystyle\sum_{k=0}^{Nt} \left(\frac{\beta_k}{x_{Nt+1} - x_k}\right)\left(H_{0D}(\omega_k) - (-1)^k \delta\right)}{\displaystyle\sum_{k=0}^{Nt} \left(\frac{\beta_k}{x_{Nt+1} - x_k}\right)}$$

with $x_k = \cos \omega_k T$

Solving for δ results in

$$\delta = \frac{a_0 H_{0D}(\omega_0) + a_1 H_{0D}(\omega_1) + \ldots + a_{Nt+1} H_{0D}(\omega_{Nt+1})}{a_0 - a_1 + a_2 - \ldots + (-1)^{Nt+1} a_{Nt+1}} \ . \tag{7.40}$$

The coefficients a_k in (7.40) are calculated as

$$a_k = \prod_{\substack{i=0 \\ i \neq k}}^{Nt+1} \frac{1}{x_k - x_i} \ .$$

Instead of solving the set of equations (7.37) in each iteration step as described in the previous section, the polynomial interpolation approach reduces the mathematical effort to the determination of the ripple δ using (7.40) and the calculation of the frequency response with high resolution using (7.39).

If the iteration is terminated, we still have to determine the filter coefficients. We avoided this calculation during the iteration process in order to reduce the computational effort of the algorithm. We evaluate $H_0(\omega)$ at $N + 1 = 2Nt + 1$ equidistant frequencies in the interval $\omega = 0 \dots 2\pi/T$. Inverse discrete Fourier transform (IDFT) according to (7.12a) yields the filter coefficients of the desired filter (refer to Sect. 7.3). Subroutine CoeffIDFT in the Appendix may be utilised for this purpose.

The following example will illustrate how, by application of the Remez exchange algorithm, the maximum deviation is reduced step by step. At the same time, the ripple becomes increasingly even.

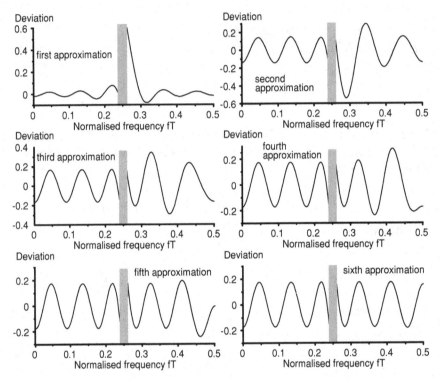

Fig. 7-21 The development of the error function in the course of the iteration for a 22nd-order low-pass filter

We consider a 22nd-order low-pass filter ($N = 22$, $Nt = 11$) with normalised edge frequencies at $f_p T = 0.24$ and $f_s T = 0.26$. According to the alternation theorem, the error curve $E(\omega)$ has at least 13 extremal frequencies with deviations of equal magnitude and alternating sign. As starting values for the iteration, we choose 13 frequencies that are equally distributed over the frequency range $fT = 0 \dots 0.24$ and $0.26 \dots 0.5$. The transition band $fT = 0.24 \dots 0.26$ is ignored. The resulting frequencies can be found in the first column (step 1) of Table 7-2. With these 13 values, and using (7.40), we calculate a first approximation of the maximum error δ, which amounts to

0.01643 in this case. Fig. 7-22 shows the first approximation of the magnitude response which has been calculated using (7.39). The corresponding error curve can be found in Fig. 7-21. This curve has exactly 13 extremal frequencies, which are entered into the second column of Table 7-2. With this set of frequencies we start the next iteration. Figure 7-21 and Fig. 7-22 illustrate how the frequency response and the error curve continuously approach the ideal case of the Chebyshev behaviour with an even ripple in the passband and stopband.

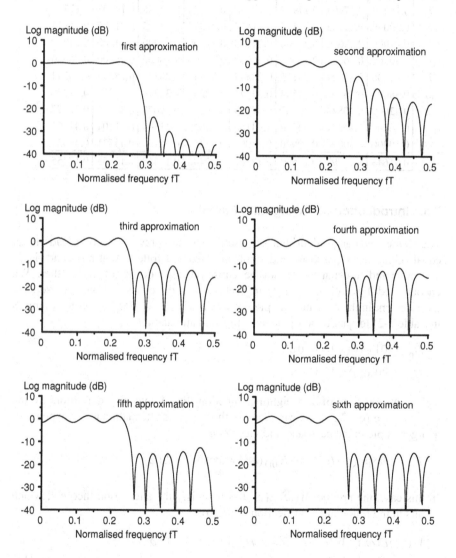

Fig. 7-22 The development of the magnitude response in the course of the iteration for a 22nd-order low-pass filter

Table 7-2 The extremal frequencies in the course of the iteration

	Step 1	Step 2	Step 3	Step 4	Step 5	Step 6	Step 7	Step 8
f_1T	0.00000	0.00000	0.00000	0.00000	0.00000	0.00000	0.00000	0.00000
f_2T	0.04000	0.04427	0.04427	0.04427	0.04427	0.04427	0.04427	0.04427
f_3T	0.08000	0.08854	0.09115	0.09115	0.08854	0.08854	0.08854	0.08854
f_4T	0.12000	0.13021	0.13542	0.13542	0.13281	0.13281	0.13281	0.13281
f_5T	0.16000	0.17708	0.17969	0.17708	0.17708	0.17708	0.17708	0.17708
f_6T	0.20000	0.22135	0.21875	0.21875	0.21875	0.21875	0.21875	0.21875
f_7T	0.24000	0.24000	0.24000	0.24000	0.24000	0.24000	0.24000	0.24000
f_8T	0.30000	0.26000	0.26000	0.26000	0.26000	0.26000	0.26000	0.26000
f_9T	0.34000	0.31729	0.29125	0.28344	0.28344	0.28344	0.28344	0.28344
$f_{10}T$	0.38000	0.36156	0.34594	0.32771	0.32250	0.32250	0.32250	0.32250
$f_{11}T$	0.42000	0.40844	0.39542	0.37979	0.36677	0.36677	0.36677	0.36677
$f_{12}T$	0.46000	0.45271	0.44750	0.43188	0.41625	0.41104	0.41104	0.41104
$f_{13}T$	0.50000	0.50000	0.50000	0.50000	0.47354	0.45792	0.45531	0.45531
δ	0.01643	0.13797	0.16549	0.17363	0.17448	0.17453	0.17453	0.17455

7.5.3 Introduction of a Weighting Function

The Remez exchange algorithm as described in the previous section leads to an equal ripple in the passband and stopband. Hence it follows that passband ripple and stopband attenuation cannot be chosen independently of each other. We encountered this problem already in Sect. 7.4.1 in the context of the least-squares error design. In the present example we obtain with $\delta = 0.17455$ (refer to column 8 in Table 7-2) for passband ripple and stopband attenuation:

$$D_p = 20\log(1+\delta) = 1.40\,\text{dB}$$
$$D_s = -20\log\delta = 15.16\,\text{dB} .$$

By introduction of a weighting function $W(\omega)$ into the definition of the Chebyshev error (7.33), we can increase the stopband attenuation at the expense of a larger ripple in the passband and vice versa.

$$E_C = \max_{\omega \in A}\left|W(\omega)(H_0(\omega)-H_{0D}(\omega))\right| = \max_{\omega \in A}\left|W(\omega)E(\omega)\right|$$

In this case, the product $W(\omega)\,E(\omega)$ has to satisfy the alternation theorem, which results in the following modification of (7.36):

$$W(\omega_i)E(\omega_i) = W(\omega_i)(H_0(\omega_i)-H_{0D}(\omega_i)) = -(-1)^i\delta$$

$$H_0(\omega_i) = H_{0D}(\omega_i)-(-1)^i\delta/W(\omega_i)$$

$$\quad (7.41)$$

$$i = 0\,.....\,Nt+1 .$$

Equation (7.41) leads to the following relations between the ripple in the passband δ_p and in the stopband δ_s:

$$\left.\begin{array}{l}\delta_p = \delta/W_p \\[2mm] \delta_s = \delta/W_s\end{array}\right\} \quad\Rightarrow\quad \delta_p/\delta_s = W_s/W_p .$$

W_p and W_s are the values of the weighting function in the passband and stopband respectively. The equations (7.39) and (7.40), which are important for carrying out the Remez algorithm in practice, assume the following form after introduction of the weighting function:

$$H_0(\omega) = \frac{\displaystyle\sum_{k=0}^{Nt}\left(\frac{\beta_k}{x-x_k}\right)\left(H_{0D}(\omega_k)-(-1)^k\,\delta/W(\omega_k)\right)}{\displaystyle\sum_{k=0}^{Nt}\left(\frac{\beta_k}{x-x_k}\right)} \qquad (7.42)$$

with $x = \cos\omega T$

and $x_k = \cos\omega_k T$.

The coefficients β_k are calculated as

$$\beta_k = \prod_{\substack{i=0 \\ i\neq k}}^{Nt}\frac{1}{x_k - x_i} .$$

$$\delta = \frac{a_0 H_{0D}(\omega_0)+a_1 H_{0D}(\omega_1)+.....+a_{Nt+1}H_{0D}(\omega_{Nt+1})}{a_0/W(\omega_0)-a_1/W(\omega_1)+.....+(-1)^{Nt+1}a_{Nt+1}/W(\omega_{Nt+1})} \qquad (7.43)$$

The coefficients a_k in (7.43) are determined as

$$a_k = \prod_{\substack{i=0 \\ i\neq k}}^{Nt+1}\frac{1}{x_k - x_i} .$$

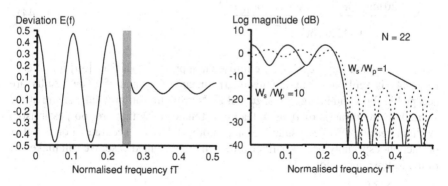

Fig. 7-23   Magnitude response and error curve of a 22nd-order low-pass filter with a weighting ratio $W_s/W_p = 10$

In order to demonstrate the effect of weighting on the ripple of the magnitude response in the passband and in the stopband, we consider again the 22nd-order low-pass filter ($N = 22$, $Nt = 11$) with normalised edge frequencies at $f_p T = 0.24$ and $f_s T = 0.26$. The weighting factor in the passband is chosen as $W_p = 1$, and that in the stopband as $W_s = 10$. Compared to the example in Sect. 7.5.2, we can expect a larger ripple in the passband and a reduced ripple in the stopband. Figure 7-23 shows the frequency response and the error curve which we obtain using the relations (7.42) and (7.43). The ripple is 10 times larger in the passband than in the stopband, since $W_s/W_p = \delta_p/\delta_s = 10$. The magnitude curve is plotted against the one with equal weighting in both bands.

The mathematical implementation of the Remez exchange algorithm, which we have introduced in this section, forms the core of the optimum equiripple design as utilised in many commercial filter design programs. Rabiner, McClellan and Parks published a popular FORTRAN version in [51, 52].

The number of values that have to be calculated for the high resolution representation of the frequency response $H_0(\omega)$ and of the error curve $E(\omega)$ depends on the filter order. A number of $16 \times Nt$ in A is a good compromise between achievable design accuracy and computational effort.

The edge frequencies of the filter, which determine the approximation region A, the filter order N and the gain and the weighting factors in the passband and stopband, are the natural design parameters of the Remez algorithm. The quantities δ_p and δ_s, which determine the ripple in the passband and the attenuation in the stopband, are a result of the Remez iteration. Since filter specifications are in general given in the form of tolerance schemes, it would be desirable to have a relation available that allows the calculation of the required filter order in terms of the ripple and the edge frequencies. Unfortunately there exists no analytical relation between these quantities. Shen and Strang [68] derived the following semi-empirical formula (7.44), which provides the most precise results that can be found in literature for the unweighted case ($\delta = \delta_p = \delta_s$).

$$N \approx \frac{20 \log_{10}(\pi\delta)^{-1} - 10 \log_{10}\log_{10}\delta^{-1}}{4.343 \ln \cot\left(\pi\frac{1-2\Delta}{4}\right)} \tag{7.44}$$

For the weighted case ($\delta_p \neq \delta_s$), we utilise an approach by Kaiser [39] in which δ in (7.44) is replaced with the geometric mean of δ_p and δ_s. The results are acceptable for a wide range of Δ, δ_p and δ_s. Narrow transition widths Δ and at the same time small ratios of δ_s to δ_p lead to estimates of N that are too pessimistic. With increasing N, the results become more and more accurate. For a better prediction, especially of lower filter orders in the weighted case, we can make use of the following relation [31].

$$N \approx \frac{D_\infty(\delta_p, \delta_s)}{\Delta F} - f(\delta_p, \delta_s)\Delta F \tag{7.45}$$

The functions $D_\infty(\delta_p,\delta_s)$ and $f(\delta_p,\delta_s)$ are defined as

$$D_\infty(\delta_p,\delta_s) = \left(a_1(\log_{10}\delta_p)^2 + a_2\log_{10}\delta_p + a_3\right)\log_{10}\delta_s$$
$$+ \left(a_4(\log_{10}\delta_p)^2 + a_5\log_{10}\delta_p + a_6\right)$$

with
$$a_1 = 5.309\cdot10^{-3} \qquad a_4 = -2.66\cdot10^{-3}$$
$$a_2 = 7.114\cdot10^{-2} \qquad a_5 = -5.941\cdot10^{-1}$$
$$a_3 = -4.761\cdot10^{-1} \qquad a_6 = -4.278\cdot10^{-1}$$

and

$$f(\delta_p,\delta_s) = b_1 + b_2\log_{10}(\delta_p/\delta_s)$$

with
$$b_1 = 11.01217$$
$$b_2 = 0.51244$$

7.5.4 Case 2, 3 and 4 Filters

The equiripple design algorithm that we have derived in this chapter is based on case 1 filters, which have a zero-phase frequency response of the form

$$H_0(\omega) = \sum_{n=0}^{Nt} B(n)\cos n\omega T .$$

For the practical implementation of filter design software, it would be desirable if this algorithmic kernel could also be used for case 2, case 3 and case 4 filters, which cover the variants of odd filter orders and antisymmetric unit-sample responses. This is made possible by the fact that the frequency responses for these symmetry cases can be expressed as the product of a function $Q(\omega)$ and a linear combination of cosine functions $P(\omega)$.

Case 2

$$H_0(\omega) = \sum_{n=1}^{Nt} B(n)\cos(n-1/2)\omega T = \cos\omega T/2 \sum_{n=0}^{Nt_{mod}} C(n)\cos n\omega T \qquad (7.46a)$$
$$\text{with } Nt_{mod} = Nt-1 = (N-1)/2$$

Case 3

$$H_0(\omega) = \sum_{n=1}^{Nt}\sin n\omega T = \sin\omega T \sum_{n=0}^{Nt_{mod}} C(n)\cos n\omega T \qquad (7.46b)$$
$$\text{with } Nt_{mod} = Nt-1 = N/2-1$$

Case 4

$$H_0(\omega) = \sum_{n=1}^{Nt} B(n)\sin(n-1/2)\omega T = \sin \omega T/2 \sum_{n=0}^{Nt_{\mathrm{mod}}} C(n)\cos n\omega T \qquad (7.46c)$$

$$\text{with } Nt_{\mathrm{mod}} = Nt - 1 = (N-1)/2$$

These relations can be generalised to the form

$$H_0(\omega) = Q(\omega)\,P(\omega)\ , \qquad\qquad (7.47)$$

with

$$
\begin{aligned}
Q(\omega) &= \cos \omega T/2 &\quad \text{for case 2}\\
Q(\omega) &= \sin \omega T &\quad \text{for case 3}\\
Q(\omega) &= \sin \omega T/2 &\quad \text{for case 4 .}
\end{aligned}
$$

Substitution of (7.47) in (7.33) results in the following expression for the Chebyshev error E_C:

$$E_C = \max_{\omega \in A}\left|W(\omega)\big(Q(\omega)P(\omega) - H_{0D}(\omega)\big)\right|$$

$$E_C = \max_{\omega \in A}\left|W(\omega)Q(\omega)\big(P(\omega) - H_{0D}(\omega)/Q(\omega)\big)\right|\ .$$

With the substitution

$$
\begin{aligned}
W_{\mathrm{mod}}(\omega) &= W(\omega)\,Q(\omega) &\qquad \text{and}\\
H_{0D\,\mathrm{mod}}(\omega) &= H_{0D}(\omega)/Q(\omega)\ ,
\end{aligned}
\qquad (7.48)
$$

we finally arrive at a form of the Chebyshev error that can be treated completely by analogy with the case 1 design problem.

$$E_C = \max_{\omega \in A}\left|W_{\mathrm{mod}}(\omega)\big(P(\omega) - H_{0D\,\mathrm{mod}}(\omega)\big)\right|$$

With regard to the quotient in (7.48), we merely have to make sure that the frequencies at which $Q(\omega)$ vanishes are excluded from the approximation interval A.

As a result of the Remez algorithm, we obtain the frequency response $P(\omega)$, which, multiplied by $Q(\omega)$, yields the frequency response $H_0(\omega)$ of the desired filter. In order to determine the filter coefficients, we evaluate $H_0(\omega)$ at $N+1$ equidistant frequencies in the interval $\omega = 0\ ...\ 2\pi/T$ and apply to this set of values one of the variants of the IDFT (7.12a–d), which are tailored to the respective symmetry cases.

7.5.5 Minimum-Phase FIR Filters

All the filter design methods that we have introduced in this chapter up to now
are related to linear-phase filters. Apart from the simple filter structure and the
guaranteed stability, the possibility of realising strictly linear-phase filters,
featuring excellent performance with respect to signal distortion, is one of the
outstanding properties of FIR filters. In the case of analog or discrete-time IIR
filters, this behaviour can only be approximated by means of complex
implementations. The design algorithms are simplified by the fact that we only
have to deal with real-valued approximation problems. The constant group delay
of $NT/2$ corresponding to half the register length of the FIR filter might be a
disadvantage in some applications. If delay is a problem, minimum-phase filters,
which exhibit in general a lower but frequency-dependent delay, become
attractive. Minimum-phase FIR filters require fewer coefficients than FIR filters
to obtain the same degree of approximation, however, a saving with respect to
implementation complexity is in general not achievable, since the coefficients of
linear-phase filters, on the other hand, occur in pairs, so that the number of
multiplications can be halved.

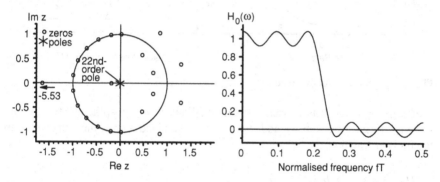

Fig. 7-24   Pole-zero plot and zero-phase frequency response of a 22nd-order low-pass
filter

Figure 7-24 shows the pole-zero plot and the zero-phase frequency response of
a 22nd-order linear-phase low-pass filter. This filter possesses zeros on the unit
circle in the stopband and zero pairs with reciprocal distance from the origin in the
passband. In Sect. 5.2.4 we explained that minimum-phase filters only have zeros
on or within the unit circle. The fact that zeros with reciprocal distance from the
origin have identical magnitude responses leads to the idea of decomposing the
linear-phase frequency response into two partial frequency responses, each having
a magnitude corresponding to the square root of the magnitude of the linear-phase
filter. One comprises the zeros within, the other the zeros outside the unit circle.
The zeros on the unit circle and the poles in the origin are equally distributed over
both partial filters. This would only work perfectly, however, if the zeros on the

unit circle occurred in pairs, which is unfortunately not the case according to Fig. 7-24. A trick [32] helps to create the desired situation. If we add a small positive constant to the zero-phase frequency response shown in Fig. 7-24, the whole curve is shifted slightly upwards. The zeros on the unit circle move closer together in pairs. If the added constant equals exactly the stopband ripple δ_s, we obtain the desired zero pairs on the unit circle (Fig. 7-25). The starting point of the following mathematical considerations is the general form of the transfer function of a FIR filter.

$$H_1(z) = \sum_{n=0}^{N} b_n z^{-n} = z^{-N} \sum_{n=0}^{N} b_n z^{N-n}$$

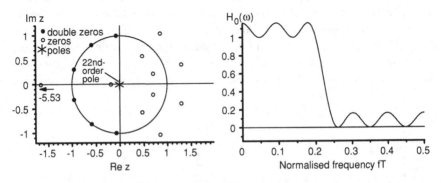

Fig. 7-25 Pole-zero plot of a modified 22nd-order low-pass filter

The filter order N is chosen even to facilitate the later split into two partial systems. In the first step we add, for reasons explained above, the constant δ_s to the transfer function.

$$H_2(z) = z^{-N} \sum_{n=0}^{N} b_n z^{N-n} + \delta_s$$

The polynomial in z can be expressed as the product of the zero terms. In addition we introduce a correction term $1/(1+\delta_s)$ that normalises the gain to unity again.

$$H_2(z) = \frac{b_0}{1+\delta_s} \frac{(z - z_{01})(z - z_{02})..(z - z_{0N})}{z^N}$$

If z_0 is a zero within or on the unit circle, the corresponding zero with reciprocal radius can be expressed as $z_0' = 1/z_0^*$. The linear-phase transfer function $H_2(z)$ can therefore be grouped into two products.

$$H_2(z) = \frac{b_0}{(1+\delta_s)z^N} \prod_{i=1}^{N/2}(z - z_{0i}) \prod_{i=1}^{N/2}\left(z - 1/z_{0i}^*\right)$$

Equivalent rearrangement yields

$$H_2(z) = K z^{-N/2} \prod_{i=1}^{N/2}(z - z_{0i}) \prod_{i=1}^{N/2}\left(z^{-1} - z_{0i}^*\right),$$

(7.49)

with

$$K = \frac{b_0(-1)^{N/2}}{(1 + \delta_s) \displaystyle\prod_{i=1}^{N/2} z_{0i}}.$$

K is, in any case, real-valued since the zeros in the product term are real or occur in complex-conjugate pairs. By the substitution of $z = e^{j\omega T}$ in (7.49), we obtain the corresponding frequency response.

$$H_2(e^{j\omega T}) = K e^{-j\omega N T/2} \prod_{i=1}^{N/2}(e^{j\omega T} - z_{0i}) \prod_{i=1}^{N/2}(e^{-j\omega T} - z_{0i}^*)$$

(7.50)

Both products in (7.50) have identical magnitude since the product terms are complex-conjugate with respect to each other. If we realise a new transfer function $H_3(z)$ from the left product in (7.49), we obtain a minimum-phase filter of order $N/2$ and a magnitude response according to the square root of the magnitude response of the modified linear-phase filter $H_2(z)$.

$$H_3(z) = \sqrt{K} z^{-N/2} \prod_{i=1}^{N/2}(z - z_{0i})$$

(7.51)

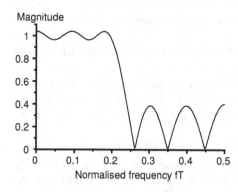

Fig. 7-26
Magnitude response of an eleventh-order minimum phase filter

Figure 7-26 depicts the magnitude response of the eleventh-order minimum-phase filter that we derived from the linear-phase filter according to Fig. 7-24. The ripple in the passband is smaller, and the one in the stopband is larger, compared to the original linear-phase filter with double filter order. The exact relationships between these ripple values can be expressed as

$$\delta_{p3} = \sqrt{1 + \frac{\delta_{p1}}{1 + \delta_{s1}} - 1} \qquad\qquad\qquad (7.52a)$$

$$\delta_{s3} = \sqrt{\frac{2\delta_{s1}}{1 + \delta_{s1}}} \; . \qquad\qquad\qquad (7.52b)$$

Multiplying out the product in (7.51) yields the transfer function in form of a polynomial in z and thus the coefficients of the resulting eleventh-order FIR filter. These coefficients are equivalent to the unit-sample response of the filter, which is graphically represented in Fig. 7-27.

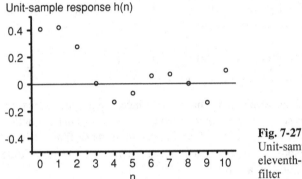

Fig. 7-27
Unit-sample response of the eleventh-order minimum-phase FIR filter

For the practical design of a minimum phase filter, we can summarise the following design steps:
- Specify the target filter in terms of the ripple in the passband δ_{p3}, the ripple in the stopband δ_{s3} and the edge frequencies of the transition band f_pT and f_sT with $\Delta = (f_s - f_p)T$.
- Calculate the ripples δ_{p1} and δ_{s1} of the linear-phase reference filter using (7.48).
- Estimate the required filter order of the reference filter using (7.44) or (7.45).
- Design the linear-phase reference filter using the Remez exchange algorithm.
- Add δ_{s1} to the obtained transfer function and determine the zeros of the polynomial.
- Choose those zeros of $H_2(z)$ which are inside the unit circle and a simple zero at those points on the unit circle where $H_2(z)$ has a pair of zeros. Multiplying out the zero terms yields the desired transfer function of the minimum-phase filter. The additional factor $\sqrt{K}$ normalises the gain to unity.

7.6 Maximally Flat (MAXFLAT) Design

According to (7.6a), we have the following zero-phase frequency response of a linear-phase filter with even filter order N:

$$H_0(\omega) = \sum_{n=0}^{Nt} B(n)\cos n\omega T \qquad \text{with } Nt = N/2. \qquad (7.53)$$

We can determine a set of coefficients $B(n)$ in closed form that leads to a monotonic decay of the magnitude in the frequency range $\omega = 0 \ldots \pi/T$ similar to the behaviour of a Butterworth low-pass filter. For the determination of the $Nt + 1$ coefficients $B(0) \ldots B(Nt)$, we need an equal number of equations. Some of these equations, say a number L, is used to optimise the frequency response in the vicinity of $\omega = 0$; the remaining equations, say a number K, dictates the behaviour around $\omega = \pi/T$. The design parameters L, K and Nt are related as

$$K + L = Nt + 1 . \qquad (7.54)$$

One of the L equations is used to normalise the gain to unity at $\omega = 0$. The remaining $L - 1$ equations enforce $L - 1$ derivatives of $H_0(\omega)$ to vanish at $\omega = 0$.

$$\left. |H_0(\omega)| \right|_{\omega = 0} = 1 \qquad (7.55a)$$

$$\left. \frac{\mathrm{d}^v H_0(\omega)}{\mathrm{d}\omega^v} \right|_{\omega = 0} = 0 \qquad v = 1 \ldots L - 1 \qquad (7.55b)$$

K equations are used to enforce $H_0(\omega)$ and $K - 1$ derivatives of $H_0(\omega)$ to vanish at $\omega = \pi/T$.

$$\left. |H_0(\omega)| \right|_{\omega = \pi/T} = 0 \qquad (7.55c)$$

$$\left. \frac{\mathrm{d}^\mu H_0(\omega)}{\mathrm{d}\omega^\mu} \right|_{\omega = \pi/T} = 0 \qquad \mu = 1 \ldots K - 1 \qquad (7.55d)$$

For the time being, we solve the problem on the basis of a polynomial of degree Nt in the variable x.

$$P_{Nt,K}(x) = \sum_{v=0}^{Nt} a_v x^v$$

The coefficients a_v of the polynomial are determined such that $P_{Nt,K}(0) = 1$ and $P_{Nt,K}(1) = 0$. $L - 1$ derivatives of $P_{Nt,K}(x)$ at $x = 0$ and $K - 1$ derivatives at $x = 1$

are forced to vanish. The closed-form solution of this problem can be found in [30]:

$$P_{Nt,K}(x) = (1-x)^K \sum_{v=0}^{Nt-K} \binom{K+v-1}{v} x^v \quad . \tag{7.56}$$

The term in parenthesis is a binomial coefficient [9], which is defined as

$$\binom{n}{m} = \frac{n!}{m!(n-m)!} \quad .$$

Equation (7.56) can thus be written in the form

$$P_{Nt,K}(x) = (1-x)^K \sum_{v=0}^{Nt-K} \frac{(K-1+v)!}{(K-1)!v!} x^v \quad . \tag{7.57}$$

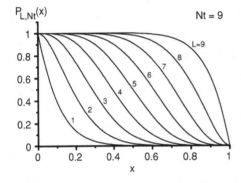

Fig. 7-28
Graphs of the polynomial family $P_{Nt,L}(x)$ for $Nt = 9$ and $L = 1 \dots 9$

Figure 7-28 shows plots of this polynomial family for $Nt = 9$ in the interval $x = 0 \dots 1$. For $Nt = 9$, equivalent to a filter order of $N = 18$, there are exactly 9 different curves and hence 9 possible cutoff frequencies. In order to obtain the frequency response of a realisable FIR filter, a transformation of the variable x to the variable ω has to be found that meets the following conditions:

- The transformation must be a trigonometric function.
- The range $x = 0 \dots 1$ has to be mapped to the frequency range $\omega = 0 \dots \pi/T$.
- The relation between x and ω must be a monotonic function in the interval $x = 0 \dots 1$ in order to maintain the monotonic characteristic of the magnitude response.

The relationship

$$x = (1 - \cos \omega T)/2 \tag{7.58}$$

meets these conditions. Substitution of (7.58) in (7.57) yields the following representation:

$$P_{Nt,K}(\omega) = \left((1+\cos\omega T)/2\right)^K \sum_{v=0}^{Nt-K} \frac{(K-1+v)!}{(K-1)!v!}\left((1-\cos\omega T)/2\right)^v \qquad (7.59)$$

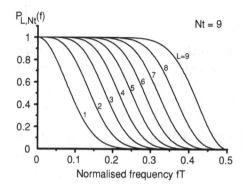

Fig. 7-29
Graphs of the polynomial family
$P_{Nt,L}(\cos\omega T)$ for $Nt = 9$ and $L = 1 \ldots 9$

Figure 7-29 shows the graphs of (7.59) again for the case $Nt = 9$. The monotonic curves are preserved by application of transformation (7.58); only the cutoff frequencies are slightly shifted.

Multiplying out (7.59) and rearrangement of the terms with respect to powers of $\cos\omega T$ leads to a polynomial of the form

$$P_{Nt,K}(\omega) = \sum_{v=0}^{Nt} c_v \left(\cos\omega T\right)^v \ .$$

Application of trigonometric identities finally yields a representation of the zero-phase transfer function according to (7.53).

$$H_0(\omega) = P_{Nt,K}(\omega) = \sum_{v=0}^{Nt} b_v \cos v\omega T$$

It is possible to derive a relation that allows the calculation of the coefficients b_v in closed form [38], but the resulting formula is rather unwieldy. A more elegant way to determine the filter coefficients [40] is to evaluate the frequency response $H_0(\omega_i) = P_{Nt,K}(\omega_i)$ at $N+1$ equidistant frequencies,

$$\omega_i = i\frac{2\pi}{(N+1)T} \qquad i = 0\ldots N \ ,$$

using (7.59), and to calculate the coefficients by inverse discrete Fourier transform (7.12a) on this obtained set of values.

The actual filter design process consists of providing a set of parameters L, K and Nt that leads to an optimum approximation of the prescribed filter specification. Nt can be approximately calculated from the transition band of the filter [40], which is defined here as the range where the magnitude drops from 95% (passband edge) down to 5% (stopband edge).

$$Nt > \frac{1}{[2(f_{0.05} - f_{0.95})T]^2} \tag{7.60}$$

Inspection of Fig. 7-28 shows that the cutoff points $x_{0.5}$ of the polynomial, defined as the values of x where the gain equals 0.5, are regularly distributed. $x_{0.5}$ can therefore be approximated as

$$x_{0.5} \approx L/(Nt+1) . \tag{7.61}$$

Substitution of (7.58) involves the half-decay cutoff frequency $f_{0.5}$ instead of $x_{0.5}$ in this design relation.

$$L/(Nt+1) \approx (1 - \cos 2\pi f_{0.5}T)/2 \tag{7.62a}$$

$$L = \text{int}((Nt+1)(1 - \cos 2\pi f_{0.5}T)/2) \tag{7.62b}$$

The practical filter design is thus performed in the following 4 steps:
- Calculation of the required filter order $N = 2 \cdot Nt$ from the width of the transition band (95%–5%) according to (7.60).
- Calculation of the design parameters L and $K = Nt + 1 - L$ using (7.62b). The input parameter is the half-decay cutoff frequency $f_{0.5}$. The result of this calculation has to be rounded to the nearest integer.
- Evaluation of the frequency response $H_0(\omega_i) = P_{Nt,K}(\omega_i)$ at $N + 1$ equidistant frequencies using (7.59).
- Calculation of the filter coefficients by inverse DFT on the $N + 1$ samples of the frequency response using (7.12a).

The presented design method is not very accurate. The calculations of the parameter Nt (7.60) and equation (7.61) are only approximations. The rounding of L in (7.62b) is a further reason for inaccuracies. The latter is especially critical in the case of low filter orders since, with Nt given, only Nt different cutoff frequencies $f_{0.5}$ can be realised. An improvement of the rounding problem can be achieved by allowing Nt to vary over a certain range, say from a value calculated using (7.60) to twice this value [40]. We finally choose that Nt which leads to the best match between the fraction $L/(Nt + 1)$ and the right-hand side of equation (7.62a).

The problem that (7.60) and (7.61) are only approximations is not completely solved by this procedure. The design results will still not be very accurate. A further drawback of the design procedure described above is the fact that the filter cannot be specified in terms of a tolerance scheme but only by the unusual parameters $f_{0.5}$ and a transition band defined by the edge frequencies $f_{0.95}$ and $f_{0.05}$.

More flexibility and, in general, more effective results with respect to the required filter order are obtained by the method described in [63], which is essentially based on a search algorithm. Specified are passband edge (v_p, ω_p) and stopband edge (v_s, ω_s) of a tolerance scheme. Starting with a value of 2, Nt is incremented step by step. At each step, the algorithm checks whether there exists an L with $L = 1 \ldots Nt$ and hence a frequency response $H_0(\omega)$ according to (7.59) that complies with the tolerance scheme. If a combination of L and Nt is found that meets the specification, the algorithm is stopped. Because of certain properties of the polynomial $P_{Nt,L}$, the search algorithm can be implemented very efficiently, since for each Nt only three values of L need to be checked.

Two C subroutines are included in the Appendix that support the design of MAXFLAT filters. The function MaxFlatPoly is a very efficient implementation of the polynomial calculation according to (7.57). The procedure MaxFlatDesign determines the parameters Nt and hence the required filter order $N = 2 \cdot Nt$, as well as the parameter L, according to the algorithm introduced in [63] from a given tolerance scheme.

Example 7-4

A maximally flat FIR filter with the following specification is to be realised:

Passband edge: $f_p T = 0.2$ Max. passband attenuation: 1 dB
Stopband edge: $f_s T = 0.3$ Min. stopband attenuation: 40 dB

We make use of the C routines given in the Appendix. On a linear scale, the minimum passband gain is 0.891, the maximum stopband gain 0.01. The procedure MaxFlatDesign calculates the optimum design parameters as $Nt = 32$ ($N = 64$) and $L = 15$. With these parameters, we evaluate the maximally flat magnitude response at $N + 1 = 65$ equally spaced frequency points

$$\omega_i = i \frac{2\pi/T}{65} \qquad \text{for } i = 0 \ldots 64$$

using the procedure MaxFlatPoly. For this purpose, the frequencies ω_i need to be mapped to the range 0 ... 1 of the input variable x_i using (7.58).

$$x_i = (1 - \cos \omega_i T)/2 = (1 - \cos(i2\pi/65))/2 \qquad \text{for } i = 0 \ldots 64$$

The filter coefficients $b(n)$ are obtained by inverse DFT of the obtained $N + 1$ samples of the frequency response.

$b(0,64) = 1.44 \cdot 10^{-11}$	$b(11,53) = -5.68 \cdot 10^{-06}$	$b(22,42) = 5.52 \cdot 10^{-03}$
$b(1,63) = 8.90 \cdot 10^{-11}$	$b(12,52) = -2.19 \cdot 10^{-05}$	$b(23,41) = 6.79 \cdot 10^{-03}$
$b(2,62) = -2.15 \cdot 10^{-10}$	$b(13,51) = 1.50 \cdot 10^{-05}$	$b(24,40) = -1.02 \cdot 10^{-02}$
$b(3,61) = -2.44 \cdot 10^{-09}$	$b(14,50) = 9.73 \cdot 10^{-05}$	$b(25,39) = -1.71 \cdot 10^{-02}$
$b(4,60) = -2.30 \cdot 10^{-10}$	$b(15,49) = -1.56 \cdot 10^{-05}$	$b(26,38) = 1.62 \cdot 10^{-02}$
$b(5,59) = 3.17 \cdot 10^{-08}$	$b(16,48) = -3.48 \cdot 10^{-04}$	$b(27,37) = 3.89 \cdot 10^{-02}$
$b(6,58) = 3.54 \cdot 10^{-08}$	$b(17,47) = -8.01 \cdot 10^{-05}$	$b(28,36) = -2.26 \cdot 10^{-02}$
$b(7,57) = -2.57 \cdot 10^{-07}$	$b(18,46) = 1.03 \cdot 10^{-03}$	$b(29,35) = -8.90 \cdot 10^{-02}$
$b(8,56) = -4.79 \cdot 10^{-07}$	$b(19,45) = 5.80 \cdot 10^{-04}$	$b(30,34) = 2.75 \cdot 10^{-02}$
$b(9,55) = 1.44 \cdot 10^{-06}$	$b(20,44) = -2.58 \cdot 10^{-03}$	$b(31,33) = 3.12 \cdot 10^{-01}$
$b(10,54) = 3.83 \cdot 10^{-06}$	$b(21,43) = -2.27 \cdot 10^{-03}$	$b(32) = 4.71 \cdot 10^{-01}$

With the subroutine **ZeroPhaseResp**, the magnitude response of the designed filter can be calculated with high resolution. The attenuation amounts to 0.933 dB at the passband edge and to 40.285 dB at the stopband edge. The magnitude response which is graphically represented in Fig. 7-30 therefore fully complies with the prescribed tolerance scheme which is also depicted.

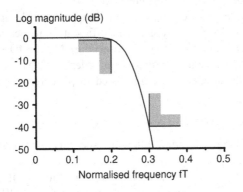

Log magnitude (dB)

Normalised frequency fT

Fig. 7-30
Magnitude of a maximally flat
FIR filter with the parameters
$Nt = 32$ and $L = 15$

Figure 7-31 shows a graphical representation of the obtained filter coefficients and hence of the unit-sample response of the filter. It is worthy of note that the values of the coefficients span an enormous dynamic range, which is a property typical of MAXFLAT filters. Additionally, the filter order is much larger than for a standard equiripple design with identical filter specification. However, maximally flat FIR filters have the advantage of having a monotonous frequency response, which is desirable in certain applications. An attenuation exceeding 100 dB almost everywhere in the stopband can be easily achieved.

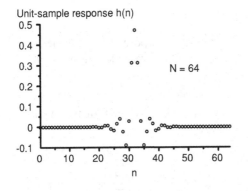

Unit-sample response h(n)

N = 64

n

Fig. 7-31
Unit-sample response of a maximally
flat FIR filter with the parameters
$Nt = 32$ and $L = 15$

The large dynamic range of the coefficients that would be needed in our example is certainly not realisable in fixed-point representation with usual wordlengths of 8, 16 or 24 bit. If we map the calculated filter coefficients to the named fixed-point formats, many coefficients will become zero, which reduces the filter order. While the filter has 65 coefficients in floating-point representation, the filter length is reduced to 51 coefficients with 24-bit, 41 coefficients with 16-bit and 21 coefficients with 8-bit coefficient wordlength. Figure 7-32 shows the magnitude responses for these coefficient resolutions.

If the magnitude response obtained with the limited coefficient wordlength is not satisfactory, other filter structures may be used that are less sensitive to coefficient inaccuracies. One possibility is the cascade structure, where the whole filter is decomposed into first- and second-order sections. For the design, one has to determine the zeros of the transfer function and combine complex-conjugate zero pairs to second-order sections with real coefficients. A further alternative is the structure presented in [71], which is especially tailored to the realisation of maximally flat filters. This structure features very low sensitivity with respect to coefficient deviations and has the additional advantage of requiring fewer multiplications than the direct form.

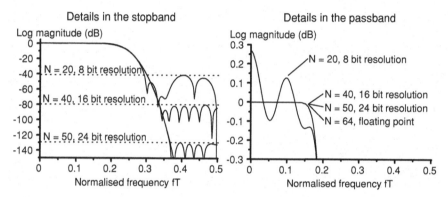

Fig. 7-32 Magnitude response of the example filter with different coefficient wordlengths

8 Effects of Finite Register Lengths

8.1 Classification of the Effects

All practical implementations of digital signal processing algorithms have to cope with finite register lengths. These lead to impairments of the filter performance that become visible in different ways.

The filter coefficients, which can be determined by means of the presented design algorithms with arbitrary precision, can only be approximated by numbers with a given limited wordlength. This implies that the characteristics of the resulting filter deviate more or less from the original specification.

Further effects such as the generation of noise and unstable behaviour are caused by the truncation of bits which becomes necessary after multiplication. If we multiply, for instance, a signal sample with a resolution of a bits by a coefficient with a wordlength of b bits, we obtain a result with $a+b$ bits. In recursive filters without that truncation, the number of digits would increase by b after each clock cycle. Under certain circumstances, the effects caused by the truncation of digits can be treated mathematically like an additive noise signal. Into the same category falls, by the way, the noise generated in analog-to-digital converters by the quantisation of the analog signal. Significant differences can be observed between floating-point and fixed-point arithmetic with respect to the order of magnitude of the additive noise and the dependencies of these effects on the signal to be processed.

The occurrence of so-called limit cycles is another effect that can be observed in the context of recursive filters realised with fixed-point arithmetic. A limit cycle is a constant or periodic output signal which does not vanish even if a constant zero is applied to the input of the filter. The filter thus shows a certain unstable behaviour.

The size of this chapter, about 85 pages, points to the large complexity of this topic. Nevertheless, it is not possible to present all aspects of finite wordlengths in detail in this textbook. The explanations should be sufficient enough, however, to avoid the grossest design mistakes and to finally arrive at filters which exhibit the desired performance.

8.2 Number Representation

Fixed-point arithmetic offers a bounded range of numbers which requires careful scaling of the DSP algorithms in order to avoid overflow and achieve an optimum

D. Schlichthärle, *Digital Filters*, DOI: 10.1007/978-3-642-14325-0_8,
© Springer-Verlag Berlin Heidelberg 2011

signal-to-noise ratio. The implementation of floating-point arithmetic is more complex but avoids some of the drawbacks of the fixed-point format. In the following sections we introduce in detail the properties of both number formats.

8.2.1 Fixed-Point Numbers

As in the case of decimal numbers, we can distinguish between integer and fractional numbers. Each position of a digit in a number has a certain weight. In case of binary numbers, these weights are powers of 2. On the left of the binary point we find the coefficients of the powers with positive exponent, including the zero; on the right are placed the coefficients of the powers with negative exponent, which represent the fractional part of the fixed point number.

$$Z = a_4\, a_3\, a_2\, a_1\, a_0\, .\, a_{-1}\, a_{-2}\, a_{-3}$$

The a_i can assume the values 0 or 1. The value of this number is

$$Z = a_4\, 2^4 + a_3\, 2^3 + a_2\, 2^2 + a_1\, 2^1 + a_0\, 2^0 + a_{-1}\, 2^{-1} + a_{-2}\, 2^{-2} + a_{-3}\, 2^{-3}$$

Let us consider the binary number 0001101.01010000 as an example. One byte (8 bits) is used to represent the integer part and again one byte for the representation of the fractional part of the fixed-point number. The corresponding decimal number is 13.3125.

```
   8.0000
 +4.0000
 +1.0000
 +0.2500
 +0.0625
 _____
 +13.3125
```

| 0 | 0 | 0 | 0 | 1 | 1 | 0 | 1 | . | 0 | 1 | 0 | 1 | 0 | 0 | 0 | 0 |

+13.3125

The most significant bit of binary numbers is commonly used as a sign bit if signed numbers have to be represented. A "0" denotes positive, a "1" negative numbers. Three different representations of negative numbers are commonly used; of these, the one's complement is only of minor importance in practice.

Sign-magnitude

The binary number following the sign bit corresponds to the magnitude of the number.

Example

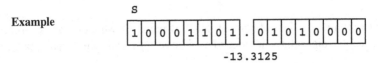

s

| 1 | 0 | 0 | 0 | 1 | 1 | 0 | 1 | . | 0 | 1 | 0 | 1 | 0 | 0 | 0 | 0 |

-13.3125

One's complement

All bits of the positive number are inverted to obtain the corresponding negative number.

Example

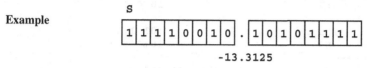

-13.3125

Two's complement

All bits of the positive number are inverted again. After that, a 1 is added to the least significant bit (LSB). A carry that may be generated by an overflow in the most significant bit (MSB) will be ignored.

Example

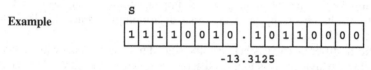

-13.3125

Table 8-1 Signed one-byte number representations

Decimal	-128	-127	-1	-0	+0	+1	+127
Sign-magnitude		11111111	10000001	10000000	00000000	00000001	01111111
One's complement		10000000	11111110	11111111	00000000	00000001	01111111
Two's complement	10000000	10000001	11111111	00000000		00000001	01111111

Table 8-1 shows the results of these rules for various exposed values of signed one-byte numbers. Both the sign-magnitude and the one's complement format have the special property that there are two representations of the zero (+0, −0). The two's complement arithmetic has only one representation of the zero, but the value ranges for positive and negative numbers are asymmetrical. Each of these representations has its pros and cons. Their choice depends on the hardware or software implementation of the arithmetical operations. Multiplications are preferably performed in sign-magnitude arithmetic. Additions and subtractions are more easily realised in two's complement form. Positive and negative numbers can be added in arbitrary sequence without the need for any pre-processing of the numbers. An advantageous property of the complement representations becomes effective if more than two numbers have to be added in a node of the flowgraph. If it is guaranteed that the total sum will remain within the amplitude limits of the respective number representation, intermediate results may lead to overflows without causing any error in the total sum. After the last addition we obtain in any case the correct result. Possible overflows occurring in these additions are ignored. The described behaviour will be illustrated by means of a simple example using an 8 bit two's complement representation, which has an amplitude range from −128 to +127.

Example

```
    100                      01100100
  +  73                    + 01001001
  ───────                  ──────────
    173                      10101101   (-83)

    173                      10101101
  + -89                    + 10100111
  ───────                  ──────────
     84                   (1)01010100
```

After the first addition, the permissible number range is exceeded. A negative result is pretended. By addition of the third, a negative number, we return back into the valid number range and obtain a correct result. The occurring overflow is ignored. Care has to be taken in practical implementations that such "false" intermediate results never appear at the input of a multiplier since this leads to errors that cannot be corrected.

A significant difference exists between the signed number formats with respect to the overflow behaviour. This topic is discussed in detail in Sect. 8.7.1. Special cases of fixed-point numbers are integers and fractions.

Programming languages like C and Pascal offer various types of signed and unsigned integers with different wordlengths. In practice, number representations with 1, 2, 4 and 8 bytes are used. Table 8-2 shows the valid ranges of common integer types together with their names in C and Pascal terminology. Negative numbers are in all cases realised in two's complement arithmetic.

Table 8-2 The range of various integer types

Number representation	C	Pascal	Range
1 byte signed	char	ShortInt	$-128 \dots 127$
1 byte unsigned	unsigned char	Byte	$0 \dots 255$
2 bytes signed	int	Integer	$-32768 \dots 32767$
2 bytes unsigned	unsigned int	Word	$0 \dots 65535$
4 bytes signed	long int	LongInt	$-2147483648 \dots 2147483647$
4 bytes unsigned	unsigned long		$0 \dots 4294967295$
8 bytes signed		Comp	$-9.2 \, 10^{18} \dots 9.2 \, 10^{18}$

In digital signal processing, it is advantageous to perform all calculations with pure fractions. The normalisation of signal sequences and coefficients to values less than unity has the advantage that the multiplication of signals by coefficients always results in fractions again, which makes overflows impossible. The return to the original wordlength which is necessary after multiplications can be easily performed by truncation or rounding of the least significant bits. This reduces the accuracy of the number representation, but the coarse value of the number remains unchanged.

Example

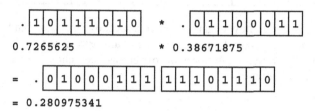

0.7265625   * 0.38671875

= .| 0 | 1 | 0 | 0 | 0 | 1 | 1 | 1 | | 1 | 1 | 1 | 0 | 1 | 1 | 1 | 0 |

= 0.280975341

Truncation to the original wordlength of 8 bits yields

= .| 0 | 1 | 0 | 0 | 0 | 1 | 1 | 1 | truncated

= 0.27734375

The error that we make by the truncation of the least significant bits can be reduced on an average by rounding. For rounding we consider the MSB in the block of bits that we want to remove. If it is 0, we simply truncate as before. If it is 1, we truncate and add a 1 to the LSB of the remaining number.

= .| 0 | 1 | 0 | 0 | 1 | 0 | 0 | 0 | rounded

= 0.28125

8.2.2 Floating-Point Numbers

If the available range of integers is not sufficient in a given application, which is often the case in technical and scientific problems, we can switch to the floating point representation. A binary floating-point number is of the general form

$$F = Mantissa \times 2^{Exponent}$$

An important standard, which defines the coding of mantissa and exponent and details of the arithmetic operations with floating-point numbers, has been developed by the group P754 of the IEEE (Institute of Electrical and Electronical Engineers). The widely used floating-point support processors of Intel (80287, 80387, from the 80486 processor on directly integrated in the CPU) and Motorola (68881, 68882), as well as most of the floating-point signal processors, obey this standard.

According to IEEE P754, the exponent is chosen in such a way that the mantissa (M) is always in the range

$$2 > M \geq 1.$$

This normalisation implies that the digit on the left of the binary point is always set to 1. Applying this rule to the above example of the decimal number 13.3125 yields the following bit pattern for a representation with 2 bytes for the mantissa and one byte for the exponent:

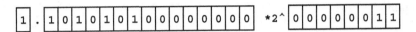

We still need to clarify how negative numbers (negative mantissa) and numbers with magnitude less than unity (negative exponent) are represented.

For the exponent, a representation with offset has been chosen. In case of an one-byte exponent we obtain

$$F = M \times 2^{E-127}$$

The mantissa is represented according to IEEE P754 in the sign-magnitude format.

Since we guarantee by the normalisation of the number that the bit on the left of the binary point is always set to 1, this digit needs not be represented explicitly. We thus gain a bit that can be used to increase the accuracy of the mantissa. The normalisation rule makes it impossible, however, to represent the number zero. A special agreement defines that $F = 0$ is represented by setting all bits of the mantissa (M) and of the exponent (E) to zero. The exponent $E = 255$ (all bits of the exponent are set) plays another special role. This value identifies infinite and undefined numbers, which may result from divisions by zero or taking the square root of negative numbers.

According to IEEE P754, there are 3 floating-point representations, with 4, 8 and 10 bytes. An additional format with 6 bytes is realised in Pascal by software. This format is not supported by mathematical coprocessors. In the following, we give details of how the bits are distributed over sign bit, mantissa and exponent. Table 8-3 shows the valid ranges of these floating-point types together with their names in C and Pascal terminology.

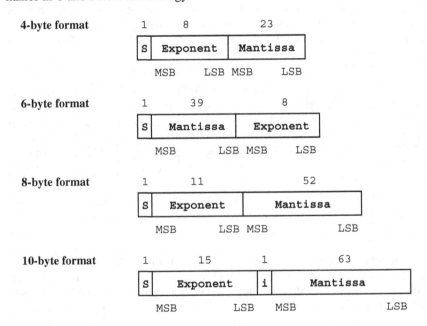

The coprocessor's internal calculations are based on the 10-byte format, which is the only one to explicitly show the bit (i) on the left of the binary point. All floating-point numbers are converted to this format if transferred to the coprocessor.

Table 8-3 The ranges of various floating-point types

Number representation	C	Pascal	Range
4 bytes, 8-bit exponent	float	Single	$1.5 \ 10^{-45} \dots 3.4 \ 10^{+38}$
6 bytes, 8-bit exponent		Real	$2.9 \ 10^{-39} \dots 1.7 \ 10^{+38}$
8 bytes, 11-bit exponent	double	Double	$5.0 \ 10^{-324} \dots 1.7 \ 10^{+308}$
10 bytes, 15-bit exponent	long double	Extended	$3.4 \ 10^{-4932} \dots 1.1 \ 10^{+4932}$

8.3 Quantisation

The multiplication of numbers yields results with a number of bits corresponding to the sum of the bits of the two factors. Also in the case of the addition of floating-point numbers, the number of bits increases depending on the difference between the exponents. In DSP algorithms, there is always a need to truncate or round off digits to return to the original wordlength. The errors introduced by this quantisation manifest themselves in many ways, as we will see later.

8.3.1 Fixed-Point Numbers

If $x_q(n)$ is a signal sample that results from $x(n)$ by quantisation, the quantisation error can be expressed as

$$e(n) = x_q(n) - x(n) \ . \tag{8.1}$$

If $b1$ is the number of bits before truncation or rounding and b the reduced number of bits, then

$$N = 2^{b1-b}$$

steps of the finer resolution before quantisation are reduced to one step of the new coarser resolution. The error $e(n)$ thus can assume N different values. The range of the error depends on the one hand on the chosen number representation (sign-magnitude or two's complement), on the other hand on the type of the quantiser (truncation or rounding).

In the case of truncation, the corresponding number of least significant bits is simply dropped.

Example

Figure 8-1a shows how the grid of possible values of the previous finer resolution is related to the coarser grid after quantisation. The details depend on the method of representation of negative numbers.

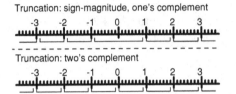

Fig. 8-1a
Quantisation characteristics in the case of the truncation of 3 bits.

Rounding is performed by substituting the nearest number that can be represented by the reduced number of bits. If the old value lies exactly between two possible rounded values, the number is always rounded up without regard of sign or number representation. This rule simplifies the implementation of the rounding algorithm, since the MSB of the dropped block of bits has to be taken into account only. In the following examples, the resolution is reduced by two bits.

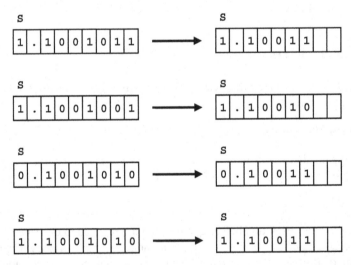

The relation between fine and coarse grid of values is depicted in Fig. 8-1b, where the resolution is again reduced by three bits. We omit the consideration of the one's complement representation, since rounding leads to large errors in the range of negative numbers.

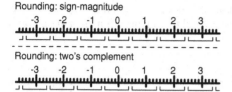

Fig. 8-1b

Quantisation characteristics in the case of rounding of 3 bits.

Mathematical rounding is similar. Rounding is also performed by substituting the nearest number that can be represented by the reduced number of bits. If the old value lies exactly between two possible rounded values, however, the number is always rounded in such a way that the LSB of the rounded value becomes zero. This procedure has the advantage that in this borderline case rounding up and rounding down is performed with equal incidence.

Example

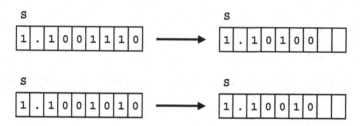

Figure 8-1c shows the quantisation characteristic, which is identical for sign-magnitude and two's complement representation.

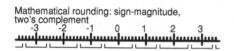

Fig. 8-1c

Quantisation characteristic in the case of mathematical rounding

There are further rounding methods with the same advantage, where rounding up and rounding down is performed randomly in the borderline case.

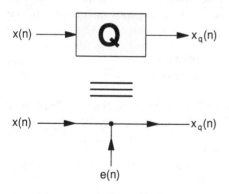

Fig. 8-2

Equivalent representation of a quantiser

By rearrangement of (8.1), the output signal of the quantiser $x_q(n)$ can be interpreted as the sum of the input signal $x(n)$ and the quantisation error $e(n)$.

$$x_q(n) = x(n) + e(n)$$

According to this relation, a quantiser can be represented by an equivalent flowgraph, as shown in Fig. 8-2. Of special interest in this context is the mean value of the error $e(n)$, which has the meaning of an offset at the output of the quantiser, as well as the mean-square error, which corresponds to the power of the superimposed error signal. Assuming a concrete sequence of samples, the error signal $e(n)$ can be determined exactly. If the signal sequence is sufficiently complex, as in the case of speech or music, we can characterise the error signal by statistical parameters such as the mean value or the variance. In these cases, $e(n)$ is a noiselike signal, for which all N possible error values occur with equal probability. If the latter is fulfilled at least approximately, the mean error can be calculated as

$$\overline{e(n)} = \frac{1}{N} \sum_{k=1}^{N} e_k \; . \tag{8.2}$$

The e_k represent the N possible error values. The mean-square error can be expressed accordingly.

$$\overline{e^2(n)} = \frac{1}{N} \sum_{k=1}^{N} e_k^2 \tag{8.3}$$

The quantity

$$\sigma_e^2 = \overline{e^2(n)} - \overline{e(n)}^2 \tag{8.4}$$

is commonly referred to as the variance of the error, which has the meaning of the mean power of the superimposed noise. As an example, we calculate offset and noise power for the case of sign-magnitude truncation. Inspection of Fig. 8-1a shows that the range of the occurring error values depends on the sign of the signal $x(n)$.

$$e_k = k \frac{\Delta x_q}{N} \qquad \begin{array}{ll} 0 \le k \le N-1 & \text{for } x(n) < 0 \\ -N+1 \le k \le 0 & \text{for } x(n) > 0 \; . \end{array} \tag{8.5}$$

Δx_q is the quantisation step at the output of the quantiser. If the signal sequence $x(n)$ is normalised to the range $-1 \le x \le +1$, then Δx_q can be expressed in dependence of the wordlength b as

$$\Delta x_q = 2 \times 2^{-b} \; . \tag{8.6}$$

We substitute (8.5) in (8.2) to calculate the mean error. For negative signal values we obtain

$$\overline{e(n)} = \frac{1}{N}\sum_{k=0}^{N-1} k\frac{\Delta x_q}{N}$$

$$\overline{e(n)} = \frac{\Delta x_q}{N^2}\sum_{k=0}^{N-1} k = \frac{\Delta x_q}{N^2}\frac{N(N-1)}{2}$$

$$\overline{e(n)} = \frac{\Delta x_q}{2}\left(1 - \frac{1}{N}\right) \qquad \text{for } x(n) < 0 \ .$$

Positive signal values accordingly yield

$$\overline{e(n)} = -\frac{\Delta x_q}{2}\left(1 - \frac{1}{N}\right) \qquad \text{for } x(n) > 0 \ .$$

For the calculation of the mean-square error, we substitute (8.5) in (8.3). In this case, the result is independent of the sign of the signal $x(n)$.

$$\overline{e^2(n)} = \frac{1}{N}\sum_{k=0}^{N-1} k^2\frac{\Delta x_q^2}{N^2}$$

$$\overline{e^2(n)} = \frac{\Delta x_q^2}{N^3}\sum_{k=0}^{N-1} k^2 = \frac{\Delta x_q^2}{N^3}\left(\frac{N^3}{3} - \frac{N^2}{2} + \frac{N}{6}\right)$$

$$\overline{e^2(n)} = \frac{\Delta x_q^2}{3}\left(1 - \frac{3}{2N} + \frac{1}{2N^2}\right)$$

The variance, equivalent to the mean power of the superimposed noise, is calculated according to (8.4) as

$$\sigma_e^2 = \overline{e^2(n)} - \overline{e(n)}^2 = \frac{\Delta x_q^2}{12}\left(1 - \frac{1}{N^2}\right) \ .$$

The mean and mean-square values of the error for the other number representations and quantiser types are summarised in Table 8-4. A comparison of the various results shows that the terms to calculate the noise power are identical in all but one case. For large N, the variances of the error approach one common limiting value in all considered cases:

$$\sigma_e^2 = \frac{\Delta x_q^2}{12} \tag{8.7}$$

For a reduction in the wordlength of 3 or 4 bits, (8.7) is already an excellent approximation. A result of this reasoning is that the noise power is not primarily dependent on the number of truncated bits but rather on the resulting accuracy of the signal representation after truncation or rounding off. A special quantiser in this context is the A/D converter which realises the transition from the practically infinite resolution of the analog signal to a resolution according to the number of bits of the chosen number representation. Thus the conversion from analog to digital generates a noise power according to (8.7). Substitution of (8.6) in (8.7) yields the noise power in terms of the wordlength.

$$\sigma_e^2 = \frac{4 \times 2^{-2b}}{12} = \frac{2^{-2b}}{3} \tag{8.8}$$

Table 8-4 Mean values and variances for various quantiser types

Type of quantiser	$\overline{e(n)}$	$\overline{e^2(n)}$	σ_e^2
Truncation: Sign-magnitude One's complement	$\pm\frac{\Delta x_q}{2}\left(1-\frac{1}{N}\right)$	$\frac{\Delta x_q^2}{3}\left(1-\frac{3}{2N}+\frac{1}{2N^2}\right)$	$\frac{\Delta x_q^2}{12}\left(1-\frac{1}{N^2}\right)$
Truncation: Two's complement	$-\frac{\Delta x_q}{2}\left(1-\frac{1}{N}\right)$	$\frac{\Delta x_q^2}{3}\left(1-\frac{3}{2N}+\frac{1}{2N^2}\right)$	$\frac{\Delta x_q^2}{12}\left(1-\frac{1}{N^2}\right)$
Rounding: Sign-magnitude	$\pm\frac{\Delta x_q}{2}\frac{1}{N}$	$\frac{\Delta x_q^2}{12}\left(1+\frac{2}{N^2}\right)$	$\frac{\Delta x_q^2}{12}\left(1-\frac{1}{N^2}\right)$
Rounding: Two's complement	$+\frac{\Delta x_q}{2}\frac{1}{N}$	$\frac{\Delta x_q^2}{12}\left(1+\frac{2}{N^2}\right)$	$\frac{\Delta x_q^2}{12}\left(1-\frac{1}{N^2}\right)$
Mathematical rounding: Sign-magnitude Two's complement	0	$\frac{\Delta x_q^2}{12}\left(1+\frac{2}{N^2}\right)$	$\frac{\Delta x_q^2}{12}\left(1+\frac{2}{N^2}\right)$

8.3.2 Floating-Point Numbers

Floating point numbers behave in a different way. We consider a floating-point number x of the form

$$x = M\, 2^E \;,$$

where the mantissa M assumes values between 1 and 2. After reduction of the resolution of the mantissa by truncation or rounding we obtain the quantised value

$$x_q = M_q\, 2^E \;.$$

The error with respect to the original value amounts to

$$e = x_q - x$$

$$e = (M_q - M) 2^E .$$ (8.9)

Because of the exponential term in (8.9), the quantisation error, caused by rounding or truncation of the mantissa, is proportional to the magnitude of the floating-point number. It makes therefore sense to consider the relative error

$$e_{rel} = \frac{e}{x} = \frac{M_q - M}{M}$$

$$e_{rel} = \frac{e_M}{M} .$$

The calculation of the mean and the mean-square error is more complex in this case because the relative error is the quotient of two random numbers, the error of the mantissa and the value of the mantissa itself. The mean-square error is calculated as

$$\overline{e_{rel}^2} = \sum_{k=1}^{N} \sum_{j=1}^{2^b} \frac{e_{Mk}^2}{M_j^2} p(j,k) .$$ (8.10)

In this double summation, the error of the mantissa e_M progresses with the parameter k through all possible values which can occur by rounding or truncation. The mantissa M progresses with the parameter j through all possible values that it can assume in the range between 1 and 2 with the given wordlength. $p(j,k)$ is a so-called joint probability which defines the probability that an error value e_{Mk} coincides with a certain value of the mantissa M_j. Assuming that the error and mantissa values are not correlated with each other and all possible values e_{Mk} and M_j occur with equal likelihood, (8.10) can be essentially simplified. These conditions are satisfied by sufficient complex signals like voice and music. Under these assumptions, the joint probability $p(j,k)$ can be approximated by a constant. The value corresponds to the reciprocal of the number of all possible combinations of the e_{Mk} and M_j.

$$p(j,k) = \frac{1}{N \, 2^b}$$

N is the number of possible different error values e_{Mk}, 2^b is the number of mantissa values M_j that can be represented by b bits.

$$\overline{e_{rel}^2} = \frac{1}{N \, 2^b} \sum_{k=1}^{N} \sum_{j=1}^{2^b} \frac{e_{Mk}^2}{M_j^2}$$

Since the joint probability is a constant, the two summations can be decoupled and calculated independently.

$$\overline{e_{\text{rel}}^2} = \frac{1}{N} \sum_{k=1}^{N} e_{\text{M}k}^2 \frac{1}{2^b} \sum_{j=1}^{2^b} \frac{1}{M_j^2} \tag{8.11}$$

With increasing length of the mantissa b, the second sum in (8.11) approaches the value $1/2$. $b = 8$ for instance yields a value of 0.503. We can therefore express the relative mean-square error in a good approximation as

$$\overline{e_{\text{rel}}^2} = \frac{1}{2N} \sum_{k=1}^{N} e_{\text{M}k}^2 \quad .$$

The mean value of the relative error is also of interest and can be calculated as

$$\overline{e_{\text{rel}}} = \sum_{k=1}^{N} \sum_{j=1}^{2^b} \frac{e_{\text{M}k}}{M_j} p(j,k) \quad . \tag{8.12}$$

For uniformly distributed and statistically independent error and mantissa values, the mean value becomes zero in any case as the mantissa assumes positive and negative values with equal probability. According to (8.4), the variance of the relative error and hence the relative noise power can be calculated as

$$\sigma_{\text{rel}}^2 = \overline{e_{\text{rel}}^2} - \overline{e_{\text{rel}}}^2 \quad .$$

Since the mean value vanishes, we have

$$\sigma_{\text{rel}}^2 = \overline{e_{\text{rel}}^2} = \frac{1}{2N} \sum_{k=1}^{N} e_{\text{M}k}^2 \quad .$$

The mean-square error for various number representations and quantiser types has been derived in the previous section and can be directly taken from Table 8-4. The variance of the relative error of floating-point numbers is thus simply obtained by multiplying the corresponding value of the mean-square error in Table 8-4 by $1/2$. For sufficient large N, these values approach two characteristic values:

$$\sigma_{\text{rel}}^2 = \frac{1}{6} \Delta x_{\text{q}}^2 = \frac{1}{6} 2^{-2(b-1)} = \frac{2}{3} 2^{-2b} \qquad \text{for truncation} \tag{8.13a}$$

$$\sigma_{\text{rel}}^2 = \frac{1}{24} \Delta x_{\text{q}}^2 = \frac{1}{24} 2^{-2(b-1)} = \frac{1}{6} 2^{-2b} \qquad \text{for rounding} \quad . \tag{8.13b}$$

Truncation generates a quantisation noise power that is four times (6 dB) higher than the noise generated by rounding.

8.4 System Noise Behaviour

Before we analyse in detail the noise generated in the various filter structures, we discuss basic concepts in the context of noise. We introduce the signal-to-noise ratio, which is an important quality parameter of signal processing systems, and consider the filtering of noise in LSI systems. Finally we touch on the problem of signal scaling, which is necessary to obtain an optimum signal-to-noise ratio in fixed-point implementations.

8.4.1 Signal-to-Noise Ratio

The signal-to-noise ratio (SNR) is calculated as the ratio of the signal power, equivalent to the variance of $x(n)$, to the noise power, equivalent to the variance of the error signal $e(n)$.

$$SNR = \frac{\sigma_x^2}{\sigma_e^2} \tag{8.14}$$

The signal-to-noise ratio is often expressed on a logarithmic scale in dB.

$$10 \log SNR = 10 \log\frac{\sigma_x^2}{\sigma_e^2} = 10 \log \sigma_x^2 - 10 \log \sigma_e^2 \tag{8.15}$$

In order to calculate the SNR, we have to relate the noise power generated in the quantiser, as derived in Sect. 8.3, to the power of the signal. In this context we observe significant differences between the behaviour of fixed-point and floating-point arithmetic.

We start with the simpler case of floating-point representation. By rearrangement of (8.14), the SNR can be expressed as the reciprocal of the relative noise power σ_{erel}^2.

$$SNR = \frac{1}{\sigma_e^2/\sigma_x^2} = \frac{1}{\sigma_{erel}^2}$$

Substitution of (8.13a) and (8.13b) yields the following logarithmic representations of the SNR:

$$10 \log SNR = 1.76 + 6.02b \quad \text{(dB)} \qquad \text{for truncation} \tag{8.16a}$$

$$10 \log SNR = 7.78 + 6.02b \quad \text{(dB)} \qquad \text{for rounding} \tag{8.16b}$$

Comparison of (8.16a) and (8.16b) shows that we gain about 6 dB, equivalent to an increase of the wordlength by one bit, if we apply the somewhat more complex rounding rather than simple truncation.

In contrast with fixed-point arithmetic, which we will consider subsequently, the signal-to-noise ratio is only dependent on the wordlength of the mantissa and not on the level of the signal. Floating-point arithmetic thus avoids the problem that care has to be taken regarding overload and scaling of the signals in all branches of the flowgraph in order to optimise the *SNR*. It is only necessary that we do not leave the allowed range of the number format used, which is very unlikely to happen in normal applications. A one byte exponent already covers the range from 10^{-38} to 10^{+38}, equivalent to a dynamic range of about 1500 dB.

In the case of fixed-point arithmetic, the noise power amounts to $\sigma_e^2 = 2^{-2b}/3$ according to (8.8). Substitution in (8.15) yields the relation:

$$10 \log SNR = 10 \log \sigma_x^2 - 10 \log \left(2^{-2b}/3\right)$$

$$10 \log SNR = 10 \log \sigma_x^2 + 4.77 + 6.02\, b \qquad \text{(dB)} . \qquad (8.17)$$

According to (8.17), the variance and thus the power of a signal $x(n)$ must assume the largest possible value in order to obtain the best possible signal-to-noise ratio when this signal is quantised. This is accomplished if the available number range is optimally utilised and hence the system operates as closely as possible to the amplitude limits. The maximum signal power, which can be achieved within the amplitude limits of the used number representation, strongly depends on the kind of signal. In the following, we assume a purely fractional representation of the signals with amplitude limits of ±1. For sinusoids, the rms value is $1/\sqrt{2}$ times the peak value. The maximally achievable signal power or variance thus amounts to 1/2. An equally distributed signal, which features equal probability for the occurrence of all possible signal values within the amplitude limits, has a maximum variance of 1/3. For music or voice, the variance is approximately in the range from 1/10 to 1/15. For normally distributed (Gaussian) signals, it is not possible to specify a maximum variance, as this kind of signal has no defined peak value. If overload is allowed with a certain probability p_0, however, it is still possible to state a maximum signal power. The corresponding values are summarised in Table 8-5, which is derived using the Gaussian error function.

p_0	σ_x^2
10^{-1}	0.3696
10^{-2}	0.1507
10^{-3}	0.0924
10^{-4}	0.0661
10^{-5}	0.0513
10^{-6}	0.0418
10^{-7}	0.0352
10^{-8}	0.0305
10^{-9}	0.0268

Table 8-5
Maximum variance of the normally distributed signal for various probabilities of overload

Substitution of the named maximum variances for the various signal types in (8.17) yields the following signal-to-noise ratios that can be achieved with a given wordlength of b bits.

sinusoids	: $6.02\,b + 1.76$	(dB)	(8.18a)
equal distribution	: $6.02\,b$	(dB)	(8.18b)
audio signals	: $6.02\,b + (-5 \ldots -7)$	(dB)	(8.18c)
Gaussian ($p_o = 10^{-5}$)	: $6.02\,b - 8.13$	(dB)	(8.18d)
Gaussian ($p_o = 10^{-7}$)	: $6.02\,b - 9.76$	(dB)	(8.18e)

The results obtained in this section can be directly applied to a special quantiser, the analog-to-digital converter. The A/D converter realises the transition from the practically infinite resolution of the analog signal to the chosen resolution of the digital signal. So even at the beginning of the signal processing chain, a noise power according to (8.8) is superimposed on the digital signal. Equation (8.18) can be used to determine the least required number of bits from the desired signal-to-noise ratio. If audio signals are to be processed with an *SNR* of 80 dB, for instance, a resolution of about 15 bits is required according to (8.18c). Each multiplication or floating-point addition in the further course of signal processing generates additional noise, which in the end increases the required wordlength beyond the results of (8.18).

The digital filter as a whole behaves like a quantiser which can be mathematically expressed by generalisation of the equations (8.16) and (8.17) and introduction of two new constants C_{fx} and C_{fl}. If we assume a noiseless input signal, the *SNR* at the output of the filter can be calculated in the case of fixed-point arithmetic as

$$10 \log SNR_{fx} = 10 \log \sigma_x^2 + 6.02\,b + C_{fx} \quad \text{(dB)}. \tag{8.19a}$$

In the case of floating-point arithmetic, the *SNR* is independent of the signal level, which leads to the following relation:

$$10 \log SNR_{fl} = 6.02\,b + C_{fl} \quad \text{(dB)}. \tag{8.19b}$$

The constants C_{fx} and C_{fl} are determined by the realised transfer function, the order of the filter and the chosen filter structure. In case of the cascade structure, the pairing up of poles and zeros as well as the sequence of second-order partial systems is of decisive importance. In general we can say that an increase of the wordlength by one bit leads to an improvement of the signal-to-noise ratio by 6 dB.

8.4.2 Noise Filtering by LSI Systems

The noise that is superimposed on the output signal of the filter originates from various sources. Besides the errors produced in the analog-to-digital conversion process, all quantisers in the filter required after multiplications and floating-point additions contribute to the noise at the output of the filter. These named noise sources do not have a direct effect on the output of the system. For the noise that is already superimposed on the input signal, the transfer function between the input and output of the filter is decisive. For the quantisation noise generated within the filter, the transfer functions between the locations of the quantisers and the filter output have to be considered. In this context, it is of interest to first of all establish general relations between the noise power at the input and output of LSI filters, which are characterised by their transfer functions $H(e^{j\omega T})$ or unit-sample responses $h(n)$.

We characterise the input signal $x(n)$ by its power-density spectrum $P_{xx}(\omega T)$, which is the square magnitude of the Fourier transform of $x(n)$ according to (3.13). The term $P_{xx}(\omega T)\,d\omega T$ specifies the contribution of the frequency range $\omega T \ldots (\omega + d\omega)T$ to the total power of the signal. The total power is calculated by integration of the power-density spectrum.

$$\sigma_x^2 = \int_{-\pi}^{+\pi} P_{xx}(\omega T)\,d\omega T \tag{8.20}$$

The power-density spectrum $P_{yy}(\omega T)$ at the output of the filter is obtained by the relation

$$P_{yy}(\omega T) = P_{xx}(\omega T)\left|H(e^{j\omega T})\right|^2 . \tag{8.21}$$

Combining (8.20) and (8.21) yields the following relationship to calculate the variance and thus the power of the output signal:

$$\sigma_y^2 = \int_{-\pi}^{+\pi} P_{xx}(\omega T)\left|H(e^{j\omega T})\right|^2 d\omega T . \tag{8.22}$$

For the estimation of the signal power at the output of the filter, the distribution of the signal power at the input must be known. If this is not the case, we can only give an upper limit which would be reached if the signal power is concentrated in the frequency range where the filter has its maximum gain.

$$\sigma_y^2 \leq \int_{-\pi}^{+\pi} P_{xx}(\omega T)\left|H_{max}\right|^2 d\omega T$$

$$\sigma_y^2 \leq \sigma_x^2 \left|H_{max}\right|^2 \tag{8.23}$$

If the input signal is a noise-like signal with a constant spectrum, as approximately generated in a quantiser, we also obtain relatively simple relationships. The total power $\sigma_x^2 = N_0$ is equally distributed over the frequency range $-\pi \leq \omega T \leq +\pi$, which results in the constant power density

$$P_{xx}(\omega T) = N_0/2\pi .$$

Substitution of this equation in (8.22) yields a relationship for the noise power at the output of the system.

$$\sigma_y^2 = \int\limits_{-\pi}^{+\pi} \frac{N_0}{2\pi} \left| H(e^{j\omega T}) \right|^2 d\omega T$$

$$\sigma_y^2 = N_0 \frac{1}{2\pi} \int\limits_{-\pi}^{+\pi} \left| H(e^{j\omega T}) \right|^2 d\omega T = \sigma_x^2 \frac{1}{2\pi} \int\limits_{-\pi}^{+\pi} \left| H(e^{j\omega T}) \right|^2 d\omega T \qquad (8.24)$$

The noise power at the output is calculated in this case as the product of the input power and the mean-square gain of the filter.

$$\sigma_y^2 = \sigma_x^2 \overline{\left| H(e^{j\omega T}) \right|^2}$$

According to (3.30), the mean-square gain of the system is equal to the sum over the squared unit-sample response.

$$\sum_{n=-\infty}^{+\infty} h^2(n) = \frac{1}{2\pi} \int\limits_{-\pi}^{+\pi} \left| H(e^{j\omega T}) \right|^2 d\omega T$$

Substitution in (8.24) yields

$$\sigma_y^2 = \sigma_x^2 \sum_{n=-\infty}^{+\infty} h^2(n) . \qquad (8.25)$$

If the noise power of a quantiser $N_0 = \Delta x_q^2/12$ is superimposed on the input signal of a filter, the superimposed noise power at the output amounts to

$$\sigma_y^2 = \frac{\Delta x_q^2}{12} \sum_{n=-\infty}^{+\infty} h^2(n) = \frac{\Delta x_q^2}{12} \overline{\left| H(e^{j\omega T}) \right|^2} . \qquad (8.26)$$

8.4.3 Optimum Use of the Dynamic Range

With floating-point arithmetic, overflow of the system is usually not a problem, since the number representations used in practice provide a sufficient dynamical range. The situation is different with fixed-point arithmetic. On the one hand, it is important to avoid overload, which as a rule leads to severe deterioration of the signal quality. On the other hand, the available dynamic range has to be used in an optimum way in order to achieve the best possible signal-to-noise ratio. This does not only apply to the output but also to each node inside the filter. The transfer functions from the input of the filter to each internal node and to the output are subject to certain constraints in the time or frequency domain, that avoid the occurrence of overflow. We first derive the most stringent condition, that definitely avoids overflow. The starting point is the convolution sum (3.3). The magnitude of the output signal does not exceed unity if the following relation applies:

$$|y(n)| = \left| \sum_{m=-\infty}^{+\infty} x(n-m)h(m) \right| \leq 1 \ .$$

In the following, we derive an upper bound for the magnitude of the above convolution term. The magnitude of a sum is in any case smaller than or equal to the sum of the magnitudes (triangle inequality):

$$\left| \sum_{m=-\infty}^{+\infty} x(n-m)h(m) \right| \leq \sum_{m=-\infty}^{+\infty} |x(n-m)| |h(m)| \leq 1 \ .$$

If we replace in the right sum the magnitude of the signal by the maximally possible value, which is unity, the above inequality certainly still holds.

$$\left| \sum_{m=-\infty}^{+\infty} x(n-m)h(m) \right| \leq \sum_{m=-\infty}^{+\infty} |h(m)| \leq 1$$

If the magnitudes of input and output signal are limited to unity, we thus obtain the following condition, which guarantees the absence of overflow.

$$\sum_{m=-\infty}^{+\infty} |h(m)| \leq 1 \qquad\qquad (8.27)$$

For an optimum use of the dynamic range, the equals sign has to be aimed at. If a given filter does not satisfy this condition, the input signal of the filter has to be multiplied by a scaling factor S, which equals the reciprocal of (8.27).

$$S = \frac{1}{\displaystyle\sum_{m=-\infty}^{+\infty} |h(m)|} \qquad\qquad (8.28)$$

A scaling according to (8.28) is too pessimistic in practice, and leads in most cases to a bad utilisation of the available dynamic range. Other scalings are therefore applied, that on the one hand allow occasional overload, but on the other hand lead to far better signal-to-noise ratios. Relation (8.29) defines a scaling factor which, as a rule, is too optimistic.

$$S = \frac{1}{\sqrt{\displaystyle\sum_{m=-\infty}^{+\infty} h^2(m)}} \tag{8.29}$$

Relation (8.29) is a good choice, especially for wideband signals. A compromise often used in practice follows from the inequality

$$\sqrt{\sum_{m=-\infty}^{+\infty} h^2(m)} \le \max_{\omega T}|H(\omega T)| \le \sum_{m=-\infty}^{+\infty}|h(m)| \ .$$

The resulting scaling sees to it that the maximum gain of the filter is restricted to unity. This guarantees absence of overflow in the case of sinusoidal input signals.

$$S = \frac{1}{\max_{\omega T}|H(\omega T)|} \tag{8.30}$$

Multiplication of the input signal of a filter by one of the scaling factors (8.28–8.30) reduces or avoids the probability of overflow. In the case of a cascade structure, it may be more advantageous to distribute the scaling factor over the partial filters. This is done in such a way that overflow is avoided at the output of each partial filter block.

8.5 Noise Optimisation with Fixed-Point Arithmetic

The noise at the output of a cascade filter realisation depends on various factors. The pairing of poles and zeros to form second-order filter blocks has a decisive influence. Furthermore, the sequence of the filter blocks plays an important role. In order to give quantitative estimations of the noise power at the output of the filter, we first have to determine the noise generated by the individual first- and second-order filter sections. In the next step, we will optimise the pole/zero pairing and the sequence of the partial filters in such a way that the noise of the overall filter is minimised.

8.5.1 Noise Behaviour of First- and Second-Order Filter Blocks

The noise power generated in first- and second-order filter sections and its spectral distribution mainly depends on three factors:
- the location of the poles and zeros of the filter,
- the filter structure and
- the number and positioning of the quantisers.

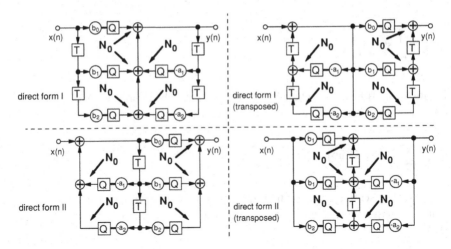

Fig. 8-3 Noise sources in the second-order direct-form structures

Figure 8-3 shows the four possible block diagrams of second-order direct-form filters with each multiplier followed by a quantiser. The arrows, labelled with N_0, indicate the locations where a noise signal with the power $N_0 = \Delta x_q^2/12$ is fed into the circuit. With the results of section 8.4.2, it is easy to determine the noise power at the output of the filter. For that purpose, me must calculate the mean-square gain of the paths from the noise sources to the output. If $H(e^{j\omega T})$ denotes the frequency response of the filter and $H_R(e^{j\omega T})$ the frequency response of the recursive part, the noise power can be calculated by means of the following relations.

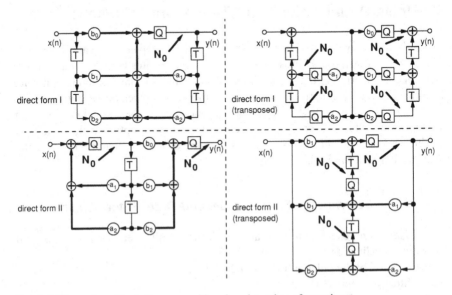

Fig. 8-4 Direct-form block diagrams with reduced number of quantisers

direct form I and

direct form II (transposed)

$$N_{\text{qout}} = 5N_0 \left| H_R(e^{j\omega T}) \right|^2 \qquad (8.30a)$$

direct form I (transposed) and

direct form II

$$N_{\text{qout}} = 3N_0 + 2N_0 \left| H(e^{j\omega T}) \right|^2 \qquad (8.30b)$$

The number of quantisers and hence the number of noise sources can be reduced if not all multipliers are directly followed by a quantiser. If adders or accumulators with double wordlength are available in the respective implementation, the quantisers can be shifted behind the adders as depicted in Fig. 8-4. The biggest saving can be achieved in the case of the direct form I, where five quantisers can be replaced by one. The paths in the block diagrams that carry signals with double wordlength are printed in bold. The general strategy for the placement of the quantisers is that a signal with double wordlength is not allowed at the input of a coefficient multiplier. The noise power at the output of the various direct-form variants is now calculated as follows:

direct form I

$$N_{\text{qout}} = N_0 \left| H_R(e^{j\omega T}) \right|^2 \qquad (8.31a)$$

direct form II

$$N_{\text{qout}} = N_0 + N_0 \left| H(e^{j\omega T}) \right|^2 \qquad (8.31b)$$

direct form I (transposed)

$$N_{\text{qout}} = 3N_0 + 2N_0 \left| H(e^{j\omega T}) \right|^2 \qquad (8.31c)$$

direct form II (transposed)

$$N_{\text{qout}} = 3N_0 \left| H_R(e^{j\omega T}) \right|^2 . \qquad (8.31d)$$

With direct form I and direct form II, we have found the configurations with the lowest possible number of noise sources. The transposed structures can only be further optimised if registers with double wordlength are made available for an interim storage of state variables. The big advantage of the non-transposed structures becomes obvious, as most of the mathematical operations take place in the accumulator. The transposed structures, on the other hand, require more frequent memory accesses, which must happen even with double wordlength if noise is to be minimised. Fig. 8-5 shows the corresponding block diagrams with only one or two quantisers respectively.

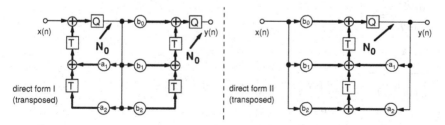

Fig. 8-5 Reduction of the quantisers in the case of the transposed structures

With double precision storage of state variables, the noise power at the output is calculated as follows:

direct form I (transposed) $N_{\text{qout}} = N_0 + N_0 \overline{\left| H(e^{j\omega T}) \right|^2}$ (8.32a)

direct form II (transposed) $N_{\text{qout}} = N_0 \overline{\left| H_R(e^{j\omega T}) \right|^2}$.   (8.32b)

Direct form I and direct form II look quite different if considered separately. In a cascade arrangement of direct form I filter blocks, however, the various parts of the block diagram can be newly combined in such a way that we obtain cascaded direct form II blocks and vice versa. The direct form II has the disadvantage that an additional scaling coefficient has to be placed in front of the first block in order to optimally exploit the dynamic range of the filter chain (Fig. 5-18). In direct form I, this scaling can be done by an appropriate choice of the coefficients b_ν.

According to (8.30), (8.31) and (8.32), the mean-square gain of the path between the noise source and the output of the filter has to be determined in order to explicitly calculate the total noise power at the output. The most elegant way to evaluate the occurring integrals of the form

$$\overline{\left| H(e^{j\omega T}) \right|^2} = \frac{1}{2\pi/T} \int_0^{2\pi/T} \left| H(e^{j\omega T}) \right|^2 d\omega = \frac{1}{2\pi} \int_0^{2\pi} \left| H(e^{j\omega T}) \right|^2 d\omega T \qquad (8.33)$$

is to use the residue method, a powerful tool of functions theory. For that purpose, the integral (8.33) has to be converted into an equivalent integral over the complex variable z. First of all, we make use of the fact that the square magnitude can be expressed as the product of the frequency response and its complex-conjugate counterpart.

$$\overline{\left| H(e^{j\omega T}) \right|^2} = \frac{1}{2\pi} \int_0^{2\pi} H(e^{j\omega T}) H(e^{-j\omega T}) \, d\omega T$$

Using the differential relationship

$$\frac{d e^{j\omega T}}{d\omega T} = j e^{j\omega T} ,$$

we have

$$\overline{\left| H(e^{j\omega T}) \right|^2} = \frac{1}{2\pi j} \int_0^{2\pi} H(e^{j\omega T}) H(e^{-j\omega T}) e^{-j\omega T} \, d e^{j\omega T} .$$

The substitution $e^{j\omega T} = z$ results in the integral

$$\overline{\left|H(e^{j\omega T})\right|^2} = \frac{1}{2\pi j} \oint_{z=e^{j\omega T}} H(z)H(1/z)z^{-1}dz .$$ (8.34)

Because of the relation $z = e^{j\omega T}$ and $0 \le \omega \le 2\pi/T$, the integration path is a full unit circle in the z-plane. All the filter structures that we consider in the following have noise transfer functions of the general form

$$H(z) = \frac{1 + k_1 z^{-1} + k_2 z^{-2}}{1 + a_1 z^{-1} + a_2 z^{-2}} .$$ (8.35)

Substitution of (8.35) in (8.34) yields:

$$\frac{1}{2\pi j} \oint_{z=e^{j\omega T}} \frac{1 + k_1 z^{-1} + k_2 z^{-2}}{1 + a_1 z^{-1} + a_2 z^{-2}} \frac{1 + k_1 z + k_2 z}{1 + a_1 z + a_2 z} \frac{1}{z} dz .$$ (8.36)

Within the integration contour, which is the unit circle, the integrand possesses three singularities: two at the complex-conjugate poles of the transfer function and one at $z = 0$. Integral (8.36) can therefore be represented by three residues [45]. We obtain the following expression as a result [12]:

$$\sigma_y^2 = N_0 \frac{2k_2\left(a_1^2 - a_2^2 - a_2\right) - 2a_1\left(k_1 + k_1 k_2\right) + \left(1 + k_1^2 + k_2^2\right)\left(1 + a_2\right)}{\left(1 + a_1 + a_2\right)\left(1 - a_1 + a_2\right)\left(1 - a_2\right)} .$$ (8.37)

In order to calculate the noise power at the output of a direct form I filter block, we have to determine the mean-square gain of the recursive filter part according to (8.31a).

$$H_R(z) = \frac{1}{1 + a_1 z^{-1} + a_2 z^{-2}}$$

So we have to set $k_1 = 0$ and $k_2 = 0$ in (8.35). With (8.37), we obtain the following variance of the noise:

$$\sigma_y^2 = N_0 \frac{1 + a_2}{1 - a_2} \frac{1}{(1 + a_2)^2 - a_1^2} .$$ (8.38a)

The coefficients a_1 and a_2 can also be represented by the polar coordinates of the complex-conjugate pole pair

$$\left(z - r \cdot e^{j\varphi}\right)\left(z - r \cdot e^{-j\varphi}\right) = z^2 - 2r\cos\varphi z + r^2$$

$$a_1 = -2r\cos\varphi \qquad\qquad a_2 = r^2 .$$

The noise power can thus be expressed as a function of the radius and the angle of the pole pair.

$$\sigma_y^2 = N_0 \frac{1+r^2}{1-r^2} \frac{1}{\left(1+r^2\right)^2 - 4r^2 \cos^2 \varphi} \tag{8.38b}$$

The noise power increases more the closer the pole pair comes to the unit circle ($r \to 1$). Additionally, there is a dependence on the angle and hence on the frequency of the pole. The minimum noise occurs when $\varphi = \pm\pi/2$. With respect to noise, the direct form is apparently especially appropriate for filters whose poles are located around a quarter of the sampling frequency.

First-order filter blocks can be treated in the same way. In the block diagrams in Fig. 8-4, the branches with the coefficients a_2 and b_2 simply have to be deleted. In order to calculate the noise power, the coefficients in (8.37) have to be set as follows:

$$k_1 = 0 \qquad\qquad k_2 = 0 \qquad\qquad a_2 = 0 \;.$$

The resulting noise power can be expressed as

$$\sigma_y^2 = N_0 \frac{1}{1-a_1^2} \;. \tag{8.39}$$

Equation (8.39) shows that the noise power increases also for first-order filter blocks if the pole approaches the unit circle.

Normal-form and wave digital filter have been introduced in Chap. 5 as alternatives to the direct form. Under the idealised assumption, that signals and coefficients can be represented with arbitrary precision, all these filter structures exhibit completely identical behaviour. They only differ with respect to the complexity. In practical implementations with finite wordlengths, however, significant differences arise with respect to noise and coefficient sensitivity.

Figure 8-6 shows the block diagram of a second-order normal form filter including a nonrecursive section to realise the numerator polynomial of the transfer function. The structure requires at least two quantisers. The transfer functions between the noise sources and the output of the filter can be expressed as

$$H_1(z) = \frac{\beta z^{-2}}{1 - 2\alpha z^{-1} + \left(\alpha^2 + \beta^2\right) z^{-2}} = \frac{\beta z^{-2}}{1 + a_1 z^{-1} + a_2 z^{-2}}$$

$$H_2(z) = \frac{z^{-1} - \alpha z^{-2}}{1 - 2\alpha z^{-1} + \left(\alpha^2 + \beta^2\right) z^{-2}} = \frac{z^{-1}\left(1 - \alpha z^{-1}\right)}{1 + a_1 z^{-1} + a_2 z^{-2}} \;.$$

The delay terms in the numerator can be neglected as they do not influence the noise power at the output. The filter coefficients and the coefficients of the transfer function are related as follows:

$$\alpha = r\cos\varphi = -a_1/2$$

$$\beta = r\sin\varphi = \sqrt{a_2 - a_1^2/4} \;.$$

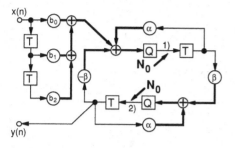

Fig. 8-6
Noise sources in a second-order normal-form structure

The two contributions to the overall noise at the output can be calculated using relation (8.37).

$$\sigma_y^2 = N_0 \frac{\beta^2}{\left(1+a_2\right)^2 - a_1^2} + N_0 \frac{2a_1\alpha + \left(1+\alpha^2\right)\left(1+a_2\right)}{\left(1+a_2\right)^2 - a_1^2}$$

$$\sigma_y^2 = N_0 \frac{1}{1-a_2} = N_0 \frac{1}{1-r^2} \tag{8.40}$$

The spectral distribution of the noise power, generated by the two sources, complement each other at the output of the filter in such a way that the total noise power becomes independent of the angle and hence of the frequency of the pole. Again, the noise power increases if the pole approaches the unit circle.

Finally, we consider the second-order wave digital filter block. Figure 8-7 shows that two quantisers are required if an accumulator with double wordlength is available.

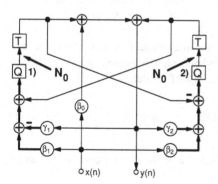

Fig. 8-7
Noise sources in a second-order wave digital filter block

In this case, the transfer functions between the noise sources and the output of the filter can be expressed as

$$H_1(z) = \frac{z^{-1}\left(1-z^{-1}\right)}{1+\left(\gamma_1-\gamma_2\right)z^{-1}+\left(1-\gamma_1-\gamma_2\right)z^{-2}} = \frac{z^{-1}\left(1-z^{-1}\right)}{1+a_1 z^{-1}+a_2 z^{-2}}$$

$$H_2(z) = \frac{z^{-1}\left(1 + z^{-1}\right)}{1 + \left(\gamma_1 - \gamma_2\right) z^{-1} + \left(1 - \gamma_1 - \gamma_2\right) z^{-2}} = \frac{z^{-1}\left(1 + z^{-1}\right)}{1 + a_1 z^{-1} + a_2 z^{-2}} \ .$$

Using (8.37), the resulting noise power is given by

$$\sigma_y^2 = N_0 \frac{2}{1 - a_2} \frac{1}{1 + a_2 - a_1} + N_0 \frac{2}{1 - a_2} \frac{1}{1 + a_2 + a_1}$$

$$\sigma_y^2 = N_0 \frac{1 + a_2}{1 - a_2} \frac{4}{\left(1 + a_2\right)^2 - a_1^2} = N_0 \frac{1 + r^2}{1 - r^2} \frac{4}{\left(1 + r^2\right)^2 - 4 r^2 \cos^2 \varphi} \ . \qquad (8.41)$$

The filter structure according to Fig. 8-7 thus generates four times the quantisation noise power of the direct form I structure. The dependence on radius and angle of the poles is the same.

If the filter implementation allows the storage of the state variables with double wordlength, a filter with only one quantiser could be realised (Fig. 8-8). This structure tends to demonstrate unstable behaviour, however, as will be shown in Sect. 8.7. Therefore, we will not investigate this variant any further.

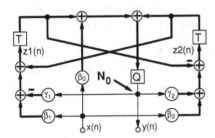

Fig. 8-8
Realisation of the wave digital filter structure with only one quantiser

In the following, we compare the various filter structures with respect to their noise behaviour. We derive the achievable signal-to-noise ratio under the assumption of a sinusoid signal and a 16-bit signal resolution.

$$10 \log SNR = 10 \log \frac{\sigma_{\text{signal}}^2}{\sigma_e^2} = 10 \log \frac{\sigma_{\text{signal}}^2}{\nu N_0} \quad \text{(dB)} \qquad (8.42)$$

The output noise is represented as a multiple ν of the noise power of the single quantiser N_0.

$$10 \log SNR = 10 \log \sigma_{\text{signal}}^2 - 10 \log \nu N_0 = 10 \log \sigma_{\text{signal}}^2 - 10 \log \nu - 10 \log N_0$$

$$10 \log SNR = 10 \log \sigma_{\text{signal}}^2 - 10 \log \nu - 10 \log 2^{-2b}/3$$

$$10 \log SNR = 10 \log \sigma_{\text{signal}}^2 - 10 \log \nu + 4.77 + 6.02 b \quad \text{(dB)} \qquad (8.43)$$

With $\sigma_{\text{signal}}^2 = 1/2$ and $b = 16$, it follows that

$$10 \log SNR = 98.1 - 10 \log v \quad \text{(dB)} .$$

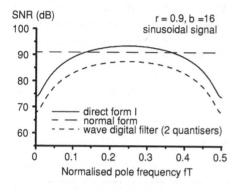

Fig. 8-9
Comparison of the quantisation noise of various filter structures with $r = 0.9$, $b = 16$, sinusoid signal

Figure 8-9 compares the achievable signal-to-noise ratios of the direct-form, normal-form and wave digital filter structure for a pole radius of $r = 0.9$. In the frequency range around $f_s/4$, the direct form exhibits the best performance. The normal form represents a compromise which features constant noise power over the whole frequency range.

8.5.2 Shaping of the Noise Spectrum

The explanations in the previous section have shown that the noise power at the output decisively depends on the filter structure. The transfer function between the location of the quantiser and the output of the filter is the determining factor. The quantisation noise passes through the same paths as the signal to be filtered. For the shaping of the noise spectrum, additional paths are introduced into the filter structure that affect the noise but leave the transfer function for the signal unchanged. For that purpose, it is necessary to get access to the error signal which develops from the quantisation process. So the surplus digits in the case of rounding or truncation must not be dropped but made available for further processing. Fig. 8-10 shows the equivalent block diagram of a quantiser with access to the error signal.

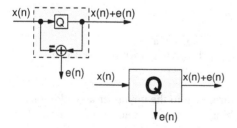

Fig. 8-10
Quantiser with access to the error signal

The relation between the dropped bits and the error signal $e(n)$ depends on the number representation (sign-magnitude or two's complement), the type of quantiser (truncation or rounding) and the sign of the signal value. The value $x(n)$ to be quantised is composed of three parts: the sign (S), the part that is to be preserved after quantisation (MSB, most significant bits), and the bits to be dropped (LSB, least significant bits). In addition to that, we have to distinguish in the case of rounding if the most significant bit of the LSB part is set or not. This leads to the two following cases that have to be considered separately:

Case 1:   S | MSB | 0 . . . . .

Case 2:   S | MSB | 1 . . . . . .

Table 8-6 Rules to derive the error signal $e(n)$

Number representation	Quantiser	Sign	Rules to derive $e(n)$
Sign-magnitude	Truncation	positive	1 | 00 ... 00 | LSB
		negative	0 | 00 ... 00 | LSB
	Rounding	positive	Case1: 1 | 00 ... 00 | LSB Case2: 0 | 00 ... 00 | LSB/2C
		negative	Case1: 0 | 00 ... 00 | LSB Case2: 1 | 00 ... 00 | LSB/2C
Two's complement	Truncation		1 | 11 ... 11 | LSB
	Rounding		Case1: 1 | 11 ... 11 | LSB/2C Case2: 0 | 00 ... 00 | LSB/2C

The rules to derive $e(n)$ from the LSB are summarised in Table 8-6. LSB/2C means the two's complement of the dropped bits.

We apply the method of noise shaping to the direct form I according to Fig. 8-4. The simplest way to influence the quantisation noise is to delay and feed back $e(n)$ to the summation node. This can be done with positive or negative sign as shown in Fig. 8-11a.

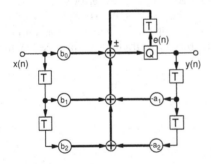

Fig. 8-11a
Direct form I with first-order noise feedback

The noise is now fed in twice, once directly into the quantiser and additionally in delayed form (Fig. 8-11b). The resulting effect is the one of a first-order FIR filter with the transfer function $H(z) = 1 \pm z^{-1}$. The noise is weighted by this transfer function before it is fed into the recursive part of the direct form structure. If the sign is negative, the transfer function has a zero at $\omega = 0$. This configuration is thus able to compensate more or less for the amplification of quantisation noise in the case of low frequency poles. In the reverse case, we obtain a filter structure that can be advantageously used for poles close to $\omega = \pi/T$. The transfer function between the point where the noise is fed in and the output of the filter can be expressed as

$$H(z) = \frac{1 \pm z^{-1}}{1 + a_1 z^{-1} + a_2 z^{-2}} \cdot \tag{8.44}$$

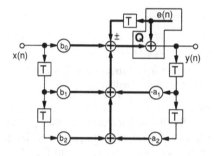

Fig. 8-11b
Alternative representation of the first-order noise feedback in the direct form I

Using (8.37), we obtain the following expressions for the noise power as a function of the radius and the angle of the pole pair. For the negative sign we have

$$\sigma_y^2 = N_0 \frac{2}{1 - a_2} \frac{1}{1 + a_2 - a_1} = N_0 \frac{2}{1 - r^2} \frac{1}{1 + r^2 + 2r \cos\varphi} \cdot \tag{8.45}$$

The noise power becomes minimum at $\varphi = 0$. For the positive sign, the noise assumes a minimum for $\varphi = \pi$ (8.46):

$$\sigma_y^2 = N_0 \frac{2}{1 - a_2} \frac{1}{1 + a_2 - a_1} = N_0 \frac{2}{1 - r^2} \frac{1}{1 + r^2 - 2r \cos\varphi} \cdot \tag{8.46}$$

The next logical step is to introduce another delay in Fig. 8-11b and to weight the delayed noise by coefficients k_1 and k_2 (Fig. 8-12). This leads to a noise transfer function between $e(n)$ and the output $y(n)$ of the following form:

$$H(z) = \frac{1 + k_1 z^{-1} + k_2 z^{-2}}{1 + a_1 z^{-1} + a_2 z^{-2}} \cdot \tag{8.47}$$

If we choose $k_1 = a_1$ and $k_2 = a_2$, the effect of the pole is completely compensated. The noise signal $e(n)$ would reach the output without any spectral boost. k_1 and k_2 cannot be chosen arbitrarily, however. Both coefficients can only assume integer values, as otherwise quantisers would be needed in this noise feedback branch, generating noise again. It must be noted that $e(n)$ must be represented in double precision already. After another multiplication by a fractional coefficient, triple precision would be required.

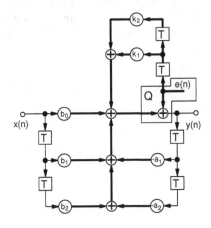

Fig. 8-12
Direct form I structure with second-order noise feedback

As a_1 and a_2 can only be approximated by the nearest integers k_1 and k_2 in the numerator polynomial, a perfect compensation for the poles is not possible. Useful combinations of k_1 and k_2 within or on the stability triangle are shown in Fig. 8-13. Simple feedback of the quantisation error, which results in the noise transfer function (8.44), corresponds to the cases A and B. Second order feedback is required in the cases C to H. The point O inside the triangle represents the case that no noise shaping takes place.

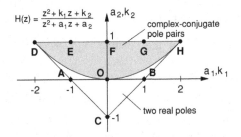

$$H(z) = \frac{z^2 + k_1 z + k_2}{z^2 + a_1 z + a_2}$$

Fig. 8-13
Possible combinations of integer coefficients in the feedback path of the quantisation noise

Figure 8-14 shows the achievable signal to noise ratio for the nine possible coefficient combinations (k_1, k_2) as a function of the pole frequency. We assume a sinusoid signal with 16-bit resolution. The nine cases can be roughly subdivided in three groups:

- The case C yields a signal-to-noise ratio that is independent of the frequency of the pole pair. The values are in general below those that we can achieve in the other configurations. A look at Fig. 8-13 shows that the coefficient pair C will be more suitable to optimise real poles.

- The cases A and B correspond to the first order error feedback. These lead at high and low frequencies to signal-to-noise ratios that are comparable to those that we achieve in the case O around $f_s/4$ without any noise shaping.
- The best signal-to-noise ratios are obtained by second-order error feedback (cases D to H). The five pole frequencies for which the maximum is reached are $fT = 0$, 1/6, 1/4, 1/3 and 1/2.

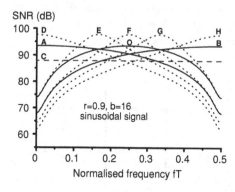

Fig. 8-14
Achievable signal-to-noise ratio for the coefficient pairs (k_1, k_2), sinusoid signal, $r = 0.9$, $b = 16$

Figure 8-15a shows a transposed direct form II with second-order error feedback. The access to the error signal $e(n)$ is explicitly realised by taking the difference between the quantised and unquantised signals. This structure is fully equivalent to Fig. 8-12. Additionally, the dependencies of the signal-to-noise ratio on the coefficients k_1 and k_2 are identical. Redrawing leads to the alternative block diagram Fig. 8-15b. The access to the error signal and the subsequent weighting by integer coefficients is automatically realised in this structure, even if it is not immediately visible. The complexity is not significantly higher compared to the corresponding structure without noise shaping. The coefficients k_1 and k_2 assume the values -2, -1, 0, 1 or 2, which in the worst case means a shifting of the signal value to the right by one digit. Furthermore, two extra additions are required.

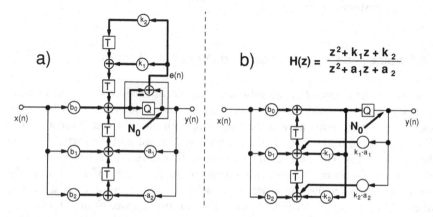

Fig. 8-15 Transposed direct form II with second-order error feedback

A particularly simple structure is obtained in the cases A and B, which yield decisive improvements for low and high frequency poles. The coefficients k_1 and k_2 assume the values

$$k_1 = \pm 1 \qquad\qquad k_2 = 0 \ .$$

One single extra addition or subtraction is required compared to the original transposed direct form II. Fig. 8-16 shows the corresponding block diagram. In addition, we show the modified direct form I which, beside the extra addition/subtraction, requires an additional buffering of the sum in the accumulator.

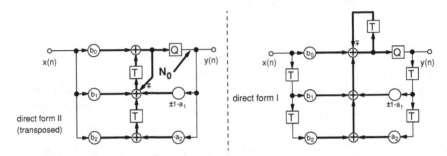

Fig. 8-16 Optimised structures for high or low frequency poles

Filters with poles at low frequencies are often needed for the processing of audio signals. The spectrum of voice and many musical instruments is mainly concentrated in the frequency range from 20 Hz up to about 4 ... 6 kHz. If we want to influence the sound using filters, poles and zeros also have to be placed into this frequency range. The standardised sampling frequencies are much higher, however. Signals are stored on compact discs at a sampling rate of 44.1 kHz, and DAT recorders use a sampling rate of 48 kHz. In order to achieve a good signal-to-noise ratio, filter structures according to Fig. 8-15 or Fig. 8-16 are therefore of special interest for this field of applications [77].

8.5.3 Noise in a Cascade of Second-Order Filter Blocks

In a cascade arrangement of second-order filter blocks, we have a number of ways to optimise the filter with respect to the noise behaviour. In the case of Chebyshev and Cauer filters, zeros have to be combined with poles to create second-order filter sections. A further degree of freedom is the sequence of the partial filters in the cascade. This leads, especially for higher filter orders, to a large number of possible combinations. Even with computer support, the effort to find the optimum is considerable. There are $NB!$ possible combinations of poles and zeros, where NB is the number of filter blocks. Also for the sequence of filter blocks, we have $NB!$ possibilities again. For higher filter orders, dynamic programming [50] is an effective way to find the optimum, but application of simple rules of thumb leads to usable suboptimal results as well.

Partially competing requirements have to be met in order to arrive at the optimum pole/zero pairing and sequence of filter blocks:

1. In the previous section, we have demonstrated that the generated noise power increases more, the closer the pole pair approaches the unit circle, and thus the higher the quality factor Q of the pole is. Obviously, it seems to be advisable to place the sections with high Q at the beginning of the cascade since the noise is filtered in the subsequent filter blocks and the contribution of the high-Q sections to the overall noise is reduced.

2. The sequence of the filter blocks has to be chosen in such a way that we obtain the flattest possible magnitude response in the passband between the input and each output of the individual filter blocks. A large ripple in the front stages would lead to small scaling coefficients between these stages in order to avoid overload. This would have to be compensated by a large gain in the back stages, with the consequence that the noise generated in the front stages would be amplified.

In the case of filters with finite zeros (inverse Chebyshev and Cauer), an appropriate combination of poles and zeros will strongly support the second requirement. A large overshot of the magnitude response is a distinctive feature of poles with high quality factor. This overshot can be reduced by combining these poles with nearby zeros. Since the most critical case is the pole with the largest quality factor, we start with assigning to this pole the nearest zero. We continue this procedure with decreasing Q by assigning the respective nearest of the remaining zeros.

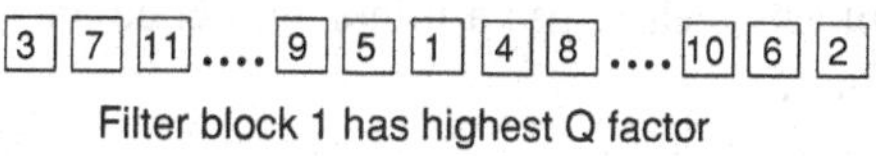

$\boxed{3}\ \boxed{7}\ \boxed{11}\ ....\ \boxed{9}\ \boxed{5}\ \boxed{1}\ \boxed{4}\ \boxed{8}\ ....\ \boxed{10}\ \boxed{6}\ \boxed{2}$

Filter block 1 has highest Q factor

Fig. 8-17
Rule of thumb for the optimum sequence of filter blocks

In the next step, the sequence of the partial filter blocks has to be optimised. For low-pass filters, a rule of thumb is given in [49], leading to suboptimal arrangements which only deviate by a few per cent from the optimum signal-to-noise ratio. The pole with the highest quality factor is placed in the middle of the filter chain, which is a compromise between the first and second requirement. The pole with the second highest Q is placed at the end, the one with the third highest Q at the beginning of the chain. Fig. 8-17 shows how the further filter blocks are arranged.

Table 8-7 shows examples for the noise power at the output of some concrete low-pass implementations. We deal in all cases with tenth-order filters realised by the first canonical structure. The cutoff frequency corresponds to a quarter of the sampling rate. The figures in the table give the output noise power as multiples v of the power N_0 of a single quantiser. Using (8.43), it is easy to calculate the achievable signal-to-noise ratio at the output of the filter. We compare Butterworth, Chebyshev and Cauer filters.

Table 8-7 Output noise power for various filter types (tenth-order, first canonical structure, $f_c T = 0.25$), normalised to N_0

Sequence	Butterworth	Chebyshev 1-dB	Cauer 1-dB/40-dB, pairing	
			optimum	wrong
Optimum	②③④①⑤ 4.9	③④①⑤② 47	③④①⑤② 258	③④①⑤② 368
Worst case	⑤④③②① 10.9	①②③④⑤ 2407	⑤④①②③ 696	①②③④⑤ 9.6·10^6
③⑤①④② Rule of thumb	5.1	55	275	455
①②③④⑤	10.2	2407	474	9.6·10^6
⑤④③②①	10.9	1189	376	6.0·10^6
①⑤④③②	6.8	178	351	20050
②④③⑤①	6.2	53	270	660
③④①②⑤	5.2	77	261	13028
③④②⑤①	6.8	62	306	439
③④⑤①②	7.2	272	542	103057

If the optimum sequence is assumed, the Butterworth filter generates a noise power that exactly corresponds to the noise of the five coefficient multipliers ($v = 5$). A larger noise can be expected for Chebyshev filters since the Q-factor of the poles is higher for this filter type. For the Cauer filter, we show the results for both, the optimum pole/zero pairing and the contrary case, where the pole with the highest Q-factor is combined with the most distant zero etc. The noise power is again increased compared to the Chebyshev filter. The wrong pole/zero pairing may yield completely useless arrangements. If poles and zeros are well combined, the choice of the sequence of the filter blocks clearly has a minor influence only on the noise performance of the overall filter.

Table 8-8 Noise power of a tenth-order 1-dB Chebyshev filter ($f_c T = 0.1$) for various filter structures, normalised to N_0

	Direct form I	Normal form	Noise shaping (first-order)	Noise shaping (second-order)
Maximum	25995	6424	2909	972
Optimum	502	119	55	16
Rule of thumb	603	144	95	28

The noise figures in Table 8-7 are calculated for filters with the normalised cutoff frequency $f_c T = 0.25$, for which direct form I filters have optimum noise behaviour. As an example, application of the rule of thumb yields a noise power of 55 N_0 for a tenth-order 1-dB Chebyshev filter. A cutoff frequency of $f_c T = 0.1$, however, leads to a value about ten times higher. Transition to other filter structures is a way out of this situation. In Table 8-8, we compare the noise power of direct form I, normal form and first- and second-order error feedback for a

tenth-order 1-dB Chebyshev filter with $f_c T = 0.1$. With the help of Fig. 8-13 and Fig. 8-14, we choose the following integer coefficients for noise shaping:
-   first-order error feedback: case A ($k_1 = 1, k_2 = 0$)
-   second-order error feedback: case D ($k_1 = -2, k_2 = 1$) .

It turns out that, in all cases, matched against the most unfavourable configuration, the rule of thumb leads to fairly good results. The deviation from the optimum is in general well below 3 dB. By analogy, all the previous considerations apply to high-pass filters, too. For bandpass and bandstop filters, however, the rule has to be modified.

The rule for the sequence of filter blocks according to Fig. 8-17 leads to disappointing results if applied to bandpass and bandstop filters. For the following considerations, we choose a twelfth-order bandpass filter with mid-band frequency $f_m T = 0.25$ and bandwidth $(f_u - f_l)T = 0.1$ as an example. For a filter with Chebyshev characteristics, the sequence ⑤④①⑥②③ produces minimum noise. No obvious systematics can be identified from this sequence. We gain more informative results if the same calculations are performed for a corresponding Cauer filter with optimum combination of poles and zeros. Using direct form filter blocks yields the following noise figures:

maximum noise:	$4123 N_0$	(sequence ①③⑤②④⑥)
minimum noise:	$12 N_0$	(sequence ⑥⑤①②④③)
rule according to Fig. 8-17:	$1322 N_0$	(sequence ③⑤①④⑥②) .

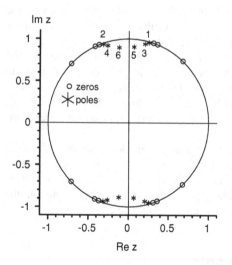

Fig. 8-18
Pole/zero plot for a twelfth-order band-pass filter with Cauer characteristics

From the optimum sequence ⑥⑤①②④③, the following rule can be derived:

The filter blocks are combined in pairs with decreasing Q-factor: highest with second highest, third highest with fourth highest Q-factor etc. These pairs are arranged again according to the well known rule as depicted in Fig. 8-17.

Figure 8-18 shows the location of the poles and zeros of our exemplary Cauer bandpass filter. These are arranged symmetrically about the mid-band frequency. The named rule thus combines in pairs the poles that are located symmetrically about the mid-band frequency and have about the same distance from the unit circle and consequently the same Q-factor. Table 8-9 shows how successfully this rule can be applied to other filter types.

Table 8-9 Noise power of twelfth-order band-pass filters (direct form I),
$f_m T = 0.25$, $(f_u - f_l)T = 0.1$, normalised to N_0

Sequence	Butterworth	Chebyshev 1-dB	Cauer 1-dB/40-dB
Optimal	⑤④①②③⑥ 13	⑤④①⑥②③ 77	⑥⑤①②④③ 12
Worst case	①③⑤⑥④② 204	①③⑤⑥④② 17702	①③⑤②④⑥ 4123
⑤⑥①②③④ Bandpass rule	15	102	12
⑥⑤②①④③ Pairs in reverse order	15	103	12
③⑤①④⑥② Low-pass rule	95	6109	1322
⑤①③⑥②④ HP-LP	139	10595	260
⑥②④④①③ LP-HP	137	10458	260

Matched against the most unfavourable and the most optimum case, the modified rule for bandpass filters yields excellent results. The sequence within the pairs plays practically no role. Direct application of the rule for low- and high-pass filters leads to useless results. Bandpass filters can also be interpreted as the cascade of a low-pass and a high-pass filter. A logical procedure would be to optimise the low-pass (LP) and high-pass (HP) pole/zero pairs separately according to the low-pass/high-pass rule. Table 8-9 shows, however, that this approach goes completely wrong. By analogy, all the previous considerations apply to bandstop filters, too.

The considerations in this section show that observing some simple rules with respect to the pairing of poles and zeros and the sequence of partial second-order filter blocks leads to quite low-noise filter implementations. The deviation from the optimum is, in general, well below 3 dB. Furthermore, depending on the location of the poles, an appropriate filter structure should be chosen.

8.6 Finite Coefficient Wordlength

The coefficients of numerator and denominator polynomial or the poles and zeros of the transfer function (8.48) can be determined with arbitrary accuracy using the design procedures that we derived in Chap. 6 and Chap. 7.

$$H(z) = \frac{\displaystyle\sum_{r=0}^{M} b_r z^{-r}}{\displaystyle\sum_{i=0}^{N} a_i z^{-i}} = \frac{b_0}{a_0} \frac{\displaystyle\prod_{m=1}^{M}\left(1 - z_{0m} z^{-1}\right)}{\displaystyle\prod_{n=1}^{N}\left(1 - z_{\infty n} z^{-1}\right)} = \frac{N(z)}{D(z)} \tag{8.48}$$

Because of the finite register lengths in practical implementations, the filter coefficients can only be represented with limited accuracy, leading to deviations from the filter specification. Poles and zeros lie on a grid of discrete realisable values.

8.6.1 Coefficient Sensitivity

We first look at the sensitivity of the zero and pole location with respect to changes of the filter coefficients. We investigate in detail the behaviour of poles, but the obtained results completely apply to zeros, too.

The denominator polynomial $D(z)$, which determines the location of the poles, can be expressed by the filter coefficients a_i or by the poles $z_{\infty n}$:

$$D(z) = 1 + \sum_{i=1}^{N} a_i z^{-i} = \prod_{n=1}^{N}\left(1 - z_{\infty n} z^{-1}\right) . \tag{8.49}$$

The starting point for the determination of the sensitivity of the ith pole with respect to the inaccuracy of the kth filter parameter a_k is the partial derivative of the denominator polynomial, taken at the pole location $z_{\infty i}$.

$$\left. \frac{\partial D(z)}{\partial a_k} \right|_{z = z_{\infty i}}$$

Expansion of this term leads to a relation that includes the desired sensitivity of the poles to the coefficient quantisation.

$$\left. \frac{\partial D(z)}{\partial a_k} \right|_{z = z_{\infty i}} = \left. \frac{\partial D(z)}{\partial z_{\infty i}} \right|_{z = z_{\infty i}} \frac{\partial z_{\infty i}}{\partial a_k} \tag{8.50}$$

Equation (8.49) can be differentiated with respect to the filter coefficients a_k and the poles $z_{\infty i}$.

$$\left. \frac{\partial D(z)}{\partial a_k} \right|_{z = z_{\infty i}} = z^{-k} \Big|_{z = z_{\infty i}} = z_{\infty i}^{-k}$$

$$\left. \frac{\partial D(z)}{\partial z_{\infty i}} \right|_{z = z_{\infty i}} = -z^{-1} \prod_{\substack{n=1 \\ n \neq i}}^{N}\left(1 - z_{\infty n} z^{-1}\right) \Bigg|_{z = z_{\infty i}}$$

$$\left.\frac{\partial D(z)}{\partial z_{\infty i}}\right|_{z=z_{\infty i}} = -z_{\infty i}^{-1} \prod_{\substack{n=1 \\ n \neq i}}^{N} \left(1 - z_{\infty n}\, z_{\infty i}^{-1}\right)$$

$$\left.\frac{\partial D(z)}{\partial z_{\infty i}}\right|_{z=z_{\infty i}} = -z_{\infty i}^{-N} \prod_{\substack{n=1 \\ n \neq i}}^{N} \left(z_{\infty i} - z_{\infty n}\right)$$

Substitution of these differential quotients in (8.50) yields an equation that gives a measure of the sensitivity of the ith pole to a change of the kth coefficient in the denominator polynomial of $H(z)$:

$$\frac{\partial z_{\infty i}}{\partial a_k} = -\frac{z_{\infty i}^{N-k}}{\displaystyle\prod_{\substack{n=1 \\ n \neq i}}^{N} \left(z_{\infty i} - z_{\infty n}\right)} \,. \tag{8.51}$$

Equation (8.51) shows that, if the poles (or zeros) are tightly clustered, it is possible that small errors in the coefficients can cause large shifts of the poles (or zeros). Furthermore, it is evident that the sensitivity increases with the number of poles (or zeros). Clusters of poles occur in the case of sharp cutoff or narrow bandpass/bandstop filters. For these filter types, it has to be carefully considered whether a chosen coefficient representation is sufficient in a given application. An improvement of the situation can be achieved in a cascade arrangement of second-order filter blocks which leads to clusters consisting of two (complex-conjugate) poles only.

The deviations of the poles from the exact value leads to consequences for the transfer function in three respects. Affected are:

- the frequency of the pole,
- the Q-factor or the 3-dB bandwidth of the resonant curve respectively (e.g. Fig. 6-11),
- the gain of the pole.

From the transfer function of a simple pole

$$H(z) = \frac{1}{z - z_{\infty}} = \frac{1}{z - r_{\infty}\, e^{j\varphi_{\infty}}} \,,$$

we can derive the three named quantities as

frequency of the pole : $\omega_{\infty} T = \varphi_{\infty}$

3 - dB bandwidth : $\Delta f_{3\mathrm{dB}} T \approx (1 - r_{\infty})/\pi$

gain : $\left| H(e^{j\omega_{\infty} T}) \right| = 1/(1 - r_{\infty})$.

Gain and bandwidth are especially critical, if the magnitude of the pole r_∞ approaches unity. The smallest deviations from the target value may lead to considerable deviations in the frequency response. At least the unpleasant distortion of the gain can be avoided by the choice of an appropriate filter structure. The boundedness of wave digital filters prevents an unlimited growth of the gain if the pole approaches the unit circle. We demonstrate this fact by means of a simple first-order low-pass filter. For the direct form, the transfer function is

$$H(z) = \frac{1}{1+a_1 z^{-1}} \qquad\qquad H(z=1, \omega=0) = \frac{1}{1+a_1} .$$

Figure 8-19 shows the analog reference filter for a wave digital filter implementation which only requires a symmetrical three-port parallel adapter and a delay.

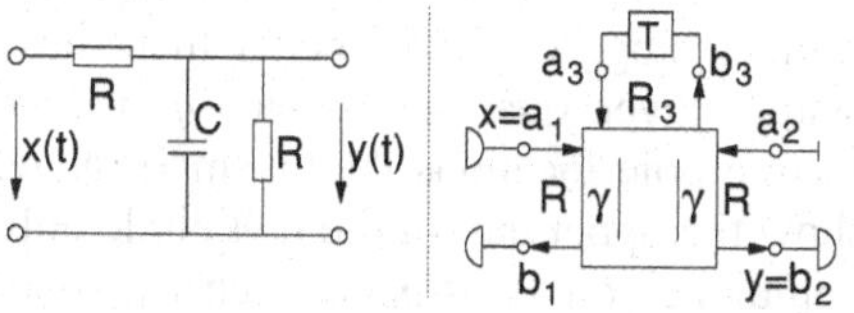

Fig. 8-19
Wave digital filter implementation of a first-order low-pass filter

The transfer function of the wave digital filter implementation according to Fig. 8-19 can be expressed as

$$H(z) = \frac{\gamma\left(1+z^{-1}\right)}{1-\left(1-2\gamma\right) z^{-1}} \qquad\qquad H(z=1, \omega=0) = 1 .$$

For $\omega = 0$ or $z = 1$, the gain amounts exactly to unity, independent of the choice of the parameter γ. By way of contrast, the gain approaches infinity as $a_1 \rightarrow -1$ in the case of the direct form.

Deviations of the pole frequency and the Q-factor cannot be avoided, however, even if wave digital filters are employed. In the following, we investigate the various types of second-order filter blocks with respect to the coefficient sensitivity. Criterion is the distribution of allowed pole locations in the z-plane.

8.6.2 Graphical Representation of the Pole Density

The coefficient sensitivity can be graphically illustrated by plotting all possible poles in the z-plane that can be represented with the given coefficient wordlength. In regions where poles are clustered, the sensitivity is low as many candidates are available in the vicinity to approximate a given exact pole location. In the reverse case, large deviations from the target frequency response are possible as the distance to the nearest realisable pole location may be considerable.

The resolution of the coefficients can be defined in different ways depending on the characteristics of the implemented arithmetical operations:

- by the quantisation step q of the coefficients which is determined by the number of available fractional digits
- by the total number of digits available for the coefficient representation

In the second case, the quantisation step q depends on the value range of the coefficients. If the magnitude is less than unity, all bits but the sign bit are used for the fractional digits. There are filter structures, however, whose coefficients may assume values up to 4. These require up to two integer digits which reduces the number of available fractional digits accordingly. In the case of pure fractional arithmetic, coefficients larger than 1 are realised by shifting the result of the multiplication by an appropriate number of bits to the left.

In the following graphical representations of possible discrete pole locations, we assume in all cases a coefficient format with four fractional bits. This corresponds to a quantisation step of $q = 1/16$. If we compare the various filter structures with respect to coefficient sensitivity with the aforementioned assumption, structures with a large coefficient range come off better than those with a small range. The reason is that, with a given quantisation step q, a large coefficient range yields more possible discrete pole locations than a small range. The following graphical representations show the upper half of the unit circle only because the pattern of allowed complex-conjugate pole pairs is symmetrical about the real axis.

Second-order direct form

For this filter type, the coefficients a_k of the denominator polynomial of the transfer function are identical with the filter coefficients in the recursive part of the block diagram. The direct form I structure (Fig. 8-3) is an example. The real and imaginary parts of the poles z_∞ and the filter coefficients a_1 and a_2 are related as follows:

$$\operatorname{Re} z_\infty = -a_1/2 \qquad\qquad (8.52a)$$

$$\operatorname{Im} z_\infty = \sqrt{a_2 - a_1^2/4} \ . \qquad\qquad (8.52b)$$

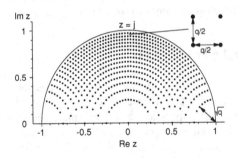

Fig. 8-20
Pattern of possible pole locations for direct form filters (coefficients with four fractional bits, $q = 1/16$)

The stability range of the coefficient pair (a_1,a_2) is the interior of the triangle according to Fig. 5-20. The coefficient a_1 assumes values in the range −2 to +2 while a_2 lies in the range -1 to $+1$. If we plot all possible complex poles in the z-plane, using (8.52), that lie within the unit circle and result from coefficients a_1 and a_2 with 4 fractional bits, we obtain the pattern according to Fig. 8-20. The possible pole locations are unevenly distributed over the unit circle. At low frequencies ($z = 1$) and high frequencies ($z = -1$), the pole density is rather low, and the poles lie on circular arcs. In the vicinity of $f_s/4$, the density is highest. The pole pattern is almost quadratic.

The distance of a pole from the point $z = 1$ can be calculated as

$$D_{z=1} = \sqrt{\left(1 - \operatorname{Re} z_\infty\right)^2 + \left(\operatorname{Im} z_\infty\right)^2} = \sqrt{\left(1 + a_1/2\right)^2 + a_2 - a_1^2/4}$$

$$D_{z=1} = \sqrt{1 + a_1 + a_2} \ . \tag{8.53}$$

For $z_\infty = 1$ we have $a_1 = -2$ and $a_2 = 1$. About this point, a_1 and a_2 take on the following discrete values:

$$a_1 = -2 + l_1 \, q$$
$$a_2 = 1 - l_2 \, q \ .$$

q is the quantisation step of the coefficients. l_1 und l_2 are integers that assume small positive values in the region about $z = 1$.

$$D_{z=1} = \sqrt{1 - 2 + l_1 \, q + 1 - l_2 \, q} = \sqrt{\left(l_1 - l_2\right) q}$$

The smallest possible distance of a pole from the point $z = 1$ thus amounts to $\sqrt{q}$.

For $z = j$ we have $a_1 = 0$ und $a_2 = 1$. About this point, a_1 und a_2 take on the following discrete values:

$$a_1 = l_1 \, q$$
$$a_2 = 1 - l_2 \, q \ .$$

Using (8.52a), the real part of the resulting realisable poles can be expressed as

$$\operatorname{Re} z_\infty = -a_1/2 = -l_1 \, q/2 \ .$$

Use of (8.53b) yields the following expression for the imaginary part:

$$\operatorname{Im} z_\infty = \sqrt{a_2 - a_1^2/4} = \sqrt{1 - l_2 \, q - l_1^2 \, q^2/4}$$
$$\operatorname{Im} z_\infty \approx \sqrt{1 - l_2 \, q} \approx 1 - l_2 \, q/2 \ .$$

In the region about $z = j$, the distance between adjacent poles amounts to $q/2$ in the horizontal and vertical direction. It becomes obvious that approximately double the coefficient wordlength is needed to achieve the same pole density in the vicinity of $z = 1$ as we find it about the point $z = j$ with the original wordlength.

As an example, we consider the case of eight fractional bits for the coefficient representation, where q assumes the value $1/256 = 0.0039$. In the region about

$z = j$, the distance between adjacent poles amounts to $q/2 = 0.002$, wheras the smallest distance of a pole from the point $z = 1$ is $\sqrt{q} = 0.0625$. Fig. 8-20 shows the result of the corresponding example with $q = 1/16$.

The normal form

The normal-form structure according to Fig. 5-27 features an even distribution of the poles. This is due to the fact that the filter coefficients α and β are identical with the real and imaginary part of the complex-conjugate pole pair. A disadvantage of the normal form is the double number of required multiplications compared to the direct form. Based on the filter coefficients α and β, the transfer function can be expressed as

$$H(z) = \frac{1}{1 - 2\alpha\, z^{-1} + \left(\alpha^2 + \beta^2\right) z^{-2}} \; .$$

The stability region of the coefficient pair (α,β) is the interior of the unit circle. Both coefficients thus assume values in the range -1 to $+1$. If we plot all possible complex poles in the z-plane that lie within the unit circle and result from coefficients α and β with four fractional bits, we obtain the pattern according to Fig. 8-21. Compared to the direct form, we observe a higher pole density about $z = 1$ and $z = -1$ and a lower one about $z = \pm j$. For wideband filters whose poles are spread over a large range of frequencies, the normal form is a good compromise.

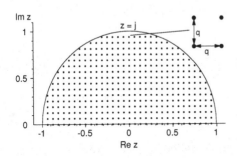

Fig. 8-21
Pattern of possible pole locations for normal-form filters (coefficients with four fractional bits, $q = 1/16$)

Second-order wave digital filter

Second-order wave digital filter sections are of interest due to their excellent stability properties. In Sect. 8.7 we will show that, under certain circumstances, unstable behaviour such as limit cycles can be avoided completely. At the beginning of this section, we also demonstrated that variations in the coefficients do not influence the maximum gain of the pole. The coefficients γ_1 and γ_2 in the block diagram (Fig. 5-61) are responsible for the location of the poles. The stability range of the coefficient paires (γ_1,γ_2) is again the interior of a triangle, as depicted in Fig. 8-22.

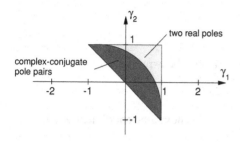

Fig. 8-22
Stability region of the coefficients γ_1 and γ_2 of second-order wave digital filter sections

Based on the coefficients γ_1 and γ_2, the transfer function of the wave digital filter section can be expressed as

$$H(z) = \frac{1}{1+(\gamma_1-\gamma_2)\,z^{-1}+(1-\gamma_1-\gamma_2)\,z^{-2}} \ . \tag{8.54}$$

By making use of (8.52) again, the poles can be calculated as

$$\mathrm{Re}\,z_\infty = -a_1/2 = (\gamma_1-\gamma_2)/2 \tag{8.55a}$$

$$\mathrm{Im}\,z_\infty = \sqrt{a_2 - a_1^2/4} = \sqrt{1-\gamma_1-\gamma_2-(\gamma_1-\gamma_2)^2/4} \ . \tag{8.55b}$$

According to Fig. 8-22, both coefficients γ_1 and γ_2 have a value range of -1 to $+1$. If we plot all possible complex poles in the z-plane that lie within the unit circle and result from coefficients γ_1 and γ_2 with four fractional bits, we obtain the pattern according to Fig. 8-23. As with the direct form, the possible pole locations are unevenly distributed over the unit circle. At low frequencies ($z = 1$) and high frequencies ($z = -1$), the pole density is rather low, and the poles lie on circular arcs. In the vicinity of $f_s/4$, the density is highest. The pole pattern is almost diamond shaped.

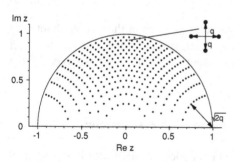

Fig. 8-23
Pattern of possible pole locations for second-order wave digital filter sections (coefficients with four fractional bits, $q = 1/16$)

According to (8.53), the distance of a pole from the point $z = 1$ can be calculated as

$$D_{z=1} = \sqrt{1+a_1+a_2} \ .$$

From (8.54), we can derive the following relations between the coefficients a_v and γ_v:

$$a_1 = \gamma_1 - \gamma_2$$
$$a_2 = 1 - \gamma_1 - \gamma_2 \;.$$

The radius of the circular arcs about $z = 1$ can thus be expressed as

$$D_{z=1} = \sqrt{1 + \gamma_1 - \gamma_2 + 1 - \gamma_1 - \gamma_2} = \sqrt{2 - 2\gamma_2} \;.$$

For $z_\infty = 1$ we have $\gamma_1 = -1$ and $\gamma_2 = 1$. About this point, the coefficient γ_2 takes on the following discrete values:

$$\gamma_2 = 1 - l_2\, q \;.$$

$$D_{z=1} = \sqrt{2 - 2\left(1 - l_2\, q\right)} = \sqrt{2 l_2\, q}$$

The smallest possible distance of a pole from the point $z = 1$ thus amounts to $\sqrt{(2q)}$.

For $z = j$ we have $\gamma_1 = 0$ und $\gamma_2 = 0$. About this point, the coefficients γ_1 und γ_2 assume the following discrete values:

$$\gamma_1 = l_1\, q$$
$$\gamma_2 = l_2\, q \;.$$

Substitution in (8.55) yields the following discrete pole locations in the vicinity of $z = j$.

$$\mathrm{Re}\, z_\infty = \left(\gamma_1 - \gamma_2\right)/2 = \left(l_2 - l_1\right) q/2 \tag{8.56a}$$

$$\mathrm{Im}\, z_\infty = \sqrt{1 - \gamma_1 - \gamma_2 - \left(\gamma_1 - \gamma_2\right)^2/4} = \sqrt{1 - \left(l_1 + l_2\right) q - \left(l_1 - l_2\right)^2 q^2/4}$$

$$\mathrm{Im}\, z_\infty \approx \sqrt{1 - \left(l_1 + l_2\right) q} \approx 1 - \left(l_1 + l_2\right) q/2 \tag{8.56b}$$

The minimum distance between possible pole locations in the region about $z = j$ amounts to $\sqrt{2} \times q/2$, which is a factor of $\sqrt{2}$ larger than in the case of the direct form. Also the radius of the circular arcs about $z = 1$ and $z = -1$ is larger, by the same factor.

8.6.3 Increased Pole Density at Low and High Frequencies

In Sect. 8.5.2, we already pointed to the importance of filters with low-frequency poles for the processing of audio signals. But we are still missing a filter structure with increased pole density about $z = 1$. The elementary structures discussed above do not get us any further. Therefore, we have to construct specialised structures with the desired property.

A primary drawback of the direct form with respect to low-frequency poles is the fact that the radius of the pole locations about the point $z_\infty = 1$ is proportional to the square root of the coefficient quantisation q. Due to this dependence, the pole density is rather low in the vicinity of $z_\infty = 1$. We will overcome this situation by choosing a new pair of filter coefficients e_1/e_2. We try the following approach:

- The coefficient e_1 determines the radius of the circular arcs about the point $z_\infty = 1$ on which the poles lie. Using (8.53), e_1 can be expressed as

$$e_1 = \sqrt{1 + a_1 + a_2} \; . \tag{8.57}$$

- The coefficient e_2 is chosen equal to a_2 and thus determines the radius of the pole location about the origin.

The filter coefficients e_1 and e_2 and the coefficients of the transfer function a_1 and a_2 are related as follows:

$$e_1 = \sqrt{1 + a_1 + a_2} \qquad\qquad a_1 = e_1^2 - 1 - e_2$$
$$e_2 = a_2 \qquad\qquad\qquad\qquad a_2 = e_2 \; .$$

By substitution of these relationships, the transfer function of the pole can be rewritten with the new coefficients e_1 and e_2 as parameters.

$$H(z) = \frac{1}{1 + a_1 z^{-1} + a_2 z^{-2}} = \frac{1}{1 + \left(e_1^2 - 1 - e_2\right) z^{-1} + e_2 z^{-2}} \tag{8.58}$$

Using (8.52), the real and imaginary parts of the pole location can also be expressed in terms of e_1 and e_2.

$$\mathrm{Re}\, z_\infty = -a_1/2 = \left(1 + e_2 - e_1^2\right)/2$$
$$\mathrm{Im}\, z_\infty = \sqrt{a_2 - a_1^2/4} = \sqrt{e2 - \left(1 + e_2 - e_1^2\right)^2/4} \tag{8.59}$$

Figure 8-24 depicts the range of the new coefficients leading to stable poles.

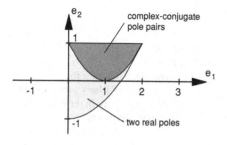

Fig. 8-24
Stability region of the coefficients e_1 and e_2 of second-order filter sections with increased pole density about $z = 1$

If we plot all possible complex poles in the z-plane that lie within the unit circle and result from coefficients e_1 and e_2 with four fractional bits, we obtain the pattern according to Fig. 8-25. The pole pattern shows the desired property. The highest pole density appears in the vicinity of $z = 1$. The poles lie on circular arcs about $z = 1$ with equidistant radii, which are determined by the coefficient e_1.

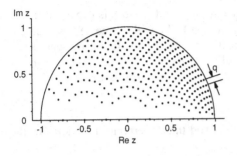

Fig. 8-25
Pattern of possible pole locations for
second-order filter sections with
increased pole density about $z = 1$
(coefficients with four fractional bits,
$q = 1/16$)

Figure 8-26 shows the prototype of a block diagram which is especially suited
to the implementation of filters with low coefficient sensitivity in the case of high
and low pole frequencies. By an appropriate choice of the filter coefficients s, t, u
and v, we can realise a variety of transfer functions from the literature which
optimise the coefficient wordlength requirements for the placement of poles near
$z = 1$ and $z = -1$. The general transfer function of this structure is given by

$$H(z) = \frac{d_0 - \left(2d_0 + d_1 + d_2\right) z^{-1} + \left(d_0 + d_1\right) z^{-2}}{1 - \left(2 - vs - uvt\right) z^{-1} + \left(1 - vs\right) z^{-2}} . \tag{8.60}$$

For the realisation of the transfer function (8.58), the filter coefficients take on the
following values:

$$s = 1 - e_2 \qquad\qquad u = e_1 \qquad\qquad t = e_1 \qquad\qquad v = 1 .$$

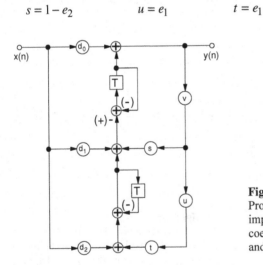

Fig. 8-26
Prototype block diagram for the
implementation of filters with reduced
coefficient sensitivity in the case of low
and high pole frequencies

In the following, we introduce further filter structures with low coefficient
sensitivity for low and high frequency poles. These feature in general a higher
concentration of poles near $z = 1$ and $z = -1$ than we could achieve with the above
approach. For all these structures, we show the valid range of filter coefficients
and the respective pole distribution pattern in the z-plane.

Avenhaus [1]

$$H(z) = \frac{d_0 - (2d_0 + d_1 + d_2)\, z^{-1} + (d_0 + d_1)\, z^{-2}}{1 - (2 - c_1)\, z^{-1} + (1 - c_1 + c_1\, c_2)\, z^{-2}}$$

$$c_1 = a_1 + 2 \qquad\qquad d_0 = b_0$$

$$c_2 = \frac{1 + a_1 + a_2}{2 + a_1} \qquad\begin{aligned} d_1 &= b_2 - b_0 \\ d_2 &= -b_0 - b_1 - b_2 \end{aligned}$$

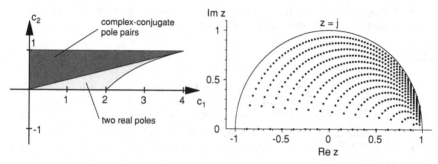

Fig. 8-27 Range of stable coefficients c_1 and c_2 and pole distribution pattern according to Avenhaus (four fractional bits, $q = 1/16$)

The coefficients in block diagram Fig. 8-26 assume the values

$$s = 1 - c_2 \qquad\qquad u = 1$$

$$t = c_2 \qquad\qquad\quad v = c_1 \ .$$

In [1], a block diagram is given that is especially tailored to this special case, saving one multiplication.

An increased pole density at high frequencies about $z = -1$ can be achieved if we choose the signs in parenthesis in Fig. 8-26. With these signs, the pole pattern in Fig. 8-27 is mirrored about the imaginary axis (Fig. 8-28). The same is valid by analogy for the two structures introduced in the following.

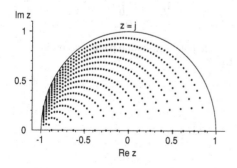

Fig. 8-28
Increased pole density about $z = -1$

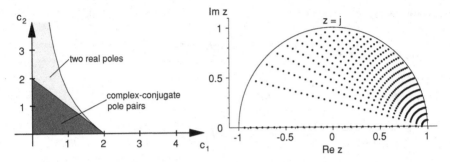

Fig. 8-29 Range of stable coefficients c_1 and c_2 and pole distribution pattern according to Kingsbury (four fractional bits, $q = 1/16$)

Kingsbury [42]

$$H(z) = \frac{d_0 - \left(2d_0 + d_1 + d_2\right) z^{-1} + \left(d_0 + d_1\right) z^{-2}}{1 - \left(2 - c_1 c_2 - c_1^2\right) z^{-1} + \left(1 - c_1 c_2\right) z^{-2}}$$

$$c_1 = \sqrt{1 + a_1 + a_2}$$
$$c_2 = \left(1 - a_2\right)/c_1$$

$$d_0 = b_0$$
$$d_1 = b_2 - b_0$$
$$d_2 = -b_0 - b_1 - b_2$$

The coefficients in block diagram Fig. 8-26 assume the values

$$s = c_2 \qquad\qquad u = 1$$
$$t = c_1 \qquad\qquad v = c_1 \ .$$

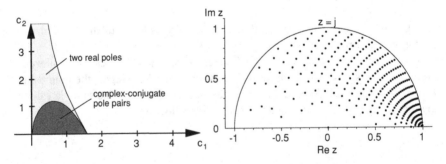

Fig. 8-30 Range of stable coefficients c_1 and c_2 and pole distribution pattern according to Zölzer (four fractional bits, $q = 1/16$)

Zölzer [77]

$$H(z) = \frac{d_0 - \left(2d_0 + d_1 + d_2\right) z^{-1} + \left(d_0 + d_1\right) z^{-2}}{1 - \left(2 - c_1 c_2 - c_1^3\right) z^{-1} + \left(1 - c_1 c_2\right) z^{-2}}$$

$$c_1 = \sqrt[3]{1 + a_1 + a_2}$$
$$c_2 = (1 - a_2)/c_1$$

$$d_0 = b_0$$
$$d_1 = b_2 - b_0$$
$$d_2 = -b_0 - b_1 - b_2$$

The coefficients in block diagram Fig. 8-26 assume the values

$s = c_2$ $u = c_1$

$t = c_1$ $v = c_1$.

8.6.4 Further Remarks Concerning Coefficient Sensitivity

There is a relationship between coefficient sensitivity and sampling frequency, which is not so obvious at first glance. Let us imagine the pole and zero distribution of an arbitrary filter. If we increase the sampling frequency without changing the frequency response, poles and zeros migrate towards $z = 1$ and concentrate on an ever-decreasing area. In order to still provide a sufficient pole density in this situation, the wordlength of the filter parameters has to be increased accordingly. The choice of a filter structure with low coefficient sensitivity about $z = 1$ would keep the required wordlength within reasonable limits.

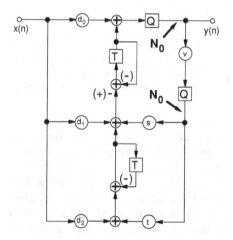

Fig. 8-31
Noise sources in the structure according to Kingsbury

A further interesting relationship exists between coefficient sensitivity and quantisation noise. It is striking that the noise of a normal-form filter is independent of the angle of the pole, and the poles are evenly distributed over the unit circle. Direct-form and wave digital filter structure exhibit the largest coefficient sensitivity at high ($z = -1$) and low ($z = 1$) pole frequencies, which are the pole locations that create the largest quantisation noise. In order to confirm a general rule, we consider in the following a filter with low coefficient sensitivity at low frequencies, such as the Kingsbury structure. Figure 8-31 shows a block diagram with the relevant noise sources. The quantisation after the multiplication by v is needed to return back to the original wordlength before the signal in this path is multiplied by the coefficients s and t. The transfer functions between the noise sources and the output of the filter are

$$H_1(z) = \frac{\left(1 - z^{-1}\right)^2}{1 + a_1 z^{-1} + a_2 z^{-2}}$$

$$H_2(z) = -(s + t)\, z^{-1}\, \frac{1 - s/(s + t)\, z^{-1}}{1 + a_1 z^{-1} + a_2 z^{-2}}\ .$$

Using the relations that we derived in Sect. 8.5.1, the generated quantisation noise power can be calculated as a function of the angle of the pole. Figure 8-32 shows the result together with the curves for the noise power of the direct and normal forms. It proves true that the smallest quantisation noise is created for those pole locations where the filter exhibits the lowest coefficient sensitivity.

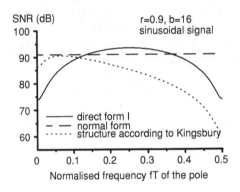

Fig. 8-32
Comparison of the quantisation noise of various filter structures (pole radius $r = 0.9$, wordlength $b = 16$, sinusoidal signal)

The coincidence of low quantisation noise and low coefficient sensitivity can be easily understood if the process of quantisation is interpreted in a somewhat different way. The effect of rounding or truncation could also be achieved by modifying the filter coefficients slightly in such a way that the rounded or truncated values appear as a result of the coefficient multiplications. Noiselike variations are superimposed onto the nominal values of the coefficients, which finally results in a shift-variable system. Gain, Q-factor and frequency of the poles and zeros change irregularly within certain limits and thus modulate the quantisation noise onto the signal. The lower the coefficient sensitivity is, the slighter are the variations of the transfer function and the smaller is the resulting quantisation noise.

8.6.5 A Concluding Example

The decomposition of transfer functions into second-order sections is a proven measure to reduce coefficient sensitivity. The mathematical background of this fact has been given in Sect. 8.6.1. A further possibility to reduce coefficient sensitivity is the transition to wave digital filter structures (refer to Sect. 5.7.4), whose properties are based on the choice of analog lossless networks inserted between resistive terminations as prototypes for the digital filter design. The following example is to show what benefits there are in practice from choosing a cascade or wave digital filter structure. Figure 8-33a shows the magnitude response of a third-order Chebyshev filter with a ripple of 1.256 dB and a cutoff frequency at $f_c T = 0.2$. The transfer function of the filter is

$$H(z) = 0.0502 \; \frac{\left(1 + z^{-1}\right)^3}{1 - 1.3078 \; z^{-1} + 1.0762 \; z^{-2} - 0.3667 \; z^{-3}} \; .$$

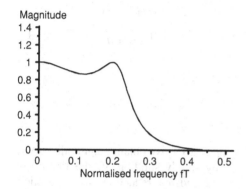

Magnitude

Fig. 8-33a
Third-order Chebyshev filter (1.256 dB ripple, $f_c T = 0.2$), ideal curve

In the case of the direct form, the filter coefficients are equal to the coefficients of the transfer function.

$$H(z) = V \; \frac{\left(1 + z^{-1}\right)^3}{1 + a_1 z^{-1} + a_2 z^{-2} + a_3 z^{-3}}$$

with
$$V = 0.0502$$
$$a_1 = -1.3078$$
$$a_2 = 1.0762$$
$$a_3 = -0.3667$$

We will check in the following how the magnitude curve is influenced by variations of the coefficients. Figure 8-33b illustrates the result for the case that each of the three decisive filter coefficients a_1, a_2 and a_3 assumes its nominal as well as a 3% higher or lower value, which leads to 27 different curves in total. We observe considerable variations in the curves and large deviations from the ideal Chebyshev (equiripple) characteristic.

We decompose now the third-order direct-form structure into the series arrangement of a first- and a second-order filter block, resulting in the following expression for the transfer function:

Magnitude

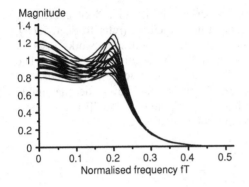

Fig. 8-33b
Third-order Chebyshev filter (1.256 dB ripple, $f_cT = 0.2$), direct form, 3% coefficient variation

$$H(z) = V\frac{1+2\,z^{-1}+z^{-2}}{1+d\,z^{-1}+e\,z^{-2}}\frac{1+z^{-1}}{1+f\,z^{-1}} = \frac{V\left(1+z^{-1}\right)^3}{1+(d+f)\,z^{-1}+(e+df)\,z^{-2}+ef\,z^{-3}}$$

with $V = 0.0502$

$d = -0.7504$

$e = 0.6579$ (8.61)

$f = -0.5575$

If the filter coefficients d, e and f vary by up to $\pm 3\%$, we obtain magnitude curves as shown in Fig. 8-33c. It turns out that, by simply splitting off the first-order section from the third-order filter, we achieve a visible reduction of the variation range by more than a factor of two. The coefficient V, which determines the overall gain of the filter, was not included in these considerations since it has no direct influence on the shape of the magnitude curve. It must not be forgotten, however, that the gain can only be adjusted with an accuracy according to the wordlength of the coefficient V.

Magnitude

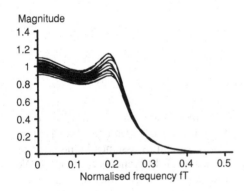

Fig. 8-33c
Third-order Chebyshev filter (1.256 dB ripple, $f_cT = 0.2$), cascade structure, 3% coefficient variation

In the case of filter structures with low coefficient sensitivity, the coefficients of the transfer function are calculated as products and sums of the filter coefficients, as (8.61) shows for the cascade structure. A more extreme behaviour in this respect is shown by the wave digital filter. The relations between the coefficients of the transfer function a_1, a_2 and a_3 and the filter coefficients γ_1, γ_2 and γ_3 according to the block diagram Fig. 5-55 is far more complex.

$$H(z) = \frac{\gamma_1\,\gamma_3\left(1+z^{-1}\right)^3}{1+a_1 z^{-1}+a_2 z^{-2}+a_3{}^{z-3}}$$

$$a_1 = 2\gamma_1 + 2\gamma_2 + 4\gamma_3 - \gamma_1\gamma_3 - \gamma_2\gamma_3 - 3$$

$$a_2 = 6\gamma_1\gamma_3 + 6\gamma_2\gamma_3 + 4\gamma_1\gamma_2 - 4\gamma_1\gamma_2\gamma_3 - 4\gamma_1 - 4\gamma_2 - 4\gamma_3 + 3$$

$$a_3 = 4\gamma_1\gamma_2\gamma_3 - 4\gamma_1\gamma_2 - \gamma_2\gamma_3 - \gamma_1\gamma_3 + 2\gamma_1 + 2\gamma_2 - 1$$

with
$$\gamma_1 = 0.2213$$
$$\gamma_2 = 0.2213$$
$$\gamma_3 = 0.2269$$

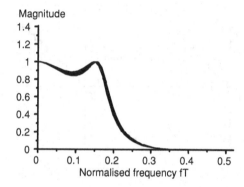

Fig. 8-33d
Third-order Chebyshev filter (1.256 dB ripple, $f_c T = 0.2$), wave digital filter, 3% coefficient variation

Variations of the coefficients by up to ±3% yield the magnitude curves in Fig. 8-33d. We find here nine different curves only, since the coefficients γ_1 and γ_2 assume identical values. The maximum gain is strictly limited to unity. The low variation range of the magnitude response proves the superiority of the wave digital filter structure with respect to coefficient sensitivity.

8.7 Limit Cycles

In the previous sections, we already identified roundoff noise as one of the deviations from the behaviour of an ideal discrete-time system which are due to the finite precision of signal representation. In recursive filters we can observe another phenomenon, which manifests itself by constant or periodical output signals which remain still present even when the input signal vanishes. These oscillations are called limit cycles and have their origin in nonlinearities in the feedback branches of recursive filter structures. Two different nonlinear mechanisms can be distinguished in this context:

- Rounding or truncation that is required to reduce the number of digits to the original register length after multiplications or after additions in the case of floating-point arithmetic.
- Overflow which occurs if the results of arithmetic operations exceed the representable number range.

Overflow cycles are especially unpleasant since the output signal may oscillate between the maximum amplitude limits. Quantisation limit cycles normally exhibit amplitudes of few quantisation steps only. But they may assume considerable values if the poles of the filter are located in close vicinity to the unit circle. The named unstable behaviour is mainly influenced by the choice of the filter structure and by the quantisation and overflow characteristics of the implemented arithmetic. As a consequence, the stable coefficient regions of the various filter structures as depicted in Fig. 8-38 are in most cases reduced compared to the ideal linear case.

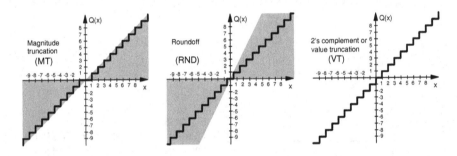

Fig. 8-34 Quantisation characteristics

8.7.1 Nonlinearities in Actual Implementations

The most common quantisation characteristics are:
- Magnitude truncation (MT)
- Two's complement truncation, also called value truncation (VT)
- Rounding (RND)

Fig. 8-34 shows graphical illustrations of the nonlinear characteristics of the three named types of quantisers. For all types of number representations, rounding is performed by substituting the nearest possible number that can be represented by the reduced number of bits. If binary digits are truncated in case of the two's complement format, the value is always rounded down irrespectively of the sign of the number (value truncation). In the case of the sign-magnitude representation, the magnitude of the number is always rounded down (magnitude truncation), which may be in effect a rounding up or a rounding down depending on the sign.

For rounding and magnitude truncation we can define sectors, the shaded regions in Fig. 8-34, that are bounded by two straight lines and that fully enclose the steplike quantisation curves. The slopes of the bounding lines amount to 0 and k_q respectively, where k_q assumes the value 1 for magnitude truncation and 2 for rounding. The quantisation characteristics $Q(x)$ thus satisfy the following sector conditions:

$$0 \le Q(x)/x \le k_q \qquad \text{with} \qquad \begin{cases} k_q = 1 & \text{for magnitude truncation} \\ k_q = 2 & \text{for rounding} \end{cases} \qquad (8.62)$$

The sector condition (8.62) is a simple means to characterise the quantiser. The smaller the aperture angle of the sector is, the less the nonlinearity deviates from the ideal linear behaviour. For the truncation in the two's complement number format, we cannot define a corresponding sector that is bounded by straight lines with finite slope. This fact makes the further mathematical treatment of this format more difficult.

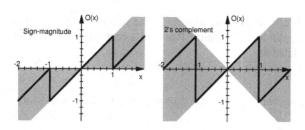

Fig. 8-35a
Natural overflow characteristics of two's complement and sign-magnitude arithmetic

Overflow nonlinearities can be treated in the same way. They may also be characterised by an overflow characteristic $O(x)$ and a certain sector condition. We start with considering the natural overflow characteristics of the sign-magnitude and the two's complement representation. Overflow in the case of two's complement arithmetic leads to a behaviour where the result of a summation jumps over the whole representable number range to the value with the opposite sign (Fig. 8-35a). For the sign-magnitude arithmetic, the magnitude jumps to zero in the case of an overflow. The sector, which comprises the nonlinear characteristic, is bounded by straight lines with the slopes 1 and k_o, where k_o assumes the value -1 for the two's complement and 0 for the sign-magnitude representation.

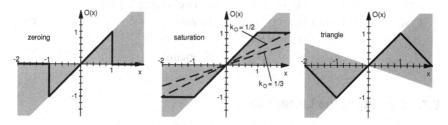

Fig. 8-35b Special overflow characteristics: zeroing, saturation, triangle

Special measures taken in the adders have the goal of narrowing down the sectors of the nonlinear characteristics in order to minimise the deviation from the ideal linear case as far as possible. Common measures are zeroing, saturation and triangle characteristic as depicted in Fig. 8-35b. All three methods have advantages in the context of two's complement arithmetic since the sector is clearly narrowed down. The slope of the bounding line amounts to $k_o = 0$ for zeroing and saturation; for the triangle characteristic we have $k_o = -1/3$. The behaviour of the saturation and triangle characteristics can be further improved by making special assumptions concerning the degree of overflow occurring in concrete implementations.

If we guarantee in the case of the triangle characteristic that the sum is always less than 2, the slope k_o will always be positive. This applies for instance to normal-form filters. The signal amplitude is limited to unity. Also the filter coefficients α and β have magnitudes of less than 1 for stable filters. In case of a vanishing input signal, only two signal values, weighted by these coefficients, are added up in the normal-form structure resulting in a sum whose magnitude will always be less than 2.

In case of saturation, k_o is always positive as long as the sum is finite. Denoting the maximum possible magnitude of the sum that can occur in a given filter structure as S_{max}, k_o is calculated as

$$k_o = 1/S_{max} \ .$$

Assuming that the input signal vanishes, we can thus estimate for the normal-form structure

$$k_o = 1/\left(|\alpha| + |\beta|\right) > 1/2 \ .$$

As for the direct form, the magnitude of the coefficient a_1 is limited to 2, the magnitude of a_2 to 1. If the input signal vanishes, we can thus give a lower limit for k_o:

$$k_o = 1/\left(|a_1| + |a_2|\right) > 1/3 \ .$$

Equation (6.83) summarises all the overflow sector conditions that we have discussed above.

$$k_o \leq O(x)/x \leq 1 \ \text{ with } \begin{cases} k_o = 0 & \text{for sign - magnitude} \\ k_o = -1 & \text{for two' s complement} \\ k_o = 0 & \text{for zeroing} \\ k_o = 1/S_{max} & \text{for saturation} \\ k_o = 2/S_{max} - 1 & \text{for triangle} \end{cases} \tag{8.63}$$

8.7.2 Stability of the Linear Filter

Up to now, we have considered stability from the point of view that a bounded input signal should always result in a bounded output signal. This so-called BIBO stability requires in the time domain that the unit-sample response is absolutely summable (refer to Sect. 3.4). We derived a related condition in Sect. 5.3.4 for the frequency domain which states that all poles of the system must lie within the unit circle. For the discussion of limit cycles we need a different definition of stability, however, since we investigate the behaviour of filters under the assumption of vanishing input signals.

In the state-space representation, a general second-order system is described by the following vector relationship (5.42):

$$z(n+1) = A\ z(n) + b\ x(n)$$

$$y(n) = c^T\ z(n) + d\ x(n)\ .$$

The vector $z(n)$ combines the state variables which correspond to the contents of the delay memories of the system. If the input signal $x(n)$ vanishes, the state variables of the respective next clock cycle $z(n+1)$ are simply calculated by multiplication of the current state vector $z(n)$ by the system matrix A.

$$z(n+1) = A\ z(n) \tag{8.64}$$

Since the output signal $y(n)$ is obtained by linear combination of the state variables (multiplication of the state vector $z(n)$ by c^T), it is sufficient to prove stability for $z(n)$. In this context it is of interest to investigate the behaviour of the system if a starting vector z_0 is continuously multiplied by the system matrix A. The curve that the arrowhead of the state vectors passes in the state space by these multiplications is called a trajectory. The following considerations are restricted to second-order filter blocks but can be easily extended to higher-order systems. The state vector consists of the two components $z_1(n)$ and $z_2(n)$, which fully describe the state of the system in the time domain. Fig. 8-36 shows four examples of possible trajectories in the z_1/z_2-plane.

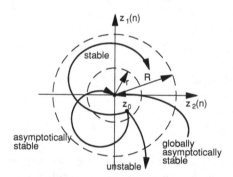

Fig. 8-36
Stability in the sense of Lyapunov

The stability criterion according to Lyapunov [17, 43] considers the behaviour of systems in the neighbourhood of equilibrium points which can be characterised in the linear case by the relation

$$z(n+1) = A\ z(n) = z(n)\ . \tag{8.65}$$

The origin $z = 0$ that should be reached with a vanishing input signal $x(n)$ is such an equilibrium point, which can be easily verified using (8.65). Stability can be defined in this context as follows:

For a neighbourhood of the origin with the radius r there exists another neighbourhood with radius R such that the following applies: If the starting point of a trajectory $z_0 = z(0)$ lies within the neighbourhood with radius r, then the entire trajectory $z(n)$ as $n \rightarrow \infty$ remains within the neighbourhood with radius R (Fig. 8-36).

A system is asymptotically stable if it is stable in the above sense and the trajectory converges to the origin as $n \rightarrow \infty$.

A system is globally asymptotically stable if it is asymptotically stable and z_0 can be chosen arbitrarily in the state space.

The eigenvalues λ_i of the matrix A, which are identical to the poles of the transfer function, have a decisive influence on the stability behaviour of the system. If z_0 is the state of the system for $n = 0$, we obtain the state after N clock cycles by means of N times multiplying z_0 by the Matrix A.

$$z(N) = A^N z_0 \tag{8.66}$$

Each nonsingular matrix can be written in the form

$$A = P^{-1} L P \ ,$$

where L is a diagonal matrix with the eigenvalues λ_i of A appearing as the diagonal elements. The matrix P contains the eigenvectors of A as the columns. Equation (8.66) can thus be expressed as follows:

$$z(N) = \left(P^{-1} L P\right)\left(P^{-1} L P\right)\left(P^{-1} L P\right)...\left(P^{-1} L P\right) z_0$$

$$z(N) = P^{-1} L L L ... L P z_0$$

$$z(N) = P^{-1} L^N P z_0 \ .$$

For second-order systems, L^N can be written as

$$L^N = \begin{pmatrix} \lambda_1^N & 0 \\ 0 & \lambda_2^N \end{pmatrix} \ .$$

If the magnitude of both eigenvalues λ_1 and λ_2 is less than unity, $z(n)$ approaches the origin with increasing N. The system is thus asymptotically stable. Matrices whose eigenvalues have magnitudes of less than unity are therefore called stable matrices. If single eigenvalues lie on the unit circle, $z(n)$ remains bounded. The system is stable. Multiple eigenvalues on the unit circle and eigenvalues with a magnitude larger than unity lead to unstable systems.

Stability as discussed above is closely related to the existence of a Lyapunov function for the considered system. A Lyapunov function v is a scalar function of the state vector $z(n)$ or of the two components $z_1(n)$ and $z_2(n)$ respectively. In a certain neighbourhood of the origin, this function has the following properties:

A. v is continuous.
B. $v(0) = 0$.
C. v is positive outside the origin.
D. v does not increase, mathematically expressed as $v[z(n+1)] \leq v[z(n)]$.

If one succeeds in finding such a function, then the region in the neighbourhood of the origin is stable. For a given system, the fact that no Lyapunov function has been found yet does not allow us to draw any conclusions concerning the stability of the system. If the more stringent condition D' applies instead of D, then the origin is asymptotically stable.

D'. v decreases monotonically, mathematically expressed as $v[z(n+1)] < v[z(n)]$.

Condition D' guarantees that the trajectories $z(n)$ which originate in the neighbourhood of the origin will converge to the origin.

Lyapunow functions are an extension of the energy concept. It is evident that the equilibrium state of a physical system is stable if the energy decreases continuously in the neighbourhood of this equilibrium state. The Lyapunov theory allows us to draw conclusions concerning the stability of a system without any knowledge of the solutions of the system equations. The concept of Lyapunov functions will be illustrated by means of the example of a simple electrical resonant circuit (Fig. 8-37). The reader is presumably more familiar with energy in the context of electrical circuits than with the energy of number sequences processed by a mathematical algorithm.

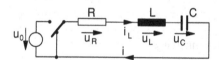

Fig. 8-37
Electrical resonant circuit

The energy stored in the circuit according to Fig. 8-37 depends on the state quantities i_L and u_C, which can be combined to the state vector $z(t)$.

$$z(t) = \begin{pmatrix} u_C(t) \\ i_L(t) \end{pmatrix}$$

We choose the total energy stored in the inductor and in the capacitor as a candidate for a Lyapunov function:

$$v(t) = 1/2\, C u_C^2(t) + 1/2\, L i_L^2(t) \; ,$$

or in quadratic vector form

$$v(t) = \begin{pmatrix} u_C(t) & i_L(t) \end{pmatrix} \begin{pmatrix} 1/2C & 0 \\ 0 & 1/2L \end{pmatrix} \begin{pmatrix} u_C(t) \\ i_L(t) \end{pmatrix} . \tag{8.67}$$

The conditions A to C for the existence of a Lyapunov function are surely met by (8.67). v is continuous, positive and vanishes in the origin. Condition D requires that v must not increase. In order to check the observance of this condition, we form the derivative of (8.67) with respect to the time.

$$
\begin{aligned}
\mathrm{d}v/\mathrm{d}t &= C\, u_C\, \mathrm{d}u_C/\mathrm{d}t + L\, i_L\, \mathrm{d}i_L/\mathrm{d}t \\
&= u_C\, i + i_L\, u_L = u_C\, i + u_L\, i = i\left(u_C + u_L\right) \\
&= i\left(-u_R\right) = -i^2\, R
\end{aligned}
$$

For $R \geq 0$, v does not increase, which proves (8.67) to be a Lyapunov function. The stability analysis by means of the chosen Lyapunov function correctly predicts that the system according to Fig. 8-37 is stable for $0 \leq R \leq \infty$. It is worth mentioning that we did not have to solve a differential equation to arrive at this result.

Also for linear discrete-time systems, quadratic forms are a promising approach for the choice of Lyapunov functions:

$$
v(n) = z(n)^{\mathrm{T}}\, V\, z(n) \geq 0 \ . \tag{8.68}
$$

If the matrix V is positive definite, $v(n)$ will be positive for any arbitrary state vector $z(n) \neq 0$. A general 2x2 matrix

$$
\begin{pmatrix} A & C \\ D & B \end{pmatrix}
$$

possesses this property if

$$
A > 0, \qquad B > 0 \qquad \text{and} \qquad 4AB > \left(C + D\right)^2 \ .
$$

We still have to make sure that $v(n)$ does not increase:

$$
\begin{aligned}
& v(n+1) \leq v(n) \\
& z(n+1)^{\mathrm{T}}\, V\, z(n+1) \leq z(n)^{\mathrm{T}}\, V\, z(n) \\
& z(n)^{\mathrm{T}}\, A^{\mathrm{T}}\, V\, A\, z(n) \leq z(n)^{\mathrm{T}}\, V\, z(n) \\
& z(n)^{\mathrm{T}}\, V\, z(n) - z(n)^{\mathrm{T}}\, A^{\mathrm{T}}\, V\, A\, z(n) \geq 0 \\
& z(n)^{\mathrm{T}}\left(V - A^{\mathrm{T}}\, V\, A\right) z(n) \geq 0 \ .
\end{aligned} \tag{8.69}
$$

Relation (8.68) is thus a Lyapunov function if both matrices V and $V - A^{\mathrm{T}}\, V\, A$ are positive definite. From (8.68) and (8.69) we can derive a number of relations that allow us to check whether a matrix V leads to a Lyapunov function, if the system matrix A is given. With a given matrix V, conversely, we can determine the ranges of the coefficients of the system matrix A in which stability is guaranteed. For the most important second-order structures (direct-form, normal-form, wave digital filter), appropriate V matrices are given in [10].

$$
V_{\text{direct}} = \begin{pmatrix} 1 + a_2 & a_1 \\ a_1 & 1 + a_2 \end{pmatrix} \qquad\qquad A_{\text{direct}} = \begin{pmatrix} -a_1 & -a_2 \\ 1 & 0 \end{pmatrix}
$$

$$V_{\text{normal}} = \begin{pmatrix} 1 & 0 \\ 0 & 1 \end{pmatrix} \qquad\qquad A_{\text{normal}} = \begin{pmatrix} \alpha & -\beta \\ \beta & \alpha \end{pmatrix}$$

$$V_{\text{WDF}} = \begin{pmatrix} 1-\gamma_2 & 0 \\ 0 & 1-\gamma_1 \end{pmatrix} \qquad A_{\text{WDF}} = \begin{pmatrix} -\gamma_1 & 1-\gamma_1 \\ \gamma_2-1 & \gamma_2 \end{pmatrix}$$

With these V matrices, (8.68) is always a Lyapunov function if the coefficients of the system matrix A lie in the stable regions as shown in Fig. 8-38.

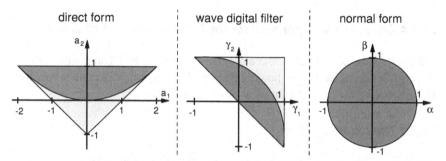

Fig. 8-38 Regions of stable coefficient pairs as derived in Chap. 5 and 8

8.7.3 Stability with Quantisation and Overflow Correction

Lyapunow functions that are based on the quadratic form can also be constructed for nonlinear systems. However, these are usually too pessimistic with regards to the predicted stable range of filter coefficients. In the following we introduce two algorithms which allow the investigation of stability in the nonlinear case: the "constructive algorithm" by Brayton and Tong [7, 8] and an "exhaustive search algorithm" in the state space [3, 59].

The constructive algorithm is based on the search for an appropriate Lyapunov function. Global asymptotical stability can be proved for those ranges of the filter coefficients for which a Lyapunov function can be found. For the remaining ranges of the coefficients we cannot make any statement, as pointed out in Sect. 8.7.2. A prerequisite for applying this method is the possibility of specifying sectors which comprise the respective quantisation and overflow characteristics as depicted in Fig. 8-34 and Fig. 8-35. The case of two's complement truncation can therefore not be treated by this method. The discussion of the constructive algorithm will give the reader a good insight into the dynamic behaviour of second-order filter blocks.

The exhaustive search algorithm is only suitable for the investigation of quantisation limit cycles. The case of two's complement truncation is covered by this method. Unlike the constructive algorithm which leaves some coefficient ranges undefined with respect to stability, the exhaustive search method provides clear results concerning stable and unstable regions in the coefficient plane. In the context of the introduction of this algorithm, we will also present formulas for the estimation of the amplitude of limit cycles.

The two aforementioned algorithms are presented in this chapter since they are universally valid in a certain sense. By way of contrast, numerous publications on this topic such as [2, 5, 6, 13, 14, 15, 18, 35, 36, 46] only consider partial aspects such as special quantisation or overflow characteristics, number representations or filter structures. They generally do not provide any new information. In most cases, the results of theses papers are more conservative with respect to stable ranges of filter coefficients than those of the constructive or search algorithm.

8.7.4 A Constructive Approach for the Determination of Stability

In the following, we introduce the algorithm by Brayton and Tong, which allows the construction of Lyapunov functions for nonlinear systems. We start, however, with illustrating the principle for the case of linear systems.

8.7.4.1 The Linear System

The starting point of our considerations is the definition of stability in the sense of Lyapunov, as introduced in a previous section. We choose a diamond-shaped region W_0 in the state space x_1/x_2 according to Fig. 8-39 as the initial neighbourhood in which all trajectories start. Our goal is to determine the neighbourhood W^* of the origin which all trajectories that start in W_0 pass through. If the system is stable, this neighbourhood W^* must be finite. In the first step we had to multiply, in principle, each point (x_1/x_2) of W_0 by the system matrix A. Since W_0 possesses the property of being convex, only a few points out of the infinite number of points of W_0 actually have to be calculated.

A set K is convex if the following applies to any arbitrary pair of points a and b in K: If a and b are in K, then all points on the straight connecting line between a and b are also in K.

With this property, only the corner points or extremal points of the convex set have to be multiplied by the matrix A. If we connect in our case the four resulting points to a new quadrangle, we obtain the region into which all trajectories move from W_0 after the first clock cycle. The four new extremal points are again multiplied by A which yields the region into which all trajectories move after the second clock cycle. This procedure is repeated until certain termination criteria are fulfilled which identify the system as stable or unstable.

The system is stable if the nth quadrangle, calculated by n successive multiplications by the system matrix A, does not increase the "total area" of all quadrangles calculated up to the $(n-1)$th step. "Total area" in this context means the convex hull which can be imagined as the area formed by a rubber band spanning all these quadrangles as indicated by the dotted lines in Fig. 8-39. If the new calculated quadrangle (drawn in grey) lies completely within the convex hull over all previous quadrangles (drawn in black), the algorithm can be terminated. The system is stable. Fig. 8-39 illustrates this algorithm by means of a system, that turns out to be stable after the 5th step because the new grey quadrangle lies completely within the convex hull W_5. With W_5 we have also found the neighbourhood W^* which all trajectories that start in W_0. pass through.

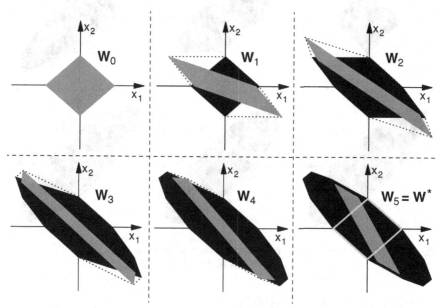

Fig. 8-39 Proof of the stability of the denominator polynomial
$N(z) = 1 + 1.67\, z^{-1} + 0.9\, z^{-2}$, realised in direct form

Figure 8-39 also shows that the convex hull W^* remains in touch with the initial region W_0, which is a general property of stable systems. This means illustratively that the W_k must not expand in all directions. If W_0 lies after the kth step completely within the convex hull W_k, the algorithm can be stopped with the result "unstable". Figure 8-40 shows an example in which instability is proved after the 5th iteration step. After the first step, W_0 touches the convex hull with the upper and lower corner only. After the 5th step, W_0 lies completely within W_5. The complex-conjugate pole pair, realised by $N(z)$, is unstable.

If the system is stable, the size of the region W^* is finite. W^* possesses an interesting property: All trajectories that start within W^* will completely run within or on the boundary of W^*. It is true that we have constructed W^* by trajectories which all have their origin in W_0, but each point that a trajectory passes can be used as starting point for a new trajectory which will follow the same curve as the original one that passed through this point. So all trajectories that start outside W_0 but within W^* will never leave W^*. The exact proof is given in [7].

With this property of W^*, we can define a Lyapunov function which can be later applied to nonlinear systems, too. As a Lyapunov function $v[z(n)]$ we choose a positive constant α that scales the region W^* in such a way that $z(n)$ lies on the boundary of αW^*, or otherwise expressed as

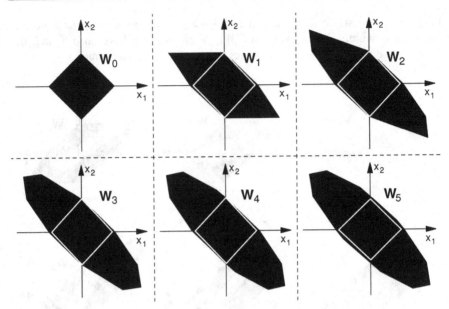

Fig. 8-40 Proof of the instability of the denominator polynomial
$N(z) = 1 + 1.67 z^{-1} + 1.02 z^{-2}$, realised in direct form

$$\nu[z(n)] = \left\{ \alpha \colon \alpha > 0, z(n) \in \partial \left(\alpha W^* \right) \right\},$$

∂ stands for the boundary of the region.

It is easy to realise that ν is a Lyapunov function. ν is always positive and vanishes for $z(n) = 0$. Since the boundary of W^* is a continuous curve, ν will also be continuous. As a state $z(n)$ on the boundary αW^* cannot migrate outwards, ν cannot increase. Hence all criteria of a Lyapunov function are met.

By means of the presented algorithm, we can only prove stability and not asymptotic stability as we do not further investigate whether the trajectories converge into the origin. If we want to exclude, however, any occurrence of limit or overflow cycles, the system has to be asymptotically stable. A trick helps in this case. All coefficients of the system matrix A are multiplied by a factor a little larger than 1 such as 1.00001. The system becomes a little more unstable, since the eigenvalues of the matrix increase accordingly. If the system matrix, modified in this way, still turns out to be stable, then the original matrix is asymptotically stable [8].

8.7.4.2 Stability of the Nonlinear System

The method to prove stability that we illustrated in the previous section for linear systems is also suitable to investigate limit and overflow cycles in real implementations. The constructive approach enables us to consider any combination of the following features:

- structure of the filter
- type of quantiser (rounding or truncation)
- placing of the quantisers (after each coefficient multiplication or after summation)
- inclusion of arbitrary overflow characteristics

According to (8.63), the behaviour of autonomous systems (systems in the situation that no signal is applied to the input), can be described in vector form by the state equation

$$z(n+1) = A\, z(n) \ .$$

In component form we can write for a second-order system

$$z_1(n+1) = a_{11}\, z_1(n) + a_{12}\, z_2(n)$$
$$z_2(n+1) = a_{21}\, z_1(n) + a_{22}\, z_2(n) \ .$$

Overflows, which result in overflow corrections $O(x)$ as illustrated in Fig. 8-35, can occur in practice after the summation of several signals.

$$z_1(n+1) = O\big(a_{11}\, z_1(n) + a_{12}\, z_2(n)\big)$$
$$z_2(n+1) = O\big(a_{21}\, z_1(n) + a_{22}\, z_2(n)\big)$$

For quantisation, characterised by $Q(x)$ according to Fig. 8-34, we have two possibilities: the quantisation may take place after each coefficient multiplication or only after summation of the signals from several branches:

$$z_1(n+1) = O\big[Q\big(a_{11}\, z_1(n) + a_{12}\, z_2(n)\big)\big]$$
$$z_2(n+1) = O\big[Q\big(a_{21}\, z_1(n) + a_{22}\, z_2(n)\big)\big]$$

or

$$z_1(n+1) = O\big[Q\big(a_{11}\, z_1(n)\big) + Q\big(a_{12}\, z_2(n)\big)\big]$$
$$z_2(n+1) = O\big[Q\big(a_{21}\, z_1(n)\big) + Q\big(a_{22}\, z_2(n)\big)\big] \ .$$

In the following, we will interpret the influence of the quantisation characteristic $Q(x)$ and overflow characteristic $O(x)$ in a different way: we assume that the filter coefficients a_{ij} are modified by $Q(x)$ and $O(x)$ in such a way that we obtain the respective quantised or overflow corrected values of $z_1(n+1)$ and $z_2(n+1)$. This means in particular that the coefficients a_{ij} vary from cycle to cycle and hence become a function of n.

$$z_1(n+1) = a'_{11}(n)\, z_1(n) + a'_{12}(n)\, z_2(n)$$
$$z_2(n+1) = a'_{21}(n)\, z_1(n) + a'_{22}(n)\, z_2(n)$$

The nonlinear system is thus converted into a shift-variant system with the state equation

$$z(n+1) = A'(n)\, z(n).$$

The range of values that the coefficients of $A'(n)$ may assume depends immediately on the parameters k_q as specified in (8.62) and k_o as specified in (8.63).

Examples

Quantisation after a coefficient multiplication

$$y = Q(ax) = \frac{Q(ax)}{ax} ax$$

The quotient $Q(ax)/ax$ assumes values between 0 and k_q according to (8.62).

$$y = \left(0 \ldots k_q\right) ax = a'x$$

In the case of magnitude truncation, the coefficient a' takes on values between 0 and a, for rounding between 0 and $2a$.

Overflow correction after an addition

$$y = O(ax_1 + bx_2) = \frac{O(ax_1 + bx_2)}{ax_1 + bx_2}(ax_1 + bx_2)$$

The quotient $O(ax_1 + bx_2)/(ax_1 + bx_2)$ assumes values between k_o and 1 according to (8.63).

$$y = \left(k_o \ldots 1\right)\left(ax_1 + bx_2\right) = a'x_1 + b'x_2$$

This leads to the following ranges for a' and b':

$$k_o a \le a' \le a \qquad \text{and} \qquad k_o b \le b' \le b \quad .$$

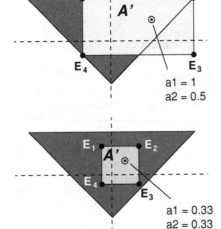

Fig. 8-41
Two examples for the scattering range of the coefficients of a direct form filter with triangular overflow characteristic and rounding after each coefficient multiplication

A stable filter may become unstable by these modifications of the coefficients a_{ij}, since eigenvalues of $A'(n)$ with magnitude larger than unity may occur. Figure 8-41 shows the example of a direct-form filter with two possible stable coefficient pairs. The applied overflow correction is triangle, the quantisation characteristic is rounding after each coefficient multiplication. In the one case, the scattering range of the coefficients extends beyond the stability triangle. In the other case, the coefficients remain completely within the stability triangle.

As for linear systems, the trajectories are simply calculated by continued multiplication by the system matrix A. In the case of nonlinear systems with rounding and possibly overflow, we obtain arbitrary sequences of matrices in the set A'. In our example, the set of matrices A' is defined as

$$A' = \begin{pmatrix} -a_1' & -a_2' \\ 1 & 0 \end{pmatrix},$$

where the pair a_1'/a_2' may assume any value within the grey regions in Fig. 8-41. In order to guarantee stability, each possible sequence of matrices in A' has to satisfy the stability criterion that we introduced in the previous section. This requires that, from some finite iteration step, the neighbourhood of the origin passed by the trajectories does not grow any further. If some sequence satisfies the instability criterion (complete enclosure of W_0), stability is no longer guaranteed.

Our problem seems to be insoluble at first glance since an infinite number of possible matrix sequences of A' would have to be checked. Brayton and Tong [7] showed, however, that stability can be proven in a finite number of steps:
1. For proof of stability, it is sufficient to consider the extreme matrices E_i only, which form the corner points of the scattering region of the set of matrices A' as illustrated in Fig. 8-41. The proof is given in [7] that if the set of extreme matrices of A' is stable, then also A' is stable.
2. We start the algorithm with the extreme matrix E_1. The initial neighbourhood W_0 represented by its extremal points is continuously multiplied by the matrix E_1 until one of the known termination criteria is fulfilled. If the result is "unstable", the whole procedure can be terminated with the result "unstable". If the result is "stable", the algorithm is continued. We denote the convex hull that we obtain as a result of the described first step of the algorithm as W_1. In the second step, we multiply W_1, represented by its extremal points, continuously by the extreme matrix E_2 until one of the termination criteria is fulfilled. If the result is "unstable", we terminate. If the result is "stable", we obtain a new convex hull that we denote as W_2. We continue with the matrix E_3 accordingly. In the following, all m extreme matrices E_i are applied cyclically (E_1, E_2, E_3, ..., E_m, E_1, E_2, ...) in the described manner. The algorithm is terminated with the result "unstable" if the criterion for instability is fulfilled for the first time. We terminate with the result "stable" if we do not observe any further growth of the convex hull after a complete pass through all m extreme matrices. This termination criterion can be expressed as $W_k = W_{k+m} = W^*$.

In the same way as we did for the linear system in Sect. 8.7.4.2, the finite neighbourhood W^* of the origin, which we finally obtain as a result of the algorithm, can be used to define a Lyapunov function for the nonlinear system.

In the context of the described algorithm, instability means that at least one sequence of the E_i leads to the result "unstable". Since these unstable sequences may not be run through in practical operation, the actual behaviour of the filter must be considered undetermined. If the result is "stable", global stability is guaranteed in any case.

The constructive algorithm allows us to determine the range of coefficients in a given implementation for which the filter is globally asymptotically stable. The involved nonlinearities are characterised by the parameters k_o and k_q and hence by the sectors which enclose the quantisation or overflow characteristics. The details of the mechanisms that lead to limit or overflow cycles are not considered with this algorithm. Since the sectors for quantisation and overflow overlap to a great extend, it is not possible to conclude whether limit or overflow cycles or both will occur in the filter under consideration when the algorithm says "unstable". If we investigate both effects separately in the following, then we do that under the assumption that in each case only one of the nonlinearities – quantisation or overflow – is implemented.

The constructive algorithm yields in most cases comparable or even larger coefficient ranges with guaranteed stability than have been previously seen in the literature. In a few cases, the results are more pessimistic, however, which is a consequence of the fact that the algorithm is only based on the sectors of the nonlinear characteristics, and the detailed mechanisms of quantisation and overflow are not considered [19, 20].

In the following, we test direct form, normal form and wave digital filters for stability. We investigate quantisation and overflow characteristics in various combinations. In the graphical representations, which show the results of the constructive algorithm, black areas in the coefficient plane indicate that at least one extreme matrix lies outside the stability region of the respective filter structure (refer to Fig. 8-38). In this case, we need not apply the algorithm, as stability is not guaranteed. There is in any case one sequence of matrices in A' that is unstable, that is, the sequence consisting of only this unstable extreme matrix. In the grey areas, potential instability is proven by the algorithm. It is interesting to note that the algorithm may come to the result "unstable" even if all extreme matrices are formed by coefficient pairs that lie within the respective stability regions. In the white areas, globally asymptotical stability is guaranteed.

Direct-form filters

With only one quantiser, placed behind the adder as shown in Fig. 8-42, the two possible extreme matrices take on the form

$$\begin{pmatrix} -\alpha_i a_1 & -\alpha_i a_2 \\ 1 & 0 \end{pmatrix} \quad i = 1,2$$

with $\alpha_1 = k_q$, $\alpha_2 = \min(0, k_o)$.

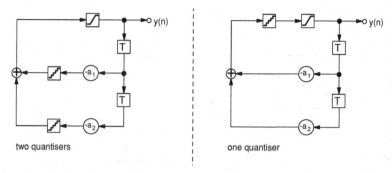

two quantisers   one quantiser

Fig. 8-42 Placement of nonlinearities in the direct form structure

If the sector of the overflow characteristic completely overlaps with the sector of the quantiser, the overflow correction does not contribute anything else to the variation range of the coefficients. Only for negative values of k_o, overflow may reduce the stability range compared to pure quantisation. The latter applies to two's complement overflow and, in certain coefficient ranges, to the triangle characteristic. If we intend to investigate overflow alone, the parameters α_1 and α_2 assume the following values:

$$\alpha_1 = 1, \quad \alpha_2 = k_o .$$

It is assumed in this case that adders, multipliers and memories operate as if they were analog.

If we quantise directly after each coefficient multiplication (Fig. 8-42), we obtain four extreme matrices:

$$\begin{pmatrix} -\alpha_i a_1 & -\beta_j a_2 \\ 1 & 0 \end{pmatrix} \quad i,j = 1,2$$

$$\text{with } \alpha_1 = \beta_1 = k_q, \quad \alpha_2 = \beta_2 = \min\left(0, k_o k_q\right) .$$

Figure 8-43 shows the stability regions for magnitude truncation and rounding with one or two quantisers respectively and without considering overflow. Fig. 8-44 depicts the results for overflow without quantisation. In addition to the stable regions of the coefficients a_1 and a_2, we also show the regions within the unit circle in the z-plane, where the corresponding poles (realised with the coefficients a_1 and a_2) are stable. For reasons of symmetry it is sufficient to show the upper half of the unit circle only. In some cases, the stability regions are considerably restricted compared to the linear filter, which is stable within the whole unit circle. What recommendation can be derived from the figures? Sign-magnitude arithmetic with truncation after summation, which corresponds to the case of one MT quantiser, yields the largest stability region. The two's complement overflow characteristic should be avoided in any case. Saturation is the best choice.

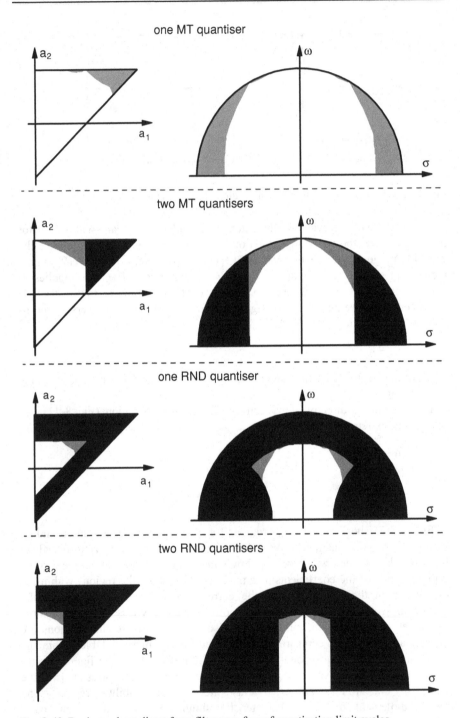

Fig. 8-43 Regions where direct form filters are free of quantisation limit cycles

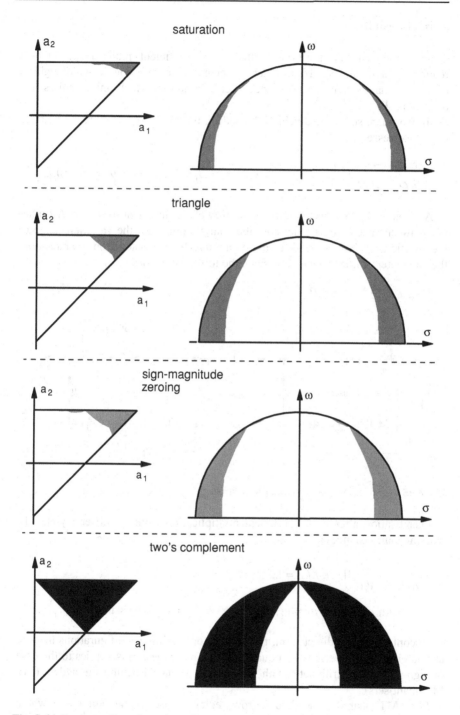

saturation

triangle

sign-magnitude
zeroing

two's complement

Fig. 8-44 Regions where direct form filters are free of overflow limit cycles

Normal-form filters

Quantisation may be performed after each coefficient multiplication which requires four quantisers. If an adder or accumulator with double wordlength is available, quantisation can take place after summation, which only requires two quantisers (Fig. 8-45).

With two quantisers placed behind the adders, the four possible extreme matrices can be expressed as

$$\begin{pmatrix} -\alpha_i\sigma & -\alpha_i\omega \\ \beta_j\omega & \beta_j\sigma \end{pmatrix} \quad i,j=1,2 \quad \text{with } \alpha_1=\beta_1=k_{\mathrm{q}}, \quad \alpha_2=\beta_2=\min(0,k_{\mathrm{o}}).$$

A negative k_{o} only occurs in the case of two's complement overflow. All other overflow characteristics do not contribute anything else to the variation range of the coefficients. If we intend to investigate overflow isolated from quantisation, the parameters α_1, α_2, β_1 and β_2 assume the following values:

$$\alpha_1=\beta_1=1, \quad \alpha_2=\beta_2=k_{\mathrm{o}}.$$

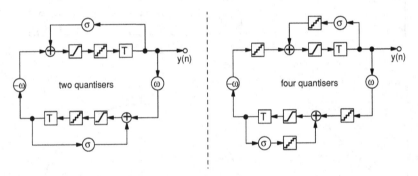

Fig. 8-45 Nonlinearities in the normal-form structure

Quantisation after each coefficient multiplication (four quantisers) yields 16 extreme matrices in total:

$$\begin{pmatrix} -\alpha_i\sigma & -\beta_j\omega \\ \mu_k\omega & \nu_l\sigma \end{pmatrix} \quad i,j,k,l=1,2$$

$$\text{with } \alpha_1=\beta_1=\mu_1=\nu_1=k_{\mathrm{q}}, \quad \alpha_2=\beta_2=\mu_2=\nu_2=\min(0,k_{\mathrm{o}}k_{\mathrm{q}}).$$

In contrary to the direct form, the normal form exhibits configurations that are absolutely free of overflow and quantisation limit cycles. This applies to the case of sign-magnitude arithmetic with the quantisers placed behind the adders (two MT quantisers).

Four MT quantisers lead to narrow regions close to the unit circle where instabilities my occur for saturation, zeroing and triangle overflow characteristics

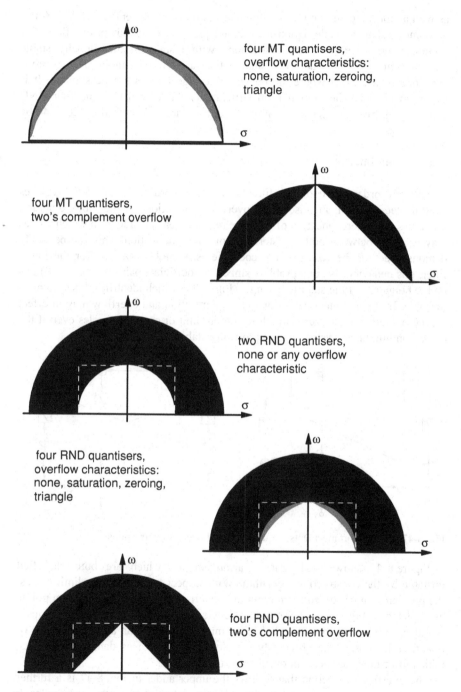

Fig. 8-46 Regions where normal form filters are free of overflow and quantisation limit cycles

as well as for the case that no overflow correction is implemented (see Fig. 8-46). A diamond-shaped stable region is obtained for two's complement overflow. For rounding, we observe severe restrictions with respect to the globally stable regions, which are all equal to or smaller than a circle with radius 1/2. Here we have one of the cases where the constructive algorithm is too pessimistic. It is shown in [2], that the normal form structure is stable with two and four RND quantisers if the coefficients lie within the unit square as indicated in Fig. 8-46 by dashed lines.

Wave digital filter

The second-order wave digital filter structure according to Fig. 5-52 requires careful placement of quantisers and overflow corrections in the block diagram. The crosswise interconnection of the state registers leads to a loop in which values may oscillate between both registers without any attenuation. This can be easily demonstrated for the case that the coefficients γ_1 and γ_2 vanish. For the linear filter, this behaviour is not a problem since the coefficient pair ($\gamma_1 = 0$, $\gamma_2 = 0$) lies on the boundary of the stability triangle (Fig. 8-31) which identifies this system as unstable. In the nonlinear case, however, quantisation and overflow may in effect set these coefficients to zero, which results in limit or overflow cycles even if the corresponding linear system is asymptotically stable.

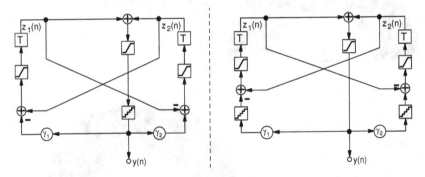

Fig. 8-47 Unstable arrangements of quantisers and overflow corrections

Figure 8-47 shows such critical arrangements, which are both classified unstable by the constructive algorithm. With respect to quantisation limit cycles, the problem can be solved by moving the quantisers behind the summation points as depicted in Fig. 8-48. This requires adders that are able to handle the double wordlength resulting from the coefficient multiplication. If an appropriate type of quantiser is chosen, the previously mentioned loop receives damping elements, which force quantisation limit oscillations to die down.

The overflow correction that follows the upper adder in Fig. 8-47 is a further critical element that has to be considered in the context of overflow limit cycles. It turns out that none of the five overflow characteristics (8.63) leads to filters that ensure freedom from overflow limit cycles. Investigation of various alternatives using the constructive algorithm leads to the conclusion that stable conditions can only be achieved if this overflow correction is completely omitted. So in order to preserve the sum $z_1(n) + z_2(n)$, an extra front bit is required which the coefficient multipliers and the subsequent adders have to cope with.

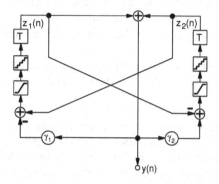

Fig. 8-48
Optimum positioning of quantisers and overflow corrections

With the mentioned special properties of the adders and multipliers, the structure according to Fig. 8-48 possesses four extreme matrices:

$$\begin{pmatrix} -\gamma_1\alpha_i & (1-\gamma_1)\,\alpha_i \\ (-1+\gamma_2)\,\beta_j & \gamma_2\beta_j \end{pmatrix} \quad i,j = 1,2$$

$$\text{with } \alpha_1 = \beta_1 = k_q, \quad \alpha_2 = \beta_2 = \min(0,k_o) \;.$$

Application of the constructive algorithm using these extreme matrices yields very contrary results. In the case of rounding, no combination of the coefficients γ_1 and γ_2 exists that is globally asymptotically stable. The situation with MT quantisation is totally different. Stability is guaranteed in combination with any overflow characteristic in the whole coefficient triangle. The only exception is two's complement overflow, which only allows stable filters if both coefficients are positive (Fig. 8-49).

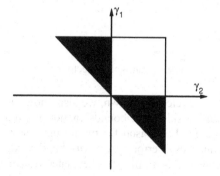

Fig. 8-49
Stable coefficient range with MT quantisation and two's complement overflow

8.7.5 A Search Algorithm in the State Plane

The constructive algorithm, which we introduced in the previous section, proved to be a powerful and universal tool to identify coefficient ranges in real filter implementations that are globally asymptotically stable and thus free of any limit cycles. The computation times to produce graphical representations of the stable regions such as in Fig. 8-43 are moderate. Overflow mechanisms can be included in the stability analysis. The following aspects are disadvantageous, however:

- The method cannot be applied to two's complement (value) truncation since no sector condition is valid in this case.
- The method identifies potentially unstable ranges of coefficients, but is not able to give any indication if limit cycles really occur in actual filter implementations or not.

The method described below overcomes these limitations, but is applicable to quantisation limit cycles only. Two's complement truncation is covered, and the algorithm clearly terminates with the result "stable" or "unstable". In order to come to these concrete results, we finally have no choice but to check each point in the state plane to see if a trajectory starting in this point

- converges into the origin,
- lies on a limit cycle, or
- leads into a limit cycle (Fig. 8-50).

If we find a point that ends up in a cycle, the respective pair of filter coefficients is unstable. Because of the finite precision of the signal representation, the state plane consists of a grid of discrete points (z_1/z_2). If we multiply a state vector $z(n)$ by the system matrix A and apply the respective rounding or truncation rules, we obtain a state vector $z(n+1)$ that again fits into the grid of discrete state values. Before starting the "exhaustive search" [3, 59] for unstable points in the state plane, we must have some idea of the size of the region around the origin that we have to scan and of the maximum period of the limit cycle.

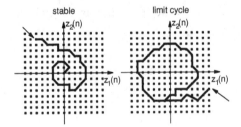

Fig. 8-50
Stable and unstable trajectories

If the bounds on the amplitude of the limit cycles are known, we also know the size of the region to be scanned. Any point outside these bounds cannot lie on a limit cycle and hence must not be considered. The reason for the occurrence of limit cycles is the existence of periodic sequences of error values $e(n)$ that develop from the quantisation process. The mechanism is similar to the generation of quantisation noise as demonstrated in [41], where the behaviour of a filter is described that exhibits limit cycles in the absence of an input signal. The amplitude of the input signal $x(n)$ is raised step by step from zero up by a few quantisation steps. The initial periodic limit cycle gradually turns into a random sequence which is superimposed onto the signal. The amplitudes of the initial limit cycle and of the finally superimposed noise are of the same order of magnitude.

The quantisation error satisfies the following condition for the various types of quantisers:

$$|e(n)| \le \varepsilon \, \Delta x_q$$

with $\varepsilon = 1/2$ for RND quantisation and $\varepsilon = 1$ for MT and VT quantisation.

The maximum amplitude that a limit cycle may assume depends on the transfer function $H(z)$ between the place of the respective quantiser and the state registers (delay elements) of the filter whose output are the state variables $z_1(n)$ and $z_2(n)$. In Sect. 8.4.3, we have derived the following relationship between the maximum values of input and output signal of a LSI system, which is characterised by its unit-sample response $h(n)$:

$$|y|_{\max} = |x|_{\max} \sum_{n=0}^{\infty} |h(n)| \; .$$

A bound M on the amplitude of limit cycles can therefore expressed as

$$M = \varepsilon \Delta x_q \sum_{n=0}^{\infty} |h(n)|$$

with $\varepsilon = 1/2$ for RND quantisation and $\varepsilon = 1$ for MT and VT quantisation.

If a filter possesses NQ quantisers, the partial contributions of these quantisers to the overall limit cycle have to be summed up at the output.

$$M = \varepsilon \Delta x_q \sum_{i=1}^{NQ} \sum_{n=0}^{\infty} |h_i(n)| \tag{8.70}$$

According to the results of Sect. 8.5.1, the internal transfer functions between a quantiser and a state register can be represented in the following general form:

$$H(z) = \frac{z^{-1}\left(1 - b_1 z^{-1}\right)}{1 + a_1 z^{-1} + a_2 z^{-2}} = \frac{z - b_1}{z^2 + a_1 z + a_2} = \frac{z - b_1}{\left(z - z_{\infty 1}\right)\left(z - z_{\infty 2}\right)} \; . \tag{8.71}$$

The transfer function (8.71) may be rewritten in three different ways:

$$H_a(z) = \frac{z - b_1}{z - z_{\infty 1}} \frac{1}{z - z_{\infty 2}}$$

$$H_b(z) = \frac{1}{z - z_{\infty 1}} \frac{z - b_1}{z - z_{\infty 2}}$$

$$H_c(z) = \frac{z_{\infty 1} - b_1}{z_{\infty 1} - z_{\infty 2}} \frac{1}{z - z_{\infty 1}} + \frac{z_{\infty 2} - b_1}{z_{\infty 2} - z_{\infty 1}} \frac{1}{z - z_{\infty 2}} \quad .$$

$H_a(z)$ and $H_b(z)$ are based on the assumption of a cascade arrangement of the pole terms with two different assignments of the zero to one of the poles. $H_c(z)$ is based on the assumption of a parallel arrangement of the pole terms. For each of the three forms, we will determine an upper bound M on the absolute sum over the respective unit-sample response. The smallest of these three values is then used to calculate the bound on the amplitude of the limit cycles using (8.70).

$$H_a(z) = \frac{z - b_1}{z - z_{\infty 1}} \frac{1}{z - z_{\infty 2}}$$

$$H_a(z) = \left[1 + \frac{z_{\infty 1} - b_1}{z - z_{\infty 1}}\right] \frac{1}{z - z_{\infty 2}}$$

The unit-sample responses of the two partial systems are easily obtained as

$$h_I(n) = \delta(n) + \left(z_{\infty 1} - b_1\right) z_{\infty 1}^{n-1} u(n-1) \quad ,$$

$$h_{II}(n) = z_{\infty 2}^{n-1} u(n-1) \quad .$$

The summation is performed in both cases over a geometric progression which will converge in any case as the magnitude of the poles of stable filters is less then unity.

$$\sum_{n=0}^{\infty} |h_I(n)| = 1 + |z_{\infty 1} - b_1| \sum_{n=0}^{\infty} |z_{\infty 1}|^{n-1} = 1 + \frac{|z_{\infty 1} - b_1|}{1 - |z_{\infty 1}|}$$

$$\sum_{n=0}^{\infty} |h_I(n)| = \frac{1 - |z_{\infty 1}| + |z_{\infty 1} - b_1|}{1 - |z_{\infty 1}|}$$

$$\sum_{n=0}^{\infty} |h_{II}(n)| = \sum_{n=0}^{\infty} |z_{\infty 2}|^{n-1} = \frac{1}{1 - |z_{\infty 2}|}$$

$$M_a = \sum_{n=0}^{\infty} |h_I(n)| \sum_{n=0}^{\infty} |h_{II}(n)| = \frac{1 - |z_{\infty 1}| + |z_{\infty 1} - b_1|}{\left(1 - |z_{\infty 1}|\right)\left(1 - |z_{\infty 2}|\right)}$$

The corresponding result for the transfer function $H_b(z)$ can be directly written down since $z_{\infty 1}$ and $z_{\infty 2}$ simply reverse roles.

$$M_b = \frac{1 - |z_{\infty 2}| + |z_{\infty 2} - b_1|}{\left(1 - |z_{\infty 1}|\right)\left(1 - |z_{\infty 2}|\right)}$$

$H_c(z)$ is a partial fraction representation of (8.71) with the following unit-sample response:

$$h(n) = \frac{z_{\infty 1} - b_1}{z_{\infty 1} - z_{\infty 2}} z_{\infty 1}^{n-1} u(n-1) + \frac{z_{\infty 2} - b_1}{z_{\infty 2} - z_{\infty 1}} z_{\infty 2}^{n-1} u(n-1) \ .$$

Using the triangle inequality we have

$$h(n) \leq \frac{|z_{\infty 1} - b_1|}{|z_{\infty 1} - z_{\infty 2}|} |z_{\infty 1}|^{n-1} u(n-1) + \frac{|z_{\infty 2} - b_1|}{|z_{\infty 2} - z_{\infty 1}|} |z_{\infty 2}|^{n-1} u(n-1) \ .$$

Summation over the geometric progressions finally results in

$$M_c = \sum_{n=0}^{\infty} |h(n)| = \frac{1}{|z_{\infty 1} - z_{\infty 2}|} \left(\frac{|z_{\infty 1} - b_1|}{1 - |z_{\infty 1}|} + \frac{|z_{\infty 2} - b_1|}{1 - |z_{\infty 2}|} \right) \ .$$

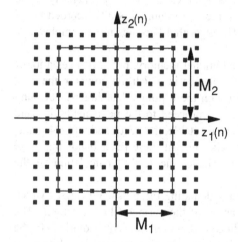

Fig. 8-51
Region to be scanned with respect to the occurrence of limit cycles

For direct-form and normal-form filters, the bounds on the amplitude of the limit cycles M_1 and M_2 are equal for both state variables $z_1(n)$ and $z_2(n)$, which yields a square in the state plane that has to be scanned (Fig. 8-51). For the wave digital filter structure, M_1 and M_2 differ depending on the values of the coefficients γ_1 and γ_2, resulting in a rectangle. The number of points that we have to check for stability thus amounts to $4 \times M_1 \times M_2$.

One point after the other in the bounded region is chosen as a starting point of a trajectory, which is obtained by continued multiplication by the system matrix A and application of the respective rounding or truncation operations. After at least $4 \times M_1 \times M_2$ operations, one of the following termination criteria applies:

- The trajectory leaves the bounded region. This means that the starting point does not lie on a limit cycle. The further calculation of the trajectory can be stopped, and the next point is checked.

- The trajectory returns back to the starting point. This means that a limit cycle has been found within the bounded region. The corresponding pair of filter coefficients is unstable. Further points need not be checked.
- The trajectory converges into the origin. The next point is checked.
- After $4 \times M_1 \times M_2$ operations, none of the above events has taken place. This means that the trajectory has entered a limit cycle since at least one point must have been reached for a second time. The corresponding pair of filter coefficients is unstable. Further points need not be checked.

If all $4 \times M_1 \times M_2$ point are scanned and no limit cycle has been found, then the corresponding pair of filter coefficients is stable.

There are ways to make the described algorithm more efficient. One possibility is to mark all starting points that finally end up in the origin. If such a point is hit later by another trajectory, the further pursuit of the trajectory can be stopped. This requires the establishment of a map of the scanned region and hence the provision of memory. A further improvement is to record each trajectory temporarily in an interim buffer. If the trajectory converges into the origin, all points recorded in the interim buffer are copied into the map. The algorithm becomes more complex, and still more memory is required.

A strategy that does not require any additional memory, but still saves a lot of computation time, is based on a special sequence in which the points in the state plane are checked for stability. We start with the 8 points that directly surround the origin. If all these points are stable, we proceed to the next quadratic band which consists of 16 points. If these points are stable, the next band with 24 points is checked, and so on. If a trajectory reaches one of these inner bands that have already been checked and identified as stable, the further calculation of the respective trajectory can be stopped.

The latter method has been applied in calculating the following figures which show the definitely stable coefficient ranges of the various filter structures. The coefficients are chosen in steps according to the VGA screen resolution. In the black regions, quantisation limit cycles have been found, whereas the white areas are asymptotically stable.

Direct-form filter

For RND and MT quantisation, the coefficient ranges that are free of quantisation limit cycles are in all cases larger than predicted by the constructive approach. For one MT and for two RND quantisers, the unstable regions coincide with regions where at least one extreme matrix is unstable (Fig. 8-52). For two MT quantisers, there are even small regions free of quantisation limit cycles in which unstable extreme matrices occur. With one exception, all grey regions in Fig. 8-43 turn out to be stable. We remember that these are regions where finite products of stable extreme matrices exist that are unstable. The mentioned exception is the case of one MT quantiser, where small irregular regions are observed along the upper edge of the stability triangle that exhibit limit cycles. These fall into the grey regions in Fig. 8-43 which predict potential instability. A magnified detail of this area is depicted in Fig. 8-53. The stable regions for value truncation are asymmetrical and have extremely irregular boundaries.

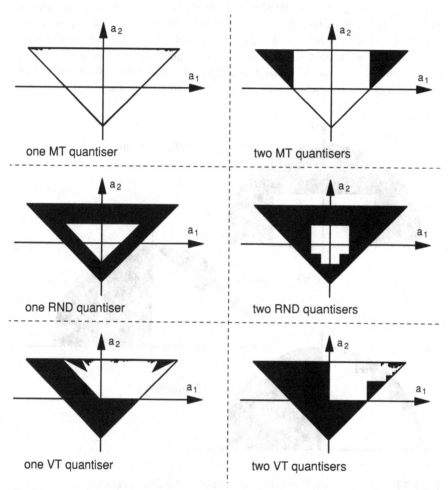

one MT quantiser

two MT quantisers

one RND quantiser

two RND quantisers

one VT quantiser

two VT quantisers

Fig. 8-52 Coefficient regions of direct-form filters that are free of limit cycles

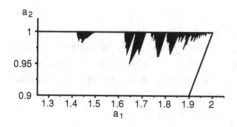

Fig. 8-53
Magnified detail of the case with one MT quantiser (direct form)

Normal-form filter

For the normal-form filter too, the results of the constructive approach are more conservative compared to those of the search algorithm (Fig. 8-54). The normal-form structure, if realised with two or four MT quantisers, is in any case free of quantisation limit cycles. For RND quantisers, the unit square is stable, which is in contrast to the constructive algorithm which predicts the circle with radius ½ to be free of limit cycles. For value truncation, we again obtain asymmetrical regions with irregular boundaries. In the case of two VT quantisers, we observe a narrow linear region that is stable. Under the angle of 45° where both filter coefficients have equal magnitude, quantisation effects obviously cancel out.

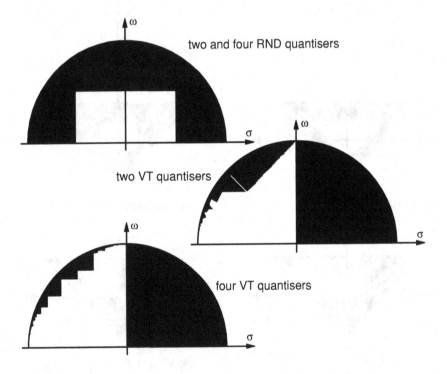

Fig. 8-54 Coefficient regions of normal-form filters that are free of limit cycles

Wave digital filter

The wave digital filter structure according to Fig. 8-48 proves to be free of quantisation limit cycles within the whole stability triangle if MT quantisation is applied. For RND and VT quantisation, however, wide areas of the coefficient plane are unstable. Merely some rectangular regions show asymptotic stability (Fig. 8-55). For VT quantisation, we again observe as a special feature a narrow linear region along the γ_1-axis that is free of limit cycles.

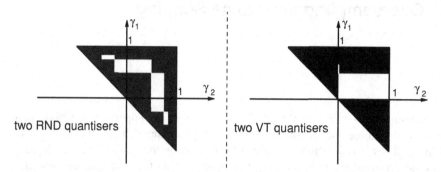

two RND quantisers   two VT quantisers

Fig. 8-55 Coefficient regions of the wave digital filter structure that are free of limit cycles

The computational effort of the search algorithm is higher compared to the constructive approach. For the application in filter design programs, this is not important, however, since only a few pairs of coefficients have to be checked for stability. In contrast to that, about 100,000 coefficient pairs have been calculated to obtain the graphs in Fig. 8-52, Fig. 8-54 and Fig. 8-55.

8.7.6 Summary

Numerous combinations of filter structures and quantisation or overflow characteristics have been checked with respect to their stability properties. It turns out that MT quantisation is superior. Normal form and wave digital filter structure are definitely free of quantisation limit cycles if MT quantisation is applied. As for the direct form, the variant with one quantiser should be chosen because only very small coefficient regions are unstable.

As for overflow limit cycles, the two's complement overflow characteristic should be avoided. For all other overflow characteristics, normal form and wave digital filter structures are free of overflow oscillations. In the case of the direct form, saturation yields the largest coefficient region without overflow limit cycles.

The direct form is the only structure that cannot be made free of quantisation or overflow limit cycles by any combination of arithmetic, quantiser and overflow correction. The direct form is nevertheless attractive, as it features the lowest implementational complexity of all structures.

9 Oversampling and Noise Shaping

Analog-to-digital and digital-to-analog conversion are important functions that greatly determine the quality of the signal processing chain. We showed in Chap. 4 that a certain effort is required for filtering in the continuous-time domain in order to avoid errors in the transition process from the analog to the discrete-time world and vice versa. The complexity of the involved filters increases more, the better the frequency range $0 \le |\omega| < \omega_s/2$, the baseband, is to be utilised and the higher the requirements are with respect to the avoidance of signal distortion. The resolution of the converters determines the achievable signal-to-noise ratio (*SNR*). Each additional bit improves the *SNR* by about 6 dB. The costs of the converters, however, increase overproportionally with the resolution.

Since modern semiconductor technologies facilitate ever-increasing processing speeds and circuit complexities, a trend can be observed to reduce the complexity of analog circuitry as far as possible and to shift the effort into the digital domain. In particular, the truncation of the spectra before A/D and after D/A conversion is carried out by digital filters. Additionally, the resolution of the converters can be reduced without degrading the *SNR*. This can be achieved by choosing sampling frequencies much higher than required to meet Shannon's sampling theorem.

9.1 D/A Conversion

From the viewpoint of the frequency domain, the periodically continued portions of the spectrum of the discrete-time signal have to be filtered away in order to reconstruct the corresponding analog signal. Figure 9-1 shows an example of such a periodic spectrum. In case of a converter that operates without oversampling, the spectrum has to be limited to half the sampling frequency by means of sharp cutoff analog low-pass filters. Converters functioning according to this principle are sometimes referred to as Nyquist converters. The name originates from the fact that half the sampling frequency is also called the Nyquist frequency in the literature.

The disadvantages of sharp cutoff analog filters in this context were already discussed in Chap. 4. In order to avoid the drawbacks of this filtering, oversampling is applied. All today's CD players, for instance, use this technique. On a compact disc, signals are stored at a sampling rate of 44.1 kHz and a resolution of 16 bits. By appropriate signal processing, the sampling rate is increased by a factor of 4, 8 or even more before the signal is applied to the D/A converter. We denote this factor in the following as the oversampling ratio R.

D. Schlichthärle, *Digital Filters*, DOI: 10.1007/978-3-642-14325-0_9,
© Springer-Verlag Berlin Heidelberg 2011

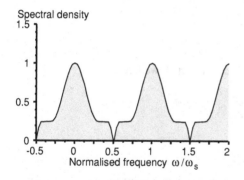

Fig. 9-1
Periodic spectrum of a discrete-time signal

If we aim at reducing the complexity of the analog filter only, oversampling ratios of 4, 8 or 16 are common. But oversampling can also be used to simplify or even eliminate completely the D/A converter. In this case, oversampling ratios of 128 or higher are needed.

9.1.1 Upsampling and Interpolation

In order to arrive at the higher sampling rate, we start with inserting $R-1$ zeros between two adjacent samples of the present sequence $f(n)$ (e.g. 3 in the case of four-times and 7 in the case of eight-times oversampling). This procedure is commonly referred to as upsampling. Figure 9-2 shows the block diagram symbol of the upsampling function. Assuming a sampling rate f_s at the input of the block, the output signal is sampled at a rate of Rf_s. The sampling period is reduced by a factor of R.

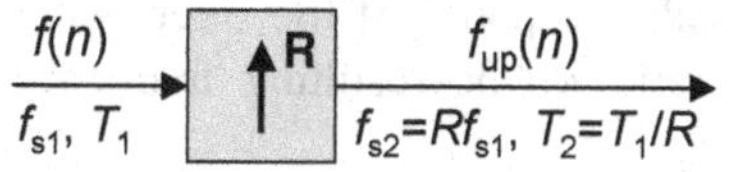

$$\underrightarrow{\begin{array}{c} f(n) \\ f_{s1},\ T_1 \end{array}} \boxed{\uparrow R} \underrightarrow{\begin{array}{c} f_{up}(n) \\ f_{s2}{=}Rf_{s1},\ T_2{=}T_1/R \end{array}}$$

Fig. 9-2
Block diagram representation of upsampling

We will show, in the following, how upsampling manifests itself in the frequency domain. According to (3.8), the discrete-time sequence $f(n)$ to be converted to analog has the spectrum

$$F(e^{j\omega T_1}) = \sum_{n=-\infty}^{+\infty} f(n)\,e^{-j\omega T_1 n}\ .$$

The sequence $f_{up}(n)$ that results from the insertion of zeros as described above has the spectrum

$$F_{up}(e^{j\omega T_2}) = \sum_{n=-\infty}^{+\infty} f_{up}(n)\,e^{-j\omega T_2 n}\ , \tag{9.1}$$

accordingly. Since only every Rth sample of $f_{up}(n)$ is nonzero, we can replace n by Rn in (9.1) without affecting the result.

$$F_{\mathrm{up}}(\mathrm{e}^{\mathrm{j}\omega T_2}) = \sum_{n=-\infty}^{+\infty} f_{\mathrm{up}}(Rn)\,\mathrm{e}^{-\mathrm{j}\omega T_2 Rn}$$

Since $f_{\mathrm{up}}(Rn)$ corresponds to the original sequence $f(n)$, we have

$$F_{\mathrm{up}}(\mathrm{e}^{\mathrm{j}\omega T_2}) = \sum_{n=-\infty}^{+\infty} f(n)\,\mathrm{e}^{-\mathrm{j}\omega T_2 Rn} = \sum_{n=-\infty}^{+\infty} f(n)\,\mathrm{e}^{-\mathrm{j}\omega T_1 n}$$

$$F_{\mathrm{up}}(\mathrm{e}^{\mathrm{j}\omega T_2}) = F(\mathrm{e}^{\mathrm{j}R\omega T_2}) = F(\mathrm{e}^{\mathrm{j}\omega T_1}) \ . \tag{9.2}$$

The shape of the spectrum is, in principle, not influenced by the insertion of the zeros. Only the scale of the frequency axis has been changed. Figure 9-3 illustrates this result under the assumption of a four-times oversampling.

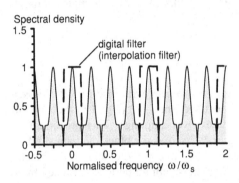

Spectral density

digital filter (interpolation filter)

Normalised frequency $\omega/\omega_{\mathrm{s}}$

Fig. 9-3
Spectrum of the digital signal after insertion of zeros to achieve four-times oversampling. Frequency response of interpolation filter indicated as a dashed line

This scaling has an interesting consequence. The frequency response $H(\mathrm{e}^{\mathrm{j}\omega T})$ of a filter is equivalent to the spectrum of the impulse response of the filter. After insertion of an upsampler with oversampling ratio R, the resulting frequency response can be expressed as

$$H(\mathrm{e}^{\mathrm{j}R\omega T}) = H\!\left((\mathrm{e}^{\mathrm{j}\omega T})^R\right).$$

Since $z = \mathrm{e}^{\mathrm{j}\omega T}$ on the unit circle, a z-transfer function $H(z)$ is transformed by the upsampler to $H(z^R)$. Figure 9-4 illustrates this situation.

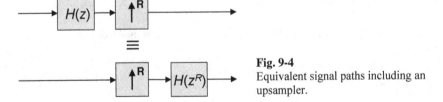

Fig. 9-4
Equivalent signal paths including an upsampler.

Both signal paths in Fig. 9-4 are fully equivalent. A simple delay may serve as an example. Let us assume that a delay by one sample with the transfer function z^{-1} is arrangend at the left low-frequency side of the upsampler. Clearly, if we intend to realise the same delay time at the right high-frequency side, we need a delay by R samples with the transfer function z^{-R}.

A digital filter can be used to filter out the desired baseband after upsampling. The periodic frequency response of such a filter is indicated in Fig. 9-3. As a result of this filtering, we observe a large spacing between the baseband and the periodically continued portions of the spectrum (Fig. 9-5). A relatively simple analog filter is now sufficient to recover the baseband and thus the desired continuous-time signal. A possible frequency response of this analog interpolation filter is also shown in Fig. 9-5. The stopband edge frequency of this filter may be chosen as high as $7/8\,\omega_s$ while still avoiding aliasing.

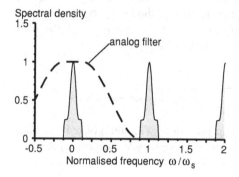

Fig. 9-5
Spectrum of the oversampled signal after low-pass filtering using a digital interpolation filter (dashed line indicates frequency response of a subsequent analog filter)

As a result of the digital low-pass filtering, we obtain a new discrete-time sequence featuring a higher temporal resolution than the original. The previously inserted zeros are replaced by values that appropriately interpolate the original samples. One can easily imagine that, with increasing R, we achieve an increasingly better approximation of the desired analog signal in the discrete-time domain, which finally reduces the effort for filtering in the continuous-time domain.

Example 9-1

Determine the impulse response of an ideal interpolation filter for arbitrary oversampling ratios R.

The required frequency response of the interpolation filter is defined as

$$H_{\text{int}}(\omega) = \begin{cases} 1 & \text{for } |\omega| < \pi/RT \\ 0 & \text{for } \pi/RT \le |\omega| \le \pi/T \end{cases}.$$

The impulse response of the filter is calculated as the inverse Fourier transform of the frequency response.

$$h_{\text{int}}(n) = \frac{T}{2\pi} \int_{-\pi/T}^{+\pi/T} H_{\text{int}}(\omega)\, e^{j\omega nT}\, d\omega = \frac{T}{2\pi} \int_{-\pi/RT}^{+\pi/RT} e^{j\omega nT}\, d\omega$$

$$h_{\text{int}}(n) = \frac{T}{2\pi} \frac{e^{j\omega nT}}{jnT} \Bigg|_{-\pi/RT}^{+\pi/RT}$$

$$h_{\text{int}}(n) = \frac{1}{2\pi} \frac{e^{jn\pi/R} - e^{-jn\pi/R}}{jn} = \frac{\sin(n\pi/R)}{n\pi}$$

$$h_{\text{int}}(n) = \frac{1}{R} \frac{\sin(n\pi/R)}{n\pi/R} \tag{9.3}$$

For practical purposes, e.g. for an FIR filter implementation, this infinite non-causal impulse response needs to be truncated to a finite length using one of the windows introduced in Sect. 7.4.2 and shifted to the right to assure causality. It must be noted that the decay time of the filter increases with the oversampling ratio R. In order to achieve sufficient steepness of the filter slopes, the filter length must be chosen proportionally to R.

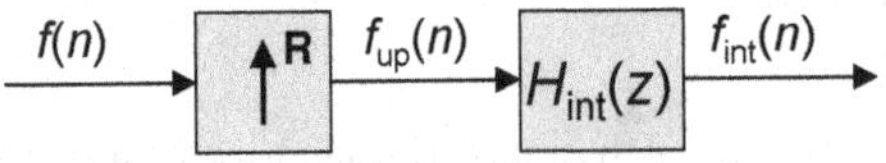

Fig. 9-6
Upsampling and interpolation

Figure 9-6 summarizes the oversampling procedure. The interpolated sequence $f_{\text{int}}(n)$ is obtained by discrete-time convolution of the upsampled sequence $f_{\text{up}}(n)$ and the impulse response of the interpolation filter $h_{\text{int}}(n)$.

$$f_{\text{int}}(n) = \sum_{m=-\infty}^{+\infty} f_{\text{up}}(m) \, h_{\text{int}}(n-m)$$

Assuming an ideal interpolation filter, we can insert the impulse response (9.3).

$$f_{\text{int}}(n) = \frac{1}{R} \sum_{m=-\infty}^{+\infty} f_{\text{up}}(m) \frac{\sin((n-m)\pi/R)}{(n-m)\pi/R}$$

Since only every Rth sample of $f_{\text{up}}(m)$ is nonzero, we can substitute m by mR without changing anything.

$$f_{\text{int}}(n) = \frac{1}{R} \sum_{m=-\infty}^{+\infty} f_{\text{up}}(mR) \frac{\sin((n-mR)\pi/R)}{(n-mR)\pi/R}$$

$f_{\text{up}}(mR)$ is the original signal sequence $f(m)$ before upsampling. So we can finally write

$$f_{\text{int}}(n) = \frac{1}{R} \sum_{m=-\infty}^{+\infty} f(m) \frac{\sin((n/R-m)\pi)}{(n/R-m)\pi} \quad . \tag{9.4}$$

Equation (9.4) constitutes the interpolation rule for R times oversampling assuming an ideal interpolation filter. Note that the interpolated sequence is attenuated by a factor of $1/R$ which must be compensated to yield the original amplitude.

As an additional side-effect of oversampling, also the equalisation of the sin$x$/x-distortion caused by the digital-to-analog converter becomes easier, since the frequency range of interest only covers the first quarter of the frequency range depicted in Fig. 4-7 for 4 times oversampling, leaving only a loss of at most 0.25 dB to be compensated. For common oversampling ratios of eight or higher, this aspect of signal reconstruction can be neglected in practice.

9.1.2 The CIC Interpolator Filter

High oversampling ratios require interpolation filters with very narrow passbands and steep cutoff slopes at the high-frequency side of the upsampler. Depending on the requirements concerning the suppression of the undesired images in the upsampled spectrum, the realisation of such filters may be a challenge. IIR filters require poles with high quality factors which are clustered in a verry narrow frequency range. Such filter designs are difficult to implement because of roundoff noise and coefficient sensitivity problems. If phase distortion shall be avoided in a given application, linear-phase FIR filters are the preferred solution. But extremely high filter orders may be required depending on the realised oversampling ratio.

An efficient way of interpolation and decimation filtering was introduced by Hogenauer [33]. This technique facilitates the implementation of multiplier-free filters which complete a great deal of the job. A conventional FIR filter of reasonable order arranged at the low-frequency side of the upsampler makes for a flat response in the passband and an adequate cutoff slope.

The Hogenauer CIC (Cascaded Integrator-Comb) filter is characterised by a transfer function of the form

$$H(z) = \left(\frac{1 - z^{-R}}{1 - z^{-1}} \right)^N = \frac{(1 - z^{-R})^N}{(1 - z^{-1})^N} \quad . \tag{9.5}$$

The denominator term represents the series connection of N integrators whose block diagram is shown in Fig. 9-7. The transfer function of the integrator is expressed as

$$H(z) = \frac{1}{1 - z^{-1}} \quad .$$

The numerator term represents the series connection of N comb filters whose block diagram is also depicted in Fig. 9-7. The transfer function of the comb filter section is expressed as

$$H(z) = 1 - z^{-R} \quad .$$

The whole filter thus would require $(R+1)N$ storage elements. As an example, 1032 storage elements would be required for an 8th-order CIC filter with an oversampling ratio of $R = 128$.

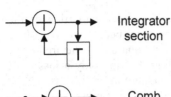

Integrator
section

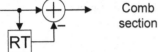

Comb
section

Fig. 9-7
Block digrams of the integrator and comb filter
sections

This number can be dramatically reduced if the comb filter sections are moved to the low-rate side of the upsampler. Making use of the rule introduced in Fig. 9-4, the transfer function of the comb section simplifies to

$$H(z) = 1 - z^{-1} .$$

The memory requirement reduces to $2N$. Only 16 storage elements would be needed in our example. Fig. 9-8 shows the optimised block diagram of a 4th-order CIC interpolator filter.

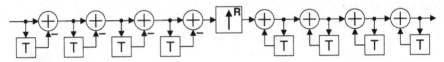

Fig. 9-8   Block diagram of a 4th-order CIC interpolator filter

Example 9-2

Derive the magnitude response of the CIC filter and determine the gain at $\omega = 0$.

The transfer function of the CIC filter is given by Eq. (9.5).

$$H(z) = \left(\frac{1 - z^{-R}}{1 - z^{-1}} \right)^{N}$$

Equivalent transformation results in

$$H(z) = z^{-\frac{N(R-1)}{2}} \left(\frac{z^{R/2} - z^{-R/2}}{z^{1/2} - z^{-1/2}} \right)^{N} .$$

The frequency response is obtained by the substitution $z = e^{j\omega T}$.

$$H(e^{j\omega T}) = e^{-j\frac{N(R-1)}{2}\omega T} \left(\frac{e^{jR\omega T/2} - e^{-jR\omega T/2}}{e^{j\omega T/2} - e^{-j\omega T/2}} \right)^{N}$$

$$H(e^{j\omega T}) = e^{-j\frac{N(R-1)}{2}\omega T}\left(\frac{\sin(R\omega T/2)}{\sin(\omega T/2)}\right)^N$$

Since the exponential term has unity magnitude, the magnitude response of the CIC filter can be expressed as

$$\left|H(e^{j\omega T})\right| = \left|\frac{\sin(R\omega T/2)}{\sin(\omega T/2)}\right|^N .$$

For ω approaching zero, the sine function can be approximated by its argument. We therefore have

$$\lim_{\omega\to 0}\left|H(e^{j\omega T})\right| = \left|\frac{R\omega T/2}{\omega T/2}\right|^N = R^N . \tag{9.6}$$

CIC magnitude response [dB]

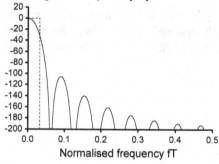

Fig. 9-9
Magnitude response of the CIC filter $(R = 16, N = 8)$. Target response in dashed line

Figure 9-9 shows the typical magnitude response of a CIC filter normalised to unity gain at $\omega = 0$. The zeros of the magnitude coincide with the centre of the mirror image spectra which are created by the upsampling process. So the mirror images are strongly attenuated. But, compared to the desired response, the cutoff slope of the filter is rather poor, and the passband shows a significant droop.

Droop compensation response [dB]

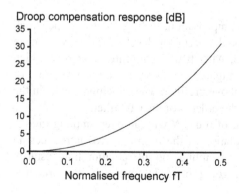

Fig. 9-10
Magnitude response of a droop compensation filter $(R = 16, N = 8)$

The droop may be compensated by an FIR filter inserted at the low-rate side of the upsampler. Fig. 9-10 shows an example of the magnitude response to be realised. As a result, the magnitude is now flat in the passband. The slope, however, is still poor (Fig. 9-11).

CIC response with droop compensation [dB]

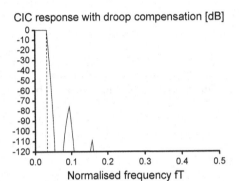

Fig. 9-11
Magnitude response of a CIC filter with droop compensation filtering prior to upsampling ($R = 16$, $N = 8$). Ideal response indicated in dashed line

The functionality of the FIR prefilter may be extended to improve the selectivity of the CIC filter. For this purpose, we apply upsampling with a low oversampling ratio to the input signal, e.g. $R = 2$, and use an FIR filter with a magnitude response as sketched in Fig 9-12.

Droop compensation + filter slope [dB]

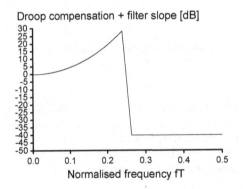

Fig. 9-12
Magnitude response of a prefilter for droop compensation, improvement of the selectivity, and mirror image suppression ($R = 16$, $N = 8$)

This filter provides the required droop compensation, realises a cutoff slope which increases the selectivity of the filter, and lowers the level of the mirror image created by upsampling by a factor of 2. After that, a CIC filter is applied which realises the desired oversampling ratio. Figure 9-13 shows the overall response of the CIC filter which makes use of the above prefilter. The remaining peaks in the stopband can be lowered by increasing the order N of the CIC filter.

According to Equ. (9.6), a CIC filter of order N with an oversampling ratio R has a gain of R^N at $\omega = 0$. The interpolation of the inserted zeros introduces an attenuation by a factor of $1 / R$ (9.4). Upsampling and interpolation thus result in an overall maximum gain of R^{N-1}. When going forward through the filter structure

CIC response with droop compensation [dB]

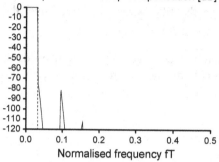

Fig. 9-13
CIC Filter response with droop compensation and cutoff slope enhancement ($R = 16$, $N = 8$)

of Fig. 9-8, the gain increases continuously after each processing block. So in each integrator or comb, the register length for the signal representation must be increased to avoid overflow. At the output,

$$G = \operatorname{ld} R^{N-1} = (N-1)\operatorname{ld} R$$

additional bits are required. For an 8th-order CIC filter with $R = 128$, for instance, 49 additional bits are required. Truncation or rounding is required in order to return to the original signal register length or to adapt to the signal resolution of the subsequent D/A converter. It is important to note that this truncation or rounding must be performed behind the last integrator. It is not possible to insert any quantisers somewhere in the chain of processing blocks. The integrator alone is an unstable filter structure because the pole is located on the unit circle. For ω approaching zero, the gain grows to infinity. Only if the number of combs preceding the integrators is equal to or higher than the number of integrators present in the chain, the overall transfer function will be stable. So seen from a quantiser somewhere in the middle of the filter structure (Fig. 9-8), the number of combs in the signal path is always lower than the number of integrators. Any DC offset or quantisation noise generated by truncation or rounding causes the error to grow without bound resulting in an unstable filter.

Example 9-3
Show that the CIC filter is stable though the integrator which is a part of the filter structure is unstable.
Applying long division to the fraction in the transfer function (9.5) yields a polynomial.

$$\frac{1-z^{-R}}{1-z^{-1}} = 1 + z^{-1} + z^{-2} + \ldots + z^{-(R-1)}$$

The transfer function of the CIC filter can therefore be expressed as

$$H(z) = \left(1 + z^{-1} + z^{-2} + \ldots + z^{-(R-1)}\right)^{N} .$$

$H(z)$ is a polynomial in z^{-1} and therefore the transfer function of a FIR filter which is inherently stable. Though the filter structure is recursive, the transfer function could also be realised by a FIR filter.

The FIR filter would have to be completely implemented on the high-rate side of the upsampler. An optimisation with respect to the memory requirements as in the case of the structure in Fig. 9-8 is not possible.

9.1.3 Noise Filtering

The concept of oversampling is not only helpful in reducing the effort for filtering in the signal conversion process. By making use of a side-effect of oversampling, also the complexity of the D/A converter may be reduced.

According to (8.8), quantisation noise with a flat spectrum and a power of

$$N_0 = 2^{-2b_2}/3 \tag{9.7}$$

is created when the output register length of the interpolation filter is reduced to the b_2 bit resolution of the D/A converter. The higher the oversampling ratio is chosen, the larger is the frequency range over which the quantisation noise power is spread. If the baseband up to the original Nyquist frequency is filtered out of the oversampled quantised signal by means of an appropriate low-pass filter, a large portion of the wideband quantisation noise is removed.

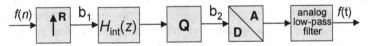

Fig. 9-14 Functional blocks of oversampling D/A conversion

The principle described can be applied to the D/A conversion process. Figure 9-14 shows the involved functional blocks. The digital input sequence $f(n)$ is represented with a resolution of b_1 bits. Depending on the chosen oversampling ratio, $R-1$ zeros are inserted between the samples of the original sequence. In the subsequent digital low-pass filter, the baseband is filtered out of the oversampled signal. The quantiser reduces the wordlength to the b_2-bit resolution of the D/A converter, resulting in a wideband quantisation noise power of

$$\sigma_e^2 = N_0 = 2^{-2b'}/3$$

which is also superimposed onto the stepped signal at the output of the D/A converter. The final analog low-pass filter can be optimised in two respects:

- If low effort for filtering is an objective, a low-pass filter is chosen with a passband edge at $f_s(0.5/R)$ and a stopband edge at $f_s(1-0.5/R)$. The resolution b_2 of the D/A converter has to be chosen to be at least equal to the original resolution b_1 of the input signal $f(n)$ if loss of signal quality is to be avoided.

- If the complexity of the converter is to be minimised, on the other hand, an analog low-pass filter is required that cuts off sharply at the Nyquist frequency. Besides the unwanted out-of-band portions of the periodic spectrum (Fig. 9-3), a large amount of the noise power outside the baseband is removed.

In case of four-times oversampling, only a quarter of the quantisation noise still exists in the signal, corresponding to a loss of 6 dB. The signal-to-noise ratio is thus improved as if the signal was present at a one-bit higher resolution. With $R = 16$, the effective resolution even rises by two bits. Generally speaking, each doubling of the sampling rate leads to a gain in resolution of half a bit, equivalent to a lowering of the quantisation noise by 3 dB. Since the signal is available at a better signal-to-noise ratio after low-pass filtering compared to ordinary Nyquist conversion, the resolution b_2 of the quantiser and thus of the D/A converter can be reduced below b_1 without a deterioration in the quality of the original signal sequence $f(n)$.

A good idea, at first glance, to further reduce the complexity of the converter would be to increase the oversampling ratio such that a resolution of only one bit would be sufficient. The D/A converter would then consist of a simple two-level digital output. Noise filtering as described above is not appropriate, however, to reach this desired goal. On the one hand, an R of about 10^9 would be required mathematically to achieve a signal-to-noise ratio equivalent to a 16-bit resolution. On the other hand, the complete information about the progress of a signal between the zero crossings gets lost and cannot be recovered no matter what R is used. It can be stated that the superimposed noise signal $e(n)$ becomes increasingly correlated with the processed signal as the resolution decreases. As a consequence, larger portions of the noise spectrum fall into the baseband. The concept of noise filtering thus becomes ineffective. Figure 9-15 illustrates this behaviour using the example of a sinewave which is sampled with 2-bit resolution. The noise spectrum is mainly composed of harmonics of the original analog sinewave. The analog low-pass filter would certainly remove some of the higher harmonics, but the harmonics falling into the passband would strongly distort the original signal.

Quantisation noise spectrum

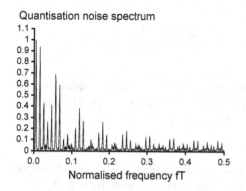

Fig. 9-15
Quantisation noise spectrum of a 116 Hz sinewave, sampled at 32 kHz, quantised with 2-bit resolution

Different is the situation if the same signal is sampled with a resolution of 16 bits. The spectrum shown in Fig. 9-16 is irregular and has roughly constant magnitude in the whole frequency range up to half the sampling rate. Harmonics of the sampled sinewave cannot be observed. The superimposed error signal is rather a noiselike signal. So a sharp cutoff filter behind the D/A converter may substantially reduce the quantisation noise.

Quantisation noise spectrum

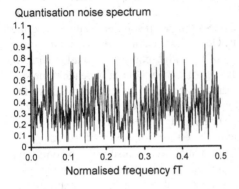

Fig. 9-16
Quantisation noise spectrum of a 116 Hz sinewave, sampled at 32 kHz, quantised with 16-bit resolution

9.1.4 Noise Shaping

The portion of the noise spectrum that falls into the baseband can be effectively reduced if the concept of noise shaping is applied which we are already familiar with in the context of noise reduction in IIR filters. It is the strength of this concept that it decorrelates signal and noise so that it is also applicable with very low level resolutions, down to even one bit. In order to implement noise shaping, the bits that have been truncated in the quantiser are delayed and subtracted from the input signal of the quantiser, as depicted in Fig. 9-5a.

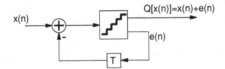

Fig. 9-17
First-order noise shaping

The dropped bits correspond to the error signal $e(n)$ which is superimposed onto the signal $x(n)$. Since this error signal is added to $x(n)$ once directly and once after delay with inverse sign, it is finally weighted by the transfer function

$$H(z) = 1 - z^{-1} \ .$$

The square magnitude of this transfer function determines the noise power density at the output of the quantiser, which is now frequency-dependent (refer to 8.21).

$$\left| H(e^{j\omega T}) \right|^2 = 2 - 2\cos \omega T = 2(1 - \cos \omega T)$$

At low frequencies, the noise spectrum is attenuated, while it is boosted close to the Nyquist frequency. It becomes obvious that noise shaping makes no sense without oversampling, since the overall noise power even increases. In case of oversampling, however, the amplified portion of the noise spectrum is later removed by the low-pass filter that decimates the spectrum to the baseband again.

The described concept can be extended to higher-order noise shaping. The error signal passes a chain of delays and is added to the input signal of the quantiser after weighting by integer coefficients. Fig. 9-18 shows a second-order noise shaper.

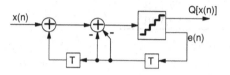

Fig. 9-18
Second-order noise shaping

In the second-order case, the quantisation noise is weighted by the transfer function

$$H(z) = (1 - z^{-1})^2 = 1 - 2z^{-1} + z^{-2} .$$

The corresponding squared magnitude can be expressed as

$$\left| H(e^{j\omega T}) \right|^2 = 4(1 - \cos \omega T)^2 = 4(1 - 2\cos \omega T + \cos^2 \omega T) .$$

We now calculate the portion of the noise power that falls into the baseband if noise shaping is applied. For first-order error feedback, the integral

$$\sigma_e^2 = N_0 \frac{1}{\pi} \int_0^{\pi/R} 2(1 - \cos \omega T) \, d\omega T$$

has to be evaluated.

$$\sigma_e^2 = N_0 2 \left[1/R - \sin(\pi/R)/\pi \right] \tag{9.8}$$

Second-order error feedback yields

$$\sigma_e^2 = N_0 \left[6/R - 8\sin(\pi/R)/\pi + \sin(2\pi/R)/\pi \right] , \tag{9.9}$$

accordingly. Table 9-1 summarises the achievable reduction of the noise power under the assumption that the spectrum is limited to the baseband $f_s/(2R)$ after quantisation, using a sharp cutoff low-pass filter. The figures for noise filtering and first- and second-order noise shaping are compared.

For large values of the oversampling ratio, the noise in the baseband (in-band noise) decreases with $1/R^3$ for first-order and with $1/R^5$ for second-order noise shaping. By appropriate series expansion of the sine function in (9.4) and (9.5), we obtain the following approximations:

Table 9-1 Possible reduction of the quantisation noise power by oversampling

R	Noise filtering	First-order noise shaping	Second-order noise shaping
Nyquist	0.0 dB	+3.0 dB	+7.8 dB
2	-3.0 dB	-4.4 dB	-3.4 dB
4	-6.0 dB	-13.0 dB	-17.5 dB
8	-9.0 dB	-22.0 dB	-32.3 dB
16	-12.0 dB	-31.0 dB	-47.3 dB
32	-15.1 dB	-40.0 dB	-62.4 dB
64	-18.1 dB	-49.0 dB	-77.4 dB
128	-21.1 dB	-58.0 dB	-92.5 dB

$$\sigma_e^2 = N_0/R \qquad\qquad \text{filtering}$$

$$\sigma_e^2 \approx N_0\pi^2/3R^3 \qquad\qquad \text{first - order noise shaping}$$

$$\sigma_e^2 \approx N_0\pi^4/5R^5 \qquad\qquad \text{second - order noise shaping} .$$

Each doubling of the oversampling ratio thus reduces the noise by 3 dB, 9 dB or 15 dB and provides 0.5, 1.5 or 2.5 bits of extra resolution respectively.

In the case of a one-bit resolution, second-order noise shaping with $R = 128$ results in an achievable signal-to-noise ratio, comparable to a 16-bit resolution. Assuming amplitude limits of ± 1, the maximum power of a sinusoidal signal amounts to 0.5. The step size of the quantiser is $\Delta x_q = 2$. According to (9.3), this results in a noise power of $N_0 = 1/3$ which is reduced by 92.5 dB by applying oversampling and noise shaping. The resulting signal-to-noise ratio can thus be calculated as

$$SNR = 10 \log(1/2)\,\text{dB} - 10 \log(1/3)\,\text{dB} + 92.5\,\text{dB} = 94.3\,\text{dB} .$$

9.2 A/D Conversion

9.2.1 Oversampling

Oversampling is also a means to simplify the A/D conversion process in terms of filter and converter complexity. The solid line in Fig. 9-19 represents the spectrum of the analog signal which shall be converted to digital. The shaded part of that spectrum marks the frequency range which shall be represented by the digital signal without being distorted by alias spectral components. Figure 9-19 illustrates the situation using the example of four-times oversampling. The periodically continued portions of the spectrum created by the oversampling process are represented by dotted lines. These extend into the desired frequency band and distort the original signal. Because of the four-times oversampling, the desired frequency band covers the range up to $|\omega| < 1/8 \, \omega_s$ as opposed to half the sampling frequency in the case of Nyquist sampling. It is easy to see that the analog antialiasing filter has to cut off at $7/8 \, \omega_s$ in order to avoid distortion of the desired signal. A possible magnitude response of this filter is shown in Fig. 9-19 as a dashed line.

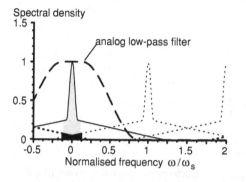

Fig. 9-19
Analog low-pass filtering in the case of four-times oversampling.
The desired baseband is shaded. The black areas represent the distorting spectral portions that are mirrored into the desired frequency band.

In the following, we compare the filter complexity required for a Nyquist converter with that required for a converter using four-times oversampling. In the frequency band of interest, the assumed Chebyshev filter shall exhibit a passband ripple of 0.1 dB. Alias spectral components in the stopband shall be attenuated by at least 72 dB. With this attenuation, these components are in the same order of magnitude as the quantisation noise of a 12-bit A/D converter. In the case of the Nyquist converter, the transition band of the filter shall cover the range $0.45 \, \omega_s$ to $0.5 \, \omega_s$. According to (2.26b), a filter order of $N = 24$ is required to fulfil this specification. In the case of four-times oversampling, the transition band extends between $1/8 \, \omega_s$ and $7/8 \, \omega_s$. This merely requires a filter order of $N = 5$.

9.2.2 Downsampling and Decimation

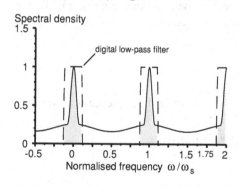

Fig. 9-20
Periodically continued spectrum after sampling

Figure 9-20 shows the periodically continued spectrum after analog low-pass filtering (cutoff frequency at $7/8 \, \omega_s$) and sampling at a rate of ω_s. The baseband remains unaffected. The actual limitation of the spectrum to the range up to $1/8\ \omega_s$ is performed in the discrete-time domain by a digital filter whose periodic magnitude response is also depicted in Fig. 9-20. For this purpose, it is preferable to use linear-phase filters which exclude the emergence of group delay distortion. Efficient realisations are again possible by the concept of CIC filters which act in this case as decimation filters. Figure 9-21 shows the resulting spectrum after digital low-pass filtering.

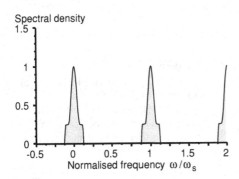

Spectral density

Fig. 9-21
Spectrum after digital low-pass filtering

In order to arrive at the actually desired sampling rate, which is reduced by a factor of four, each fourth sample is picked out of the digitally filtered signal. This process is commonly referred to as downsampling. Figure 9-22 shows the block diagram symbol of the downsampling function. Assuming a sampling rate f_{s1} at the input of the block, the output signal is sampled at a rate of f_{s1}/R. The sampling period is multiplied by a factor of R, accordingly.

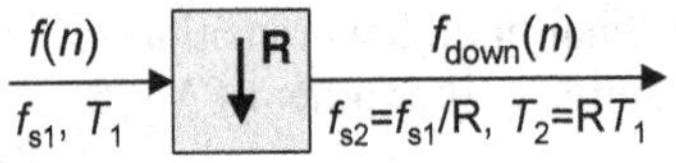

$$f(n) \quad \boxed{\downarrow R} \quad f_{\text{down}}(n)$$
$$f_{s1},\ T_1 \qquad\qquad f_{s2}=f_{s1}/R,\ T_2=RT_1$$

Fig. 9-22   Block diagram representation of downsampling

In the following, we will investigate what the consequences of this downsampling process are in the frequency domain. According to (3.14), the samples $f(n)$ and the associated spectrum are related as

$$f(n) = \frac{1}{2\pi} \int_{-\pi}^{+\pi} F(e^{j\omega T_1}) e^{j\omega T_1 n} \, d\omega T_1 \ .$$

By picking out every Rth sample out of $f(n)$, we obtain the sequence $f_{\text{down}}(n)$.

$$f_{\text{down}}(n) = f(Rn) = \frac{1}{2\pi} \int_{-\pi}^{+\pi} F(e^{j\omega T_1}) e^{j\omega T_1 Rn} \, d\omega T_1$$

$$f_{\text{down}}(n) = \sum_{i=0}^{R-1} \frac{1}{2\pi} \int_{-\pi/R}^{+\pi/R} F(e^{j(\omega T_1 + i2\pi/R)}) e^{j(\omega T_1 + i2\pi/R)Rn} \, d\omega T_1$$

$$f_{\text{down}}(n) = \frac{1}{2\pi} \int_{-\pi/R}^{+\pi/R} \sum_{i=0}^{R-1} F(e^{j(\omega T_1 + i2\pi/R)}) e^{j\omega T_1 R n} \, d\omega T_1$$

With $T_1 = T_2 / R$ we obtain

$$f_{\text{down}}(n) = \frac{1}{2\pi} \int_{-\pi}^{+\pi} \frac{1}{R} \sum_{i=0}^{R-1} F(e^{j(\omega T_2 + i2\pi)/R}) e^{j\omega T_2 n} \, d\omega T_2$$

The spectrum of the new sequence $f_{\text{down}}(n)$ can thus be expressed as

$$F_{\text{down}}(e^{j\omega T_2}) = \frac{1}{R} \sum_{i=0}^{R-1} F(e^{j(\omega T_2 + i2\pi)/R}) \ . \tag{9.10}$$

Inspection of (9.10) reveals that the gaps in the spectrum shown in Fig. 9-21 are filled up and the scale of the frequency axis is changed. Because of the preceding digital low-pass filtering, the periodically continued spectral portions do not overlap (Fig. 9-23). The described procedure finally leads to the same result as intricate analog low-pass filtering and subsequent sampling at double the Nyquist frequency.

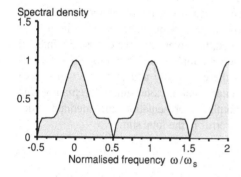

Spectral density

Normalised frequency ω / ω_s

Fig. 9-23
Spectrum of the digital signal after reduction of the sampling rate by a factor of four. The spectrum is normalised to unity.

The process of appropriate low-pass filtering and subsequent downsampling is commonly referred to as decimation (Fig. 9-24). This function is the inverse operation of interpolation which we introduced in Sect. 9.1.1.

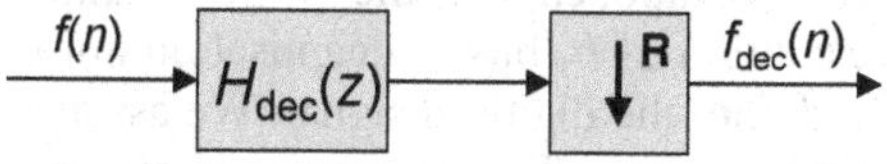

$$f(n) \rightarrow \boxed{H_{\text{dec}}(z)} \rightarrow \boxed{\downarrow R} \xrightarrow{f_{\text{dec}}(n)}$$

Fig. 9-24
Downsampling and decimation

9.2.3 CIC Decimation Filter

The decimation filter has a similar purpose as the interpolation filter in the context of upsampling: the bandwidth of an oversampled signal is to be limited to an Rth of the Nyquist bandwidth. The main difference is in the sequence of filtering and rate change. In the case of signal decimation, filtering precedes the downsampling process. An efficient implementation can be achieved again by shifting the comb filter section to the low-rate side of the downsampler. As a consequence, the filter chain starts with the integrators followed by the downsampler and the comb filter blocks as illustrated in Fig. 9-25.

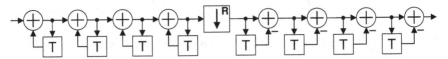

Fig. 9-25   Block diagram of a 4th-order CIC decimation filter

The reversed order of integrators and comb filters changes the situation concerning stability. In the path from any internal node to the output of the filter, the number of comb filter sections is always larger than the number of integrators. So all these partial filters are stable. It is therefore possible to perform scaling of the signal at any internal node of the filter. The quantisation noise created in this context does not lead to any problems as in the case of the CIC interpolation filter where scaling was only possible at the output of the filter.

Scaling is applied to make sure that no overflow occurs at internal nodes and at the output of the filter. Since the gain of the integrators approaches infinity as ω approaches zero, overflow cannot be avoided, however large we choose the internal signal wordlength of the filter. Here we can make use of a property of two's complement arithmetic which is valid if there are only adders in the signal path (refer to Sect. 8.2.1). As long as it is guaranteed that the total sum will remain within the amplitude limits of the the chosen number representation, intermediate nodes may run into overflow without causing any error in the total sum.

9.2.4 Noise Filtering

Oversampling, with subsequent limitation of the spectrum to the baseband, is also advantageous in the context of A/D conversion. Fig. 9-26 shows the functional block diagram of an oversampling A/D converter. The initial analog low-pass filter avoids the occurence of alias spectral components in the baseband after sampling. After A/D conversion with a resolution of b_1 bits, a quantisation noise power N_0 according to (9.7) is superimposed onto the digitised signal. We assume for the time being that the subsequent digital low-pass filter operates without any quantisers so that no additional noise is induced. Together with the undesired spectral components (Fig. 9-20), the portion of the quantisation noise power outside the desired band is also filtered out. The noise power that is superimposed onto the remaining signal thus amounts to merely N_0/R. Assuming $R = 4$, the signal-to-noise ratio will be improved as if the analog signal was sampled with a one-bit higher resolution. With $R = 16$, the effective resolution is even increased by two bits. We can say in general that each doubling of the oversampling ratio provides 0.5 bits of extra resolution, equivalent to an improvement of the signal-to-noise ratio by 3 dB.

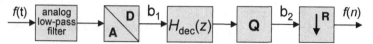

Fig. 9-26 Oversampling A/D conversion

As an example we assume that the analog signal is sampled with a 14-bit converter and an oversampling ratio of 16. The decimation low-pass filter is a CIC filter without any internal scaling. Since the noise power after filtering corresponds to a signal resolution of 16 bits, we can quantise to a b_2-bit resolution of 16 bits without noticeably degrading the signal quality. We have increased the effective resolution of the A/D converter by two bits at the expense of a 16 times higher sampling rate. The decimation low-pass filter is thus able to correctly interpolate four additional steps between the discrete output levels of the A/D converter.

A cost reduction with respect to the A/D converter can hardly be achieved using the procedure described above. In fact, a desired signal resolution can be achieved by means of a less accurate converter, but the much higher processing speed required for oversampling more-or-less nullifies this advantage. It would be far more attractive if the oversampling ratio could be chosen high enough so that a one-bit resolution and thus a simple sign detector would be sufficient for the A/D conversion. This is made possible by the introduction of noise shaping as we previously described in the context of the oversampling D/A converter. The concept has to be modified, however, since the signal to be quantised down to one bit is analog in this case, and analog circuitry has to be kept as simple as possible.

9.2.5 Oversampling ΔΣ Conversion

The A/D converter in Fig. 9-26 can be considered a special quantiser that turns analog signals with infinite amplitude resolution into signals with 2^{b_1} discrete amplitude steps. This discretisation generates a wideband quantisation noise N_0 according to (9.7). The principle of noise shaping may also be applied to this special quantiser. The procedure is to delay the quantisation error by one clock cycle and to subtract it from the input signal. For the feedback of the error signal, we cannot fall back on the truncated bits as described in the case of D/A conversion. Instead, we provide this error signal by explicitly calculating the difference between the input and output signals of the A/D converter. The circuit in Fig. 9-27 thus reduces the quantisation noise in the desired band as derived in Sect. 9.1.4. If we replace the A/D converter in Fig. 9-26 by the circuit in Fig. 9-27, we can maintain the same signal-to-noise ratio as in the case of simple oversampling conversion with an even smaller resolution b_1 of the A/D converter.

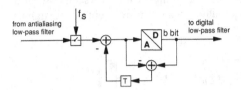

Fig. 9-27
A/D conversion with noise shaping

By choosing a sufficiently high sampling frequency, the A/D converter may be replaced by a one-bit quantiser, equivalent to a simple sign detector. In addition to that, we perform some equivalent rearrangements in the block diagram. Fig. 9-28 shows the result. The recursive structure in front of the sign detector, consisting of a delay and an adder, represents a discrete-time integrator (the value at the input equals the previous output value plus the present input value). The bold paths are carrying analog signals.

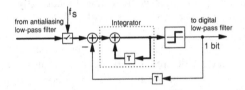

Fig. 9-28
Alternative implementation of the A/D converter with noise shaping

If, in the next step, we shift the sampler behind the quantiser, the discrete-time integrator can be replaced by a common analog integrator. Choosing the time constant of this integrator equal to the sampling period T yields a circuit with approximately identical timing conditions to those in the previous discrete-time implementation according to Fig. 9-28. Figure 9-29 depicts the resulting block diagram, which includes this modification. The delay in the feedback path may be omitted. Its role in the loop is taken over by the sampler.

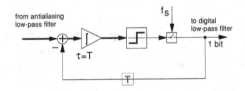

Fig. 9-29
Structure of the oversampling $\Delta\Sigma$-A/D converter

The structure shown in Fig. 9-29 is commonly referred to as a $\Delta\Sigma$ modulator [11]. The ouput of this circuit is a binary signal whose pulse width is proportional to the applied analog voltage. Figure 9-30 shows the reaction of the $\Delta\Sigma$ modulator to a sinusoid with amplitude 0.9. If positive values are applied, the comparator provides a "high" level most of the time. In case of negative input values, the output tends more to the "low" level. A zero input yields rapidly alternating "high" and "low" levels with an average value in the middle between the two extreme levels.

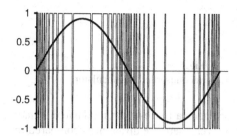

Fig. 9-30
Output of the $\Delta\Sigma$ modulator in case of a sinusoidal input

Since the $\Delta\Sigma$ modulator is still fully equivalent to block diagram 9-27, the spectral contents of the pulse train is composed of the spectrum of the analog input signal and of the spectrum of the quantisation noise, weighted by the transfer function $1-z^{-1}$. The A/D converter can thus be replaced in a practical implementation by an inverting integrator, a comparator and a D flip-flop, which takes over the role of the sampler. The circuit is highly insensitive to nonideal behaviour of the operational amplifier, which is commonly used to realise the integrator. The circuit is inherently stable as long as no extraordinary delay or gain occurs in the loop. For certain constant or slowly varying input values, the noise level may assume marked extreme values. The spectrum may also contain single audible tones. In order to avoid these unfavourable situations, a high frequency noise (dither) is added to the analog input signal, which is later removed again by the digital low-pass filter. This additional noise ensures that the $\Delta\Sigma$ modulator does not stay too long in these unfavourable operational conditions.

For a signal conversion of CD quality, a first-order oversampling $\Delta\Sigma$ converter is not sufficient. We showed in Sect. 9.1.4 that a second-order noise shaper with R of at least 128 is required for this purpose. The necessary transfer function $(1-z^{-1})^2$ to provide sufficient attenuation of the quantisation noise in the desired band can be realised by applying the algorithm according to Fig. 9-27 twice, as depicted in Fig. 9-31.

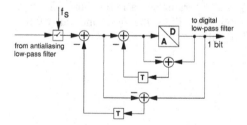

Fig. 9-31
Double application of the noise-shaping algorithm

Equivalent rearrangement of Fig. 9-31 and shift of the sampler behind the comparator leads to the block diagram of a second-order $\Delta\Sigma$ modulator. This circuit features two integrators arranged in series. The binary output signal of the modulator is fed back to the inputs of both integrators with inverse sign (Fig. 9-32).

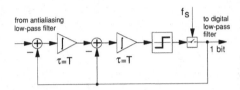

from antialiasing
low-pass filter

f_s

to digital
low-pass
filter

1 bit

$\tau = T$ $\tau = T$

Fig. 9-32
Second-order $\Delta\Sigma$ modulator

The second-order $\Delta\Sigma$ modulator also provides a pulse-width modulated signal. In the fine structure of the pulse train, however, there are differences compared to the first-order modulator, which result in the desired lowering of the noise spectrum in the baseband. As an example, a constant input value of 0.4 leads to the following periodic sequences at the output:

first-order: 1 1 1 -1 1 1 -1 1 1 -1 ...   average $= (7-3)/10$
second-order: 1 1 1 -1 1 -1 1 1 1 1 -1 1 -1 1 1 1 -1 1 1 -1 ... average $= (14-6)/20$.

Both sequences have mean values of 0.4.

The second-order $\Delta\Sigma$ modulator is somewhat more complex than the first-order variant. The sensitivity to tolerances of the components is more critical. A 30% increase of loop gain and/or additional delay in the loop will cause instability. An additional dither signal may not be needed, since the correlation between the quantisation error and the input signal is reduced by the additional feedback loop [11].

The explanations have shown that the hardware complexity of the analog front ends of A/D and D/A converters may be considerably reduced by techniques such as oversampling and noise shaping. The interfaces between digital and analog functions are simple binary interfaces operating at high clock rates. The information is coded in these interfaces by means of pulse-width modulation. In the case of high quality audio applications, the required clock frequencies are in the order of 5 to 10 MHz (44.1 kHz × 128 or 256). Considering the A/D converter, the complexity of the analog circuity is reduced to an anti-aliasing filter with moderate performance, two integrators, a comparator and a D flip-flop. The D/A converter is reduced to a low-pass filter that filters the baseband out of the spectrum of the pulse-width modulated signal.

10 Appendix

10.1 Symbols and Abbreviations

a	incident wave
A	system matrix
$a(t), a(n)$	unit-step response of the system
$a(\omega)$	logarithmic magnitude response
a_ν	coefficients of the denominator polynomial
A_ν	normalised denominator coefficients
B	bandwidth of a bandpass or bandstop filter
b	reflected wave
b	wordlength in bits
$b(\omega)$	phase response
$\boldsymbol{b}, \boldsymbol{c}$	coefficient vectors
b_ν	coefficients of the numerator polynomial
B_ν	normalised numerator coefficients
C	capacitance
Δ	normalised width of the filter transition band
$\delta(n)$	unit-sample sequence
$\delta(t)$	delta function
DFT	discrete Fourier transform
$\delta_\mathrm{p}, \delta_\mathrm{s}$	ripple in the passband and stopband
Δx_q	step size of the quantiser
ε	design parameter of Chebyshev and Cauer filters
$e(n), e(t)$	error signal (discrete-time and continuous-time)
f	frequency
FIR	finite impulse response
$f_\mathrm{s}, f_\mathrm{c}$	sampling frequency, cutoff frequency
γ	coefficient of the wave digital filter
G	conductance
$H(\mathrm{j}\omega), H(\mathrm{e}^{\mathrm{j}\omega T})$	frequency response (continuous-time, discrete-time)
$H(p), H(z)$	transfer function (continuous-time, discrete-time)
$h(t), h(n)$	impulse response, unit sample response
$I_0(x)$	zeroth-order modified Bessel function of first kind
IDFT	inverse discrete Fourier transform
IIR	infinite impulse response
Im	imaginary part of a complex number

D. Schlichthärle, *Digital Filters*, DOI: 10.1007/978-3-642-14325-0_10,
© Springer-Verlag Berlin Heidelberg 2011

$K(k)$	complete elliptic integral of the first kind
k, k_1	design parameters of elliptic filters
L	inductance
M	degree of the numerator polynomial
MT	magnitude truncation
N	degree of the denominator polynomial, order of the filter
N_0	noise power of the quantiser
$O(x)$	overflow characteristic
OSR	oversampling ratio
p	complex frequency
P	signal power
P	normalised complex frequency
p_0, P_0	zero frequency
p_∞, P_∞	pole frequency
Q	quality factor
q	quantisation step of the filter coefficients
$Q(x)$	quantisation operator
$q(x)$	unit rectangle
r	reflection coefficient
R	resistance, port resistance
$r(t), r(n)$	ramp function (continuous-time, discrete-time)
Re	real part of a complex number
RND	rounding
σ	real part of p
σ	variance
S_{11}, S_{22}	reflectance
S_{12}, S_{21}	transmittance
$\mathrm{sn}(u,k)$	Jacobian elliptic function
SNR	signal-to-noise ratio
T	sampling period
T	transformation matrix
t, τ	time variables
$\tau_g(\omega)$	group delay
$T_n(x)$	Chebyshev polynomial of the first kind of order n
$u(t), u(n)$	unit-step function (continuous-time, discrete-time)
V	gain parameter
v_p	minimum gain in the passband
v_{norm}	normalised gain
v_s	maximum gain in the stopband
VT	value truncation
ω	angular frequency, imaginary part of p
Ω	normalised angular frequency
$W(\omega)$	weighting function
ω_c	cutoff frequency
$W_n(1/x)$	Bessel polynomial

ω_p	passband edge frequency
ω_s	sampling angular frequency
ω_s	stopband edge frequency
$x(t), x(n)$	input signal (continuous-time, discrete-time)
ψ	frequency variable of the bilinear transform
$y(t), y(n)$	output signal (continuous-time, discrete-time)
z	frequency variable of discrete-time systems
$\mathbf{z}$	state vector

10.2 References

[1]   Avenhaus E (1972) A Proposal to Find Suitable Canonical Structures for the Implementation of Digital Filters with Small Coefficient Wordlength. NTZ 25:377-382

[2]   Barnes CW, Shinnaka S (1980) Stability Domains for Second-Order Recursive Digital Filters in Normal Form with "Matrix Power" Feedback. IEEE Trans. CAS-27:841-843

[3]   Bauer PH, Leclerc LJ (1991) A Computer-Aided Test for the Absence of Limit Cycles in Fixed-Point Digital Filters. IEEE Trans. ASSP-39:2400-2409

[4]   Bellanger M (1984) Digital Processing of Signals. John Wiley & Sons, New York

[5]   Bose T, Brown DP (1990) Limit Cycles Due to Roundoff in State-Space Digital Filters. IEEE Trans. ASSP-38:1460-1462

[6]   Bose T, Chen MQ (1991) Overflow Oscillations in State-Space Digital Filters. IEEE Trans. CAS-38:807-810

[7]   Brayton RK, Tong CH (1979) Stability of Dynamical Systems: A Constructive Approach. IEEE Trans. CAS-26:224-234

[8]   Brayton RK, Tong CH (1980) Constructive Stability and Asymptotic Stability of Dynamical Systems. IEEE Trans. CAS-27:1121-1130

[9]   Bronstein I, Semendjajew K (1970) Taschenbuch der Mathematik. Verlag Harri Deutsch, Frankfurt

[10] Butterweck HJ, Ritzerfeld J, Werter M (1989) Finite Wordlength Effects in Digital Filters. AEÜ 43:76-89

[11] Candy JC, Temes GC (1992) Oversampling Methods for A/D and D/A Conversion. In: Oversampling Delta-Sigma Data Converters. IEEE Press, Piscataway

[12] Chang TL (1981) Suppression of Limit Cycles in Digital Filters Designed with One Magnitude-Truncation Quantizer. IEEE Trans. CAS-28:107-111

[13] Claasen TACM, Mecklenbräuker WFG, Peek JBH (1973) Second-Order Digital Filter With Only One Magnitude-Truncation Quantiser And Having Practically No Limit Cycles. Electron Lett. 9:531-532

[14] Claasen TACM, Mecklenbräuker WFG, Peek JBH (1973) Some Remarks on the Classification of Limit Cycles in Digital Filters. Philips Res. Rep. 28:297-305

[15] Claasen TACM, Mecklenbräuker WFG, Peek JBH (1975) Frequency Domain Criteria for the Absence of Zero-Input Limit Cycles in Nonlinear Discrete-Time Systems, with Applications to Digital Filters. IEEE Trans. CAS-22:232-239

[16] Darlington S(1939) Synthesis of reactance 4-poles which produce prescribed insertion loss characteristics. J. Math. Phys. Vol.18:257-353

[17] Dorf RC (Editor) (1993) The Electrical Engineering Handbook. CRC Press, London

[18] Ebert PM, Mazo JE, Taylor MG (1969) Overflow Oscillations in Digital Filters. Bell Syst. Tech. J. 48:2999-3020

[19] Erickson KT, Michel AN (1985) Stability Analysis of Fixed-Point Digital Filters Using Computer Generated Lyapunov Functions-Part I: Direct Form and Coupled Form Filters. IEEE Trans. CAS-32:113-132

[20] Erickson KT, Michel AN (1985) Stability Analysis of Fixed-Point Digital Filters Using Computer Generated Lyapunov Functions-Part II: Wave Digital Filters and Lattice Digital Filters. IEEE Trans. CAS-32:132-142

[21] Fettweis A (1971) Digital Filter Structures Related to Classical Filter Networks. AEÜ 25:79-89

[22] Fettweis A (1972) A Simple Design of Maximally Flat Delay Digital Filter. IEEE Trans. AU-20:112-114

[23] Fettweis A (1986) Wave Digital Filters: Theory and Practice. Proc. IEEE 74: 270-327

[24] Fettweis A (1988) Passivity and Losslessness in Digital Filtering. AEÜ 42:1-8

[25] Gazsi L (1985) Explicit Formulas for Lattice Wave Digital Filters. IEEE Trans. CAS-32:68-88

[26] Gentile K (2002) The care and feeding of digital, pulse-shaping filters. RF Design April 2002:50-61

[27] Gold B, Rader CM (1969) Digital Processing of Signals. McGraw-Hill

[28] Gray AH, Markel JD (1973) Digital Lattice and Ladder Filter Synthesis. IEEE Trans. AU-21:491-500

[29] Hamming RW (1962) Numerical Methods for Scientists and Engineers. McGraw-Hill, New York

[30] Herrmann O (1971) On the Approximation Problem in Nonrecursive Digital Filter Design. IEEE Trans. CT-18:411-413

[31] Herrmann O, Rabiner LR, and Chan DSK (1973) Practical Design Rules for Optimum Finite Impulse Response Low-Pass Digital Filters. Bell Syst. Tech. J.52:769-799

[32] Herrmann O, Schuessler W (1970) Design of Nonrecursive Digital Filters with Minimum Phase. Electron. Lett. 6:185-186

[33] Hogenauer EB (1981) An Economical Class of Digital Filters for Decimation and Interpolation. IEEE Trans. ASSP-29:155-162

[34] Ikehara M, Funaishi M, Kuroda H (1992) Design of Complex All-Pass Networks Using Remez Algorithm. IEEE Trans. CASII-39:549-556

[35] Jackson LB (1969) An Analysis of Limit Cycles Due to Multiplication Rounding in Recursive Digital (Sub)Filters. Proc. 7th Annu. Allerton Conf. Circuit System Theory, pp. 69-78

[36] Jackson LB (1979) Limit Cycles in State-Space Structures for Digital Filters. IEEE Trans. CAS-26:67-68

[37] Jahnke E, Emde F, Lösch F (1966) Tafeln Höherer Funktione. Teubner, Stuttgart

[38] Jinaga BC, Dutta Roy SC (1984) Explicit Formulas for the Weighting Coefficients of Maximally Flat Nonrecursive Digital Filters. Proc. IEEE 72:1092

[39] Kaiser JF (1974) Nonrecursive Digital Filter Design Using The I_0-Sinh Window Function. Proc. IEEE Int. Symp. on Circuits and Syst., pp. 20-23

[40] Kaiser JF (1979) Design Subroutine (MXFLAT) for Symmetric FIR Low Pass Digital Filters with Maximally-Flat Pass and Stop Band. In: Programs for Digital Signal Processing, IEEE Press, pp. 5.3-1 to 5.3-6

[41] Kieburtz RB (1973) An Experimental Study of Roundoff Effects in a Tenth-Order Recursive Digital Filter. IEEE Trans. COM-21:757-763

[42] Kingsbury NG (1972) Second-Order Recursive Digital Filter Element for Poles Near the Unit Circle and the Real z-Axis. Electronic Letters 8:155-156

[43] La Salle J, Lefschetz S (1967) Die Stabilitätstheorie von Ljapunov. Bibliographisches Institut, Mannheim

[44] Lang M, Laakso TI (1994) Simple and Robust Method for the Design of Allpass Filters Using Least-Squares Phase Error Criterion. IEEE Trans. CASII-41:40-48

[45] Laugwitz D (1965) Ingenieurmathematik V. Bibliographisches Institut, Mannheim

[46] Lepschy A, Mian GA, Viaro U (1986) Stability Analysis of Second-Order Direct-Form Digital Filters with Two Roundoff Quantizers. IEEE Trans. CAS-33:824-826

[47] Lightstone M, Mitra SK, Lin I, Bagchand S, Jarske P, Neuvo Y (1994) Efficient Frequency-Sampling Design of One- and Two-Dimensional FIR Filters Using Structural Subband Decomposition. IEEE Trans. CAS-41:189-201

[48] Lim YC, Lee JH, Chen CK, Yang RH (1992) A Weighted Least Squares Algorithm for Quasi-Equiripple FIR and IIR Digital Filter Design. IEEE Trans. SP-40:551-558

[49] Lueder E (1981) Digital Signal Processing with Improved Accuracy. Proc. 5th European Conf. on Circuit Theory and Design, The Hague, pp. 25-33

[50] Lueder E, Hug H, Wolf W (1975) Minimizing the Round-Off Noise in Digital Filters by Dynamic Programming. Frequenz 29:211-214

[51] McClellan JH, Parks TW, Rabiner LR (1973) A Computer Program For Designing Optimum FIR Linear Phase Digital Filters. IEEE Trans. AU-21:506-526

[52] McClellan JH, Parks TW, Rabiner LR (1979) FIR Linear Phase Filter Design Program. In: Programs for Digital Signal Processing, IEEE Press, pp. 5.1-1 to 5.1-13

[53] Meerkötter K, Wegener W (1975) A New Second-Order Digital Filter Without Parasitic Oscillations. AEÜ 29:312-314

[54] Mitra SK, Hirano K (1974) Digital All-Pass Networks. IEEE Trans. CAS-21:688-700

[55] Oppenheim AV, Schafer RW (1975) Digital Signal Processing. Prentice-Hall, Englewood Cliffs

[56] Orchard HJ, Willson AN (1997) Elliptic Functions for Filter Design. IEEE Trans. CAS-44:273-287

[57] Pfitzenmaier G (1971) Tabellenbuch Tiefpässe, Unversteilerte Tschebyscheff- und Potenz-Tiefpässe. Siemens Aktiengesellschaft, Berlin · München

[58] Pregla R, Schlosser W (1972) Passive Netzwerke, Analyse und Synthese. Teubner, Stuttgart

[59] Premaratne K, Kulasekere EC, Bauer PH, Leclerc LJ (1996) An Exhaustive Search Algorithm for Checking Limit Cycle Behavior of Digital Filters. IEEE Trans. ASSP-44:2405-2412

[60] Press WH, Teukolsky SA, Vetterling WT, Flannery BP (1992) Numerical Recipes in C. Cambridge University Press

[61] Rabiner LR, McClellan JH, Parks TW (1975) FIR Digital Filter Design Techniques Using Weighted Chebyshev Approximation. Proc. IEEE 63:595-610

[62] Rabiner LR, McGonegal CA, Paul D (1979) FIR Windowed Filter Design Program - WINDOW. In: Programs for Digital Signal Processing, IEEE Press, pp. 5.2-1 to 5.2-19

[63] Rajagopal LR, Dutta Roy SC (1989) Optimal Design of Maximally Flat FIR Filters with Arbitrary Magnitude Specifications. IEEE Trans. ASSP-37:512-518

[64] Regalia PA, Mitra SK, Vaidyanathan PP (1988) The Digital All-Pass Filter: A Versatile Signal Processing Building Block. Proc. IEEE 76:19-37

[65] Saal R (1979) Handbook of Filter Design. AEG-Telefunken, Backnang

[66] Selesnick IW (1999) Low-Pass Filters Realizable as All-Pass Sums: Design via a New Flat Delay Filter. IEEE Trans. CASII-46:40-50

[67] Shannon CE (1949) Communication in the Presence of Noise. Proc. IRE 37:10-21
[68] Shen J, Strang G (1999) The Asymptotics of Optimal (Equiripple) Filters. IEEE Trans. ASSP-47:1087-1098
[69] Sunder S, Ramachandran V (1994) Design of Equiripple Nonrecursive Digital Differentiators and Hilbert Transformers Using a Weighted Least-Squares Technique. IEEE Trans. SP-42:2504-2509
[70] Thiran JP (1971) Recursive Digital Filters with Maximally Flat Group Delay. IEEE Trans. CT-18:659-664
[71] Vaidyanathan PP (1984) On Maximally-Flat Linear-Phase FIR Filters. IEEE Trans. CAS-31:830-832
[72] Vaidyanathan PP, Mitra SK (1984) Low Pass-Band Sensitivity Digital Filters: A Generalized Viewpoint and Synthesis Procedures. Proc. IEEE 72:404-423
[73] Vaidyanathan PP, Mitra SK, Neuvo Y (1986) A New Approach to the Realization of Low-Sensitivity IIR Digital Filters. IEEE Trans. ASSP-34:350-361
[74] Vollmer M, Kopmann H (2002) A Novel Approach to An IIR Digital Filter Bank With Approximately Linear Phase, IEEE Int. Symp. On Circuits and Systems, Phoenix, May 2002, Vol. II, pp.512-515
[75] Weinberg L (1962) Network Analysis and Synthesis. McGraw-Hill
[76] Wupper H (1975) Neuartige Realisierung RC-aktiver Filter mit geringer Empfindlichkeit. Habilitationsschrift, Ruhr-University Bochum
[77] Zölzer U (1989) Entwurf digitaler Filter für die Anwendung im Tonstudiobereich. Doctorate dissertation, Technical University Hamburg-Harburg

10.3 Filter Design Tables

This section presents a selection of design tables for the filter types that we introduced in Chap. 2:
- Butterworth
- Chebyshev
- Inverse Chebyshev
- Cauer
- Bessel

All tables show the coefficients for filter orders up to $N = 12$. For Butterworth and Bessel filters, which have the filter order as the only independent design parameter, these tables are complete. Chebyshev and Cauer filters have far more degrees of freedom, so we can only include a representative choice in this textbook. In the case of design problems for which these tables are not sufficient, extensive filter catalogues are available in the literature [57, 65].

All first- and second-order filter blocks that occur in the context of the filter types named above can be expressed in the following general form:

$$H(P) = \frac{1 + B_2 P^2}{1 + A_1 P + A_2 P^2} \ .$$

Three different forms of transfer functions appear in practice.

Simple real pole

$$B_2 = 0 \quad A_2 = 0 \quad A_1 \neq 0$$

This simple first-order pole occurs in the case of odd-order low-pass filters.

Complex-conjugate pole pair

$$B_2 = 0 \quad A_2 \neq 0 \quad A_1 \neq 0$$

Butterworth, Chebyshev and Bessel filters are composed of these second-order filter blocks.

Complex-conjugate pole pair with imaginary zero pair

$$B_2 \neq 0 \quad A_2 \neq 0 \quad A_1 \neq 0$$

Inverse Chebyshev and Cauer filters have zero pairs on the imaginary $j\omega$-axis.

Filter catalogues often show poles and zeros instead of normalised filter coefficients. The filter coefficients A_1, A_2 and B_2 can be easily calculated from these poles and zeros. For second-order filter blocks we have the following relations:

$$A_1 = -2 \operatorname{Re} P_\infty / |P_\infty|^2$$
$$A_2 = 1 / |P_\infty|^2$$
$$B_2 = 1 / |P_0|^2 \ .$$

For the first-order pole, A_1 is calculated as

$$A_1 = -1 / P_\infty \ .$$

The following filter tables contain a factor V, which normalises the gain of the cascade arrangement of second-order filters blocks to unity. V equals one in most cases. In the case of even-order Chebyshev and Cauer filters, V depends on the given passband ripple.

$$V = v_p = 1 / \sqrt{1 + \varepsilon^2}$$

Butterworth filters

$V = 1$
$B_2 = 0$

N	A_1	A_2
1	1.000000	
2	1.414214	1.000000
3	1.000000	1.000000
	1.000000	
4	0.765367	1.000000
	1.847759	1.000000
5	0.618034	1.000000
	1.618034	1.000000
	1.000000	
6	0.517638	1.000000
	1.414214	1.000000
	1.931852	1.000000
7	0.445042	1.000000
	1.246980	1.000000
	1.801938	1.000000
	1.000000	
8	0.390181	1.000000
	1.111140	1.000000
	1.662939	1.000000
	1.961571	1.000000

N	A_1	A_2
9	0.347296	1.000000
	1.000000	1.000000
	1.532089	1.000000
	1.879385	1.000000
	1.000000	
10	0.312869	1.000000
	0.907981	1.000000
	1.414214	1.000000
	1.782013	1.000000
	1.975377	1.000000
11	0.284630	1.000000
	0.830830	1.000000
	1.309721	1.000000
	1.682507	1.000000
	1.918986	1.000000
	1.000000	
12	0.261052	1.000000
	0.765367	1.000000
	1.217523	1.000000
	1.586707	1.000000
	1.847759	1.000000
	1.982890	1.000000

Chebyshev filters

$B_2 = 0$

V equals one for odd filter orders. In the case of even filter orders, V depends on
the passband ripple:
- 0.25 dB: $V = 0.971628$
- 0.50 dB: $V = 0.944061$
- 1.00 dB: $V = 0.891251$
- 2.00 dB: $V = 0.794328$

Chebyshev 0.25 dB

N	A_1	A_2
1	0.243421	
2	0.849883	0.473029
3	0.573140	0.747032
	1.303403	
4	0.365795	0.860621
	2.255994	2.198549
5	0.245524	0.912880
	1.318006	1.864219
	2.288586	
6	0.175678	0.940542
	0.809742	1.586787
	3.535069	5.071206
7	0.130987	0.956845
	0.543394	1.416677
	1.959450	3.535167
	3.250989	
8	0.101215	0.967243
	0.390423	1.310156
	1.171842	2.627537
	4.783087	9.092020

N	A_1	A_2
9	0.080466	0.974280
	0.294846	1.239844
	0.775649	2.128887
	2.575534	5.762661
	4.205054	
10	0.065459	0.979263
	0.231071	1.191128
	0.553748	1.832684
	1.512316	3.972113
	6.018950	14.261358
11	0.054269	0.982923
	0.186303	1.155992
	0.417320	1.642616
	0.989240	3.031036
	3.181532	8.546939
	5.155207	
12	0.045709	0.985689
	0.153607	1.129805
	0.327217	1.512940
	0.700890	2.486660
	1.843984	5.617915
	7.248804	20.579325

Chebyshev 0.5 dB

N	A_1	A_2
1	0.349311	
2	0.940260	0.659542
3	0.548346	0.875314
	1.596280	
4	0.329760	0.940275
	2.375565	2.805743
5	0.216190	0.965452
	1.229627	2.097461
	2.759994	
6	0.151805	0.977495
	0.719120	1.694886
	3.691705	6.369532
7	0.112199	0.984148
	0.471926	1.477359
	1.818204	3.938897
	3.903658	
8	0.086211	0.988209
	0.335124	1.348920
	1.036706	2.788231
	4.980968	11.356882
9	0.068277	0.990873
	0.251348	1.266842
	0.671707	2.209747
	2.385021	6.396216
	5.040188	
10	0.055392	0.992718
	0.196118	1.211094
	0.474268	1.880381
	1.335834	4.203190
	6.259890	17.768641
11	0.045831	0.994050
	0.157651	1.171413
	0.355133	1.673930
	0.855617	3.139405
	2.943283	9.468583
	6.173405	
12	0.038544	0.995044
	0.129707	1.142112
	0.277330	1.535091
	0.599686	2.547068
	1.627510	5.935974
	7.533733	25.605025

Chebyshev 1 dB

N	A_1	A_2
1	0.508847	
2	0.995668	0.907021
3	0.497051	1.005829
	2.023593	
4	0.282890	1.013680
	2.411396	3.579122
5	0.181032	1.011823
	1.091107	2.329385
	3.454311	
6	0.125525	1.009354
	0.609201	1.793016
	3.721731	8.018803
7	0.0920921	1.007375
	0.391989	1.530326
	1.606177	4.339334
	4.868210	
8	0.070429	1.005894
	0.275575	1.382088
	0.875459	2.933762
	5.009828	14.232606
9	0.055600	1.004790
	0.205485	1.289680
	0.556611	2.280179
	2.103365	7.024249
	6.276263	
10	0.045006	1.003957
	0.159743	1.227864
	0.389282	1.921118
	1.126613	4.412333
	6.289486	22.221306
11	0.037176	1.003317
	0.128092	1.184305
	0.289916	1.700378
	0.708287	3.233739
	2.593589	10.382028
	7.681621	
12	0.031226	1.002818
	0.105202	1.152365
	0.225632	1.553671
	0.491819	2.598631
	1.371714	6.223779
	7.565021	31.985093

Chebyshev 2 dB

N	A_1	A_2	N	A_1	A_2
1	0.764783		9	0.042558	1.015849
2	0.976619	1.214978		0.157778	1.307956
3	0.416333	1.128547		0.432087	2.337937
	2.710682			1.723645	7.602867
4	0.225886	1.076803		8.289826	
	2.285707	4.513278	10	0.034388	1.012859
5	0.141700	1.050236		0.122298	1.241205
	0.898462	2.543558		0.299870	1.953980
	4.580677			0.887534	4.589612
6	0.097258	1.035249		5.913813	27.587958
	0.481606	1.876388	11	0.028368	1.010640
	3.508721	10.007393		0.097873	1.194521
7	0.070933	1.026046		0.222356	1.721517
	0.304860	1.573834		0.549392	3.311060
	1.317946	4.708418		2.124017	11.223492
	6.437500			10.140090	
8	0.054045	1.020012	12	0.023804	1.008948
	0.212583	1.408883		0.080271	1.160468
	0.690371	3.057181		0.172584	1.568437
	4.714502	17.698915		0.378610	2.640205
				1.080071	6.467695
				7.109976	39.674546

Inverse Chebyshev filters

$V = 1$

The coefficients B_2 are independent of the stopband attenuation of the filter.

N	B_2
1	---
2	0.500000
3	0.750000
4	0.853553
	0.146447
5	0.904508
	0.345492
6	0.933013
	0.500000
	0.066987
7	0.950484
	0.611260
	0.188255
8	0.961940
	0.691342
	0.308658
	0.038060

N	B_2
9	0.969846
	0.750000
	0.413176
	0.116978
10	0.975528
	0.793893
	0.500000
	0.206107
	0.024472
11	0.979746
	0.827430
	0.571157
	0.292292
	0.079373
12	0.982963
	0.853553
	0.629410
	0.370590
	0.146447
	0.017037

30 dB stopband attenuation

N	A_1	A_2
1	31.606961	
2	5.533785	15.811388
3	1.866438	4.233593
	1.866438	
4	0.943411	2.372921
	2.277597	1.665815
5	0.573384	1.765237
	1.501139	1.206220
	0.927755	
6	0.386924	1.491737
	1.057095	1.058725
	1.444018	0.625712
7	0.279334	1.344439
	0.782677	1.005215
	1.131001	0.582210
	0.627658	
8	0.211428	1.255566
	0.602097	0.984968
	0.901101	0.602284
	1.062921	0.331686
9	0.165740	1.197594
	0.477229	0.977748
	0.731157	0.640924
	0.896897	0.344725
	0.477229	
10	0.133490	1.157571
	0.387403	0.975935
	0.603395	0.682043
	0.760322	0.388150
	0.842823	0.206514
11	0.109859	1.128721
	0.320677	0.976405
	0.505516	0.720132
	0.649401	0.441267
	0.740675	0.228348
	0.385972	
12	0.092017	1.107207
	0.269779	0.977798
	0.429157	0.753654
	0.559288	0.494835
	0.651304	0.270691
	0.698936	0.141282

40 dB stopband attenuation

N	A_1	A_2
1	99.995000	
2	9.959874	50.000000
3	2.838494	8.807047
	2.838494	
4	1.337350	3.906721
	3.228648	3.199614
5	0.784536	2.515899
	2.053941	1.956882
	1.269406	
6	0.518865	1.937759
	1.417566	1.504746
	1.936431	1.071734
7	0.369949	1.641491
	1.036575	1.302267
	1.497894	0.879262
	0.831268	
8	0.277721	1.468564
	0.790882	1.197966
	1.183639	0.815282
	1.396197	0.544684
9	0.216469	1.358346
	0.623297	1.138499
	0.954947	0.801675
	1.171416	0.505477
	0.623297	
10	0.173632	1.283518
	0.503901	1.101883
	0.784844	0.807990
	0.988961	0.514097
	1.096272	0.332462
11	0.142458	1.230251
	0.415834	1.077935
	0.655521	0.821662
	0.842102	0.542797
	0.960461	0.329878
	0.500504	
12	0.119043	1.190908
	0.349015	1.061499
	0.555203	0.837355
	0.723555	0.578536
	0.842597	0.354392
	0.904218	0.224982

50 dB stopband attenuation

N	A_1	A_2
1	316.226185	
2	17.754655	158.113883
3	4.233618	18.673520
	4.233618	
4	1.842787	6.650654
	4.448881	5.943547
5	1.037435	3.722221
	2.716039	3.163204
	1.678604	
6	0.669948	2.608071
	1.830333	2.175059
	2.500281	1.742046
7	0.470579	2.068541
	1.318534	1.729317
	1.905337	1.306312
	1.057382	
8	0.349764	1.765502
	0.996045	1.494904
	1.490686	1.112221
	1.758384	0.841623
9	0.270736	1.577551
	0.779554	1.357705
	1.194346	1.020881
	1.465083	0.724683
	0.779554	
10	0.216072	1.452477
	0.627065	1.270841
	0.976676	0.976948
	1.230684	0.683056
	1.364224	0.501420
11	0.176614	1.364773
	0.515534	1.212457
	0.812689	0.956184
	1.044004	0.677319
	1.190740	0.464399
	0.620505	
12	0.147160	1.300743
	0.431453	1.171334
	0.686342	0.947190
	0.894458	0.688371
	1.041619	0.464227
	1.117795	0.334817

60 dB stopband attenuation

N	A_1	A_2
1	999.999500	
2	31.606961	500.000000
3	6.259920	39.936595
	6.259920	
4	2.501934	11.539482
	6.040202	10.832375
5	1.345577	5.644660
	3.522766	5.085643
	2.177189	
6	0.845771	3.602658
	2.310690	3.169645
	3.156462	2.736632
7	0.583966	2.672247
	1.636236	2.333023
	2.364430	1.910018
	1.312160	
8	0.429063	2.171174
	1.221868	1.900576
	1.828654	1.517893
	2.157045	1.247295
9	0.329439	1.869655
	0.948583	1.649809
	1.453313	1.312985
	1.782752	1.016787
	0.948583	
10	0.261378	1.673459
	0.758548	1.491823
	1.181466	1.197931
	1.488734	0.904038
	1.650274	0.722402
11	0.212706	1.538214
	0.620885	1.385898
	0.978764	1.129625
	1.257349	0.850760
	1.434071	0.637840
	0.747307	
12	0.176634	1.440779
	0.517863	1.311369
	0.823801	1.087225
	1.073599	0.828406
	1.250232	0.604262
	1.341665	0.474853

Cauer filters

The tables of the Cauer filters contain an additional column indicating the width of the transition band ω_s/ω_p of the respective filter.

Passband ripple: 1 dB

$V = 1$ for odd N

$V = 0.891251$ for even N

Cauer filter, 1-dB passband ripple, 30-dB stopband attenuation

N	A_1	A_2	B_2	ω_s/ω_p
1	0.508849			62.114845
2	0.962393	0.893099	0.031688	4.004090
3	0.404019	0.984054	0.262019	1.732505
	1.787125			
4	0.165088	0.995394	0.581107	1.250383
	1.754288	2.297852	0.139656	
5	0.067497	0.998247	0.801280	1.095539
	0.579010	1.440901	0.463510	
	2.219449			
6	0.027596	0.999292	0.913451	1.037991
	0.212903	1.162894	0.731929	
	1.925771	2.751629	0.169694	
7	0.011281	0.999711	0.963679	1.015356
	0.083238	1.063737	0.880383	
	0.603379	1.551368	0.501672	
	2.304618			
8	0.004611	0.999882	0.984993	1.006248
	0.033405	1.025582	0.949274	
	0.217117	1.198555	0.755958	
	1.955920	2.837962	0.175090	
9	0.001885	0.999952	0.993839	1.002549
	0.013551	1.010378	0.978949	
	0.084134	1.076951	0.892081	
	0.607275	1.571093	0.508125	
	2.319269			
10	0.000770	0.999980	0.997477	1.001041
	0.005522	1.005229	0.991342	
	0.033642	1.030770	0.954408	
	0.217706	1.204761	0.759916	
	1.961000	2.852711	0.176001	
11	0.000315	0.999992	0.998968	1.000425
	0.002254	1.001726	0.996452	
	0.013627	1.012464	0.981110	
	0.084233	1.079226	0.893987	
	0.607921	1.574425	0.509205	
	2.321729			
12	0.000129	0.999997	0.999578	1.000174
	0.000921	1.000705	0.998549	
	0.005549	1.005075	0.992236	
	0.033661	1.031660	0.995241	
	0.217801	1.205804	0.760576	
	1.961850	2.855183	0.176154	

Cauer filter, 1-dB passband ripple, 40-dB stopband attenuation

N	A_1	A_2	B_2	ω_s/ω_p
1	0.508847			196.512846
2	0.985058	0.902474	0.010126	7.044820
3	0.452106	0.994690	0.131433	2.416184
	1.909413			
4	0.210869	1.001457	0.386003	1.515484
	2.013945	2.764201	0.080466	
5	0.099952	1.001110	0.636119	1.218682
	0.733855	1.674652	0.321263	
	2.595082			
6	0.047599	1.000574	0.806052	1.098870
	0.303672	1.282199	0.585449	
	2.356753	3.734747	0.113924	
7	0.022695	1.000279	0.902280	1.045995
	0.135465	1.126194	0.775128	
	0.796829	1.934861	0.378329	
	2.796049			
8	0.010825	1.000133	0.952133	1.021682
	0.062612	1.058360	0.885644	
	0.317801	1.373784	0.632778	
	2.443707	4.012650	0.122697	
9	0.005163	1.000064	0.976873	1.010285
	0.029425	1.027430	0.943723	
	0.139262	1.163888	0.804391	
	0.811036	2.002385	0.391919	
	2.844769			
10	0.002463	1.000030	0.988900	1.004893
	0.013937	1.012992	0.972748	
	0.063819	1.075113	0.901436	
	0.320594	1.396459	0.643513	
	2.464013	4.079467	0.124755	
11	0.001175	1.000014	0.994690	1.002331
	0.006626	1.006177	0.986906	
	0.029869	1.035156	0.951713	
	0.139888	1.173013	0.810872	
	0.814260	2.018229	0.395040	
	2.856020			
12	0.000560	1.000007	0.997463	1.001111
	0.003156	1.002942	0.993732	
	0.014120	1.016619	0.976668	
	0.063975	1.079126	0.904893	
	0.321206	1.401720	0.645953	
	2.468661	4.094866	0.125227	

Cauer filter, 1-dB passband ripple, 50-dB stopband attenuation

N	A_1	A_2	B_2	ω_s/ω_p
1	0.508847			621.456171
2	0.992304	0.905568	0.003213	12.484568
3	0.475795	1.000427	0.063303	3.460613
	1.969598			
4	0.240441	1.006120	0.239692	1.908184
	2.177965	3.086060	0.045962	
5	0.125733	1.004054	0.470461	1.407231
	0.849859	1.870794	0.215871	
	2.847722			
6	0.066518	1.002272	0.670150	1.198914
	0.381611	1.398334	0.448369	
	2.711769	4.680949	0.077468	
7	0.035320	1.001225	0.808326	1.101265
	0.186351	1.195534	0.653917	
	0.967594	2.339835	0.282982	
	3.231412			
8	0.018777	1.000654	0.892993	1.052645
	0.094878	1.099700	0.797976	
	0.413323	1.575551	0.517273	
	2.890996	5.311008	0.088636	
9	0.009986	1.000348	0.941574	1.027667
	0.049332	1.051861	0.886913	
	0.196415	1.273942	0.705474	
	1.002200	2.503109	0.303867	
	3.343619			
10	0.005312	1.000185	0.968482	1.014624
	0.025930	1.027262	0.938161	
	0.098546	1.137504	0.830830	
	0.421486	1.633180	0.537209	
	2.944416	5.508338	0.091991	
11	0.002825	1.000099	0.983108	1.007753
	0.013707	1.014411	0.966614	
	0.050838	1.070945	0.906151	
	0.198629	1.298530	0.719802	
	1.012099	2.552223	0.309923	
	3.376373			
12	0.001503	1.000052	0.990978	1.004117
	0.007267	1.007641	0.982099	
	0.026610	1.037135	0.948932	
	0.099213	1.149131	0.839763	
	0.423725	1.650156	0.542875	
	2.959761	5.565840	0.092956	

Cauer filter, 1-dB passband ripple, 60-dB stopband attenuation

N	A_1	A_2	B_2	ω_s/ω_p
1	0.508847			1965.225746
2	0.994603	0.906560	0.001017	22.176712
3	0.487098	1.003271	0.029897	5.021121
	1.998311			
4	0.258381	1.009197	0.142774	2.460783
	2.276772	3.289813	0.026091	
5	0.144391	1.006476	0.330087	1.671161
	0.931913	2.019421	0.141961	
	3.072028			
6	0.082338	1.003943	0.529063	1.343537
	0.443972	1.498899	0.332159	
	2.990838	5.508885	0.052929	
7	0.047295	1.002313	0.692997	1.185474
	0.231599	1.262563	0.531840	
	1.112351	2.728504	0.210158	
	3.601562			
8	0.027238	1.001341	0.809410	1.103082
	0.126530	1.143868	0.695302	
	0.499674	1.784259	0.415274	
	3.289942	6.644266	0.065191	
9	0.015702	1.000775	0.885187	1.058218
	0.070824	1.080607	0.811008	
	0.251636	1.396267	0.604619	
	1.177950	3.042259	0.236698	
	3.806594			
10	0.009055	1.000447	0.932070	1.033179
	0.040168	1.045732	0.886205	
	0.134654	1.212283	0.748612	
	0.517389	1.900155	0.444653	
	3.398181	7.086114	0.069688	
11	0.005223	1.000258	0.960233	1.019007
	0.022949	1.026130	0.932690	
	0.074465	1.117437	0.846299	
	0.257159	1.447926	0.628906	
	1.200468	3.157678	0.245952	
	3.878535			
12	0.003013	1.000149	0.976864	1.010920
	0.013165	1.014991	0.960602	
	0.041937	1.066146	0.908245	
	0.136565	1.237964	0.765807	
	0.523176	1.941448	0.454589	
	3.435250	7.241268	0.071232	

Bessel filters

$V = 1$

Normalised to t_0

N	A_1	A_2
1	1.000000	
2	1.000000	0.333333
3	0.569371	0.154812
	0.430629	
4	0.366265	0.087049
	0.633735	0.109408
5	0.256073	0.055077
	0.469709	0.070065
	0.274218	
6	0.189781	0.037716
	0.358293	0.047955
	0.451926	0.053188
7	0.146771	0.027325
	0.281315	0.034558
	0.370779	0.038961
	0.201135	
8	0.117236	0.020647
	0.226517	0.025927
	0.306756	0.029468
	0.349492	0.031272
9	0.096041	0.016118
	0.186326	0.020085
	0.256809	0.022911
	0.302019	0.024637
	0.158805	
10	0.080289	0.012913
	0.156045	0.015968
	0.217637	0.018235
	0.261565	0.019770
	0.284462	0.020548
11	0.068245	0.010565
	0.132690	0.012969
	0.186582	0.014805
	0.227721	0.016132
	0.253569	0.016940
	0.131193	
12	0.058816	0.008797
	0.114304	0.010723
	0.161652	0.012226
	0.199526	0.013363
	0.226028	0.014132
	0.239675	0.014520

$V = 1$

Normalised to ω_{3dB}

N	A_1	A_2
1	1.000000	
2	1.361654	0.618034
3	0.999629	0.477191
	0.756043	
4	0.774254	0.388991
	1.339664	0.488904
5	0.621595	0.324533
	1.140177	0.412845
	0.665639	
6	0.513054	0.275641
	0.968607	0.350473
	1.221734	0.388718
7	0.433228	0.238072
	0.830363	0.301095
	1.094437	0.339457
	0.593694	
8	0.372765	0.208745
	0.720236	0.262125
	0.975366	0.297924
	1.111250	0.316161
9	0.325742	0.185418
	0.631960	0.231049
	0.871017	0.263562
	1.024356	0.283414
	0.538619	
10	0.288318	0.166512
	0.560356	0.205909
	0.781532	0.235149
	0.939275	0.254934
	1.021499	0.264964
11	0.257940	0.150928
	0.501515	0.185268
	0.705206	0.211495
	0.860698	0.230458
	0.958389	0.241998
	0.495859	
12	0.232862	0.137889
	0.452546	0.168086
	0.640003	0.191640
	0.789953	0.209464
	0.894879	0.221511
	0.948908	0.227595

10.4 Filter Design Routines

This part of the Appendix contains some useful subroutines written in C for the design of IIR and FIR filters. These were developed and tested using Visual C++.

Cauer

The subroutine Cauer allows the calculation of the poles and zeros of Cauer (elliptic) filters. Input parameters are the parameters k, k1, e and N, as introduced in Sect. 2.5. This parameter set must satisfy Eqn. (2.46).

```
void Cauer(double k, double k1, double e, int N, double *ZerR,
           double *ZerI, double *PolR, double *PolI)
/***************************************************************
 * Subroutine to calculate poles and zeros of elliptic filters. *
 * The subroutine delivers: Nz=N/2 conjugate complex poles and  *
 * zeros plus one real pole if N is odd, in total Np=(N+1)/2     *
 * pole values.                                                  *
 * INPUT:   k      transition width wp/ws                        *
 *          k1     vsnorm/vpnorm                                 *
 *          e      1/vpnorm                                      *
 *          N      filter degree                                 *
 * RETURN: *ZerR   array for the real part of the zeros          *
 *         *ZerI   array for +- the imaginary part of the poles  *
 *         *PolR   array for the real part of the poles          *
 *         *PolI   array for +- the imaginary part of the poles  *
 * REQUIRED ADDITIONAL ROUTINES: none                           *
 * Zer, ZeI, PolR, PolI are arrays of dimension Np. The parameters *
 * N, k and k1 cannot be chosen independently. They are related by *
 * a formula making use of complete elliptic integrals (Chap. 2.5). *
 * Subroutine CauerParam calculates these parameters from the   *
 * specification of the desired low-pass filter.                *
 ***************************************************************/
{
    double Cei1(double);
    double InverseSn(double, double);
    void EllipFunctions(double, double, double *, double *, double *);
    int Nz, Np, Lm, L;
    double kc, k1c, u0, s1, c1, d1, dx, sn, cn, dn, de;

    Nz = N / 2;
    Np = (N + 1) / 2;
    kc = sqrt(1.0-k*k);
    k1c = sqrt(1.0-k1*k1);
    u0 = Cei1(kc) / Cei1(k1c) * InverseSn(1.0 / e, k1);
    EllipFunctions(u0, kc, &s1, &c1, &d1);
    dx = Cei1(k) / N;
    for (Lm = 1; Lm<= Nz; Lm++)
    {
        L = N + 1 - Lm - Lm;
        EllipFunctions(L * dx, k, &sn, &cn, &dn);
        ZerR[Lm] = 0.0;
        ZerI[Lm] = 1.0 / k / sn;
        de = c1 * c1 + k * k * sn * sn * s1 * s1;
        PolR[Lm] = -cn * dn * s1 * c1 / de;
        PolI[Lm] = sn * d1 / de;
    }
```

```
     if (N != 2 * (N / 2))
     {
        PolR[Np] = -s1 / c1;
        PolI[Np] = 0.0;
        ZerR[Np] = 0.0;
        ZerI[Np] = 0.0;
     }
}

double Cei1(double x)
/********************************
* Complete elliptic integral    *
* of the 1st kind               *
********************************/
{
    double a, a1, b, b1, Test;
    int l;
    a = 1.0;
    b = sqrt(1.0 - x * x);
    Test = 1.0e-12;
    for (l = 1; l <= 25; l++)
    {
        a1 = 0.5 * (a + b);
        b1 = sqrt(a * b);
        a = a1;
        b = b1;
        if ((a - b) < Test) l=25;
    }
    return (0.5 * M_PI / a);
}

double InverseSn(double x, double k)
/********************************
* Inverse elliptic function of  *
* an imaginary argument         *
********************************/
{
    double p, q, tt, tw;
    double a, b, a1, b1, f, Test;
    int kq, kh, j;
    a = 1.0; b = k;
    Test = 1.0e-12;
    p = atan(x);
    tw = 1.0;
    kq = 1;
    for (j = 1; j <= 25; j++)
    {
        f = p / a / tw;
        tw = tw * 2.0;
        kq = 2 * kq - 1;
        q = b * sin(p) / a / cos(p);
        tt = atan(q);
        if(tt < 0) kq = kq + 1;
        kh = kq / 2;
        p = p + tt + M_PI * kh;
        a1 = a; b1 = b;
        a = 0.5 * (a1 + b1); b = sqrt(a1 * b1);
        if ((a-b) < Test) j=25;
    }
    return f;
}
```

```
void EllipFunctions(double x, double k, double *sn, double *cn,
                    double *dn)
/***********************************************************************
 * Calculate the elliptic functions sn,cn, and dn                      *
 ***********************************************************************/
{
    double p, q, kc, Test;
    int l, limit, j1, jk, nn;
    double a[26], b[26], c[26];
    kc = sqrt(1.0 - k * k);
    a[1] = 1.0;
    b[1] = kc;
    c[1] = k;
    Test = 1.0e-12;
    for (l = 1; l < 25; l++)
    {
        a[l + 1] = 0.5 * (a[l] + b[l]);
        b[l + 1] = sqrt(a[l] * b[l]);
        c[l + 1] = 0.5 * (a[l] - b[l]);
        if (c[l + 1] < Test)
        {
            limit = l + 1;
            l = 24;
        }
    }
    j1 = limit - 1;
    nn = j1 - 1;
    p = exp(nn * log(2.0)) * a[j1] * x;
    for (l = 1; l <= nn; l++)
    {
        jk = j1 + 1 - l;
        q = c[jk] * sin(p) / a[jk];
        p = 0.5 *(p + asin(q));
    }
    *sn = sin(p);
    *cn = cos(p);
    *dn = sqrt(1.0 - k * k * (*sn) * (*sn));
}
```

CauerParam

The subroutine **CauerParam** allows the calculation of a harmonised set of the parameters k, k1, e and N of a Cauer filter. Given are passband ripple and stopband attenuation in dB and the order N of the filter. The subroutine returns k, k1, and e.

```
void CauerParam(double ap, double as, int N, double *k, double *k1,
                double *e)
/***********************************************************************
 * Subroutine to calculate the filter parameters of elliptic          *
 * filters from filter order, passband ripple and stopband            *
 * attenuation.                                                        *
 * INPUT:  ap     passband ripple in dB                               *
 *         as     stopband ripple in dB                               *
 *         N      filter order                                         *
 * RETURN: *k     transition width wp/ws                              *
 *         *k1    vsnorm/vpnorm                                        *
 *         *e     1/vpnorm                                             *
 ***********************************************************************/
```

```
{
    double Cei1(double);
    double Modul(double);
    double vpnorm, vsnorm, q, k1c;
    vpnorm = 1.0 / sqrt(exp(log(10.0) * ap / 10.0) - 1.0);
    vsnorm = 1.0 / sqrt(exp(log(10.0) * as / 10.0) - 1.0);
    *k1 = vsnorm / vpnorm;
    k1c = sqrt(1.0 - (*k1) * (*k1));
    q = exp(-M_PI * Cei1(k1c) / Cei1(*k1) / N);
    *k = Modul(q);
    *e = 1.0 / vpnorm;
}

double Modul(double q)
{
    double Theta2, Theta3, qp, qf, Test;
    Test = 1.0e-20;
    Theta2 = 1.0;
    Theta3 = 1.0;
    qp = 1.0;
    qf = 1.0;
    while (qf > Test)
    {
        qp = qp * q;
        qf = qf * qp;
        Theta3 = Theta3 + 2.0 * qf;
        qf = qf * qp;
        Theta2 = Theta2 + qf;
    }
    Theta2 = 2.0 * sqrt(sqrt(q)) * Theta2;
    return (Theta2 * Theta2 / Theta3 / Theta3);
}
```

Bessel

Subroutine **Bessel** calculates the coefficients of analog Bessel filters (Chap. 2.6). The input parameter `Param` must be set to zero in this case. Moreover, the coefficients of analog reference filters for the design of digital Bessel filters by means of the bilinear transform can be determined by equating `Param` with the reciprocal of the design parameter μ (refer to Sect. 6.4.4).

```
void Bessel(double *Coeff, int N, double Param)
/******************************************************************
* Subroutine to calculate the filter coefficients of Bessel filters.*
* INPUT:   N          filter order                              *
*          Param      design parameter                          *
* RETURN:  Coeff      Bessel filter coefficients                *
* For Param = 0, the subroutine calculates the Bessel coefficients  *
* of analog filters (Sect. 2.6). For Param = 1/mu, the subroutine   *
* calculates the Bessel coefficients for the design of digital   *
* filters with mu = 2*t0/T and mu > N-1 (Sect. 6.4.4).          *
******************************************************************/
{
    int i, j;
    double g, h;
    Coeff[0] = 1.0;
    for (j = 1; j <= N; j++) Coeff[j] = 0.0;
    for (i = N; i > 0; i--)
```

```
{
    g = 2.0 * i - 1.0;
    h = 1.0 - (i - 1.0) * (i - 1.0) * Param * Param;
    for (j = N; j > 0; j--)
    {
        if (j != 2 * (j / 2)) Coeff[j] = Coeff[j-1] * h;
        else Coeff[j] = Coeff[j-1] + Coeff[j] * g;
    }
    Coeff[0] = Coeff[0] * g;
}
}
```

FirFreqResp

Subroutine **FirFreqResp** calculates the frequency response of FIR filters in the frequency range 0 to Range. Range is normalised to the sampling period (Range = f_{max} * T). NP is the number of equidistant points to be calculated within the specified frequency range. The array Coeff contains the filter coefficients. N specifies the filter order. The calculated zero phase frequency response is returned back to the calling program in the array H0. Type specifies the kind of symmetry of the impulse response. Type = 0 applies to Case 1 and Case 2 filters, Type = 1 to Case 3 and Case 4 filters accordingly.

```
void FirFreqResp(double *H0, int NP, double *Coeff, int N, int Type,
                 double Range)
/****************************************************************
* Subroutine to calculate the zero phase frequency response of FIR *
* filters from the coefficients of the filter with an arbitrary    *
* frequency resolution.                                            *
* INPUT    NP      number of calculated frequency points           *
*          Coeff   array containing the filter coefficients        *
*          N       degree of the filter                            *
*          Type    symmetry of the impulse response                *
*                  Type  0 : impulse response with even symmetry    *
*                  Type  1 : impulse response with odd symmetry     *
*          Range   desired normalised frequency range tmaxT         *
* RETURN   H0      arraycontaining the calculated frequency response *
****************************************************************/
{
    int imax, i, j;
    double Nhalf, dw, w, sum;
    imax = (N - 1) / 2;
    Nhalf = N / 2.0;
    dw = 2.0 * M_PI * Range / (NP - 1.0);
    if (Type == 0)
    {
        for (j = 0; j < NP; j++)
        {
            sum = 0.0;
            w = j * dw;
            for (i = 0; i <= imax; i++)
                sum = sum + Coeff[i] * cos(w * (i - Nhalf));
            sum = sum * 2.0;
            if (N == 2 * (N / 2)) sum = sum + Coeff[N/2];
            H0[j] = sum;
        }
    }
```

```
    else
    {
        for (j = 0; j < NP; j++)
        {
            sum = 0.0;
            w = j * dw;
            for (i = 0; i <= imax; i++)
                sum = sum + Coeff[i] * sin(w * (i - Nhalf));
            sum = sum * 2.0;
            H0[j] = sum;
        }
    }
}
```

CoeffIDFT

The procedure **CoeffIDFT** calculates the $N+1$ coefficients of an Nth-order FIR filter from values of the zero phase frequency response taken at $N+1$ equidistant frequency points. These values are sampled at the frequencies $\omega_i = 2*\pi*i/T/(N+1)$ with $i = 0...N$. The algorithm is based on an inverse discrete Fourier transform (IDFT) and makes use of the special symmetry conditions of linear-phase filters. The array H passes the samples of the frequency response to the subroutine. The filter coefficients are returned back in the array Coeff. type again specifies the kind of symmetry of the impulse response. type = 0 applies to Case 1 and Case 2 filters, type = 1 to Case 3 and Case 4 filters accordingly.

```
void CoeffIDFT(double *h, double *coeff, int N, int type)
/*****************************************************************
 * Procedure to calculate the N+1 coefficients of an Nth-order FIR *
 * filter from N+1 equally spaced samples of the zero-phase       *
 * frequency response. The samples are taken at the frequencies   *
 * wi=2*pi*i/T/(N+1) with i=0...N. The algorithm is based on an   *
 * inverse discrete Fourier transform (IDFT) and makes use of the *
 * special symmetry conditions of linear-phase filters.           *
 * INPUT:     h      array containing the frequency response      *
 *            N      order of the filter                          *
 *            type   symmetry of the unit sample response         *
 *                   type 0: unit sample response with even symmetry *
 *                   type 1: unit sample response with odd symmetry  *
 * RETURN:    coeff array containing the filter coefficients       *
 * REQUIRED ADDITIONAL ROUTINES: none                             *
 * h and coeff are arrays of dimension N+1                        *
 *****************************************************************/
{
    int i, j, imax;
    double sum, nhalf, dw;

    imax = N / 2;
    nhalf = N / 2.0;
    for (j = 0; j <= N; j++)
    {
        dw = 2.0 * M_PI * (j-nhalf) / (N+1);
        if (type == 0) sum = h[0];
        else
        {
            sum = 0.0;
```

```
        if ((N % 2) == 1) sum = h[imax+1];
        if (((j-imax-1) % 2) == 1) sum = -sum;
    }
    for (i = 1; i <= imax; i++)
    {
        if (type == 0) sum = sum + 2.0 * h[i] * cos(dw * i);
        else sum = sum + 2.0 * h[i] * sin(dw * i);
    }
    coeff[j] = sum / (N+1);
    }
}
```

MaxFlatPoly

The function **MaxFlatPoly** returns values of the MaxFlat polynomial according to Sect. 7.6 (Eq. 7.57). Input values are the argument x, the degree of the polynomial Nt and the parameter L.

```
double MaxFlatPoly(double x, int L, int Nt)
/**************************************************************
 * Function to calculate values of the maximum flat polynomial *
 * PNt,L(x). This algorithm is an optimised version of the formula *
 * given by Kaiser (1979). The polynomial is defined in the range *
 * x=0 ... 1. L is in the range L=1 ... Nt.                    *
 * INPUT:   x              argument of the polynomial          *
 *          Nt             Nt/N/2 (half order of the filter)   *
 *          L              degree of tangency at x=0           *
 * RETURN: MaxFlatPoly     result of the polynomial calculation *
 **************************************************************/
{
    int j, jj;
    double sum, y, z;

    if (L == 0 || L == Nt+1)
    {
        if (L == 0) sum = 0.0;
        if (L == Nt+1) sum = 1.0;
    }
    else
    {
        sum = 1.0;
        if (L != 1)
        {
            y = x;
            for (j = 1; j < L; j++)
            {
                z = y;
                if (Nt+1-L != 1)
                    for (jj = 1; jj <= Nt-L; jj++) z=z*(1.0+(double)j/jj);
                y = y * x;
                sum = sum + z;
            }
        }
        y = 1.0 - x;
        for (j = 1; j <= Nt + 1 - L; j++) sum = sum * y;
    }
    return sum;
}
```

MaxFlatDesign

The procedure **MaxFlatDesign** determines the parameters of a MAXFLAT filter from the characteristic data of a low-pass tolerance scheme. Input data to the procedure are the passband edge frequency wp, the stopband edge frequency ws, the minimum gain in the passband vp and the maximum gain in the stopband vs. As a result, we obtain half the filter order Nt (Nt = N/2) and the parameter L.

```
void MaxFlatDesign(double wp, double ws, double gp, double gs, int
*Nt, int *L)
/*********************************************************************
 * Procedure to calculate the filter parameters Nt and L for a      *
 * MAXFLAT FIR filter. This routine is based on the algorithm       *
 * introduced by Rajagopal and Dutta Roy (1989). Input is the       *
 * tolerance mask of the low-pass filter. The cutoff frequencies    *
 * are normalised to the sampling frequency of the filter.          *
 * INPUT:   wp              passband cutoff frequency               *
 *          ws              stopband cutoff frequency               *
 *          vp              minimum passband gain (linear)          *
 *          vs              maximum stopband gain (linear)          *
 * RETURN:  Nt              half the order of the filter (Nt=N/2)   *
 *          L               degree of tangency at w=0               *
 *********************************************************************/
{
    int InRange(double, double, int, int);
    int Found(double, double, double, double, int, int);
    const int lim = 200;
    double xp, xs;
    int exit;

    xp = (1.0-cos(2.0*M_PI*wp))/2.0;
    xs = (1.0-cos(2.0*M_PI*ws))/2.0;
    *L = 0; *Nt = 1; exit = 0;
    while (!exit && *Nt<lim)
    {
        *Nt += 1;
        while (!InRange(gp,xp,*L,*Nt)) *L += 1;
        if(Found(gp,gs,xp,xs,*L,*Nt)) exit = 1;
        if(Found(gp,gs,xp,xs,*L+1,*Nt))
        {
            exit = 1;
            *L += 1;
        }
    }
}


int InRange(double gp, double xp, int L, int Nt)
{
    double MaxFlatPoly(double, int, int);

    if ((gp>MaxFlatPoly(xp,L,Nt)) && (gp<=MaxFlatPoly(xp,L+1,Nt)))
        return 1;
    else return 0;
}


int Found(double gp, double gs, double xp, double xs, int L, int Nt)
```

```
{
    double MaxFlatPoly(double, int, int);
    int flg = 1;

    if (MaxFlatPoly(xp, L, Nt) < gp) flg = 0;
    if (MaxFlatPoly(xs, L, Nt) > gs) flg = 0;
    return flg;
}
```

Using the function **MaxFlatPoly** and the procedure **MaxFlat Design**, it is extremely easy to calculate the coefficients of an FIR low-pass filter with maximally flat frequency response. The following listing shows a short example how these routines may be applied to calculate the coefficients of a MAXFLAT filter.

```
main(int argc, char *argv[])
{
    void MaxFlatDesign(double,double,double,double,int *,int *);
    double MaxFlatPoly(double, int, int);
    void CoeffIDFT(double *, double *, int, int);
    int i, L, Nt, N;
    double wp, ws, gp, gs, x, s1, s2;
    double h[200], coeff[200];
    // Entry of filter characteristics
    printf("Passband edge frequency  = ");scanf("%lf", &wp);
    printf("Stopband edge frequency  = ");scanf("%lf", &ws);
    printf("Passband ripple [dB]     = ");scanf("%lf", &gp);
    gp = pow(10.0,-gp/20.0);
    printf("Stoband attenuation [dB] = ");scanf("%lf", &gs);
    gs = pow(10.0,-gs/20.0);
    // Calculation of Nt and L
    MaxFlatDesign(wp, ws, gp, gs, &Nt, &L);
    N = 2 * Nt;
    // Calculate N+1 frequency points
    for (i = 0; i <= N; i++)
    {
        x=(1.0-cos(2.0*M_PI*i/(N+1)))/2.0;
        h[i] = MaxFlatPoly(x, L, Nt);
    }
    // Filter parameters via inverse DFT
    CoeffIDFT(h, coeff, N, 0);
    // Filter coefficients can be found in array coeff
    for (i = 0; i <= N; i++) printf("%i: %e\n",i,coeff[i]);
}
```

Index

D. Schlichthärle, *Digital Filters*, DOI: 10.1007/978-3-642-14325-0,
© Springer-Verlag Berlin Heidelberg 2011